G PROTEIN-COUPLED RECEPTORS

This text provides a comprehensive overview of recent discoveries and current understandings of G protein-coupled receptors (GPCRs). Recent advances include the first mammalian non-rhodopsin GPCR structures and reconstitution of purified GPCRs into membrane discs for defined studies, novel signaling features including oligomerization, and advances in understanding the complex ligand pharmacology and physiology of GPCRs in new assay technologies and drug targeting.

The first chapters of this book illustrate the history of GPCRs based on distinct species and genomic information. This is followed by discussion of the homo- and hetero-oligomerization features of GPCRs, including receptors for glutamate, $GABA_B$, dopamine, and chemokines. Several chapters are devoted to the key signaling features of GPCRs. The authors take time to detail the importance of the pathophysiological function and drug targeting of GPCRs, specifically β-adrenoceptors in cardiovascular and respiratory diseases, metabotropic glutamate receptors in CNS disorders, S1P receptors in the immune system, and Wnt/Frizzled receptors in osteoporosis.

This book will be invaluable to researchers and graduate students in academia and industry who are interested in the GPCR field.

Dr. Sandra Siehler is a Research Investigator at the Novartis Institutes for BioMedical Research in Basel, Switzerland. Dr. Siehler is a member of the American Society for Pharmacology and Experimental Therapeutics and the British Pharmacological Society.

Dr. Graeme Milligan is Professor of Molecular Pharmacology at the University of Glasgow. He is actively involved in numerous associations, such as the Biochemical Society and the British Pharmacological Society. Dr. Milligan was awarded the Ariens Award for Pharmacology from the Dutch Pharmacological Society in 2006.

G Protein-Coupled Receptors

STRUCTURE, SIGNALING, AND PHYSIOLOGY

Edited by

Sandra Siehler
Novartis Institutes for BioMedical Research

Graeme Milligan
University of Glasgow

CAMBRIDGE
UNIVERSITY PRESS

CAMBRIDGE UNIVERSITY PRESS
Cambridge, New York, Melbourne, Madrid, Cape Town, Singapore,
São Paulo, Delhi, Dubai, Tokyo, Mexico City

Cambridge University Press
32 Avenue of the Americas, New York, NY 10013-2473, USA

www.cambridge.org
Information on this title: www.cambridge.org/9780521112086

© Cambridge University Press 2011

First published 2011

Printed in the United States of America

A catalog record for this publication is available from the British Library.

Library of Congress Cataloging in Publication data
G protein-coupled receptors : structure, signaling, and physiology / [edited by]
    Sandra Siehler, Graeme Milligan.
        p. ; cm.
    Includes bibliographical references and index.
    ISBN 978-0-521-11208-6 (hardback)
    1. G proteins.   2. Cell receptors.   3. Cellular signal transduction. I. Siehler,
    Sandra. II. Milligan, Graeme.
    [DNLM: 1. Receptors, G-Protein-Coupled – physiology.   2. Drug Delivery
    Systems.   3. Signal Transduction–physiology. QU 55.7]
    QP552.G16G1747 2011
    572'.696–dc22   2010030378

ISBN 978-0-521-11208-6 Hardback

Contents

Color plates follow page 32.

Figures

Tables

Contributors

Rosa López Almagro, Ph.D
Research and Development Center
Almirall
Barcelona, Spain

J. Kurt Chuprun, Ph.D.
Center for Translational Medicine
Thomas Jefferson University
Philadelphia, PA

Michele Ciccarelli, MD
Center for Translational Medicine
Thomas Jefferson University
Philadelphia, PA

Laetitia Comps-Agrar, Ph.D.
Department of Molecular
    Pharmacology
Institut de Génomique Fonctionnelle
Montpellier, France

John F. Cryan, Ph.D
Senior Lecturer
School of Pharmacy
Department of Pharmacology
    and Therapeutics
University College Cork
Cork, Ireland

Yehia Daaka, Ph.D.
Department of Urology
UF Prostate Disease Center
University of Florida
College of Medicine
Gainesville, FL

Etienne Doumazane, Ph.D.
Department of Molecular
    Pharmacology
Institut de Génomique
    Fonctionnelle
Montpellier, France

Thierry Durroux, Ph.D.
Department of Molecular
    Pharmacology
Institut de Génomique
    Fonctionnelle
Montpellier, France

Karin F. K. Ejendal, Ph.D.
Postdoctoral Research Associate
Department of Medicinal Chemistry
    and Molecular Pharmacology
School of Pharmacy and
    Pharmaceutical Sciences
Purdue University
West Lafayette, IN

Susan R. George
Professor
Department of Pharmacology
    and Toxicology
University of Toronto
Toronto, Ontario, Canada

Nuria Godessart, Ph.D.
Head of Autoimmunity Department
Almirall Laboratories
Llobregat, Spain

J. Silvio Gutkind, Ph.D.
Oral & Pharyngeal Cancer Branch
National Institute of Dental and
  Craniofacial Research
National Institutes of Health
Bethesda, MD

Ahmed Hasbi, Ph.D.
Postdoctoral Fellow
Department of Pharmacology
  and Toxicology
University of Toronto
Toronto, Ontario, Canada

Ralf Heilker, Ph.D.
Boehringer Ingelheim Pharma
  GmbH & Co. KG
Department of Lead Discovery
Biberach, Germany

Peter Hein, MD, Ph.D.
Postdoctoral Researcher
Department of Molecular and Cellular
  Pharmacology and Psychiatry
University of California at San
  Francisco
San Francisco, CA

Daniel Hoyer, Ph.D.
Neuropsychiatry
Neuroscience Research
Novartis Institutes for BioMedical
  Research
Basel, Switzerland

Terry Kenakin, Ph.D.
Department of Biological Reagents
  and Assay Development
Molecular Discovery
GlaxoSmithKline Research and
  Development
Research Triangle Park, NC

Brian K. Kobilka, MD
Professor
Depatment of Molecular and Cellular
  Physiology
Stanford University
Stanford, CA

Walter J. Koch, Ph.D.
Center for Translational Medicine
Thomas Jefferson University
Philadelphia, PA

Adam J. Kuszak
Department of Pharmacology
University of Michigan
Ann Arbor, MI

Carlos Martínez-A., Ph.D.
Professor
Department of Immunology and
  Oncology
Centro Nacional de Biotecnologia
Madrid, Spain

Damien Maurel, Ph.D.
Scientist
Ecole Polytechnique Fédérale de
  Lausane
Lausane, Switzerland

Mario Mellado, Ph.D.
Research Scientist
Department of Immunology
  and Oncology
Centro Nacional de
  Biotecnologia
Madrid, Spain

Graeme Milligan, Ph.D.
Professor
Neuroscience and Molecular
  Pharmacology
University of Glasgow
Scotland

Carine Monnier, Ph.D.
Department of Molecular
  Pharmacology
Institut de Génomique Fonctionnelle
Montpellier, France

Zhongzhen Nie, Ph.D.
Department of Urology
UF Prostate Disease Center
University of Florida
College of Medicine
Gainesville, FL

Richard M. O'Connor
School of Pharmacy
Department of Pharmacology
    and Therapeutics
University College Cork
Cork, Ireland

Brian F. O'Dowd
Professor
Department of Pharmacology and
    Toxicology
University of Toronto
Toronto, Ontario, Canada

Stefan Offermanns, MD
Director
Department of Pharmacology
Max-Planck Institute for Heart and
    Lung Research
Hessen, Germany

Jean Phillipe Pin, Ph.D.
Director
Department of Molecular
    Pharmacology
Institut de Génomique Fonctionnelle
Montpellier, France

Laurent Prézeau, Ph.D.
Department of Molecular
    Pharmacology
Institut de Génomique Fonctionnelle
Montpellier, France

Julie A. Przybyla
Department of Medicinal Chemistry
    and Molecular Pharmacology
School of Pharmacy and
    Pharmaceutical Sciences
Purdue University
West Lafayette, IN

Sören G. F. Rasmussen, Ph.D.
Postdoctoral Scholar
Department of Molecular and
    Cellular Physiology
Stanford University
Stanford, CA

Georges Rawadi, Ph.D.
Business Development & Alliance
    Manager
Galapagos
Romainville, France

Marie-Laure Rives, Ph.D.
Postdoctoral Research Fellow
Columbia University
New York, NY

José Miguel Rodríguez-Frade, Ph.D.
Research Scientist
Department of Immunology and
    Oncology
Centro Nacional de
    Biotecnologia
Madrid, Spain

Philippe Rondard, Ph.D.
Department of Molecular
    Pharmacology
Institut de Génomique
    Fonctionnelle
Montpellier, France

Andreas Russ, Ph.D.
Department of Biochemistry
University of Oxford
Oxford, United Kingdom

Torsten Schöneberg, Ph.D.
Molecular Biochemistry
Institute of Biochemistry
University of Leipzig
Leipzig, Germany

Kristin Schröck, Ph.D.
Molecular Biochemistry
Institute of Biochemistry
University of Leipzig
Leipzig, Germany

Sandra Siehler, Ph.D
Research Investigator II
Center for Proteomic Chemistry
Novartis Institutes for Biomedical
    Research
Basel, Switzerland

Claudia Stäubert, Ph.D.
Molecular Biochemistry
Institute of Biochemistry
University of Leipzig
Leipzig, Germany

Roger K. Sunahara, Ph.D.
Associate Professor
Department of Pharmacology
University of Michigan
Ann Arbor, MI

Gema Tarrasón, Ph.D.
Research and Developement Center
Almirall
Barcelona, Spain

Erin Trinquet, Ph.D.
Cisbio Bioassays
Parc technologique Marcel Boiteux
Bagnols/Cèze, France

José Vázquez-Prado, Ph.D.
Professor
Department of Pharmacology
Center for Research and Advanced
    Studies
National Polytechnic Institute
Mexico

Ivan Toma Vranesic, Ph.D.
Neuropsychiatry
Neuroscience Research
Novartis Institutes for BioMedical
    Research
Basel, Switzerland

Val J. Watts, Ph.D.
Department of Medicinal Chemistry
    and Molecular Pharmacology
School of Pharmacy and
    Pharmaceutical Sciences
Purdue University
West Lafayette, IN

Michael Wolff, Ph.D.
Department of Lead Discovery
Boehringer Ingelheim Pharma GmbH
    & Co. KG
Biberach, Germany

Xiao Jie Yao
Research Associate
Department of Molecular and
    Cellular Physiology
Stanford University
Stanford, CA

Introduction

Sandra Siehler and Graeme Milligan

This book provides a comprehensive overview of recent discoveries and the current understanding in the G protein-coupled receptor (GPCR) field.

A plethora of distinct GPCRs exist on the cell surface of every cell type and generate signals inside cells to regulate key physiological events. The human genome contains between 720 and 800 GPCRs with specific tissue and subcellular expression profiles. Chapter 1 of this volume illustrates the evolutionary history of GPCRs based on genomic information available from distinct species and ancient genomic information. Many GPCRs are involved in olfactory/sensory mechanisms. Three hundred sixty-seven non-sensory human GPCRs are known or predicted to be activated by native ligands; endogenous ligands for 224 human GPCRs are described currently, but remain to be identified for 143 orphan receptors. Three hundred sixty-seven ligand-activated non-sensory GPCRs consist of 284 class A (rhodopsin-like) receptors, 50 class B (secretin-like) receptors, 17 class C (metabotropic receptor-like) receptors, and 11 belong to the atypical class of frizzled-/smoothened receptors. Polymorphisms (e.g., of β adrenoceptors, see Chapter 15) and alternative splicing (e.g., of metabotropic glutamate receptors, see Chapter 16) further increase the variety of GPCR proteins. Posttranslational modifications such as N-linked glycosylation or carboxyterminal palmitoylation can influence their function.

GPCRs are integral membrane proteins containing an extracellular amino terminus of widely varying length, seven transmembrane α-helical stretches, and an intracellular carboxy terminus. The molecular understanding of GPCRs developed with the cloning of the β_2 adrenoceptor in 1986 and appreciation that it was related to the photon receptor rhodopsin. The majority of signaling events originate at the inner face of the plasma membrane and involve transactivation of one or more members of the four G protein families (G_s, $G_{i/o}$, $G_{q/11}$, $G_{12/13}$), which link GPCRs to effector cascades. Chapter 7 explains functions of mammalian G proteins elucidated using subunit- and tissue-specific gene targeting. Besides effector cascades involving G proteins, non-G protein-mediated signaling has been described for various GPCRs. Moreover, the activity of G proteins can be regulated by non-GPCR proteins such as receptor tyrosine kinases. The activity of GPCRs is further modulated by cellular signals in an auto- and transregulatory fashion. GPCRs form intra- and juxtamembrane signaling complexes

A polymer THAT CONSISTS OF 2,3, or 4 MONOMERS

comprising not only G proteins, but also other GPCRs, ion channels, membrane and cytosolic kinases and other enzymes, G protein-modulatory proteins, and interact with elements of the cell cytoskeleton. Chapters 3–6 describe homo- and hetero-oligomerization features of GPCRs including receptors for glutamate, $GABA_B$, dopamine, and chemokines. Dopamine receptors can hetero-dimerize not only with other subtypes in the same receptor family, but also with less-related GPCR members and ion channels such as NMDA or $GABA_A$ receptors. For class C receptors, which contain a large extracellular domain, oligomerization is mandatory for receptor function. For other GPCRs, oligomerization may result in altered and/or novel ligand pharmacology. Methods applied to measure GPCR complexes and oligomer signaling comprise GPCR-$G\alpha$ protein fusion constructs containing either a mutated receptor or $G\alpha$ mutant, and time-resolved fluorescence resonance energy transfer (TR-FRET).

Downstream of the cellular plasma membrane, the complexity of intracellular communication controlled by GPCRs increases dramatically. Ligand-activated GPCRs often internalize, which mostly causes desensitization of signaling events, although both prolonged signaling and even signaling initiated following receptor internalization have been described. Receptor hetero-oligomers can co-internalize, and activation and internalization of one partner can therefore silence the other interaction partner. Chapters 8–11 describe key signaling features of GPCRs better understood because of significant recent advancements. These include understanding of kinetics of receptor activation and signaling events studied using FRET and bioluminescent RET (BRET). Multiple related proteins control GPCR-mediated cell signaling processes. For example four RhoGTPase nucleotide exchange factors (Rho-GEFs) link $G_{12/13}$ to pathways controlling, for example, contractile complexes of the cytoskeleton, whereas nine mammalian adenylyl cyclases (ACs) are regulated by GPCRs in a receptor- and tissue-specific manner. These enzymes are integral membrane proteins directly regulated by G_s and $G_{i/o}$ proteins, although $G_{q/11}$-coupled GPCRs also influence AC activities via calcium and protein kinase C, and $G_{12/13}$ proteins were recently found to regulate AC activity as well. Arrestins are known to bind to agonist-stimulated phosphorylated GPCRs and promote endocytosis. Novel functions of arrestins include interactions with non-GPCR receptors or direct interaction with signaling proteins including, for example, the ERK MAP kinases. Modern assay technologies to assess GPCR signaling and ligand pharmacology are described in Chapter 12. Multiplexing subcellular readouts using high content screening allows the simultaneous capture of multiple signals, in both temporal and spatial fashion. The pharmacological complexity of orthosteric and allosteric GPCR ligands in the context of both receptor-G protein complexes and activation state models, is illustrated in Chapters 13 and 14. Functional selectivity of GPCR ligands due to receptor allosterism toward intracellular effector pathways contributes to the complex pharmacological nature.

Dysregulated ligand concentration, GPCR protein level, coupling, and/or signaling are implicated in and often causative for many pathophysiological conditions including central nervous system (CNS) disorders, cardiovascular and metabolic diseases, respiratory malfunctions, gastrointestinal disorders, immune diseases, cancer, musculoskeletal pathologies, and eye illnesses. Targeting of GPCRs is hence widely utilized for therapeutic intervention using small molecule weight ligands and, increasingly, therapeutic antibodies. About 30 percent of marketed drugs target GPCRs. Pathophysiological aspects of β-adrenoceptors in cardiovascular and respiratory diseases, of metabotropic glutamate receptors in CNS disorders, of sphingosine 1-phosphate (S1P) receptors in the immune system, and of Wnt/Frizzled receptors in osteoporosis are described in Chapters 15–18. Frizzled receptors possess a GPCR-like architecture, however, their coupling to G proteins remains controversial. Drugability of GPCRs is generally high since ligand binding pockets are found in the extracellular facing segments of GPCRs, meaning that cell permeability is not a requirement. Exceptions exist regarding drugability (e.g., for many chemokine receptors as elaborated in Chapter 6), and a few unique examples for intracellular binding sites for drugs have emerged.

Despite the high drugability and importance of this target class, drug discovery technologies for GPCRs remained limited for a long time when compared to other target classes such as kinases. Integrated lead finding strategies for cytosolic kinases and intracellular parts of membrane kinases comprise biochemical, biophysical, structural, and cellular approaches, which enable a detailed understanding of mechanisms of actions of compounds. Lead finding for GPCRs, on the other hand, was so far solely based on cellular approaches using recombinant and native systems, and either intact cells or cell membranes. Reasons included the challenges of purifying GPCRs in sufficient quantities, the stability of these as isolated membrane proteins, and the lack of structural knowledge. All three issues have been tackled, and recent successes become prominent. Expression, solubilization, and purification methods of GPCRs using eukaryotic insect or mammalian cells, prokaryotic bacterial cells, or in vitro expression systems have been significantly improved. New methods are being applied to stabilize isolated membrane proteins in semi-native lipid environments like, for example, recombinant high density lipoprotein (rHDL)-membrane discs. Functional studies of isolated GPCR-G protein complexes reconstituted in rHDLs are described in Chapter 2 and deliver novel insights that cannot be obtained from cellular systems.

The first crystal structures of a non-rhodopsin GPCR were published for the human β_2 adrenoceptor in 2007 using either a T4 lysozyme fusion replacing the third intracellular loop or a Fab antibody fragment binding to the third intracellular loop, and with the receptor in complex with an inverse agonist and stabilized in a lipid environment. The T4 lysozyme approach also facilitated the identification of the crystal structure of the human A_{2A} adenosine receptor in complex with an antagonist one year later. A novel approach for receptor stabilization uses targeted amino acid mutations in order to thermostabilize the receptor, and enabled crystal structure determination of the turkey β_1 adrenoceptor in complex with an antagonist in 2008. All GPCR structures available to date are derived from class A GPCRs and resemble inactive receptor conformations. More GPCR structures are expected to become public soon and will enable structural drug discovery approaches including fragment-based screening and ligand co-crystallizations. Stabilized purified GPCRs reconstituted in a lipid environment facilitate not only biochemical, but also biophysical methods such as surface plasmon resonance (SPR) or back-scattering interferometry (BSI) measurements. These novel advances allow confirmation of direct binding of a ligand – whether of competitive or allosteric nature – to a GPCR, and to directly study mechanisms of actions of ligands and G protein activation to determine pharmacological textures of GPCRs. This will boost further understanding of GPCR biology, biomedical research, and ultimately translation of new therapies into the clinic.

We thank all the authors for their comprehensive and professional contributions, and Amanda Smith, Katherine Tengco, Joy Mizan, Allan Ross and Monica Finley from Cambridge University Press and Newgen for assistance, final editing and formatting of the chapters, and printing of the book. From planning the outline of the book to final printing, it has been a rewarding experience. We hope the book will be exciting to read for both newcomers and professionals in the GPCR field.

PART I: ADVANCES IN GPCR PROTEIN RESEARCH

1 The evolution of the repertoire and structure of G protein-coupled receptors

Torsten Schöneberg, Kristin Schröck, Claudia Stäubert, and Andreas Russ

INTRODUCTION

With the advent of large, publicly available genomic data sets and the completion of numerous invertebrate and vertebrate genome sequences, there has been much effort to identify, count, and categorize G protein-coupled receptor (GPCR) genes.[1,2] This valuable source of large-scale genomic information also initiated attempts to identify the origin(s) and to follow the evolutionary history of these receptor genes and families. Since all recent genomes have been shaped by selective forces over millions of years, understanding structure-function relationships and the physiological relevance of individual GPCRs makes sense only in the light of evolution. Until recently, the study of natural selection has largely been restricted to comparing individual candidate genes to theoretical expectations. Genome-wide sequence and single nucleotide polymorphism (SNP) data now bring fundamental new tools to the study of natural selection. There has been much success in producing lists of candidate genes, which have potentially been under selection in vertebrate species or in specific human populations.[3–9]

Less effort has gone into a detailed characterization of the candidate genes, which comprises the elucidation of functional differences between selected and nonselected alleles, as well as their phenotypic consequences, and ultimately the identification of the nature of the selective force that produced the footprint of selection. Such further characterization creates a profound understanding of the role and consequences of selection in shaping genetic variation, thus verifying the signature of selection obtained from genome-wide data. Since GPCRs control almost every physiological process, several receptor variants are involved in adaptation to environmental changes and niches. Consistently, genomic scans for signatures of selection revealed a number of such loci containing GPCR genes. This chapter sheds light on the origin(s), rise, and fall of GPCR genes and functions, and focuses on recent advantages in elucidating selective mechanisms (still) driving this process.

GAIN AND LOSS OF GPCRs

The origin of GPCR genes

The GPCR superfamily comprises at least five structurally distinct families/subfamilies (GRAFS classification) named: **G**lutamate, **R**hodopsin, **A**dhesion, **F**rizzled/Taste2, and **S**ecretin receptor families.[2] Because there is very little sequence homology among the five families, the evolutionary origin of GPCRs and their ancestry remain a matter of debate.

The evolutionary success of the GPCR superfamily is reflected by both its presence in almost every eukaryotic organism and by its abundance in mammals, but proteins that display a seven transmembrane (7TM) topology are already present in prokaryotes. The prokaryotic light-sensitive 7TM proteins, such as proteo-, halo-, and bacteriorhodopsins, facilitate light energy harvesting in the oceans, coupled to the carbon cycle via a non-chlorophyll-based pathway. Further, there are prokaryotic sensory rhodopsins for phototaxis in halobacteria, which control the cell's swimming behavior in response to light. As in rhodopsins of bilateral animals, prokaryotic rhodopsins contain retinal covalently bound to 7TM. Moreover, 7TM proteins with a structural similarity to prokaryotic sensory rhodopsins are found in eukaryotes.[10,11] These structural and functional features shared by pro- and eukaryotic rhodopsins suggest a common ancestry. However, despite these similarities, sequence comparisons provide no convincing evidence of an evolutionary linkage between prokaryotic rhodopsins and eukaryotic G protein-coupled rhodopsins.[12] Therefore, the question about the evolutionary origin of eukaryotic GPCRs remains open. Currently, all our insights into their evolutionary history are based on the analysis of the GPCR repertoire of distantly related extant species.

Structural and functional data clearly show that **G-protein signaling via GPCRs** is present in yeast/fungi,[13] plants,[14] and primitive unicellular eukaryotes, such as the slime mold *Dictyostelium discoideum*.[15] This receptor-signaling

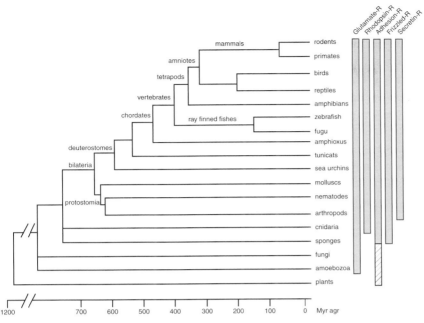

Figure 1-1: Evolutionary occurrence of the different GPCR families in eukaryotes.
GPCRs and their signal transduction probably evolved ~1.2 billion years ago, before plant/fungi/animal split. Genomes of extant plants and fungi usually contain less than ten GPCR genes. The first rhodopsin-like GPCRs, which compose the main GPCR family in vertebrates, appeared ~570–700 Myr ago. Expansion of rhodopsin-like GPCRs started ~500 Myr ago, giving rise to over 1,000 members in some mammalian genomes. The relationships of some major lineages are controversially discussed, hence a very simplified phylogenetic tree of eukaryotes together with a raw time scale are shown. There is some sequence relation between adhesion receptors and GPCRs in plants and fungi, but key features of adhesion receptors, such as the GPS domain in the N terminus, are not present in plant and fungi GPCRs ([23,24]).

complex must have evolved before the plant/fungi/animal split about 1.2 billion years ago (Figure 1.1). Signal transduction through G proteins is the most prominent and eponymous feature of GPCRs. However, one has to consider that GPCRs signal not only via G proteins but also via alternative, non-G-protein-linked signaling pathways.[16] Therefore, it remains open whether G proteins were involved in GPCR signaling from the very evolutionary beginning or if the prototypes of what we now call GPCRs initially fulfilled other functions.

In contrast to GPCR signaling as such, it is more difficult to ascertain the deep evolutionary origin of the five prototypical receptor structures we know today. Genomic data and functional evidence indicate that **glutamate-receptor-like receptors** are present in *D. discoideum*[17,18] and the sponge *Geodia cydonium*,[19,20] which diverged more than 600 million years (Myr) ago (Figure 1.1). The ligand-binding domain of glutamate-receptor-like receptors, also known as the "Venus fly trap" domain, is distantly related to the prokaryotic periplasmic-binding proteins involved in amino acid and nutrient transport in bacteria.[21] Free amino acids act at glutamate-like receptors as either direct-acting orthosteric agonists or allosteric modulators of receptor activity. In contrast to *Dictyostelium* glutamate-like receptors, the sponge receptor did show weak activation by millimolar concentrations of glutamate. This suggests that glutamate activation of glutamate-like receptors may have arisen early in metazoan evolution, with the high glutamate affinity seen in the resurrected ancestral receptor fully present by the time of the bilaterian ancestor.[22] The experimental support for glutamate-receptor-like proteins in *Dictyostelium* suggests that a prototypical receptor structure might predate the origin of metazoa.

Comparative genomic analysis also indicates that precursors of the **adhesion-GPCR** subfamily were present before the onset of metazoan evolution. Sequences distantly related to adhesion-GPCRs are found in fungi and plant genomes;[23,24] however, the homology is modest and mainly based on alignment of putative 7TM regions. Clear evidence for the ancient origin of the adhesion-GPCR subfamily comes from the analysis of the genome of a single-cell eukaryote considered to be the closest relative to metazoans, the choanoflagellate *Monosiga brevicollis*.[25,26] The *Monosiga* genome encodes proteins with the GPS (GPCR proteolytic site)-7TM domain architecture characteristic for adhesion-GPCRs, but no clearly defined homologs of **frizzled** receptors or other elements of the *wnt* pathway. Thus, like glutamate-receptor-like proteins, the signaling module used in adhesion-GPCRs might predate the origin of metazoa.

Frizzled-like receptors are identified in sponges,[27,28] jellyfishes (Cnidaria),[29,30] and placozoa (*Trichoplax adherens*)[31] and can be interpreted as a true innovation of metazoan evolution.[30] They predate the origin of *bilateria*.

A recent phylogenetic study suggests that secretin-like GPCRs descended from the family of adhesion GPCRs.[32] Several members of the ***secretin-like receptor family*** (corticotropin-releasing factor receptor, calcitonin/calcitonin gene-related peptide receptor) are found in both deuterostomia and protostomia[33] but not in Cnidaria, *Monosiga brevicollis*, and *Dictyostelium*. This suggests an evolutionary age of more than 550 Myr, concurrent with the evolution of bilaterial animals.

The structural signatures of **rhodopsin-like GPCRs** have been found in protostome Bilateria (insects, mollusks, nematodes, vertebrates, etc.) and in jellyfishes (*Cnidaria*), which suggests that rhodopsin-like receptors appeared ~570–700 Myr ago[34–37] (Figure 1.1). Within the rhodopsin-like GPCRs, glycoprotein hormone receptors and serotonin receptors appear to be among the oldest, as suggested by their presence in sea anemone (*Cnidaria*), planarians, and nematodes.[38–40] Rhodopsin is the name-giving GPCR in this family. Recent reports suggested that opsins diverged at least before the deuterostome-protostome split about 550 Myr ago.[41] There is growing evidence that opsins are abundant in jellyfishes, indicating that prototypical opsins may have existed before divergence of Cnidaria and Bilateria about 570 to 700 Myr ago,[42,43] although still long after GPCR and G-protein signaling evolved. Given that G-protein coupling to 7TM proteins evolved before the plant/fungi/animal split (about 1.2 billion years ago), and that the first rhodopsin-like receptors appeared early in metazoan evolution, one must consider the retinal-based photosensory system to be a "reinvention" (convergent evolutionary model).

Expansion of GPCR genes

The recent completion of many vertebrate and nonvertebrate genome projects has enabled us to obtain a complete inventory of GPCRs in these species and, by a comparative genomics approach, to analyze the evolution of the GPCR subfamilies. Comparison of the repertoires of GPCRs in insects (fly, mosquito, beetle) and protochordate (Ciona) to that in vertebrates (mammals, birds, fish) reveals a high level of orthology. This indicates that nonvertebrates contain the basic ancestral complement of vertebrate GPCR genes.[44–50] However, the number of GPCRs in most sequences of nonvertebrate genomes (exceptions are the chemokine receptors in worms) is substantially lower than that in vertebrate genomes.[2] This is especially true for rhodopsin-like GPCRs that constitute the most abundant GPCR family in vertebrates when compared to nonvertebrates. Most modern rhodopsin-like GPCR subfamilies expanded in the very early vertebrate evolution. Still, many nonvertebrate GPCR clusters evolved about 500 Myr ago during a time called the "Cambrian Explosion." There are interesting theories about what triggered the enormous gain of species, functionalities, and genes. It was proposed that vision triggered the Cambrian Explosion by creating a new world of organismal interactions, the evolutionary consequence of which was a race in the invention of attracting, attacking, and defending mechanisms.[43] Concomitantly, duplication events of gene and genomic blocks, including several genes of the phototransduction, were traced to the very early vertebrate evolution.

Processes of creating new genes using preexisting genes as raw materials are well characterized, such as exon shuffling, gene duplication, retroposition, gene fusion, and fission. GPCR gene expansion in vertebrates is mainly the result of a combination of species-specific gene duplications and gene or genome duplication events. Two rounds of whole-genome duplications are thought to have played an important role in the establishment of gene repertoires in vertebrates.[51] These events occurred during chordate evolution, after the split of the urochordate and cephalochordate lineages but before the radiation of extant jawed vertebrates.[52] Whole-genome duplications can contribute to prompt gene multiplication and may trigger evolutionary adaptation. One copy or even both members of a gene pair may mutate and acquire unique functionality without risking the fitness of the organism, which is ensured by the homolog. Further, gene duplications often retained overlapping expression patterns and preserved partial-to-complete redundancy consistent with a role in boosting robustness or gene doses. On the other hand, if not advantageous, continuous accumulation of mutations (neutral evolution) will eliminate one of the duplicated genes. As for other genes, disadvantageous mutations in GPCRs are removed from a population through purifying selection. Therefore, many evolutionarily old GPCR genes, including the rhodopsins, display strong features of purifying selection.[5,53,54] It is of interest that specifically GPCRs were retained in vertebrate genomes after genome duplications.[55]

Current evidence suggests that an additional whole-genome duplication occurred in the teleost lineage after it split from the tetrapod lineage, and that only a subset of the duplicates have been retained in modern teleost genomes.[56] Support of these findings comes from sequence analysis of coelacanth, one of the nearest living relatives of tetrapods. The two modern coelacanth species that are known, *Latimeria chalumnae* and *Latimeria menadoensis*, are remarkably similar to their fossil relatives, showing little morphological change over 360 Myr. Genomic sequence analyses show no evidence of whole-genome duplication, consistent with the explanation that the coelacanth genome has not experienced a recent polyploidization event.[57] Therefore, whole-genome duplications did not contribute to GPCR expansion in tetrapode evolution. However, polyploidy is common in fishes and has been determined in sturgeons up to ploidy levels of 8n.[58] GPCR duplications due to polyploidization in ancient teleost fishes have been suggested, for example, trace amine-associated (TAAR) and (purinergic) $P2Y_{12}$-like receptors.[53,59,60]

Whole-chromosome duplications have been made responsible for parallel duplications of more related GPCR. For example, parts of chromosome 4 and 5 show a very similar order of paralog receptor genes. It is assumed that a chromosomal duplication gave rise to dopaminergic receptor paralogs, DRD5 and DRD1, and adrenoceptor paralogs, ADRA2C and ADRA1, at chromosomes 4 and 5, respectively.[61]

Several GPCR subfamilies, such as olfactory,[62–64] chemokine,[65] aminergic,[66] TAAR,[59,67], vomeronasal,[68] and nucleotide receptor-like receptors,[60] cluster in vertebrate genomes and are often arranged in a tandem-like fashion. The numbers of functional receptor genes and pseudogenes of these GPCR subfamilies vary enormously among the genomes of different vertebrate species. Much of the variation in these receptor repertoires can probably be explained by the adaptation of species to different environments. For example, the platypus, a semiaquatic monotreme, has the largest repertoire of vomeronasal receptors in all vertebrates surveyed to date, with more than 300 intact genes and 600 pseudogenes in this chemosensory receptor family.[68] However, it seems that a substantial portion of variety is generated by genomic drift, which probably also has an important role in both adaptive and nonadaptive evolution.[69,70]

The molecular mechanisms of gene amplification and genomic clustering are extensively studied in prokaryotes.[71] Here, gene duplication can occur during genome replication due to an unequal sister strand exchange, producing two adjacent identical copies of a region (the amplicon) that can undergo homologous recombination. Another mechanism, termed circle-excision and reinsertion, involves creation of a circular DNA molecule and its subsequent recombination with another DNA molecule to form the duplication. Alternatively, a rolling-circle mechanism may account for instances of very rapid gene amplification. There is evidence that similar mechanisms are responsible for gene amplifications in eukaryotes.[72] However, the precise molecular mechanisms for odorant GPCR cluster formation, for example, are not yet known.

The loss of GPCR functions

Currently, most genome analyses rely on sequence information from extant species. As more than 99 percent of all species that ever lived on earth are extinct, most of the information about receptor repertoires and their evolutionary history cannot be studied. However, recent advances in DNA extraction and amplification from fossil remains[73] and new sequencing technologies have made it possible to retrieve substantial amounts of ancient DNA sequences and even almost complete genomes dating back to approximately 1 Myr ago.[74–77] Genome sequences of extant organisms contain valuable information about past functions and gene evolution. For example, pseudogenes are considered as genomic fossils and increasingly attract attention in GPCR research. In addition, selection of favorable gene variants had left footprints that can be identified by suitable bioinformatic methods (see further in this chapter).

Pseudogenes are inheritable and characterized by a homology to a known gene and nonfunctionality.[78] As a result of their nonfunctionality, most pseudogenes are released from selective pressure. Therefore, compared to functional genes, pseudogenes, if old enough, display a ratio of nonsynonymous to synonymous substitution rates (K_a/K_s) of 1,[79] and accumulate frame shifting and nonsense mutations. Depending on the mechanism by which they evolved, the majority of mammalian pseudogenes can be classified as duplicated pseudogenes or retrotransposed pseudogenes (also called processed pseudogenes). The latter are generated by the reverse-transcription of mRNAs, followed by genomic integration. The human gonadotropin releasing hormone GnRH type II receptor homolog is one well-characterized example of GPCR pseudogenization caused by retrotransposition.[80] Duplicated pseudogenes arise from local duplication or unequal crossing-over (see the subchapter on expansion of GPCR genes earlier in this chapter). Thus, they often retain the original exon–intron structures of the parental genes. However, duplication or retrotransposition events are not always mechanistically necessary for pseudogenization. Inactivating mutations in former functional genes can cause loss of functionality, as demonstrated in GPCRs[81–83] and other genes. Once released from purifying selection, it will take several million years that obvious signatures of inactivation (premature stop codon, frameshift) become fixated in the coding sequence. In primate TAAR, for example, rough estimates suggest that 7–10 Myr are required to obtain and fixate at least one of such obvious signatures of receptor inactivation (unpublished data). On the other hand, signatures of the original sequence will gradually disappear over time. As a consequence, a pseudogene may escape present-day detection depending on the date and mechanism of its pseudogenization. Rough estimates suggest that signatures of genes can be detected for more than 80 Myr of neutral evolution. For example, the neuropeptide Y receptor type 6 (Y_6R) is a pseudogene in all primate genomes investigated so far by an inactivating deletion mutation, which occurred in the common ancestor of primates. Large deletions can remove informative sequences in an even shorter period, though, as in the Y_6R gene that disappeared in rats after the mouse/rat split 14–16 Myr ago. Because of the many difficulties in identifying pseudogenes,[84] the exact number of GPCR pseudogenes within a genome can only be an estimate.

Pseudogenization is particularly frequent in odorant GPCR (OR),[85] while signatures from only thirty nonolfactory rhodopsin-like GPCR pseudogenes have been detected so far in the human genome.[86] Out of ~900 ORs, at least 63 percent appear to be pseudogenes, leaving only ~350 apparently functional OR genes.[87] Furthermore, some apparently intact human OR genes lack motifs that are very highly conserved in their mouse orthologs, suggesting that not all human OR genes with complete open reading frames encode functional OR proteins. By contrast, in the mouse genome, only about 20 percent of the OR are pseudogenes, giving mice more than three times as many intact OR genes as humans have.[88] A similar accumulation of pseudogenes has been observed in human bitter taste receptors and the vomeronasal (pheromone) receptors.[54,70] It was speculated that the evolution of trichromatic color vision in hominoids and other Old World monkeys has relaxed the functional constraints for many taste, odorant, and pheromone receptors, hence permitting their pseudogenization.[89–91]

Reduction or even a loss of GPCR function can also be restricted to distinct populations of a species. Differences in environmental conditions may relax the constraint and promote neutral evolution of a gene formerly under purifying selection. The melanocortin type 1 receptor (MC1R) gene nicely represents this situation. The MC1R controls pigmentation of melanocyte and is, therefore, a central component in determining hair, skin, and coat color. Numerous studies have shown that even single amino acid mutations in MC1R can have profound effects on pigment phenotypes in vertebrates. In many cases, these changes in MC1R function, and the resulting pigmentation pattern, are thought to be adaptive.[92] In African populations, there is strong purifying selection on the MC1R gene, whereas the great MC1R sequence diversity in European and Asian populations suggests a relaxing of constraint.[93,94] Many of the European MC1R variants show reduction or loss of functionality leading to pale skin color and red hairs.[95] This suggests that the MC1R gene has played a significant role in the maintenance of dark pigmentation in Africans and pigment variation in non-African populations. It is still a matter of debate whether the loss of MC1R functionality in vertebrates is always due to a loss of constraint as a result of adaptation to habitats in which protection from sunlight is less relevant like for example, outside of the equatorial region or for cave-adapted species.[96] Reduced pigmentation may increase fitness. The reduced MC1R activity in some Pleistocene species, for example, Neanderthals[97] and mammoths,[98] may have promoted vitamin D synthesis in skin under the extreme climate conditions during Pleistocene ice ages. Further, reduced pigmentation due to loss of MC1R function can provide an advantage against predation. Illustrating examples are the beach mice in Florida[99,100] and a lizard species in the Chihuahuan Desert.[101] Both animals have lighter-colored coats than their mainland counterparts, driven by natural selection for camouflage against the pale sand dunes. These examples demonstrate that not only the gain but also the elimination of a GPCR function (pseudogenization) may have an evolutionary advantage and may also trigger adaptation.

STRUCTURAL EVOLUTION OF GPCRS

The growing number of crystal structures of GPCRs not only allows the identification of important structural hallmarks at the molecular level,[102] but also provides some clues of the receptor activation mechanism.[103–105] Moreover, this available structural information forms the basis to generate models of other GPCRs [106] not yet crystallized. Construction of reliable receptor models, however, still requires refinements based on experimental data from comparative studies, mutagenesis, cross-linking, and/or nuclear magnetic resonance (NMR) studies, and even the best data sets are only based on mutations at a few selected positions and limited in vitro functional analyses. Therefore, we are often unable to predict mutational effects from such models, and experimental assessment is still required. Mining evolutionary diversity as an additional source of structural/functional information may direct GPCR modeling and mutagenesis studies. This approach has been successfully applied since the very early stages of GPCR structure/function analysis to predict the approximate arrangements of the 7TM helices.[107] To identify more distinct structural determinants within a given GPCR that participate in ligand recognition and signal transduction, sequence analysis has to be focused on the comparison of receptor orthologs and paralogs. The basic concept of this approach is that the structural diversity between orthologs is the result of a long evolutionary process characterized by a continuous accumulation of mutations. The maintenance of vital functions in an organism strictly requires enough structural conservation to ensure the functionality of the receptor protein. The significance of such multisequence-based analyses increases with the number of orthologs included. Studies based on large numbers of orthologs analyzed the conservation and relative orientation of the TM helixes[53] and used such an evolutionary approach to address the functional relevance of distinct residues in GPCRs.[53,60,82,108,109]

Structural shaping of the core of GPCRs

Proteins with a 7TM-core architecture are already present in prokaryotes, but it is still a matter of debate whether these prokaryotic archaeal rhodopsins are the blueprint of the eukaryotic GPCRs. Clear evidence for direct phylogenetic relation between eukaryotic and prokaryotic opsins is still lacking[12] and may not be found because of the general difficulties in establishing meaningful phylogenetic relations between pro- and eukaryotic sequences. However, recent studies showed some structural homology within the putative TM regions of haloarchaeal rhodopsins, especially to bacteriorhodopsin and sensory rhodopsin, and fungal 7TM proteins.[110,111] There are at least two alternative hypotheses how the TM core of eukaryotic GPCR evolved.[112] The **first hypothesis** proposes exon shuffling of preexisting TM domains that may have occurred in the evolution of GPCR.[113] Such scenario would suggest an exon-intron structure of a prototypical GPCR gene ("intron early" theory). The two long-standing alternative explanations for the origin of introns, the "intron-early" and "intron-late" theories, remain a matter of continuous debate not only for GPCRs.[114] Many ancestral GPCRs that have orthologs in vertebrates and invertebrates contain introns within the coding region of the 7TM core and retained their intron/exon-structure during evolution. However, evolutionary gain and loss of introns, as found in rhodopsin-like GPCRs, are frequently observed, and it was unclear why mammalian GPCR genes are characterized by a large proportion of intronless genes or a lower density of introns when compared with GPCRs of invertebrates. A specific loss of introns in mammalian rhodopsin-like GPCRs was postulated.[115] A more detailed analysis showed that there was not a major loss of introns in mammalian GPCRs, but the formation of new GPCRs among mammals explains why these have fewer introns compared to invertebrate GPCRs.[116] Several mechanisms of intron loss are suggested including recombination or gene duplication through RNA intermediates and genomic deletion.[114] As an impressive example for an RNA intermediate mechanism, the four introns found in the ancestral vertebrate rhodopsin gene were simultaneously lost in the common ancestor of modern teleost fishes.[117] However, there was also an evolutionary gain of introns in vertebrate rhodopsin-like GPCR. For example, GPR34 subtypes and the ADP-receptor $P2Y_{12}$, which are intronless in the coding regions in evolutionarily old teleostei (zebrafish) and tetrapods, gained an intron in more recent fish evolution.[53,60]

An alternate, **second hypothesis** of creation of the TM core is that duplication occurred in the evolution of an ancestral gene, such that helixes 5–7 originated as duplicates of helixes 1–3, leading to intragenic as well as intergenic similarities between helixes 1–3 and 5–7 of bacteriorhodopsin and various GPCRs. Nevertheless, there is very little sequence homology between the five present-day GPCR subfamilies and haloarchaeal rhodopsins, but also between the five families. This makes it impossible to reconstruct the GPCR prototype from extant sequences.[112]

Long evolutionary processes have shaped the TM core of the individual GPCR families over hundreds of millions of years, releasing specific structural determinants. One key feature of the TM core is a highly conserved disulfide bridge between extracellular loops (ECL) 1 and 2. This disulfide bond is found in most members of the families rhodopsin/R, glutamate/G, adhesion/A, and secretin/S. Numerous functional analyses of mutant rhodopsin-like GPCRs, in which these cysteine residues were replaced by other amino acids, have shown that this disulfide bond may be critical for receptor signaling.[118–120] In some receptors, however, this disulfide bond is required to maintain more distinct functions. Systematic mutagenesis studies of the conserved cysteine residues in several GPCRs showed that disruption of the disulfide bond does not influence the receptor's ability to activate G proteins, but interferes with high affinity ligand binding and receptor trafficking.[121–123] Despite the disruption of the disulfide bond, the receptor core structure appears to remain intact, allowing receptor function. Consistent with this notion, some GPCRs, for example, receptors for sphingosine 1-phosphate and lysophosphatidic acid, lack the conserved extracellular Cys residues.

Another hallmark of the GPCR structure is the high frequency of TM kinks, which commonly occur at proline residues. The pattern of helical kinks appears to be conserved within but differs between the individual GPCR families. This implicates conserved differences in the fine arrangement of the TM core between the GPCR families[124]. Reflecting the crystal structures of GPCR, TM kinks do not always require proline. It was proposed that in evolution, a mutation to proline initially induces the kink in a helix. The resulting packing defects are later repaired by further mutation, thereby locking the kink in the structure even in cases where the proline was removed later in evolution[124]. The relative position of kinks due to proline residues within a TM also may have played a pivotal role in the structural evolution of subfamilies. As a specific example, indels (insertion/deletion of base pairs) shaped the relative positioning of a conserved proline in TM2 of rhodopsin-like receptor subfamilies during evolution.[125]

Within their TM core, most GPCR families possess a number of highly conserved sequence motifs. In the rhodopsin-like GPCRs, for example, the E/DRY motif at the transition of TM3 and ICL2 and the N/DP(X)nY motif in 7TM are preserved during more than 570–700 Myr of evolution (see the beginning of this chapter on the origin of GPCR genes) and present in almost all family members. Numerous studies highlight the functional importance of the conserved E/DRY motif, and the crystal structures of (rhod)opsins implicate an intramotif salt bridge or multiple intramolecular hydrogen bonds of the Glu and Arg residues.[105,126] However, several family 1 GPCRs are known in which the acidic residue (Asp, Glu) within this motif is naturally substituted by His, Asn, Gln, Gly, Val, Thr, Cys, or Ser residues. This fact questions a generalization of the structural arrangement of this motif found in rhodopsin structures. The N/DP(X)nY motif within the 7TM near the cytoplasmic face of the plasma membrane is highly conserved. The functional importance of, for example, the Asn and the Pro within the N/DP(X)nY motif has been demonstrated.[127,128] It is noteworthy, however, that in a few receptors, the Asn/Asp residue within the N/DP(X)nY is naturally replaced by Ser, Thr, Lys, or His. This all indicates that even with GPCR structures in hand, highly conserved motifs may have different structures and functions and need to be individually analyzed for the respective receptor.

Structural evolution of intra- and extracellular domains of GPCRs

Several families (glutamate-like, adhesion-like GPCRs) and subfamilies (leucin-rich repeats [LRR] receptors of rhodopsin-like GPCRs) possess very long N and/or C termini. The N termini are variable in length, up to 6,000-amino-acids-long in the very large GPCR type 1 (VLGR1[129]). The receptors with very large N-terminal ectodomains are often rich in glycosylation sites and proline residues, forming what has been described as mucin-like stalks, and are likely to participate in cell adhesion.[130] The module-like structures encoded by complex intron/exon loci suggest exon shuffling as the mechanism that formed these large ectodomains.

In rhodopsin-like GPCRs, LRR-containing GPCRs (LGR), such as the classical glycoprotein hormone receptors TSHR, LHR, and FSHR, and the LGR for relaxins, constitute a unique cluster of receptors with large N termini sharing a large LRR domain for hormone binding. The early origin of LGRs is illustrated by the existence of a receptor related to human glycoprotein hormone receptors in cnidarians (sea anemone) and nematodes.[38,131] A hallmark for LGRs is the large N-terminal extracellular domain involved in selective hormone binding that is composed of tandem arrays of LRRs motifs, N- and C-terminally flanked by cysteine-rich sequences.[132] Based on the architecture of the ectodomain, that is, the number of LRRs, and the sequence similarity, LGRs can be classified into three subtypes (A, B, C).[133] The two modules, the ectodomain (or part of it) and the 7TM core, were preformed earlier and joined together by genomic rearrangement. This hypothesis is further supported by the fact that the isolated ectodomain structurally assembles and crystallizes and binds the glycoprotein hormone independently from the 7TM core.[134]

The vast majority of adhesion-GPCRs contains a GPS domain[135,136] immediately N-terminal of the 7TM domain. This GPS-7TM module is conserved from choanoflagellates to humans, suggesting that it represents a combination of core functional domains rather than an association that arose later by exon shuffling. The GPS and N-terminal TM helixes are indeed typically encoded in a single exon.

With increasing complexity of animal genomes, the adhesion-GPCR subfamily expanded to up to thirty-three members that comprise divergent long N-terminal domains followed by the conserved GPS-7TM module.[137] The N-terminal domains can be grouped into at least eight different classes again defined by their domain architectures. The number of receptors in each adhesion-GPCR subclass is variable from species to species, indicating rapid evolution by duplication and mutation. Consistent with this assumption, nearly 50 percent of adhesion-GPCR genes in mammalian genomes occur in clusters.[130,138,139]

Within the range of divergent N-termini, two highly conserved subgroups of adhesion-GPCRs appear to be highly conserved in all bilateral animals. The cadherin-like subfamily described in Drosophila as *flamingo/starry night*[140] and CELSR (Cadherin, EGF-like, LAG-like, and seven-pass receptor) in vertebrates[141,142] is an essential component of planar cell polarity and tissue polarity pathways and interacts with wnt/frizzled signaling.[143,144] The second conserved domain architecture is found in the lectin-like latrophilin/lectomedin protein family,[145] where the N terminus contains an unusual domain related to rhamnose-binding lectins found in sea urchins.[146] Latrophilin was initially described as the calcium-independent receptor for latrotoxin, the neurotoxin found in the venom of the Black Widow spider *Latrodectus mactans*. Latrophilins have been suggested to act as GPCR modulating synaptic exocytosis[145] but are very likely to also fulfill essential developmental functions. The comparison of invertebrate and vertebrate latrophilins provides a good illustration of domain shuffling, because vertebrate latrophilins have acquired an additional olfactomedin-like domain in their N termini that is not present in invertebrate homologs.[137]

Many adhesion-GPCRs contain a hormone-binding (HRM) domain in their N termini, which is homologous to the HRM of the secretin receptor subfamily. Because the 7TM regions of adhesion-GPCRs and secretin-type receptors also show significant homology, it is tempting to speculate that secretin-receptor-like proteins arose by domain loss from adhesion-GPCR-like ancestors.[32] However, functional data supporting this hypothesis have not yet been presented.

Coevolution of GPCRs and their ligands/associated factors

Comparative genomics is a valuable tool for deciphering coevolution of GPCRs and their (putative) ligands leading to novel insights in the evolutionary conservation of signaling systems. As stated previously in this chapter, gene duplication provides redundancy and therefore robustness and an increase in gene doses, but gives the chance for developing new ligand/receptor systems without risking fitness of the organism. There are a number of examples where the receptor and the ligand underwent coevolution upon gene duplications among many peptides and their receptors.[147–151] For example, gene duplication events led to the expansion of both, the subtype A LGR (see the subchapter on structural evolution of intra- and extracellular domains of GPCRs earlier in this chapter) and glycoprotein hormone subunits in vertebrates. Afterward, coevolution shaped ligand and signaling properties of the subtypes. In contrast to the high constitutive activity of fly and nematode LGR,[131,152] some LGRs have become more intramolecularly constrained during evolution in the chordate lineage. In the case of human FSHR, for example, evolutionary changes in the sequence of the 7TM core lead to a decreased basal receptor activity and, in addition, cause a decrease in the sensitivity of FSHR to structurally related glycoprotein hormones such as chorionic gonadotropin hormone (CG) and TSH. Therefore, it has been suggested that an intramolecular constraint acquired during the evolution of primates contributed to the protection of the FSHR against promiscuous activation by the primate-specific CG.[153].

SELECTION ON GPCR GENES

Determining the molecular basis of evolutionary adaptation remains largely unresolved, and the respective roles of selection and demography in shaping the genomic structure are actively debated. The recent availability of large panels of SNPs in humans[154] and other species[155] has given new momentum to both the search for signatures of recent selection and the mutations underlying variation in complex traits, with the latter being investigated in genome-wide genotype-phenotype association (GWA) studies. GPCRs, as one of the largest gene families with high relevance in almost every biological function, are highly involved in adaption processes. It is therefore not surprising that GWA studies in humans and animals revealed a number of GPCRs, at least at a statistical level, responsible for previously unappreciated phenotype variations. For instance, variations in the thyreotropin-releasing hormone receptor gene and the GPR133 gene are associated with lean body mass[156] and body height (P. Kovacs, M. Stumvoll, personal communication), respectively. This field in GPCR research just started, and the paragraphs that follow can only expose the potential power of these new methods.

Genetic signatures of selection

As in other genes, disadvantageous mutations in GPCRs are removed from a population (purifying selection). Many evolutionarily old GPCR genes, including the rhodopsins, display strong purifying selection across their entire coding regions as determined by K_a/K_s ratios [5,53,54]. A sliding-window analysis of K_a/K_s usually shows that the N- and C termini and the intracellular loop regions of GPCRs are more variable and display higher K_a/K_s ratios than the TMD region. But how can we detect more subtle adaptive changes that improve gene performance?

First, phylogenetic comparisons of several dozens of ortholog sequences can help with identifying functionally important residues and domains.[53] Fixed differences in highly conserved determinants, which are unique for a species or species group, may be indicative of adaptive change (positive selection). The modules integrated into the PAML package are well-established bioinformatic tools to test for site-specific positive selection.[157,158] It should be noted that such fixed differences are not necessarily the result of selection; therefore, testing for the functional relevance of the change is required.

Second, population genetic models predict that rapid selection should leave 'footprints' in closely linked genomic regions (selective sweeps). Complete or near-complete fixation of an allele may be indicative of adaptive changes in populations of one species, especially when accompanied by signatures of a selective sweep. There are several methods for detecting signatures of selective sweeps in genomic sequences.[159] Large genome-wide analyses have scanned the human genome for signatures of recent positive selection on the basis of species differences and allele variation within populations.[5,6]

The combination of allele frequency analyses, geographic allele distribution data, and sequence comparison to closely related species has provided evidence for the presence of selective pressure on many GPCR genes (see the subchapter immediately following). The human bitter-taste receptor TAS2R16 is an illustrative example. Sequence comparison between mammalian species and human populations revealed signatures of positive selection, as indicated by an excess of evolutionarily derived alleles at the nonsynonymous site (Lys[172]Asn) in the TAS2R16 coding region. At the functional level, the Asn[172] variant has an increased sensitivity toward harmful cyanogenic glycosides. The improved recognition of bitter natural toxins through taste may have conferred an important selective advantage and may have driven the increase of frequency of the derived Asn[172] allele at an early stage of human evolution.[160]

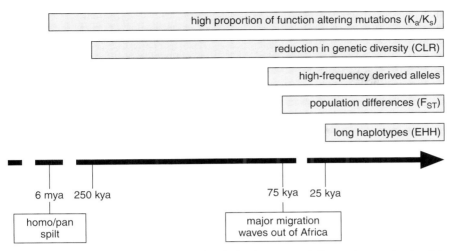

Figure 1-2: Signatures of positive selection during human evolution and examples for detection methods.

The five main signatures of selection exist over variable time scales, thus covering different periods of human evolution. The statistical tests named here are those applied in the genome-wide selection studies from Table 1.1 (adapted from Sabeti et al.[7]). K_a/K_s: rate of non-synonymous to the rate of synonymous substitutions, CLR: composite likelihood ratio, EHH: extended haplotype homozygosity, F_{ST}: fixation index, mya: million years ago, kya: kilo years ago.

Selection of genomic regions containing GPCR genes

Identifying regions in genomes that have been targets of positive selection will provide important insights into recent evolutionary history. Methods used to identify regions of recent positive selection in a given genome are based on a simple fact. Alleles that are under positive selection will increase in prevalence in a population leaving distinctive patterns of genetic variation in the DNA sequence. These signatures can be identified by comparison with the background distribution of genetic variation in the species, which is generally argued to evolve largely under neutrality (no effect on an individual's fitness).[7] Most studies are based on large-scale SNP data of selected human populations as created by the International Haplotype Map (HapMap) and Perlegen Sciences projects.[154] Many specific statistical tests have been developed to detect positive selection (see Figure 1.2), but they are all based broadly on signatures such as (1) reduction of genetic diversity, (2) frequency of derived alleles, (3) differences in allele frequency between populations, and (4) long haplotypes (for further readings see Sabeti et al.[7]). An advantage of the derived test statistics is that they are broadly not codon-based, hence carrying the potential to elucidate natural selection even on non-genic genome areas. The signatures appear and disappear in different time scales so that with the same data set, the different methods do not always produce the same results. Further, a challenging problem of such genomic scans of recent positive selection is yet to clearly distinguish selection from neutral effects because of genetic drift, demographic reasons, and bottlenecks.[161]

Table 1-1. Potentially Selected GPCR Genes in the Evolution of Modern Human Populations Identified in Genome-wide Studies.

GPCR Genes	Analyzed Data	Detection Method	Reference
LGR8, CYSLTR2, OR2B2, OR5I1, OR2W1	gene coding regions of human and chimpanzee	K_a/K_s	[5]
CXCR4*, GPR73	SNP data from HapMap1 (analysis restricted to chromosome 2)	CLR	[6]
Africans: GPR34, GPR82, OR4C2P Ψ, OR4C3, OR4C4P Ψ, OR4C5 Ψ, OR5T2, OR8J2 Ψ, OR8K5, OR8V1P Ψ, OR10V2P Ψ, OR10Y1P Ψ **Europeans**: OR2AK2, OR2L1P Ψ, OR2L2, OR2L5 Ψ, OR2L9P Ψ, OR2L13 **Asians**: GRM3*, GRM8*	SNP data from HapMap1	iHS (EHH based)	[9]
GRM6, TAS1R1, GPR111*, PTGER4, OR2A14, OR5D18, OR51D1, OR52W1	gene coding regions of human and chimpanzee	K_a/K_s	[188]
Europeans: CXCR4*, CXCR7, NPY1R, GPR65, GPR151, DRD5 **Asians**: BAI3*, GRM3*, GPR65	SNP data from HapMap2 and Perlegen Sciences project	Rsb (EHH based)	[189]
Europeans: GPR64* **Asians**: CELSR1*	SNP data from HapMap2	XP-EHH	[8]
Africans: BAI3*, OR52A1, OR52A4, OR52A5, OR4K1, OR4K5 **Europeans**: GPR64*, GPR177, TACR3, OR4K1, OR4K5 **Asians**: GPR111*, GPR115, OR4A47*	SNP data from HapMap2	LRH and/or iHS (EHH based)	[154]
Africans: OR4P4 **Asians**: OR9K2, MC2R	SNP data from Perlegen Sciences project	CLR	[190]
PTGER3, GRM7, GRM8*, LGR7, BAI3*, P2RY2, ADORA2B, CELSR1*, OR52K2	SNP data from HapMap2	F_{ST}	[191]

The following methods, mainly testing for positive selection, were applied: K_a/K_s… rate of non-synonymous to the rate of synonymous substitutions, CLR … composite likelihood ratio, EHH … extended haplotype homozygosity, iHS … integrated haplotype score, Rsb … relative integrated EHH of a site between populations, XP-EHH … cross population EHH, LRH … long-range haplotype, F_{ST} … fixation index. Please see Figure 1.2 for the signature of selection the statistical tests are based on. *… selection candidates as determined by two or more different detection methods, Ψ … pseudogenes.

The meta-analysis of genome-wide studies with human SNP data revealed a number of genomic regions with signatures of positive selection containing GPCRs. Among them are several receptors that are implicated in immune defense and in the chemosensory system [6,81,162–166] (Table 1.1). About 3.8±2.9 percent (extracted from studies in Table 1.1) of all genes in genomic regions with signatures of selection are GPCRs, indicating that there is no overrepresentation of GPCRs in potentially selected genes. It is of interest to note that several GPCR s of the adhesion receptor family (BAI3, CELSR1, GPR64/HE6) are in selected regions that were detected by more than one method (see Table 1.1). Further studies are now necessary to elucidate the functional relevance of the selected allele and the causality of selection.

Selection of individual GPCRs

As outlined earlier in the chapter, genome-wide scans can assist in identifying signatures of positive selection in a genomic sequence block. It is even more challenging to proof selection of an individual gene. Thus functional data and an idea of the selective force that worked on a specific gene variant are essentially required. There are only a few reports where selection may have favored a functional GPCR variant. Chemosensory receptors, such as odorant and taste receptors, have been found under positive selection in human and vertebrate populations suggesting adaptive functionality.[70,160,167-169] Another impressive example comes from Drosophila genetics. Methuselah, a member of the adhesion receptor family in insects, has been proposed to have major effects on stress response and longevity phenotype.[170] Methuselah has experienced an unusual high level of adaptive amino acid divergence concentrated in the intra- and extracellular loop domains of the receptor protein. This suggests the historical action of positive selection on those regions of the molecule that modulate signal transduction. Analysis of haplotypes in *D. melanogaster* populations provided further evidence for contemporary and spatially variable selection at the Methuselah locus.[171]

Domestication is an evolutionary process in which animals are exposed to specific selection pressures defined by humans. The melanocortins and their receptors play a pivotal role in skin pigmentation and regulation of the energy homeostasis. For that reason, they have been popular targets in domestication of farm animals as seen in traits of domestic pigs that have been selected because of an increase in growth and food intake. A partial inactivating mutation in the MC4R gene has been shown to account for this desired porcine phenotype.[172] Similarly, coat color phenotypes in domestic pigs result from direct human selection and not via a simple relaxation of natural selective pressures at the MC1R.[173]

As shown in Table 1.1, several genomic sequences with signatures of selection contain pseudogenes of GPCR. This raises the question of whether inactive GPCR variants may provide advantage and are, therefore, selected as shown for other genes, for example, caspase12.[85] Indeed, there are several examples where receptor inactivation may provide an advantage under distinct environmental circumstances. For example, the FY*O allele at the Duffy locus is at or near fixation in sub-Saharan Africa but rare in other parts of the world.[162] The Duffy antigen, which is a chemokine receptor, acts as a co-receptor for the cell entry of *Plasmodium vivax*. Mutation studies have shown that inactivation of Duffy leads to resistance to infections by *P. vivax*.[174]

Selection on the null-allele is probably responsible for the evolution of the chemoattractant-like receptor GPR33.[81] After the appearance of GPR33 in the mammalian genome more than 125–190 Myr ago, this receptor underwent pseudogenization in humans, other hominoids, and some rodent species. Simultaneous pseudogenization in several unrelated species within the last 1 Myr caused by neutral drift appears to be very unlikely. It was speculated that a likely cause of GPR33 inactivation was its interplay with a rodent-hominoid-specific pathogen. Although selection of the GPR33 pseudogene is still hypothetical, this consideration is supported by the fact that rats and gerbils are frequently the host of zoonotic pathogens like hanta viruses and *Yersinia pestis*, respectively, and both species share their habitat with humans.[81]

An even more impressive example was identified in TAARs. TAARs form a specific subfamily of GPCRs in vertebrates and were initially considered neurotransmitter receptors. Recent studies suggest that mouse and fish TAAR function as chemosensory receptors in the olfactory epithelium.[67,175] Two receptor subtypes, TAAR3 and TAAR4, became independently inactivated in both great apes and some New World monkeys, obviously in a comparable evolutionary time scale, on different continents (Figure 1.3). As with GPR33, given that the inactivation occurred independently and in a number of species of unrelated orders an exogenous trigger is likely to be responsible for driving the inactivation of distinct TAAR. These examples clearly show that evolutionary and population genetic data can conceptually enrich GPCR research even for those GPCRs that became inactivated during evolution.

IN VITRO EVOLUTION OF GPCRS

All approaches described previously in this chapter extract only past evolutionary information where the selective forces cannot be manipulated or remain unknown. For many biomolecules such as RNA, DNA, enzymes, and antibodies, in vitro evolution techniques have been developed, which guide structure and maturation under researcher-defined selection conditions.[176,177] Over more than one billion years of GPCR evolution, the unique TM architecture and manifold combinations of other residues assure the evolutionary adaptability toward almost any chemical structure that may serve as ligand. In keeping with the idea of a universal receptor backbone, artificial programming of the binding site by in vitro evolution approaches has led to the generation of GPCRs with new ligand binding or signal transduction properties.[178] This opened the possibility of "designing" GPCRs for special purposes, for example, as biosensors or to better study distinct GPCR signaling pathways. As a first step toward this goal, so-called RASSLs (receptors activated solely by synthetic ligands) were developed.[179] RASSLs are unresponsive to endogenous agonists but can be activated by nanomolar concentrations of pharmacologically inert, drug-like small molecules.

Until recently, most mutations converting pharmacological ligand properties were accidentally identified by site-directed mutagenesis approaches. Although

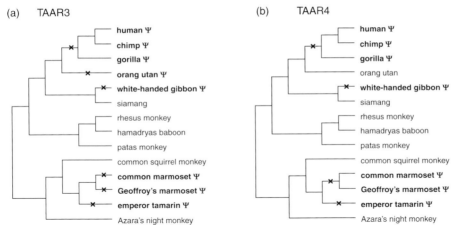

Figure 1-3: Phylogenetic trees of primate TAAR3 and TAAR4 subtypes.
Sequence information of selected primate TAAR3 (A) and TAAR4 (B) orthologs were analyzed for open reading frames. Events causing the particular pseudogenization (nucleotide insertion, deletion, or stop mutation) are displayed on the affected branches (red cross). Pseudogenes are indicated as Ψ and highlighted in red.

classical site-directed mutagenesis and alanine-scan approaches have led to many important insights into the molecular mechanisms governing GPCR function, these strategies are limited by their relative small number of mutant proteins, and combinations of mutations are rarely studied. More advanced in vitro evolution technologies in the GPCR field are mainly based on the very robust G protein/ receptor crosstalk, which is interchangeable between even distantly related species, for example, yeast and mammals. Only one minor change in the C terminus of the yeast G protein, Gpa1p, has enabled mammalian GPCRs to activate the yeast mating pathway.[180] Yeast cell growth directly depends on receptor activation by applying the following principle: The yeast pheromone receptor is replaced with a mammalian GPCR and the yeast G protein is tailored to couple mammalian GPCRs to a specifically engineered pheromone response pathway. The coupling of receptor activity to yeast growth allows rapid and economical screening of randomly modified receptor libraries. Such approaches are suitable for ligand identification, mutational maturation, saturable structure-function analysis, genetic selection of constitutively active receptors, and generation of receptors with novel ligand recognition properties.[181,182] For example, a well-established yeast mutagenesis system was used to produce hundreds of thousands of mutant M3 muscarinic receptors, and screened for signaling characteristics of an 'ideal' RASSL. After multiple rounds of mutagenesis and iterative screening, mutants were isolated that had lost the ability to respond to the natural agonist (acetylcholine) but gained the ability to respond to clozapine-N-oxide.[183]

The field of GPCR in vitro evolution has grown continuously in the past decade. An ideal RASSL is a clinically approved but physiologically almost inert drug (antibiotic or antiviral), allowing tissue engineering without the drug safety studies needed for new compounds.[184] Previous studies have already demonstrated that GPCRs with new ligand-binding and signal-transduction abilities created by sophisticated site-directed mutagenesis[179,185] and in vitro selection[183] can lead to powerful tools for biomedical in vitro and in vivo research. Recent studies extend these opportunities by showing that GPCR architecture still provides room for additional, allosteric agonist binding sites.[186] Such newly generated allosteric site, ligands can activate the receptor even when the receptor is occupied by an inverse agonist at the orthosteric binding site.[187] Receptors with designed additional agonistic binding sites may be of value as biosensors and ligand-concentration-dependent off/on bioswitches.

SUGGESTED READING

1. Fredriksson, R., and Schioth, H. B. (2005) *Mol Pharmacol* 67(5), 1414–1425.
2. Schioth, H. B., and Fredriksson, R. (2005) *Gen Comp Endocrinol* 142(1–2), 94–101.
3. Akey, J. M., Eberle, M. A., Rieder, M. J., Carlson, C. S., Shriver, M. D., Nickerson, D. A., and Kruglyak, L. (2004) *PLoS Biol* 2(10), e286.
4. Bustamante, C. D., Fledel-Alon, A., Williamson, S., Nielsen, R., Hubisz, M. T., Glanowski, S., Tanenbaum, D. M., White, T. J., Sninsky, J. J., Hernandez, R. D., Civello, D., Adams, M. D., Cargill, M., and Clark, A. G. (2005) *Nature* 437(7062), 1153–1157.
5. Nielsen, R., Bustamante, C., Clark, A. G., Glanowski, S., Sackton, T. B., Hubisz, M. J., Fledel-Alon, A., Tanenbaum, D. M., Civello, D., White, T. J. J. S., Adams, M. D., and Cargill, M. (2005) *PLoS Biol* 3(6), e170.
6. Nielsen, R., Williamson, S., Kim, Y., Hubisz, M. J., Clark, A. G., and Bustamante, C. (2005) *Genome Res* 15(11), 1566–1575.
7. Sabeti, P. C., Schaffner, S. F., Fry, B., Lohmueller, J., Varilly, P., Shamovsky, O., Palma, A., Mikkelsen, T. S., Altshuler, D., and Lander, E. S. (2006) *Science* 312(5780), 1614–1620.
8. Sabeti, P. C., Varilly, P., Fry, B., Lohmueller, J., Hostetter, E., Cotsapas, C., Xie, X., Byrne, E. H., McCarroll, S. A., Gaudet, R., Schaffner, S. F., Lander, E. S., Frazer, K. A., Ballinger, D. G., Cox, D. R., Hinds, D. A., Stuve, L. L., Gibbs, R. A., Belmont, J. W., Boudreau, A., Hardenbol, P., Leal, S. M., Pasternak, S., Wheeler, D. A., Willis, T. D., Yu, F., Yang, H., Zeng, C., Gao, Y., Hu, H., Hu, W., Li, C., Lin, W., Liu, S., Pan, H., Tang, X., Wang, J., Wang, W., Yu, J., Zhang, B., Zhang, Q., Zhao, H., Zhao, H., Zhou, J., Gabriel, S. B., Barry, R., Blumenstiel, B., Camargo, A., Defelice, M., Faggart, M., Goyette, M., Gupta, S., Moore, J., Nguyen, H., Onofrio, R. C., Parkin, M., Roy, J., Stahl, E., Winchester, E., Ziaugra, L., Altshuler, D., Shen, Y., Yao, Z., Huang, W., Chu, X., He, Y., Jin, L., Liu, Y., Shen, Y., Sun, W., Wang, H., Wang, Y., Wang, Y., Xiong, X., Xu, L., Waye, M. M., Tsui, S. K., Xue, H., Wong, J. T., Galver, L. M., Fan, J. B., Gunderson, K., Murray, S. S., Oliphant, A. R., Chee, M. S., Montpetit, A., Chagnon, F., Ferretti, V., Leboeuf, M., Olivier, J. F., Phillips, M. S., Roumy, S., Sallee, C., Verner, A., Hudson, T. J., Kwok, P. Y., Cai, D., Koboldt, D. C., Miller, R. D., Pawlikowska, L., Taillon-Miller, P., Xiao, M., Tsui, L. C., Mak, W., Song, Y. Q., Tam, P. K., Nakamura, Y., Kawaguchi, T., Kitamoto, T., Morizono, T., Nagashima, A., Ohnishi, Y., Sekine, A., Tanaka, T., Tsunoda, T., Deloukas, P., Bird, C. P., Delgado, M., Dermitzakis, E. T., Gwilliam, R., Hunt, S., Morrison, J., Powell, D., Stranger, B. E., Whittaker, P., Bentley, D. R., Daly, M. J., de Bakker, P. I., Barrett, J., Chretien, Y. R., Maller, J., McCarroll, S., Patterson, N., Pe'er, I., Price, A., Purcell, S., Richter, D. J., Sabeti, P., Saxena, R., Schaffner, S. F., Sham, P. C., Varilly, P., Altshuler, D., Stein, L. D., Krishnan, L., Smith, A. V., Tello-Ruiz, M. K., Thorisson, G. A., Chakravarti, A., Chen, P. E., Cutler, D. J., Kashuk, C. S., Lin, S., Abecasis, G.

R., Guan, W., Li, Y., Munro, H. M., Qin, Z. S., Thomas, D. J., McVean, G., Auton, A., Bottolo, L., Cardin, N., Eyheramendy, S., Freeman, C., Marchini, J., Myers, S., Spencer, C., Stephens, M., Donnelly, P., Cardon, L. R., Clarke, G., Evans, D. M., Morris, A. P., Weir, B. S., Tsunoda, T., Johnson, T. A., Mullikin, J. C., Sherry, S. T., Feolo, M., Skol, A., Zhang, H., Zeng, C., Zhao, H., Matsuda, I., Fukushima, Y., Macer, D. R., Suda, E., Rotimi, C. N., Adebamowo, C. A., Ajayi, I., Aniagwu, T., Marshall, P. A., Nkwodimmah, C., Royal, C. D., Leppert, M. F., Dixon, M., Peiffer, A., Qiu, R., Kent, A., Kato, K., Niikawa, N., Adewole, I. F., Knoppers, B. M., Foster, M. W., Clayton, E. W., Watkin, J., Gibbs, R. A., Belmont, J. W., Muzny, D., Nazareth, L., Sodergren, E., Weinstock, G. M., Wheeler, D. A., Yakub, I., Gabriel, S. B., Onofrio, R. C., Richter, D. J., Ziaugra, L., Birren, B. W., Daly, M. J., Altshuler, D., Wilson, R. K., Fulton, L. L., Rogers, J., Burton, J., Carter, N. P., Clee, C. M., Griffiths, M., Jones, M. C., McLay, K., Plumb, R. W., Ross, M. T., Sims, S. K., Willey, D. L., Chen, Z., Han, H., Kang, L., Godbout, M., Wallenburg, J. C., L'Archeveque, P., Bellemare, G., Saeki, K., Wang, H., An, D., Fu, H., Li, Q., Wang, Z., Wang, R., Holden, A. L., Brooks, L. D., McEwen, J. E., Guyer, M. S., Wang, V. O., Peterson, J. L., Shi, M., Spiegel, J., Sung, L. M., Zacharia, L. F., Collins, F. S., Kennedy, K., Jamieson, R., and Stewart, J. (2007) *Nature* 449(7164), 913–918.

9. Voight, B. F., Kudaravalli, S., Wen, X., and Pritchard, J. K. (2006) *PLoS Biol* 4(3), e72.

10. Sineshchekov, O. A., Jung, K. H., and Spudich, J. L. (2002) *Proc Natl Acad Sci U S A* 99(13), 8689–8694.

11. Waschuk, S. A., Bezerra, A. G., Jr., Shi, L., and Brown, L. S. (2005) *Proc Natl Acad Sci U S A* 102(19), 6879–6883.

12. Soppa, J. (1994) *FEBS Lett* 342(1), 7–11.

13. Elion, E. A. (2000) *Curr Opin Microbiol* 3(6), 573–581.

14. Fujisawa, Y., Kato, H., and Iwasaki, Y. (2001) *Plant Cell Physiol* 42(8), 789–794.

15. Kim, J. Y., Haastert, P. V., and Devreotes, P. N. (1996) *Chem Biol* 3(4), 239–243.

16. Violin, J. D., and Lefkowitz, R. J. (2007) *Trends Pharmacol Sci* 28(8), 416–422.

17. Prabhu, Y., and Eichinger, L. (2006) *Eur J Cell Biol* 85(9–10), 937–946.

18. Taniura, H., Sanada, N., Kuramoto, N., and Yoneda, Y. (2006) *J Biol Chem* 281(18), 12336–12343.

19. Perovic, S., Krasko, A., Prokic, I., Muller, I. M., and Muller, W. E. (1999) *Cell Tissue Res* 296(2), 395–404.

20. Pin, J. P., Galvez, T., and Prezeau, L. (2003) *Pharmacol Ther* 98(3), 325–354.

21. O'Hara, P. J., Sheppard, P. O., Thogersen, H., Venezia, D., Haldeman, B. A., McGrane, V., Houamed, K. M., Thomsen, C., Gilbert, T. L., and Mulvihill, E. R. (1993) *Neuron* 11(1), 41–52.

22. Kuang, D., Yao, Y., Maclean, D., Wang, M., Hampson, D. R., and Chang, B. S. (2006) *Proc Natl Acad Sci U S A* 103(38), 14050–14055.

23. Kulkarni, R. D., Thon, M. R., Pan, H., and Dean, R. A. (2005) *Genome Biol* 6(3), R24.

24. Gookin, T. E., Kim, J., and Assmann, S. M. (2008) *Genome Biol* 9(7), R120.

25. King, N., Hittinger, C. T., and Carroll, S. B. (2003) *Science* 301(5631), 361–363.

26. King, N., Westbrook, M. J., Young, S. L., Kuo, A., Abedin, M., Chapman, J., Fairclough, S., Hellsten, U., Isogai, Y., Letunic, I., Marr, M., Pincus, D., Putnam, N., Rokas, A., Wright, K. J., Zuzow, R., Dirks, W., Good, M., Goodstein, D., Lemons, D., Li, W., Lyons, J. B., Morris, A., Nichols, S., Richter, D. J., Salamov, A., Sequencing, J. G., Bork, P., Lim, W. A., Manning, G., Miller, W. T., McGinnis, W., Shapiro, H., Tjian, R., Grigoriev, I. V., and Rokhsar, D. (2008) *Nature* 451(7180), 783–788.

27. Adell, T., Nefkens, I., and Muller, W. E. (2003) *FEBS Lett* 554(3), 363–368.

28. Adell, T., Thakur, A. N., and Muller, W. E. (2007) *Cell Biol Int* 31(9), 939–949.

29. Momose, T., and Houliston, E. (2007) *PLoS Biol* 5(4), e70.

30. Putnam, N. H., Srivastava, M., Hellsten, U., Dirks, B., Chapman, J., Salamov, A., Terry, A., Shapiro, H., Lindquist, E., Kapitonov, V. V., Jurka, J., Genikhovich, G., Grigoriev, I. V., Lucas, S. M., Steele, R. E., Finnerty, J. R., Technau, U., Martindale, M. Q., and Rokhsar, D. S. (2007) *Science* 317(5834), 86–94.

31. Srivastava, M., Begovic, E., Chapman, J., Putnam, N. H., Hellsten, U., Kawashima, T., Kuo, A., Mitros, T., Salamov, A., Carpenter, M. L., Signorovitch, A. Y., Moreno, M. A., Kamm, K., Grimwood, J., Schmutz, J., Shapiro, H., Grigoriev, I. V., Buss, L. W., Schierwater, B., Dellaporta, S. L., and Rokhsar, D. S. (2008) *Nature* 454(7207), 955–960.

32. Nordstrom, K. J., Lagerstrom, M. C., Waller, L. M., Fredriksson, R., and Schioth, H. B. (2009) *Mol Biol Evol* 26(1), 71–84.

33. Cardoso, J. C., Pinto, V. C., Vieira, F. A., Clark, M. S., and Power, D. M. (2006) *BMC Evol Biol* 6, 108.

34. Benton, M. J., and Ayala, F. J. (2003) *Science* 300(5626), 1698–1700.

35. Feng, D. F., Cho, G., and Doolittle, R. F. (1997) *Proc Natl Acad Sci U S A* 94(24), 13028–13033.

36. Peterson, K. J., and Butterfield, N. J. (2005) *Proc Natl Acad Sci U S A* 102(27), 9547–9552.

37. Bouchard, C., Ribeiro, P., Dube, F., and Anctil, M. (2003) *J Neurochem* 86(5), 1149–1161.

38. Nothacker, H. P., and Grimmelikhuijzen, C. J. (1993) *Biochem Biophys Res Commun* 197(3), 1062–1069.

39. Saitoh, O., Yuruzume, E., Watanabe, K., and Nakata, H. (1997) *Gene* 195(1), 55–61.

40. Tierney, A. J. (2001) *Comp Biochem Physiol A Mol Integr Physiol* 128(4), 791–804.

41. Arendt, D., Tessmar-Raible, K., Snyman, H., Dorresteijn, A. W., and Wittbrodt, J. (2004) *Science* 306(5697), 869–871.

42. Suga, H., Schmid, V., and Gehring, W. J. (2008) *Curr Biol* 18(1), 51–55.

43. Alvarez, C. E. (2008) *BMC Evol Biol* 8, 222.

44. Kamesh, N., Aradhyam, G. K., and Manoj, N. (2008) *BMC Evol Biol* 8, 129.

45. Hauser, F., Cazzamali, G., Williamson, M., Park, Y., Li, B., Tanaka, Y., Predel, R., Neupert, S., Schachtner, J., Verleyen, P., and Grimmelikhuijzen, C. J. (2008) *Front Neuroendocrinol* 29(1), 142–165.

46. Gloriam, D. E., Fredriksson, R., and Schioth, H. B. (2007) *BMC Genomics* 8, 338.

47. Burke, R. D., Angerer, L. M., Elphick, M. R., Humphrey, G. W., Yaguchi, S., Kiyama, T., Liang, S., Mu, X., Agca, C., Klein, W. H., Brandhorst, B. P., Rowe, M., Wilson, K., Churcher, A. M., Taylor, J. S., Chen, N., Murray, G., Wang, D., Mellott, D., Olinski, R., Hallbook, F., and Thorndyke, M. C. (2006) *Dev Biol* 300(1), 434–460.

48. Lagerstrom, M. C., Hellstrom, A. R., Gloriam, D. E., Larsson, T. P., Schioth, H. B., and Fredriksson, R. (2006) *PLoS Comput Biol* 2(6), e54.

49. Metpally, R. P., and Sowdhamini, R. (2005) *BMC Genomics* 6, 106.

50. Vassilatis, D. K., Hohmann, J. G., Zeng, H., Li, F., Ranchalis, J. E., Mortrud, M. T., Brown, A., Rodriguez, S. S., Weller, J. R., Wright, A. C., Bergmann, J. E., and Gaitanaris, G. A. (2003) *Proc Natl Acad Sci U S A* 100(8), 4903–4908.

51. Dehal, P., and Boore, J. L. (2005) *PLoS Biol* 3(10), e314.

52. Kuraku, S., Meyer, A., and Kuratani, S. (2009) *Mol Biol Evol* 26(1), 47–59.

53. Schulz, A., and Schoneberg, T. (2003) *J Biol Chem* 278(37), 35531–35541.

54. Go, Y., Satta, Y., Takenaka, O., and Takahata, N. (2005) *Genetics* 170(1), 313–326.

55. Semyonov, J., Park, J. I., Chang, C. L., and Hsu, S. Y. (2008) *PLoS ONE* 3(4), e1903.

56. Woods, I. G., Wilson, C., Friedlander, B., Chang, P., Reyes, D. K., Nix, R., Kelly, P. D., Chu, F., Postlethwait, J. H., and Talbot, W. S. (2005) *Genome Res* 15(9), 1307–1314.

57. Noonan, J. P., Grimwood, J., Schmutz, J., Dickson, M., and Myers, R. M. (2004) *Genome Res* 14(3), 354–366.

58. Ludwig, A., Belfiore, N. M., Pitra, C., Svirsky, V., and Jenneckens, I. (2001) *Genetics* 158(3), 1203–1215.

59. Gloriam, D. E., Bjarnadottir, T. K., Yan, Y. L., Postlethwait, J. H., Schioth, H. B., and Fredriksson, R. (2005) *Mol Phylogenet Evol* 35(2), 470–482.

60. Schöneberg, T., Hermsdorf, T., Engemaier, E., Engel, K., Liebscher, I., Thor, D., Zierau, K., Römpler, H., and Schulz, A. (2007) *Purinergic Signal* 3, 255–268.

61. Perez, D. M. (2003) *Mol Pharmacol* 63(6), 1202–1205.

62. Krautwurst, D. (2008) *Chem Biodivers* 5(6), 842–852.

63. Rouquier, S., and Giorgi, D. (2007) *Mutat Res* 616(1–2), 95–102.

64. Kratz, E., Dugas, J. C., and Ngai, J. (2002) *Trends Genet* 18(1), 29–34.

65. Zlotnik, A., Yoshie, O., and Nomiyama, H. (2006) *Genome Biol* 7(12), 243.

66. Le Crom, S., Kapsimali, M., Barome, P. O., and Vernier, P. (2003) *J Struct Funct Genomics* 3(1–4), 161–176.

67. Hashiguchi, Y., and Nishida, M. (2007) *Mol Biol Evol* 24(9), 2099–2107.

68. Grus, W. E., Shi, P., and Zhang, J. (2007) *Mol Biol Evol* 24(10), 2153–2157.

69. Nei, M., Niimura, Y., and Nozawa, M. (2008) *Nat Rev Genet* 9(12), 951–963.

70. Niimura, Y., and Nei, M. (2006) *J Hum Genet* 51(6), 505–517.

71. Romero, D., and Palacios, R. (1997) *Annu Rev Genet* 31, 91–111.

72. Cohen, S., Agmon, N., Yacobi, K., Mislovati, M., and Segal, D. (2005) *Nucleic Acids Res* 33(14), 4519–4526.

73. Rompler, H., Dear, P. H., Krause, J., Meyer, M., Rohland, N., Schoneberg, T., Stiller, M., and Hofreiter, M. (2006) *Nature Protocols* 1(2), 720–728.

74. Miller, W., Drautz, D. I., Ratan, A., Pusey, B., Qi, J., Lesk, A. M., Tomsho, L. P., Packard, M. D., Zhao, F., Sher, A., Tikhonov, A., Raney, B., Patterson, N., Lindblad-Toh, K., Lander, E. S., Knight, J. R., Irzyk, G. P., Fredrikson, K. M., Harkins, T. T., Sheridan, S., Pringle, T., and Schuster, S. C. (2008) *Nature* 456(7220), 387–390.

75. Willerslev, E., Cappellini, E., Boomsma, W., Nielsen, R., Hebsgaard, M. B., Brand, T. B., Hofreiter, M., Bunce, M., Poinar, H. N., Dahl-Jensen, D., Johnsen, S., Steffensen, J. P., Bennike, O., Schwenninger, J. L., Nathan, R., Armitage, S., de Hoog, C. J., Alfimov, V., Christl, M., Beer, J., Muscheler, R., Barker, J., Sharp, M., Penkman, K. E., Haile, J., Taberlet, P., Gilbert, M. T., Casoli, A., Campani, E., and Collins, M. J. (2007) *Science* 317(5834), 111–114.

76. Noonan, J. P., Coop, G., Kudaravalli, S., Smith, D., Krause, J., Alessi, J., Chen, F., Platt, D., Paabo, S., Pritchard, J. K., and Rubin, E. M. (2006) *Science* 314(5802), 1113–1118.

77. Green, R. E., Krause, J., Ptak, S. E., Briggs, A. W., Ronan, M. T., Simons, J. F., Du, L., Egholm, M., Rothberg, J. M., Paunovic, M., and Paabo, S. (2006) *Nature* 444(7117), 330–336.

78. Balakirev, E. S., and Ayala, F. J. (2003) *Annu Rev Genet* 37, 123–151.

79. Hurst, L. D. (2002) *Trends Genet* 18(9), 486.

80. Morgan, K., Conklin, D., Pawson, A. J., Sellar, R., Ott, T. R., and Millar, R. P. (2003) *Endocrinology* 144(2), 423–436.

81. Rompler, H., Schulz, A., Pitra, C., Coop, G., Przeworski, M., Paabo, S., and Schoneberg, T. (2005) *J Biol Chem* 280(35), 31068–31075.

82. Rompler, H., Yu, H. T., Arnold, A., Orth, A., and Schoneberg, T. (2006) *Genomics* 87(6), 724–732.

83. Gaillard, I., Rouquier, S., Chavanieu, A., Mollard, P., and Giorgi, D. (2004) *Hum Mol Genet* 13(7), 771–780.

84. Torrents, D., Suyama, M., Zdobnov, E., and Bork, P. (2003) *Genome Res* 13(12), 2559–2567.

85. Wang, X., Grus, W. E., and Zhang, J. (2006) *PLoS Biol* 4(3), e52.

86. Rompler, H., Staubert, C., Thor, D., Schulz, A., Hofreiter, M., and Schoneberg, T. (2007) *Mol Interv* 7(1), 17–25.

87. Mombaerts, P. (2001) *Annu Rev Genomics Hum Genet* 2, 493–510.

88. Young, J. M., and Trask, B. J. (2002) *Hum Mol Genet* 11(10), 1153–1160.

89. Liman, E. R., and Innan, H. (2003) *Proc Natl Acad Sci U S A* 100(6), 3328–3332.

90. Zhang, J., and Webb, D. M. (2003) *Proc Natl Acad Sci U S A* 100(14), 8337–8341.

91. Wang, X., Thomas, S. D., and Zhang, J. (2004) *Hum Mol Genet* 13(21), 2671–2678.

92. Hoekstra, H. E. (2006) *Heredity* 97(3), 222–234.

93. Harding, R. M., Healy, E., Ray, A. J., Ellis, N. S., Flanagan, N., Todd, C., Dixon, C., Sajantila, A., Jackson, I. J., Birch-Machin, M. A., and Rees, J. L. (2000) *Am J Hum Genet* 66(4), 1351–1361.

94. John, P. R., Makova, K., Li, W. H., Jenkins, T., and Ramsay, M. (2003) *Ann N Y Acad Sci* 994, 299–306.

95. Rees, J. L. (2000) *Pigment Cell Res* 13(3), 135–140.

96. Gross, J. B., Borowsky, R., and Tabin, C. J. (2009) *PLoS Genet* 5(1), e1000326.

97. Lalueza-Fox, C., Rompler, H., Caramelli, D., Staubert, C., Catalano, G., Hughes, D., Rohland, N., Pilli, E., Longo, L., Condemi, S., de la Rasilla, M., Fortea, J., Rosas, A., Stoneking, M., Schoneberg, T., Bertranpetit, J., and Hofreiter, M. (2007) *Science* 318(5855), 1453–1455.

98. Rompler, H., Rohland, N., Lalueza-Fox, C., Willerslev, E., Kuznetsova, T., Rabeder, G., Bertranpetit, J., Schoneberg, T., and Hofreiter, M. (2006) *Science* 313(5783), 62.

99. Steiner, C. C., Rompler, H., Boettger, L. M., Schoneberg, T., and Hoekstra, H. E. (2009) *Mol Biol Evol* 26(1), 35–45.

100. Hoekstra, H. E., Hirschmann, R. J., Bundey, R. A., Insel, P. A., and Crossland, J. P. (2006) *Science* 313(5783), 101–104.

101. Rosenblum, E. R., Rompler, H., Schoneberg, T., and Hoekstra, H. E. (2009) *in press*

102. Kobilka, B., and Schertler, G. F. (2008) *Trends Pharmacol Sci* 29(2), 79–83.

103. Scheerer, P., Heck, M., Goede, A., Park, J. H., Choe, H. W., Ernst, O. P., Hofmann, K. P., and Hildebrand, P. W. (2009) *Proc Natl Acad Sci U S A* 106(26), 10660–10665

104. Scheerer, P., Park, J. H., Hildebrand, P. W., Kim, Y. J., Krauss, N., Choe, H. W., Hofmann, K. P., and Ernst, O. P. (2008) *Nature* 455(7212), 497–502.

105. Park, J. H., Scheerer, P., Hofmann, K. P., Choe, H. W., and Ernst, O. P. (2008) *Nature* 454(7201), 183–187.

106. Zhang, Y., Devries, M. E., and Skolnick, J. (2006) *PLoS Comput Biol* 2(2), e13.

107. Baldwin, J. M. (1993) *Embo J* 12(4), 1693–1703.

108. Staubert, C., Tarnow, P., Brumm, H., Pitra, C., Gudermann, T., Gruters, A., Schoneberg, T., Biebermann, H., and Rompler, H. (2007) *Endocrinology* 148(10), 4642–4648.

109. Tarnow, P., Schoneberg, T., Krude, H., Gruters, A., and Biebermann, H. (2003) *J Biol Chem* 278(49), 48666–48673.

110. Brown, L. S. (2004) *Photochem Photobiol Sci* 3(6), 555–565.

111. Zhai, Y., Heijne, W. H., Smith, D. W., and Saier, M. H., Jr. (2001) *Biochim Biophys Acta* 1511(2), 206–223.

112. Taylor, E. W., and Agarwal, A. (1993) *FEBS Lett* 325(3), 161–166.

113. Pardo, L., Ballesteros, J. A., Osman, R., and Weinstein, H. (1992) *Proc Natl Acad Sci U S A* 89(9), 4009–4012.

114. Roy, S. W., and Gilbert, W. (2006) *Nat Rev Genet* 7(3), 211–221.

115. Bryson-Richardson, R. J., Logan, D. W., Currie, P. D., and Jackson, I. J. (2004) *Gene* 338(1), 15–23.

116. Fridmanis, D., Fredriksson, R., Kapa, I., Schioth, H. B., and Klovins, J. (2007) *Mol Phylogenet Evol* 43(3), 864–880.

117. Venkatesh, B., Ning, Y., and Brenner, S. (1999) *Proc Natl Acad Sci U S A* 96(18), 10267–10271.

118. Kosugi, S., Ban, T., Akamizu, T., and Kohn, L. D. (1992) *Biochem Biophys Res Commun* 189(3), 1754–1762.

119. Savarese, T. M., Wang, C. D., and Fraser, C. M. (1992) *J Biol Chem* 267(16), 11439–11448.

120. Cook, J. V., and Eidne, K. A. (1997) *Endocrinology* 138(7), 2800–2806.

121. Le Gouill, C., Parent, J. L., Rola-Pleszczynski, M., and Stankova, J. (1997) *FEBS Lett* 402(2–3), 203–208.

122. Perlman, J. H., Wang, W., Nussenzveig, D. R., and Gershengorn, M. C. (1995) *J Biol Chem* 270(42), 24682–24685.

123. Schulz, A., Grosse, R., Schultz, G., Gudermann, T., and Schoneberg, T. (2000) *J Biol Chem* 275(4), 2381–2389.

124. Yohannan, S., Faham, S., Yang, D., Whitelegge, J. P., and Bowie, J. U. (2004) *Proc Natl Acad Sci U S A* 101(4), 959–963.

125. Deville, J., Rey, J., and Chabbert, M. (2009) *J Mol Evol* 68(5), 475–489.
126. Palczewski, K., Kumasaka, T., Hori, T., Behnke, C. A., Motoshima, H., Fox, B. A., Le Trong, I., Teller, D. C., Okada, T., Stenkamp, R. E., Yamamoto, M., and Miyano, M. (2000) *Science* 289(5480), 739–745.
127. Tajima, T., Nakae, J., Takekoshi, Y., Takahashi, Y., Yuri, K., Nagashima, T., and Fujieda, K. (1996) *Pediatr Res* 39(3), 522–526.
128. Gales, C., Kowalski-Chauvel, A., Dufour, M. N., Seva, C., Moroder, L., Pradayrol, L., Vaysse, N., Fourmy, D., and Silvente-Poirot, S. (2000) *J Biol Chem* 275(23), 17321–17327.
129. McMillan, D. R., Kayes-Wandover, K. M., Richardson, J. A., and White, P. C. (2002) *J Biol Chem* 277(1), 785–792.
130. Stacey, M., Lin, H. H., Gordon, S., and McKnight, A. J. (2000) *Trends Biochem Sci* 25(6), 284–289.
131. Kudo, M., Chen, T., Nakabayashi, K., Hsu, S. Y., and Hsueh, A. J. (2000) *Mol Endocrinol* 14(2), 272–284.
132. Kajava, A. V. (1998) *J Mol Biol* 277(3), 519–527.
133. Van Loy, T., Vandersmissen, H. P., Van Hiel, M. B., Poels, J., Verlinden, H., Badisco, L., Vassart, G., and Vanden Broeck, J. (2008) *Gen Comp Endocrinol* 155(1), 14–21.
134. Fan, Q. R., and Hendrickson, W. A. (2005) *Nature* 433(7023), 269–277.
135. Krasnoperov, V. G., Bittner, M. A., Beavis, R., Kuang, Y., Salnikow, K. V., Chepurny, O. G., Little, A. R., Plotnikov, A. N., Wu, D., Holz, R. W., and Petrenko, A. G. (1997) *Neuron* 18(6), 925–937.
136. Lin, H. H., Chang, G. W., Davies, J. Q., Stacey, M., Harris, J., and Gordon, S. (2004) *J Biol Chem* 279(30), 31823–31832.
137. Bjarnadottir, T. K., Fredriksson, R., and Schioth, H. B. (2007) *Cell Mol Life Sci* 64(16), 2104–2119.
138. Kwakkenbos, M. J., Kop, E. N., Stacey, M., Matmati, M., Gordon, S., Lin, H. H., and Hamann, J. (2004) *Immunogenetics* 55(10), 655–666.
139. Fredriksson, R., Gloriam, D. E., Hoglund, P. J., Lagerstrom, M. C., and Schioth, H. B. (2003) *Biochem Biophys Res Commun* 301(3), 725–734.
140. Usui, T., Shima, Y., Shimada, Y., Hirano, S., Burgess, R. W., Schwarz, T. L., Takeichi, M., and Uemura, T. (1999) *Cell* 98(5), 585–595.
141. Formstone, C. J., and Little, P. F. (2001) *Mech Dev* 109(1), 91–94.
142. Hadjantonakis, A. K., Formstone, C. J., and Little, P. F. (1998) *Mech Dev* 78(1–2), 91–95.
143. Lawrence, P. A., Struhl, G., and Casal, J. (2007) *Nat Rev Genet* 8(7), 555–563.
144. Strutt, H., and Strutt, D. (2008) *Curr Biol* 18(20), 1555–1564.
145. Sudhof, T. C. (2001) *Annu Rev Neurosci* 24, 933–962.
146. Vakonakis, I., Langenhan, T., Promel, S., Russ, A., and Campbell, I. D. (2008) *Structure* 16(6), 944–953.
147. Braasch, I., Volff, J. N., and Schartl, M. (2009) *Mol Biol Evol* 26(4), 783–799.
148. van Kesteren, R. E., and Geraerts, W. P. (1998) *Ann N Y Acad Sci* 839, 25–34.
149. Darlison, M. G., and Richter, D. (1999) *Trends Neurosci* 22(2), 81–88.
150. Goh, C. S., Bogan, A. A., Joachimiak, M., Walther, D., and Cohen, F. E. (2000) *J Mol Biol* 299(2), 283–293.
151. Wilkinson, T. N., Speed, T. P., Tregear, G. W., and Bathgate, R. A. (2005) *Ann N Y Acad Sci* 1041, 534–539.
152. Sudo, S., Kuwabara, Y., Park, J. I., Hsu, S. Y., and Hsueh, A. J. (2005) *Endocrinology* 146(8), 3596–3604.
153. Costagliola, S., Urizar, E., Mendive, F., and Vassart, G. (2005) *Reproduction* 130(3), 275–281.
154. Frazer, K. A., Ballinger, D. G., Cox, D. R., Hinds, D. A., Stuve, L. L., Gibbs, R. A., Belmont, J. W., Boudreau, A., Hardenbol, P., Leal, S. M., Pasternak, S., Wheeler, D. A., Willis, T. D., Yu, F., Yang, H., Zeng, C., Gao, Y., Hu, H., Hu, W., Li, C., Lin, W., Liu, S., Pan, H., Tang, X., Wang, J., Wang, W., Yu, J., Zhang, B., Zhang, Q.,

Zhao, H., Zhao, H., Zhou, J., Gabriel, S. B., Barry, R., Blumenstiel, B., Camargo, A., Defelice, M., Faggart, M., Goyette, M., Gupta, S., Moore, J., Nguyen, H., Onofrio, R. C., Parkin, M., Roy, J., Stahl, E., Winchester, E., Ziaugra, L., Altshuler, D., Shen, Y., Yao, Z., Huang, W., Chu, X., He, Y., Jin, L., Liu, Y., Shen, Y., Sun, W., Wang, H., Wang, Y., Wang, Y., Xiong, X., Xu, L., Waye, M. M., Tsui, S. K., Xue, H., Wong, J. T., Galver, L. M., Fan, J. B., Gunderson, K., Murray, S. S., Oliphant, A. R., Chee, M. S., Montpetit, A., Chagnon, F., Ferretti, V., Leboeuf, M., Olivier, J. F., Phillips, M. S., Roumy, S., Sallee, C., Verner, A., Hudson, T. J., Kwok, P. Y., Cai, D., Koboldt, D. C., Miller, R. D., Pawlikowska, L., Taillon-Miller, P., Xiao, M., Tsui, L. C., Mak, W., Song, Y. Q., Tam, P. K., Nakamura, Y., Kawaguchi, T., Kitamoto, T., Morizono, T., Nagashima, A., Ohnishi, Y., Sekine, A., Tanaka, T., Tsunoda, T., Deloukas, P., Bird, C. P., Delgado, M., Dermitzakis, E. T., Gwilliam, R., Hunt, S., Morrison, J., Powell, D., Stranger, B. E., Whittaker, P., Bentley, D. R., Daly, M. J., de Bakker, P. I., Barrett, J., Chretien, Y. R., Maller, J., McCarroll, S., Patterson, N., Pe'er, I., Price, A., Purcell, S., Richter, D. J., Sabeti, P., Saxena, R., Schaffner, S. F., Sham, P. C., Varilly, P., Altshuler, D., Stein, L. D., Krishnan, L., Smith, A. V., Tello-Ruiz, M. K., Thorisson, G. A., Chakravarti, A., Chen, P. E., Cutler, D. J., Kashuk, C. S., Lin, S., Abecasis, G. R., Guan, W., Li, Y., Munro, H. M., Qin, Z. S., Thomas, D. J., McVean, G., Auton, A., Bottolo, L., Cardin, N., Eyheramendy, S., Freeman, C., Marchini, J., Myers, S., Spencer, C., Stephens, M., Donnelly, P., Cardon, L. R., Clarke, G., Evans, D. M., Morris, A. P., Weir, B. S., Tsunoda, T., Mullikin, J. C., Sherry, S. T., Feolo, M., Skol, A., Zhang, H., Zeng, C., Zhao, H., Matsuda, I., Fukushima, Y., Macer, D. R., Suda, E., Rotimi, C. N., Adebamowo, C. A., Ajayi, I., Aniagwu, T., Marshall, P. A., Nkwodimmah, C., Royal, C. D., Leppert, M. F., Dixon, M., Peiffer, A., Qiu, R., Kent, A., Kato, K., Niikawa, N., Adewole, I. F., Knoppers, B. M., Foster, M. W., Clayton, E. W., Watkin, J., Gibbs, R. A., Belmont, J. W., Muzny, D., Nazareth, L., Sodergren, E., Weinstock, G. M., Wheeler, D. A., Yakub, I., Gabriel, S. B., Onofrio, R. C., Richter, D. J., Ziaugra, L., Birren, B. W., Daly, M. J., Altshuler, D., Wilson, R. K., Fulton, L. L., Rogers, J., Burton, J., Carter, N. P., Clee, C. M., Griffiths, M., Jones, M. C., McLay, K., Plumb, R. W., Ross, M. T., Sims, S. K., Willey, D. L., Chen, Z., Han, H., Kang, L., Godbout, M., Wallenburg, J. C., L'Archeveque, P., Bellemare, G., Saeki, K., Wang, H., An, D., Fu, H., Li, Q., Wang, Z., Wang, R., Holden, A. L., Brooks, L. D., McEwen, J. E., Guyer, M. S., Wang, V. O., Peterson, J. L., Shi, M., Spiegel, J., Sung, L. M., Zacharia, L. F., Collins, F. S., Kennedy, K., Jamieson, R., and Stewart, J. (2007) *Nature* 449(7164), 851–861.

155. Gibbs, R. A., Taylor, J. F., Van Tassell, C. P., Barendse, W., Eversole, K. A., Gill, C. A., Green, R. D., Hamernik, D. L., Kappes, S. M., Lien, S., Matukumalli, L. K., McEwan, J. C., Nazareth, L. V., Schnabel, R. D., Weinstock, G. M., Wheeler, D. A., Ajmone-Marsan, P., Boettcher, P. J., Caetano, A. R., Garcia, J. F., Hanotte, O., Mariani, P., Skow, L. C., Sonstegard, T. S., Williams, J. L., Diallo, B., Hailemariam, L., Martinez, M. L., Morris, C. A., Silva, L. O., Spelman, R. J., Mulatu, W., Zhao, K., Abbey, C. A., Agaba, M., Araujo, F. R., Bunch, R. J., Burton, J., Gorni, C., Olivier, H., Harrison, B. E., Luff, B., Machado, M. A., Mwakaya, J., Plastow, G., Sim, W., Smith, T., Thomas, M. B., Valentini, A., Williams, P., Womack, J., Woolliams, J. A., Liu, Y., Qin, X., Worley, K. C., Gao, C., Jiang, H., Moore, S. S., Ren, Y., Song, X. Z., Bustamante, C. D., Hernandez, R. D., Muzny, D. M., Patil, S., San Lucas, A., Fu, Q., Kent, M. P., Vega, R., Matukumalli, A., McWilliam, S., Sclep, G., Bryc, K., Choi, J., Gao, H., Grefenstette, J. J., Murdoch, B., Stella, A., Villa-Angulo, R., Wright, M., Aerts, J., Jann, O., Negrini, R., Goddard, M. E., Hayes, B. J., Bradley, D. G., Barbosa da Silva, M., Lau, L. P., Liu, G. E., Lynn, D. J., Panzitta, F., and Dodds, K. G. (2009) *Science* 324(5926), 528–532.

156. Liu, X. G., Tan, L. J., Lei, S. F., Liu, Y. J., Shen, H., Wang, L., Yan, H., Guo, Y. F., Xiong, D. H., Chen, X. D., Pan, F., Yang, T. L., Zhang, Y. P., Guo, Y., Tang, N. L., Zhu, X. Z., Deng, H. Y., Levy, S., Recker, R. R., Papasian, C. J., and Deng, H. W. (2009) *Am J Hum Genet* 84(3), 418–423.

157. Yang, Z. (2007) *Mol Biol Evol* 24(8), 1586–1591.

158. Yang, Z. (1997) *Comput Appl Biosci* 13(5), 555–556.
159. Bamshad, M., and Wooding, S. P. (2003) *Nat Rev Genet* 4(2), 99–111.
160. Soranzo, N., Bufe, B., Sabeti, P. C., Wilson, J. F., Weale, M. E., Marguerie, R., Meyerhof, W., and Goldstein, D. B. (2005) *Curr Biol* 15(14), 1257–1265.
161. Coop, G., Pickrell, J. K., Novembre, J., Kudaravalli, S., Li, J., Absher, D., Myers, R. M., Cavalli-Sforza, L. L., Feldman, M. W., and Pritchard, J. K. (2009) *PLoS Genet* 5(6), e1000500.
162. Hamblin, M. T., and Di Rienzo, A. (2000) *Am J Hum Genet* 66(5), 1669–1679.
163. Kelley, J. L., Madeoy, J., Calhoun, J. C., Swanson, W., and Akey, J. M. (2006) *Genome Res* 16(8), 980–989.
164. Liu, Y., Yang, S., Lin, A. A., Cavalli-Sforza, L. L., and Su, B. (2005) *J Mol Evol* 61(5), 691–696.
165. Maayan, S., Zhang, L., Shinar, E., Ho, J., He, T., Manni, N., Kostrikis, L. G., and Neumann, A. U. (2000) *Genes Immun* 1(6), 358–361.
166. Novembre, J., Galvani, A. P., and Slatkin, M. (2005) *PLoS Biol* 3(11), e339.
167. Moreno-Estrada, A., Casals, F., Ramirez- Soriano, A., Oliva, B., Calafell, F., Bertranpetit, J., and Bosch, E. (2008) *Mol Biol Evol* 25(1), 144–154.
168. Hashiguchi, Y., Furuta, Y., Kawahara, R., and Nishida, M. (2007) *Gene* 396(1), 170–179.
169. Shi, P., Zhang, J., Yang, H., and Zhang, Y. P. (2003) *Mol Biol Evol* 20(5), 805–814.
170. Lin, Y. J., Seroude, L., and Benzer, S. (1998) *Science* 282(5390), 943–946.
171. Schmidt, P. S., Duvernell, D. D., and Eanes, W. F. (2000) *Proc Natl Acad Sci U S A* 97(20), 10861–10865.
172. Kim, K. S., Reecy, J. M., Hsu, W. H., Anderson, L. L., and Rothschild, M. F. (2004) *Domest Anim Endocrinol* 26(1), 75–86.
173. Fang, M., Larson, G., Ribeiro, H. S., Li, N., and Andersson, L. (2009) *PLoS Genet* 5(1), e1000341.
174. Hadley, T. J., and Peiper, S. C. (1997) *Blood* 89(9), 3077–3091.
175. Liberles, S. D., and Buck, L. B. (2006) *Nature* 442(7103), 645–650.
176. Diaz Arenas, C., and Lehman, N. (2009) *Int J Biochem Cell Biol* 41(2), 266–273.
177. Joyce, G. F. (2007) *Angew Chem Int Ed Engl* 46(34), 6420–6436.
178. Ault, A. D., and Broach, J. R. (2006) *Protein Eng Des Sel* 19(1), 1–8.
179. Coward, P., Wada, H. G., Falk, M. S., Chan, S. D., Meng, F., Akil, H., and Conklin, B. R. (1998) *Proc Natl Acad Sci U S A* 95(1), 352–357.
180. Pausch, M. H. (1997) *Trends Biotechnol* 15(12), 487–494.
181. Beukers, M. W., and Ijzerman, A. P. (2005) *Trends Pharmacol Sci* 26(10), 533–539.
182. Ladds, G., Goddard, A., and Davey, J. (2005) *Trends Biotechnol* 23(7), 367–373.
183. Armbruster, B. N., Li, X., Pausch, M. H., Herlitze, S., and Roth, B. L. (2007) *Proc Natl Acad Sci U S A* 104(12), 5163–5168.
184. Conklin, B. R., Hsiao, E. C., Claeysen, S., Dumuis, A., Srinivasan, S., Forsayeth, J. R., Guettier, J. M., Chang, W. C., Pei, Y., McCarthy, K. D., Nissenson, R. A., Wess, J., Bockaert, J., and Roth, B. L. (2008) *Nat Methods* 5(8), 673–678.
185. Scearce-Levie, K., Coward, P., Redfern, C. H., and Conklin, B. R. (2001) *Trends Pharmacol Sci* 22(8), 414–420.
186. Baker, J. G., Hall, I. P., and Hill, S. J. (2003) *Mol Pharmacol* 63(6), 1312–1321.
187. Thor, D., Schulz, A., Hermsdorf, T., and Schoneberg, T. (2008) *Biochem J* 412(1), 103–112.
188. Arbiza, L., Dopazo, J., and Dopazo, H. (2006) *PLoS Comput Biol* 2(4), e38.
189. Tang, K., Thornton, K. R., and Stoneking, M. (2007) *PLoS Biol* 5(7), e171.
190. Williamson, S. H., Hubisz, M. J., Clark, A. G., Payseur, B. A., Bustamante, C. D., and Nielsen, R. (2007) *PLoS Genet* 3(6), e90.
191. Barreiro, L. B., Laval, G., Quach, H., Patin, E., and Quintana-Murci, L. (2008) *Nat Genet* 40(3), 340–345.

2 Functional studies of isolated GPCR-G protein complexes in the membrane bilayer of lipoprotein particles

Adam J. Kuszak, Xiao Jie Yao, Sören G.F. Rasmussen,
Brian K. Kobilka, and Roger K. Sunahara

INTRODUCTION

Over the past few years we have witnessed a dramatic expansion of our knowledge of G protein-coupled receptor GPCR structure and function. This increased understanding has been driven by the development of novel techniques that allow further characterization of GPCRs at an atomic level. A veritable opening of the floodgates has occurred with the reports of multiple crystal structures for GPCRs, including the β_2 adrenoceptor (AR) bound to partial inverse agonists,[1-4] the β_1AR bound to an antagonist,[5] the A_{2A} adenosine receptor,[6] and new structures of opsin.[7-9] These structures have provided insight into the various GPCR conformational states stabilized by ligands, and also suggest important mechanisms of G protein coupling and activation (see Chapter 1).

Although highly informative, one must remember that crystal structures provide static images of proteins and may be influenced by packing of the crystal lattice. However, GPCRs are dynamic molecules, undergoing both

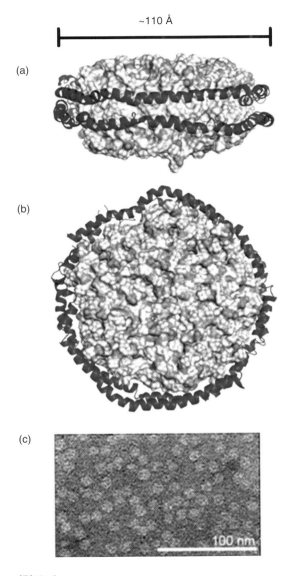

~110 Å

(a)

(b)

(c)

100 nm

`Plate 1

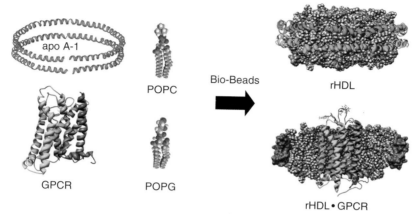

Plate 2

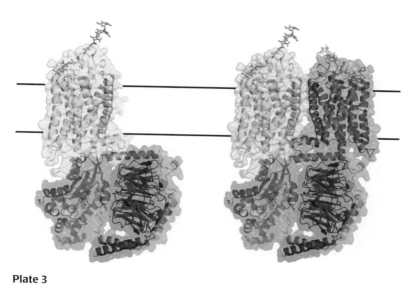

Plate 3

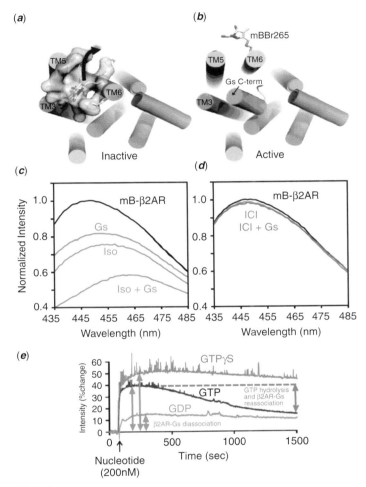

(*a*)

TM5
TM6
TM3

Inactive

(*b*)

mBBr265

TM5   TM6
Gs C-term
TM3

Active

(*c*)

mB-β2AR
Gs
Iso
Iso + Gs

Normalized Intensity

435 445 455 465 475 485
Wavelength (nm)

(*d*)

mB-β2AR
ICI
ICI + Gs

435 445 455 465 475 485
Wavelength (nm)

(*e*)

GTPγS
GTP
GTP hydrolysis
and β2AR-Gs
reassociation
GDP
β2AR-Gs diassociation

Intensity (%change)

0   500   1000   1500
Time (sec)

Nucleotide
(200nM)

Plate 4

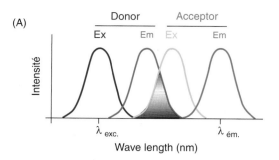

(A)

Donor   Acceptor

Ex   Em   Ex   Em

Intensité

$\lambda_{exc.}$ $\lambda_{ém.}$

Wave length (nm)

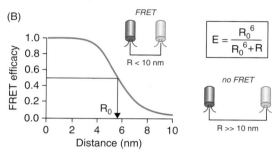

(B)

FRET efficacy

FRET

$R < 10$ nm

$$E = \frac{R_0^6}{R_0^6 + R}$$

no FRET

$R \gg 10$ nm

R_0

Distance (nm)

(C)

FRET   no FRET

Angle $\neq$ 90°   Angle = 90°

Plate 5

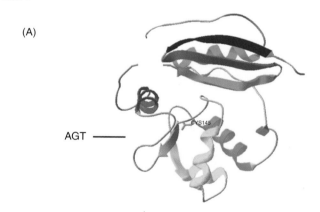

(A)

AGT ——— CYS145

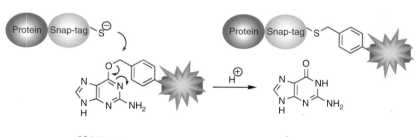

(B)

Protein   Snap-tag —S$^{\ominus}$ Protein   Snap-tag —S

$\overset{\oplus}{H}$

O^6-labeled benzylguanine   Guanine

Plate 6

(A)

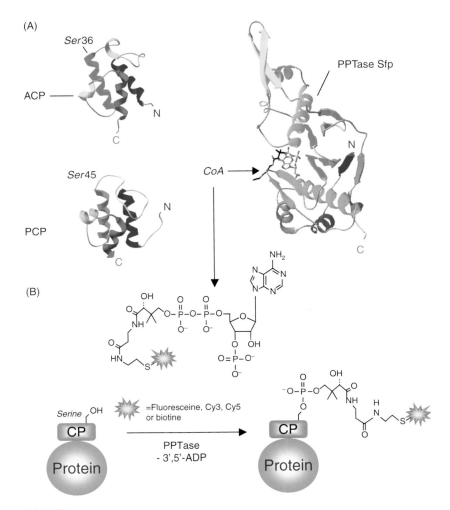

*Ser*36

ACP

*Ser*45

PCP

PPTase Sfp

CoA

(B)

=Fluoresceine, Cy3, Cy5 or biotine

Serine OH

CP

Protein

PPTase
- 3',5'-ADP

CP

Protein

Plate 7

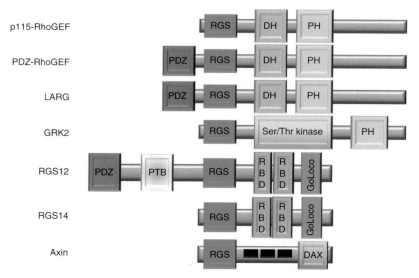

Plate 8

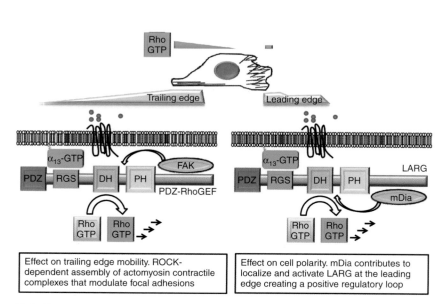

Plate 9

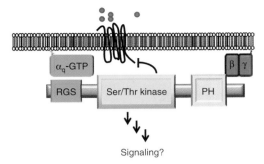

Plate 10

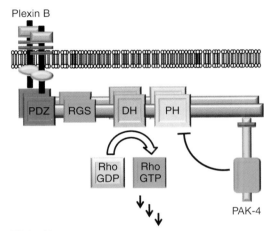

Plate 11

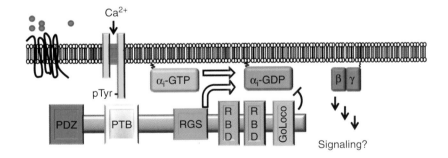

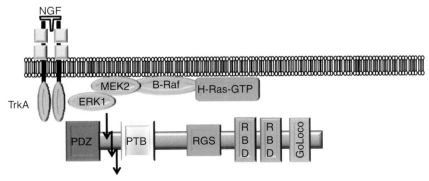

Plate 12

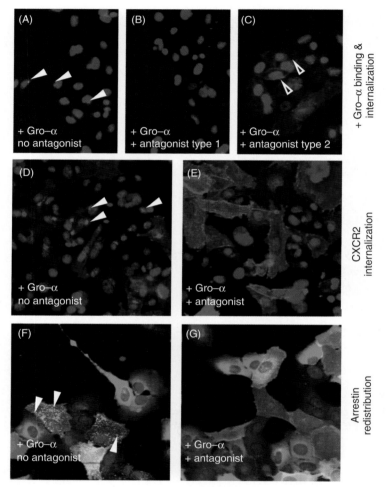

(A) + Gro–α
no antagonist

(B) + Gro–α
+ antagonist type 1

(C) + Gro–α
+ antagonist type 2

+ Gro–α binding &
internalization

(D) + Gro–α
no antagonist

(E) + Gro–α
+ antagonist

CXCR2
internalization

(F) + Gro–α
no antagonist

(G) + Gro–α
+ antagonist

Arrestin
redistribution

Plate 13

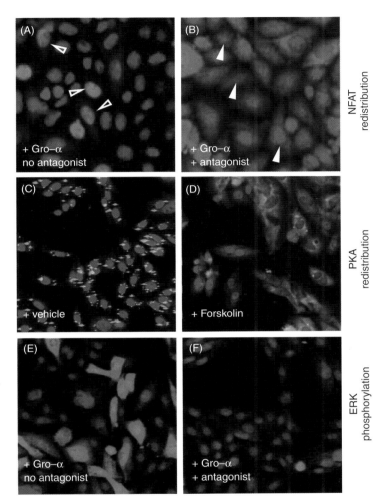

(A)

+ Gro–α
no antagonist

(B)

+ Gro–α
+ antagonist

NFAT
redistribution

(C)

+ vehicle

(D)

+ Forskolin

PKA
redistribution

(E)

+ Gro–α
no antagonist

(F)

+ Gro–α
+ antagonist

ERK
phosphorylation

Plate 14

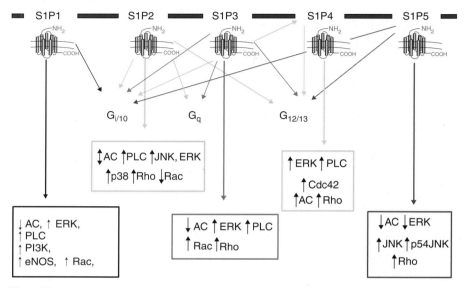

Plate 15

dramatic and subtle conformational changes during their activation process. Accordingly, crystallographic information must be supplemented with an understanding of the functional interaction between GPCRs and G protein heterotrimers. Recent technological advances in membrane protein biochemistry have led to significant enhancements in our understanding of GPCR function. These advancements have been driven by the work of multiple laboratories probing the structure and function of isolated GPCRs in reconstituted lipoprotein particles, which are derived from high density lipoprotein (HDL) particles. This novel methodology utilizes purified GPCRs isolated in a phospholipid bilayer, a mimic of their native plasma membrane environment. Reconstitution into lipoprotein particles has granted researchers the ability to create a homogeneous population of GPCRs and their cognate G proteins with a defined stoichiometry. With this unique advantage, reconstitution of GPCRs into lipoprotein particles has enabled investigators to definitively demonstrate that the minimal functional unit of GPCR required to activate G protein is a monomer. GPCR oligomerization has emerged as a fundamental aspect of receptor theory, and lipoprotein particle reconstitution may hold the key to elucidating the functional consequences of GPCR homo- and heterodimerization. Furthermore, the high efficiency of receptor reconstitution and the degree of homogeneity of the receptor-lipoprotein particle has facilitated the utilization of fluorescence-based assays to probe GPCR function. Reconstitution of receptor preparations that are covalently labeled with fluorescent probes has allowed direct, real-time measurement of conformational changes in response to ligand binding and G protein coupling, providing insight into multiple mechanisms of GPCR function.

This chapter presents an overview of the advances in our understanding of the interactions between receptors and G proteins which reconstitution into lipoprotein particles has yielded. An introduction to the methodology and properties of this novel system will first be provided, followed by a discussion of current theories regarding GPCR oligomerization that have been influenced by receptor reconstitution into lipoprotein particles. Finally, new insights into ligand binding and the mechanism of G protein activation are discussed.

STRUCTURAL AND BIOPHYSICAL CHARACTERIZATION OF RECONSTITUTED LIPOPROTEIN PARTICLES

Apolipoprotein A-1 (apo A-1) is the major protein component of high density lipoprotein (HDL), and its structure and function has received extensive scrutiny. Apo A-1 (along with apo A-2) forms nascent discoidal or spherical complexes with phospholipids, triglycerides, and cholesterol.[10] These lipoprotein particles are physiologically involved in reverse lipid transport and are also the major activator of lecithin cholesterol acyltransferase. Apo A-1 is 243 amino acids long, with the C-terminal 200 residues involved in lipid binding. This lipid binding domain is composed of eight 22-mer and two 11-mer helical repeats.[11,12] HDL particle size

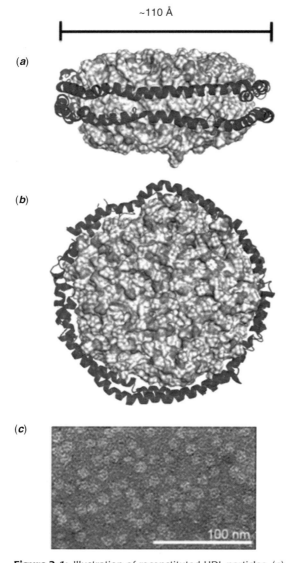

~110 Å

(*a*)

(*b*)

(*c*)

100 nm

Figure 2-1: Illustration of reconstituted HDL particles. (*a*) and (*b*) Molecular models illustrating a dimer of apo A-1 proteins wrapped around a phospholipid bilayer consisting of 160 POPC molecules. Each apo A-1 protein (blue) is depicted as a ribbon diagram. (*c*) Transmission electron micrograph of reconstituted HDL particles reveals a monodisperse and homogeneous population. Images in (*a*) and (*b*) were generated using PyMol Molecular Viewer (DeLano Scientific). *See colour plate 1.*

is heterogeneous in vivo, but specific in vitro reconstitution conditions generate a homogenous population.[13] The discoidal reconstituted HDL particle (rHDL) is composed of a phospholipid bilayer surrounded by apo A-1 protein. The first 43 amino acid residues of apo A-1 create a putative globular domain involved in scavenger receptor binding (SRB-1), and can be removed without significantly effecting the discoidal lipid bilayer formation (Δ(1–43)apo A-1; Figure 2.1).[14]

The phospholipid bilayer of reconstituted lipoprotein particles is composed of approximately 150 lipids (depending on the specific phospholipids used), with the hydrophobic tails surrounded by an apo A-1 dimer. The thickness of the lipid bilayer is approximately 40 Ångströms (Å), the same as a typical cellular plasma membrane. The inner diameter of the bilayer is constrained, to a degree, by the apo A-1 protein to ~80–85 Å. The outer diameter of the particles is ~110 Å (Figure 2.1, *a* and *b*). While the organization of the apo A-1 dimer was debated for some time to adopt either a "picket-fence" or "belt" motif, the belt model has recently gained acceptance.[15]

Sligar and coworkers have extensively studied the structural and biophysical properties of rHDL particles using an engineered form of apo A-1 (referred to as Membrane Scaffolding Protein, or MSP). Solid-state NMR studies of reconstituted particles (termed Nanodiscs) suggested an extended helical structure of the engineered apo A-1 surrounding the lipid bilayer, lending support to the double-belt model.[16] Molecular dynamics simulations were used to predict the effects of altering the length of the engineered apo A-1 and the addition of membrane proteins on the shape of the discoidal particles.[17] Unsurprisingly, truncations of the apo A-1 construct influenced the planar dimensions of the particles. Furthermore, studies of phospholipid phase transitions within HDL particles found that the outer ring of lipids adopts a more rigid state than the liquid crystalline state of the inner lipids.[18]

Reconstitution of HDL particles in vitro, with both wild-type and engineered forms of apo A-1, has historically been a mainstay technique for investigating the function of these lipid transporters, as well as the properties of lipid bilayers. However, from a membrane protein researcher's point of view, the most striking aspect of the lipoprotein particle is its structural composition and dimensions; the particles are essentially nanometer-scale, self-contained membranes. Thus in light of HDL's structural characteristics, the Sligar, Sakmar and Sunahara laboratories independently began to pursue these reconstituted lipoprotein particles as a means to study integral membrane proteins in a more physiologically relevant system compared to lipid vesicles and detergent micelles.

RECONSTITUTION OF GPCRS INTO LIPOPROTEIN PARTICLES

The incorporation of GPCRs into lipoprotein particles is relatively straightforward (Figure 2.2), an extension of years of studies on reconstitution of HDL particles.[13] Essentially, a purified GPCR of interest, an engineered apo A-1, and phospholipids are mixed together in a detergent-solubilized solution. The molar ratios between these components are critical, as they dictate proper particle formation (apo A-1:lipid) and the stoichiometry of incorporated GPCRs (apo A-1:receptor).[19,20] Removal of detergent, either via dialysis or a hydrophobic resin, results in the spontaneous formation of the discoidal particles. Indeed, the most difficult aspect of the reconstitution is generating a sample of purified and active GPCR, although there are reports of successful incorporation of membrane proteins using partially purified membrane extractions.[21] More recently, intriguing studies illustrated the combination of cell-free expression systems and

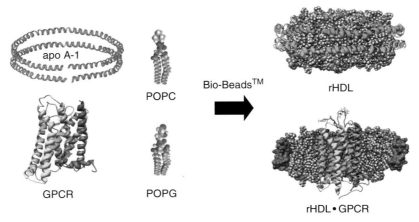

Figure 2-2: Schematic overview of GPCR reconstitution into HDL particles. Purified apolipo-protein A-1 (apo A-1)] is incubated with detergent solubilized palmitoyl-oleoyl-phosphocholine (POPC) and palmitoyl-oleoyl-phosphoglycerol (POPG) and detergent solubilized, purified GPCR. Detergent extraction with Bio-Beads™ hydrophobic resin (Bio-Rad) promotes the formation of a lipid bilayer disc containing the GPCR surrounded by an apo A-1 "belt." Reconstituted HDL's (rHDL) inner diameter of ~80 Å sterically favors the incorporation of a GPCR monomer. Typi-cal apo A-1 and GPCR molar ratios result in an excess of "empty" rHDL particles compared to rHDL•receptor particles. rHDL is shown in a side view with the apo A-1 dimer in green and blue. rHDL•GPCR is illustrated as a cross-sectional side view with apo A-1 in purple as a space-filling model. *See colour plate 2.*

lipoprotein reconstitution for direct synthesis and incorporation of membrane proteins into lipid bilayers.[22] The precise mechanism of the spontaneous particle formation in vivo and in vitro remains unclear, although theoretical models of the process have been developed based on studies of the interactions between apo A-1 and lipids,[10,23]

Several different phospholipids have been successfully used in lipoprotein par-ticle reconstitutions, including dimyristoyl phosphocholine (DMPC), dipalmi-toyl phosphocholine, (DPPC), palmitoyloleoyl phosphocholine (POPC), and palmitoyloleoyl phosphoglycerol (POPG) either as a single lipid component or as a mixture. Extracts of polar porcine brain lipids have also been utilized. Even though a variety of lipids have been successfully used, the lipid composi-tion during reconstitution is not trivial. The structural differences between the phospholipid polar head groups and hydrophobic carbon chains translates to differences in the optimal molar ratios of apo A-1 to lipid during reconstitution.[17] Furthermore, when choosing reconstitution conditions, it is wise to consider the lipid composition of the native cellular membrane where the membrane protein of interest is normally located. Perhaps the most critical aspect of the phospho-lipids used in reconstitutions is their transition temperature. Although GPCR reconstitution into lipoprotein particles requires little hands-on manipulation, it does entail lengthy incubation periods. Therefore the lipid transition tempera-ture must be compatible with maintaining stability of the GPCR. For example, DMPC's transition temperature is ~24°C,[24] which is inappropriate for multi-hour incubations with purified GPCRs. The transition temperature of POPC, however, is ~-6°C[25] and therefore suitable for reconstituting GPCRs at 0 to 4°C.

Although the reconstitution of GPCRs into lipoprotein particles is highly efficient, it does not result in a homogeneous sample without an additional rHDL•GPCR purification step. Typically the molar ratio of apo A-1 to GPCR is heavily biased on the side of apo A-1 to favor the incorporation of a single GPCR per reconstituted particle. This results in the formation of a large population of lipoprotein particles that do not contain the membrane protein. Furthermore, even at ideal ratios of apo A-1:lipid, some protein aggregates will form during the detergent removal, which may interfere with subsequent analyses. Isolation of the desired lipoprotein particles containing incorporated GPCRs is accomplished with the combination of two methods. First, size exclusion chromatography (SEC) can readily separate the reconstituted HDL particles (Stoke's diameter of ~10.5 nm) from protein aggregates and any free lipids. Second, purification of particles containing the GPCR has been accomplished in our laboratory by use of FLAG epitope-tagged receptors and subsequent affinity chromatography with M1 FLAG affinity resin gravity flow columns.[20,26,27] We have also successfully taken advantage of the carbohydrate moieties on the GPCR using concanavalinA-sepharose.[28] Similar in principle to the FLAG affinity purification, a histidine-tagged receptor can be reconstituted into lipoprotein particles and purified using immobilized metal affinity chromatography resins such as Ni-NTA or Talon™ (A.J. Kuszak, unpublished results).

The co-reconstitution of GPCRs and their cognate G proteins is also relatively straightforward. While it is possible to merely add a purified G protein heterotrimer to the reconstitution mixture, this will result in a majority of the GPCR population remaining uncoupled to heterotrimer. The low affinity between receptor and G protein in detergent micelles does not effectively support prior association between receptor and heterotrimer during reconstitution. Coupled with the large molar excess of apo A-1 (and lipid), this results in many particles containing either receptor or G protein, but not both. The approach our laboratory has taken to address these issues is to reconstitute a FLAG-tagged GPCR and purify the receptor-containing particles via FLAG affinity. Purified G protein heterotrimer (in detergent micelles) is subsequently added to the purified rHDL•GPCR particles at relatively small volumes, such that the final detergent conditions are below the critical micelle concentration, and the integrity of the lipoprotein particles is not compromised. The removal of residual detergent is accomplished either by incubation with hydrophobic resin (Bio-Beads™ SM2, BioRad) or size exclusion chromatography. It is important to note that detergent removal induces G protein aggregation and results in proportionally large losses of functional G protein. Attaining functionally satisfactory G protein:GPCR ratios therefore requires the use of high heterotrimer concentrations (Figure 2.3d).[20] Any resulting G protein aggregates may be removed via size exclusion chromatography.

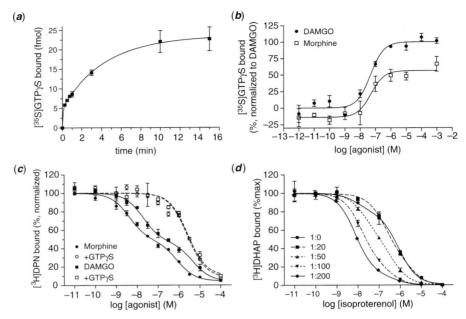

Figure 2-3: Monomeric GPCRs are capable of functional G protein coupling. (*a*) Time-course activation of $G\alpha_s$ by monomeric β_2AR. Monomeric β_2AR was reconstituted with G_s heterotrimer in lipoprotein particles and incubated with the agonist isoproterenol in the presence of $[^{35}S]GTP\gamma S$. (*b*) Agonist concentration response curves for $G\alpha_{i3}$ stimulation by monomeric MOR. Both the peptide DAMGO and the alkaloid morphine activate $G\alpha_{i3}$ with potencies similar to what is seen in cell-based assays. (*c*) Allosteric regulation of agonist binding to monomeric MOR by G_{i2} heterotrimer. Agonist competitions of antagonist [3H]diprenorphine (DPN) binding were measured in the absence and presence of GTPγS, which uncouples G protein heterotrimers from GPCRs. (*d*) A 1:1 coupling of $\beta_2AR:G_s$ results in a homogeneous population of high affinity agonist binding sites. G_s heterotrimer was added to lipoprotein particle reconstituted β_2AR in increasing molar ratios of receptor to G protein (noted as R:G in legend insert). Uncoupled β_2AR exhibits low affinity agonist binding, while β_2AR coupled to G_s displays high affinity isoproterenol binding. The method of G protein heterotrimer addition in these assays results in a significant amount of G protein aggregation,[20] necessitating high molar ratios of G protein to receptor (200:1) to achieve 100% coupling. Panels *a* and *d*, adapted with permission from Whorton et al.[20], copyright 2007 National Academy of Sciences, U.S.A. Panel *c*, adapted from Kuszak et al.[26]

In efforts to expand the utility of reconstituted lipoproteins for membrane protein study, several engineered forms of apo A-1 constructs designed to increase the effective diameter of the rHDL particle have been reported.[29,30] These larger particles have been used to reconstitute dimeric forms of rhodopsin by increasing the ratio of receptor:lipoprotein during the reconstitution,[19] although the topological orientation of the two rhodopsin molecules was undefined. Subsequent electron microscopy analysis of lipoprotein particles reconstituted with rhodopsin under similar conditions suggests that the two rhodopsins within the particle reconstitute in a random fashion.[31] Approximately 50% of the two rhodopsins were orientated as "parallel" dimers (i.e., N-termini of both receptors on the same side of the bilayer) and 50% were in an "antiparallel" orientation, the functional significance of the latter being questionable.

INSIGHTS INTO GPCR FUNCTION USING LIPOPROTEIN PARTICLES

GPCR Oligomerization

Isolation and analysis of monomeric and dimeric forms of GPCRs in lipoprotein particles

Historically GPCRs were thought to bind agonists and stimulate G proteins as a monomeric unit. However, a large body of literature over the past decade has reported the oligomerization of GPCRs in membrane fractions. The ability of GPCRs to form homo- or heterodimers, or even higher-order oligomers, may provide another level of signaling diversity and specificity, since the formation of larger oligomeric receptor complexes is postulated to generate unique ligand binding and G protein coupling surfaces. To date, our best understanding of the function of GPCR oligomerization comes from Class C GPCRs such as the metabotropic glutamate and GABA$_B$ receptors. These receptors function as obligate dimers in which the agonist binds to one member of the dimer pair whereas the G protein binding and activation is conferred through the other.[32]

A vast amount of evidence supporting the notion that Class A GPCRs oligomerize, including rhodopsin, dopamine receptors, and opioid receptors, has accumulated since the mid 1990s. Multiple studies have reported biochemical measurements of the interaction between two receptors at the plasma membrane, as well as functional consequences of dimerization.[33] The number of GPCRs found to oligomerize has increased significantly following the pioneering work by the laboratories of Blumer and Bouvier to develop fluorescence resonance energy transfer (FRET) and bioluminescence resonance energy transfer (BRET) methods to observe interactions between fluorescently-labeled GPCRs.[34,35] To date, a large proportion of defined hormone GPCRs have been shown to homo- and heterodimerize through the use of a variety of approaches.[36,37]

Unfortunately, the majority of reports supporting GPCR oligomerization were performed in heterologous overexpression systems, where receptors are expressed at hundreds- to thousands-fold higher levels then observed endogenously. Therefore, a caveat to these findings is the potential for forced or unnatural oligomerization as a result of overcrowding of receptors at the membrane. Thus many of the in vivo functional consequences of GPCR oligomerization remain unclear. More to the point, the potential for false-positive dimerization in these systems raised the question of whether class A GPCRs functioned as obligate dimers or oligomers, or if a monomeric receptor could bind ligands and activate G proteins. Answering this question required solving a significant technical problem, requiring the isolation of a population of GPCR monomers, in a lipid bilayer, from populations of receptor oligomers.

The reconstituted HDL particle system fulfills this requirement exquisitely. The structural characteristics of the particles allows for preferential incorporation of a GPCR monomer into a phospholipid bilayer that mimics its natural plasma membrane environment. Indeed, the reconstitution of several GPCRs in lipoprotein particles has been demonstrated and the restoration of their functional activities reported.[19,20,26,28,31,38] Reconstitution of detergent-solubilized β_2AR into lipoprotein particles, for example, restores the antagonist and agonist binding properties and more importantly displays functional coupling to G proteins.[20] Likewise reconstitution of the photoreceptor, rhodopsin, displays biophysical properties of activation and G protein coupling similar to the behavior of rhodopsin in native rod outer segment membranes.[28]

A critical analytical step in the study of these reconstituted receptor preparations was the determination of the receptor:lipoprotein particle stoichiometry. In one approach to addressing receptor stoichiometry, investigators took advantage of the unique buoyant properties of lipoprotein particles and the fact that their density changes dramatically with the incorporation of protein in the place of lipids. Utilizing this characteristic, rhodopsin can be reconstituted at higher GPCR:lipoprotein ratios and particles containing one or two rhodopsin molecules are resolved from each other by sucrose density centrifugation.[19] It is important to note that the orientation of the two GPCRs with respect to one another in the reconstituted lipoprotein particle may be dependent upon the specific conditions of reconstitution. Indeed, as discussed further below, heterogeneous populations of reconstituted receptor dimers have been observed,[31] and caution should therefore be taken when speculating about the contributions of GPCR oligomerization to function when studying these reconstituted particles containing multiple receptors.

Alternatively, single-molecule spectroscopy approaches may be employed to determine how many receptors are reconstituted in particle preparations that display functional G protein coupling. GPCRs fluorescently labeled with either Cy3 or Cy5 may be reconstituted and assessed for whether both fluorophores are visualized in a single particle. The successful reconstitution of a Cy3-GPCR together with a Cy5-GPCR, as a dimer, can be visualized as co-localization of the fluorophores by total internal reflection fluorescence microscopy (TIRFM). Analyses of the reconstitution of Cy3- and Cy5-labeled receptors at concentrations and stoichiometries used in functional assays indicated that greater than 95–98% of the receptors were monomeric.[20,26]

Analysis of the oligomeric state of rhodopsin is relatively easier, since active rhodopsin has a unique spectral fingerprint with a known extinction coefficient.[39] Ensemble measurements of populations of rhodopsin in lipoproteins may be made using the extinction coefficients of apo A-1 (two molecules per particle) and rhodopsin (one molecule or more) to determine the rhodopsin:lipoprotein stoichiometries.[28]

G protein coupling to GPCRs reconstituted in lipoprotein particles

Using reconstituted lipoproteins, monomeric β_2AR was shown to be capable of functional coupling to its cognate G protein, G$_s$. Following the addition of G$_s$ heterotrimer to rHDL•β_2AR, the βAR agonist isoproterenol stimulated [^{35}S]GTPγS binding to Gα_s with rapid kinetics (Figure 2.3a). The addition of G$_s$ heterotrimer also induced a high affinity isoproterenol binding site, the proportion of which is dependent on the concentration of the G protein. As illustrated in Figure 2.3d, the addition of G protein at high G:R ratios completely left-shifted the concentration dependency of isoproterenol inhibition of antagonist ([³H] dihydroalprenolol) binding.

Similarly, monomeric forms of the μ-opioid receptor (MOR) have been studied in lipoprotein particles. Unlike the β_2AR and rhodopsin, opioid receptors couple to multiple isoforms of the inhibitory $G_{i/o}$ protein family and bind both small molecule and relatively large peptide ligands.[40,41] Homo- and heterodimerization of opioid receptors has been reported,[42–44] and heterodimerization has been suggested as the cause of altered pharmacology in cells expressing multiple opioid receptor subtypes.[45,46] Furthermore, some literature suggests that opioid receptors are constitutive dimers at the plasma membrane, and perhaps even obligate dimers in terms of functional G protein coupling.[47] However, monomeric MOR displays strong coupling to $G_{i/o}$ proteins.[26,48] Reconstituted MOR demonstrated [³H] diprenorphine binding characteristics similar to those observed in membranes, a dramatic improvement over detergent-solubilized receptor preparations. Analysis of the G protein coupling revealed that the monomeric MOR was fully capable of activation of both $G\alpha_i$ and $G\alpha_o$ (as measured by [³⁵S]GTPγS binding) by the peptide DAMGO and the alkaloid morphine (Figure 2.3b).[26] Furthermore, both G_i and G_o heterotrimers demonstrated the capacity to allosterically regulate DAMGO and morphine binding to MOR (Figure 2.3c).

Reconstitutions of monomeric bovine rhodopsin appear to function in a similar manner to rhodopsin in native rod outer segments (ROS).[28,31] Many studies suggested that rhodopsin may function as a monomer over the years,[49] yet striking images using atomic force microscopy implied that rhodopsin is organized in a higher order arrays of oligomers.[50] Monomeric rhodopsin in lipoprotein particles undergoes photoactivation and exhibits formation and subsequent decay of the meta-II state at rates comparable to ROS membranes. In addition, photoactivation of rhodopsin promotes transducin binding and induces nucleotide exchange at a comparable rate as in ROS membranes.[28,51,52]

Thus it is apparent that the monomeric forms of representative GPCRs bind agonists and functionally couple to G protein heterotrimers, resulting in ligand affinities and Gα activation with the same pharmacological properties as those observed in the plasma membrane where receptor dimers are thought to exist. Yet these results are not irreconcilable with the notion of GPCR oligomerization. Rather, they offer additional insight into our understanding of biochemical data observed for GPCRs over the past few decades. Allosteric regulation of agonist binding to GPCRs by coupled G proteins is classically observed as a biphasic competition of antagonist binding by an agonist. The biphasic binding is interpreted as two populations of GPCRs, one coupled to G protein (high affinity) and one uncoupled (low affinity). Importantly, in assays performed with cell membrane preparations, the observed fraction of high affinity binding sites is usually not higher than approximately 50%.[53,54] Considering that 100% high affinity agonist binding is observed when a monomeric β_2AR population is completely coupled to G_s in HDL particles (Figure 2.3d),[20] this suggests that only ~50% of GPCRs may be coupled to G proteins in vivo. Indeed, theoretical models based on the crystal

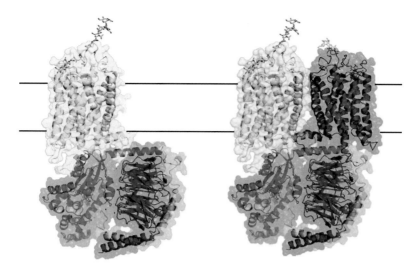

Figure 2.4: Theoretical models of monomeric (left) and dimeric (right) rhodopsin coupling to transducin (G_t). Proteins are colored as follows: rhodopsin – yellow and red, $G\alpha_t$ – purple, $G\beta$ – blue, $G\gamma$ – green. Gray lines indicate the relative position of the receptors in the plasma membrane. Each receptor condition is theorized to couple to a single transducin heterotrimer. The rhodopsin molecule coupled to the $G\alpha$ subunit of G_t (yellow) is thought to be the "activating" receptor. The functional role of the partner rhodopsin (red), for example whether it can add a level of selectivity to G protein coupling, remains unclear. *See colour plate 3.*

structure of rhodopsin and G proteins have suggested that a GPCR dimer may be unable to bind two G protein heterotrimers based on steric conflicts, and rather proposed a model of rhodopsin dimers where only one of the two rhodopsin molecules couples G_t.[55] GPCR dimers are therefore likely to couple to G proteins as a functional pentamer complex (Figure 2.4).[28,56] This hypothetical pentameric form of the receptor-G protein complex is gaining acceptance, and as a result a need to better define the concepts of receptor organization and function has emerged.[57]

Still, our understanding of the details of the receptor-G protein complex's function remains rudimentary. Importantly, it is unclear exactly how two receptor protomers may influence conformational changes between each other and how those changes influence ligand binding and G protein coupling. Although recent evidence suggests that cross-talk between receptors can influence ligand binding,[58–60] for some GPCRs it appears that this receptor-receptor interaction does not entail a transactivation mechanism in which one receptor binds agonist and the other receptor activates the G protein.[61] It is also unclear if GPCR dimerization creates unique G protein interacting sites, adding a level of specificity to receptor-G protein coupling that influences the signaling pathways activated by agonists. Heterodimerization of GPCRs that traditionally couple to two distinct G proteins may result in diverse signaling outcomes for such a complex. Furthermore, although oligomerization of GPCRs is not required for arrestin binding,[62] it is possible that receptor monomers, homodimers, and heterodimers may interact with arrestin differently. To best elucidate these potential differences between GPCR monomers and oligomers, a distinct population of each is required. Lipid vesicles and cellular expression systems are inappropriate for generating such samples due to their inability to control receptor complex stoichiometry. Reconstituted lipoprotein particles, however, have great potential for direct comparison of GPCR monomers and oligomers.

On the basic level, generating lipoprotein particles with multiple GPCRs is relatively simple; one merely has to increase the ratio of receptor to apo A-1 during the reconstitution process. Indeed, reconstitution of rhodopsin in monomeric and dimeric forms may shed some light on the relevance of the G protein-receptor pentamer. Reconstitution of two photoreceptors per lipoprotein particle resulted in rates that are approximately half as fast per rhodopsin molecule compared to the monomeric form.[19] Banerjee and colleagues also found that lipoprotein particles containing two rhodopsins exhibited similar decay rates as the monomeric receptor, but activate transducin at approximately one-half the rate observed for the monomer.[31]

However, as noted by Banerjee and colleagues, the rates of activation of the "dimer" in lipoprotein particles should be cautioned since the orientation of each rhodopsin, with respect to one another, was found to be either parallel or antiparallel in the phospholipid bilayer. While previous reports suggested that incorporation of multiple bacteriorhodopsins into lipoprotein particles results in a majority of parallel dimers due to association of the proteins prior to reconstitution,[30] Banerjee and colleagues observed mixtures of both parallel and antiparallel rhodopsins. The authors used immuno-gold staining and electron microscopy to visualize the N-terminus of rhodopsin, demonstrating that this method of incorporating multiple rhodopsins results in a near 50–50 population of parallel and antiparallel dimers.

The fact that particles containing two rhodopsins coupled to transducin half as rapidly and activate approximately one-half as many transducin molecules than particles containing one rhodopsin supports the notion of a functional pentamer.[19] However, it is unclear if lower stoichiometry of transducin coupling to rhodopsin is indicative of a pentameric complex, or if the conformation of rhodopsin was in some way disrupted such that it could not maximally couple G protein. Furthermore, the parallel or antiparallel orientation of the rhodopsin dimers is a critical characteristic that likely influences their function. Based on the premise that only the parallel dimers contribute to the observed activity whereas the antiparallel dimers were incapable of coupling G protein, Banerjee and colleagues concluded that the monomer and dimer display nearly identical properties. However this assumption of all-or-nothing activity for the parallel and antiparallel dimers, and by extension for parallel dimers composed of receptors interacting via different transmembrane helixes, may be flawed. It is plausible that multiple parallel and antiparallel receptor oligomeric conformations are capable of activity resulting in transducin activation, even if that activity is limited or altered, complicating the interpretation of transducin activation rates when comparing monomeric rhodopsin to these proposed dimer populations.

Clearly definitive analysis of the pharmacology of GPCR monomers versus dimers (or oligomers) therefore not only requires the isolation of these complexes in a membrane bilayer, but also conclusive demonstration of a homogeneous population of dimers in a parallel orientation. It may be possible to construct a "dimerization motif" as a fusion to the receptor, allowing the direct purification and reconstitution of GPCR dimers. This approach could then be optimized to generate homogeneous samples of GPCR homo- and heterodimers in lipoprotein particles. The evolving theory of GPCR function posits that while a monomer may be the minimal functional, dimers or oligomers are the predominant species in vivo. The lipoprotein reconstitution system will prove to be very powerful for studies on the pharmacology of these complexes, since no other current approach appears to be capable of definitive elucidation of GPCR dimer pharmacology.

Single-molecule analysis of ligand binding to monomeric GPCRs

The capacity to isolate a functional GPCR in a membrane-bound and mono-dispersed form in an aqueous solution has also allowed the direct visualization of ligand binding to a GPCR-G protein heterotrimer complex. We have visualized direct binding of a fluorescently-labeled μ-opioid receptor agonist, dermorphin, to a reconstituted and immobilized MOR-G_{i2} complex at a single-molecule level with TIRFM.[26] Analysis of fluorescent step-photobleaching confirmed that the agonist bound MOR-G_{i2} in a 1:1 ratio. These methods are currently being optimized to allow a direct measurement of the kinetics of ligand binding to GPCRs. Such an analysis will be incredibly useful for determining absolute ligand dissociation and association rates as opposed to the rate constants obtained from equilibrium binding studies.

These fluorescent ligands and microscopy techniques are being further developed to visualize ligand binding to living cells using TIRFM. With these single molecule approaches, a large library of opioid ligands, differentially labeled with fluorescent probes and selective for the different opioid receptor subtypes, could potentially visualize receptors on the cell surface and reveal subcellular organization. It may even be possible to discriminate receptor dimers and oligomers on the cell surface if sufficient resolution of the fluorescent probes is achieved. With the potential power to definitively probe oligomerization in native cell membranes, the future of fluorescently labeled GPCR ligands is indeed bright.

Conformational changes conferred by ligand binding and G protein coupling

A major goal of GPCR research has always been to understand the structural changes the receptor undergoes in response to agonist binding, and how these changes are conferred to the G protein heterotrimer to promote nucleotide exchange and consequent activation. Through the work or multiple laboratories, great strides have been taken toward this goal over the past decade. Using several approaches, investigators have elucidated that receptor activation involves rigid body movements of transmembrane helices 3 and 6 (TM3 and TM6).[63–67]

Moreover, recent crystallographic evidence suggests that in the inactive state TM3 and TM6 form a strong interaction between the arginine residue of the highly conserved DRY motif and a conserved glutamate residue outside TM6.[1,2,68] Combined with the finding that mutations of these residues in several receptor systems results in elevated basal activity and G protein activation, it is postulated that this putative "ionic lock" between the arginine and glutamate is broken upon receptor activation.

The significance of TM3 and TM6 for activation of the β_2AR was further revealed using environmentally sensitive fluorescent probes such as tetramethylrhodamine (TMR), fluorescein, and monobromobimane (bimane, mBBr) to monitor their movements in response to ligands.[69,70] Agonist-mediated conformational changes were observed in both detergent-solubilized[71] and lipid vesicle-reconstituted β_2AR.[72] Alterations in fluorescence intensity upon ligand binding revealed distinct kinetic steps in receptor activation,[71] as well as distinct receptor conformations in response to full and partial agonists.[70,72,73]

Combining the sensitivity of real-time fluorescence readout of receptor conformation with reconstitution into the lipoprotein mimic of the native lipid environment has provided great advances in the ability to monitor receptor-G protein interactions. In this approach, Yao and colleagues strategically placed bimane on Cys265 of the β_2AR, located on the C-terminal portion of TM6.[27] In this labeled receptor, conformational changes in TM6 alter the environment around bimane-Cys265, resulting in large changes in both the fluorescence intensity and the maximal wavelength at which bimane fluoresces (λ_{max}). Striking was the observation that a complex between G_s heterotrimer and receptor in lipoprotein particles can be observed as a decrease in bimane fluorescence intensity and an increase in λ_{max} (Figure 2.5c). The decrease in fluorescence is rapidly reversed with the addition of nucleotide, the pattern of which is dependent on the type of nucleotide (Figure 2.5e). GTPγS induced a rapid increase in bimane fluorescence (i.e., reversal of the G protein effect) sustained in a manner characteristic of the uncoupling capacity of this non-hydrolyzable GTP analogue. The rapidity of the increase in fluorescence implies that the α-subunit of the G_s heterotrimer is likely nucleotide-free. GDP induces a qualitatively similar rapid rise in bimane fluorescence, albeit with a significantly smaller maximal response. GTP, on the other hand, induces a rapid uncoupling event similar to GTPγS followed by a slow but steady reversal (decrease in fluorescence). The decrease in fluorescence is consistent with the re-coupling of G_s to β_2AR following hydrolysis of GTP to GDP, as incubation with either GDP or GTP eventually induced the same level of fluorescence.

The addition of the agonist isoproterenol induces a conformational change in bimane fluorescence of receptor alone that is comparable to that of G protein (Figure 2.5c). The addition of both isoproterenol and G_s heterotrimer appears to invoke an even further change in fluorescence, suggesting that agonist and G protein together stabilize a distinct receptor conformation. The capacity of G protein to induce a distinct agonist-bound state supports the notion that G proteins allosterically modulate agonist binding affinity. More precisely, G proteins

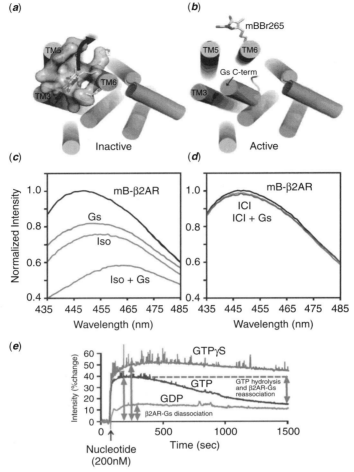

Figure 2-5: (**a** & **b**) Molecular models of the conformational changes in β$_2$AR following G$_s$ coupling: movement of TM6 away from TM3. (**c, d** & **e**) Conformational changes in β$_2$AR upon ligand binding and G$_s$ coupling monitored by a fluorescent probe (monobromobimane, mBBr) on TM6. (**c** & **d**) Wavelength scans of mBBrβ$_2$AR follow excitation (340 nm) under basal (black), Isoproterenol (light blue), G$_s$ (red), or with inverse agonist ICI-118,551 (green) treatment, or the co-application of G$_s$ with isoproterenol (magenta) or G$_s$ with ICI 118,551 (purple). (**e**) Coupling dynamics of mBBrβ$_2$AR to G$_s$ illustrated by time-scans in the presence of 200 nM concentrations of various nucleotides: GDP (light blue), GTPγS (purple) or GTP (black). Adapted with permission from Yao et al.[27] *See colour plate 4.*

in their nucleotide-free form, a state stabilized by receptor, induce a receptor conformation that displays a slow rate of agonist dissociation.

The capacity of G$_s$ to induce a coupled state of the receptor in lipoprotein particles also revealed the mechanism by which inverse agonists stabilize the inactive state of the receptor. As anticipated, the inverse agonist ICI-118,551 (ICI) had little effect on bimane fluorescence on the β$_2$AR alone (Figure 2.5d). Somewhat surprising was the capacity of ICI to prevent G protein from inducing a conformational change and thus interacting productively with the receptor. A similar

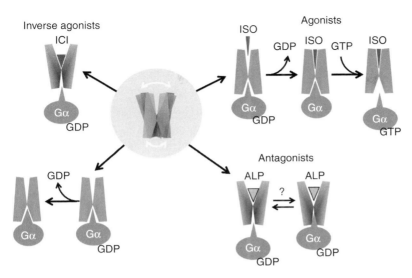

Figure 2-6: Schematic model of the β_2AR conformational changes induced by agonist, inverse agonist, neutral antagonists and G_s heterotrimer interactions. GPCRs under basal conditions exist in various conformational states (center), each of which may be stabilized by ligands and/ or G proteins. ISO (isoproterenol), ALP (alprenolol), ICI (ICI-118,551).

effect was observed with a less efficacious inverse agonist, carazolol.[27] These data strongly suggest that the mechanism by which inverse agonists decrease basal activity is to stabilize the receptor in an inactive state, one that is incapable of interacting with G protein. This effect is in stark contrast to more neutral antagonists such as alprenolol, which does not impede G protein interaction with the receptor.[27] Thus in the absence of agonist or G protein, receptors exist as an ensemble of conformational states. Inactive states may be stabilized by inverse agonists whereas "basally active" and active states may be stabilized by agonists and/or G proteins (Figure 2.6).[73,74]

These in vitro data collected using the lipoprotein reconstitution system may have considerable ramifications to our understanding of the mechanism of receptor activation in vivo. Several recent biophysical analyses of receptor coupling and activation have revealed that receptors may be "pre-coupled" to G proteins in a complex.[75–77] These investigations utilized resonance energy transfer measurements between G protein and receptor to suggest that they may be stably associated within 100 Å of each other (the approximate Förster distance for fluorescent proteins). Agonist activation results in perturbation of this complex, a process that appears to be reversible depending on the reporter system used. The in vitro data using reconstituted receptors do not refute this interpretation in that the receptor may still be associated with the G protein in the absence of agonists. This "pre-coupled" state of the receptor is insensitive to guanine nucleotides, in contrast to the exquisite nucleotide sensitivity of the nucleotide-free form of the G protein-receptor complex described earlier. Since the cellular concentrations of GDP and GTP (~10–20 mM and 100–200 mM, respectively) are sufficient to "uncouple" this state, we interpret the nucleotide- and agonist-free receptor G protein complex as a "coupled" state. It is also feasible that the in situ biophysical approaches are not necessarily measuring alterations in receptor conformation, but rather they are a measure of the consequences of the conformational change to the spatial relationship between the receptor and G protein. In contrast, the reconstitution studies reveal discrete conformational changes in the receptor following ligand binding or G protein interaction but cannot, or rather have not, revealed any changes in the spatial relationship between the receptor and G protein.

Taken together these data support a model where, under basal conditions, GPCRs may be pre-associated with G proteins in a complex that that does not alter the conformation of TM6. While it is possible to isolate and stabilize a receptor complex with G proteins in lipoprotein particles that is spectroscopically distinct and reveals rearrangement of TM6, it is dependent upon the removal of nucleotide. The presence of high levels of cellular nucleotides during in situ studies would therefore disfavor this receptor conformation but not preclude the formation of a "pre-coupled" form of β_2AR and G_s in the absence of agonists. Agonist activation of the "pre-coupled" form of the receptor-G protein complex, in the presence of cellular nucleotides, would first induce a conformational rearrangement of TM6. One would then observe a rapid uncoupling event due to the binding of GTP, as is observed in the lipoprotein particle studies with the β_2AR. Following effector activation and subsequent G protein inactivation through hydrolysis of GTP to GDP, a step that is also observable using the lipoprotein particles and monobromobimane, the G protein may again form a "pre-coupled" state with the receptor.

CONCLUSIONS

Considerable advances have been made in our understanding of GPCR function in the past few years, and key to many of these was the development of a new reconstitution system in the form of lipoprotein particles. The advantages of the reconstituted lipoprotein system for studying GPCR function are twofold. First, the capacity to isolate homogeneous populations of receptors and coupled G proteins of defined stoichiometry, specifically GPCR monomers, represents a significant improvement over lipid vesicle reconstitution. Second, the phospholipid bilayer contained within the particles adopts dimensions that are generally similar to the cellular plasma membrane. The lipoprotein reconstitution system was thus able to definitively illustrate that monomeric class A GPCRs are the minimal functional unit in terms of G protein activation. Furthermore, this approach offers the potential for confident analysis of the function of defined GPCR homo- and heterodimers. The monodispersed molecular nature of lipoprotein reconstituted GPCRs also facilitated the development of new methods for studying ligand binding to receptors, and future development may provide advances in the study of ligand binding kinetics. Finally, the development of a defined GPCR-G protein complex in the reconstitution system has yielded significant advances in our understanding of the dynamics of receptor structure during ligand binding and G protein coupling.

Thus the lipoprotein reconstitution approach has proved highly versatile for GPCR research. The isolation of defined GPCR-G protein complexes at the single molecule level is unique to this system, providing new opportunities for elucidating structure and function of the complex. Furthermore, the reconstituted lipoprotein particles are highly amendable to experimental manipulations, allowing extensive analysis of the effects of ligands and nucleotides on the GPCR-G protein complex. In conclusion, the reconstitution of GPCR-G protein complexes in lipoprotein particles has made an impressive entrance into GPCR research. Its continued use will most certainly reveal new insights into the highly dynamic nature of these signaling complexes.

REFERENCES

1. Cherezov, V., D. M. Rosenbaum, et al. (2007). "High-resolution crystal structure of an engineered human beta2-adrenergic G protein-coupled receptor." *Science* 318(5854): 1258–65.
2. Rasmussen, S. G., H. J. Choi, et al. (2007). "Crystal structure of the human beta2 adrenergic G-protein-coupled receptor." *Nature* 450(7168): 383–7.
3. Rosenbaum, D. M., V. Cherezov, et al. (2007). "GPCR engineering yields high-resolution structural insights into beta2-adrenergic receptor function." *Science* 318(5854): 1266–73.
4. Hanson, M. A., V. Cherezov, et al. (2008). "A specific cholesterol binding site is established by the 2.8 A structure of the human beta2-adrenergic receptor." *Structure* 16(6): 897–905.
5. Warne, T., M. J. Serrano-Vega, et al. (2008). "Structure of a beta1-adrenergic G-protein-coupled receptor." *Nature* 454(7203): 486–91.
6. Jaakola, V. P., M. T. Griffith, et al. (2008). "The 2.6 angstrom crystal structure of a human A2A adenosine receptor bound to an antagonist." *Science* 322(5905): 1211–7.
7. Park, J. H., P. Scheerer, et al. (2008). "Crystal structure of the ligand-free G-protein-coupled receptor opsin." *Nature* 454(7201): 183–7.
8. Scheerer, P., J. H. Park, et al. (2008). "Crystal structure of opsin in its G-protein-interacting conformation." *Nature* 455(7212): 497–502.
9. Shimamura, T., K. Hiraki, et al. (2008). "Crystal structure of squid rhodopsin with intracellularly extended cytoplasmic region." *J Biol Chem* 283(26): 17753–6.
10. Brouillette, C. G., G. M. Anantharamaiah, et al. (2001). "Structural models of human apolipoprotein A-I: a critical analysis and review." *Biochim Biophys Acta* 1531(1–2): 4–46.
11. Jonas, A., J. H. Wald, et al. (1990). "Apolipoprotein A-I structure and lipid properties in homogeneous, reconstituted spherical and discoidal high density lipoproteins." *J Biol Chem* 265(36): 22123–9.
12. Klon, A. E., M. K. Jones, et al. (2000). "Molecular belt models for the apolipoprotein A-I Paris and Milano mutations." *Biophys J* 79(3): 1679–85.
13. Jonas, A. (1986). "Reconstitution of high-density lipoproteins." *Methods Enzymol* 128: 553–82.
14. Rogers, D. P., C. G. Brouillette, et al. (1997). "Truncation of the amino terminus of human apolipoprotein A-I substantially alters only the lipid-free conformation." *Biochemistry* 36(2): 288–300.
15. Klon, A. E., J. P. Segrest, et al. (2002). "Comparative models for human apolipoprotein A-I bound to lipid in discoidal high-density lipoprotein particles." *Biochemistry* 41(36): 10895–905.

16. Li, Y., A. Z. Kijac, et al. (2006). "Structural analysis of nanoscale self-assembled discoidal lipid bilayers by solid-state NMR spectroscopy." *Biophys J* 91(10): 3819–28.

17. Shih, A. Y., I. G. Denisov, et al. (2005). "Molecular dynamics simulations of discoidal bilayers assembled from truncated human lipoproteins." *Biophys J* 88(1): 548–56.

18. Shaw, A. W., M. A. McLean, et al. (2004). "Phospholipid phase transitions in homogeneous nanometer scale bilayer discs." *FEBS Lett* 556(1–3): 260–4.

19. Bayburt, T. H., A. J. Leitz, et al. (2007). "Transducin activation by nanoscale lipid bilayers containing one and two rhodopsins." *J Biol Chem* 282(20): 14875–81.

20. Whorton, M. R., M. P. Bokoch, et al. (2007). "A monomeric G protein-coupled receptor isolated in a high-density lipoprotein particle efficiently activates its G protein." *Proc Natl Acad Sci U S A* 104(18): 7682–7.

21. Civjan, N. R., T. H. Bayburt, et al. (2003). "Direct solubilization of heterologously expressed membrane proteins by incorporation into nanoscale lipid bilayers." *Biotechniques* 35(3): 556–60, 562–3.

22. Katzen, F., J. E. Fletcher, et al. (2008). "Insertion of membrane proteins into discoidal membranes using a cell-free protein expression approach." *J Proteome Res* 7(8): 3535–42.

23. Segrest, J. P. (1977). "Amphipathic helixes and plasma lipoproteins: thermodynamic and geometric considerations." *Chem Phys Lipids* 18(1): 7–22.

24. Munford, M. L., Lima, V.R., Vieira, T.O., Heinzelmann, G., Creczynski-Pasa, T.B., Pasa, A.A. (2005). "AFM *In-Situ* Characterization of Supported Phospholipid Layers Formed by Vesicle Fusion." *Microscopy and Microanalysis* 11(Suppl. 3): 90–3.

25. Keough, K. (2003). "How thin can glass be? New ideas, new approaches." *Biophys J* 85(5): 2785–6.

26. Kuszak, A. J., S. Pitchiaya, et al. (2009). "Purification and functional reconstitution of monomeric mu-opioid receptors: allosteric modulation of agonist binding by Gi2." *J Biol Chem* 284(39): 26732–41.

27. Yao, X. J., G. Velez Ruiz, et al. (2009). "The effect of ligand efficacy on the formation and stability of a GPCR-G protein complex." *Proc Natl Acad Sci U S A* 106(23): 9501–6.

28. Whorton, M. R., B. Jastrzebska, et al. (2008). "Efficient coupling of transducin to monomeric rhodopsin in a phospholipid bilayer." *J Biol Chem* 283(7): 4387–94.

29. Denisov, I. G., Y. V. Grinkova, et al. (2004). "Directed self-assembly of monodisperse phospholipid bilayer Nanodiscs with controlled size." *J Am Chem Soc* 126(11): 3477–87.

30. Bayburt, T. H., Y. V. Grinkova, et al. (2006). "Assembly of single bacteriorhodopsin trimers in bilayer nanodiscs." *Arch Biochem Biophys* 450(2): 215–22.

31. Banerjee, S., T. Huber, et al. (2008). "Rapid incorporation of functional rhodopsin into nanoscale apolipoprotein bound bilayer (NABB) particles." *J Mol Biol* 377(4): 1067–81.

32. Pin, J. P., J. Kniazeff, et al. (2005). "Allosteric functioning of dimeric class C G-protein-coupled receptors." *Febs J* 272(12): 2947–55.

33. Hebert, T. E. and M. Bouvier (1998). "Structural and functional aspects of G protein-coupled receptor oligomerization." *Biochem Cell Biol* 76(1): 1–11.

34. Angers, S., A. Salahpour, et al. (2000). "Detection of beta 2-adrenergic receptor dimerization in living cells using bioluminescence resonance energy transfer (BRET)." *Proc Natl Acad Sci U S A* 97(7): 3684–9.

35. Overton, M. C. and K. J. Blumer (2000). "G-protein-coupled receptors function as oligomers in vivo." *Curr Biol* 10(6): 341–4.

36. Rios, C. D., B. A. Jordan, et al. (2001). "G-protein-coupled receptor dimerization: modulation of receptor function." *Pharmacol Ther* 92(2–3): 71–87.

37. Dalrymple, M. B., K. D. Pfleger, et al. (2008). "G protein-coupled receptor dimers: functional consequences, disease states and drug targets." *Pharmacol Ther* 118(3): 359–71.

38. Leitz, A. J., T. H. Bayburt, et al. (2006). "Functional reconstitution of Beta2-adrenergic receptors utilizing self-assembling Nanodisc technology." *Biotechniques* 40(5): 601–2, 604, 606, passim.

39. Wald, G. (1968). "Molecular basis of visual excitation." *Science* 162(850): 230–9.

40. Waldhoer, M., S. E. Bartlett, et al. (2004). "Opioid receptors." *Annu Rev Biochem* 73: 953–90.

41. Pineyro, G. and E. Archer-Lahlou (2007). "Ligand-specific receptor states: implications for opiate receptor signalling and regulation." *Cell Signal* 19(1): 8–19.

42. Cvejic, S. and L. A. Devi (1997). "Dimerization of the delta opioid receptor: implication for a role in receptor internalization." *J Biol Chem* 272(43): 26959–64.

43. Jordan, B. A. and L. A. Devi (1999). "G-protein-coupled receptor heterodimerization modulates receptor function." *Nature* 399(6737): 697–700.

44. Wang, D., X. Sun, et al. (2005). "Opioid receptor homo- and heterodimerization in living cells by quantitative bioluminescence resonance energy transfer." *Mol Pharmacol* 67(6): 2173–84.

45. George, S. R., T. Fan, et al. (2000). "Oligomerization of mu- and delta-opioid receptors. Generation of novel functional properties." *J Biol Chem* 275(34): 26128–35.

46. Martin, N. A. and P. L. Prather (2001). "Interaction of co-expressed mu- and delta-opioid receptors in transfected rat pituitary GH(3) cells." *Mol Pharmacol* 59(4): 774–83.

47. Levac, B. A., B. F. O ' Dowd, et al. (2002). "Oligomerization of opioid receptors: generation of novel signaling units." *Curr Opin Pharmacol* 2(1): 76–81.

48. Kuszak, A. J. (2009). The function of the monomeric form of the mu-opioid receptor: G protein-mediated allosteric regulation of agonist binding and stimulation of nucleotide exchange. *Department of Pharmacology*. Ann Arbor, University of Michigan. Ph.D. Thesis: 122 p.

49. Chabre, M. and M. le Maire (2005). "Monomeric G-protein-coupled receptor as a functional unit." *Biochemistry* 44(27): 9395–403.

50. Liang, Y., D. Fotiadis, et al. (2003). "Organization of the G protein-coupled receptors rhodopsin and opsin in native membranes." *J Biol Chem* 278(24): 21655–62.

51. Fahmy, K. and T. P. Sakmar (1993). "Light-dependent transducin activation by an ultraviolet-absorbing rhodopsin mutant." *Biochemistry* 32(35): 9165–71.

52. Jastrzebska, B., D. Fotiadis, et al. (2006). "Functional and structural characterization of rhodopsin oligomers." *J Biol Chem* 281(17): 11917–22.

53. De Lean, A., J. M. Stadel, et al. (1980). "A ternary complex model explains the agonist-specific binding properties of the adenylate cyclase-coupled beta-adrenergic receptor." *J Biol Chem* 255(15): 7108–17.

54. Kenakin, T. (2004). "G-protein coupled receptors as allosteric machines." *Receptors Channels* 10(2): 51–60.

55. Park, P. S., S. Filipek, et al. (2004). "Oligomerization of G protein-coupled receptors: past, present, and future." *Biochemistry* 43(50): 15643–56.

56. Baneres, J. L. and J. Parello (2003). "Structure-based analysis of GPCR function: evidence for a novel pentameric assembly between the dimeric leukotriene B4 receptor BLT1 and the G-protein." *J Mol Biol* 329(4): 815–29.

57. Ferre, S., R. Baler, et al. (2009). "Building a new conceptual framework for receptor heteromers." *Nat Chem Biol* 5(3): 131–4.

58. Park, P. S., C. S. Sum, et al. (2002). "Cooperativity and oligomeric status of cardiac muscarinic cholinergic receptors." *Biochemistry* 41(17): 5588–604.

59. Park, P. S. and J. W. Wells (2003). "Monomers and oligomers of the M2 muscarinic cholinergic receptor purified from Sf9 cells." *Biochemistry* 42(44): 12960–71.

60. Park, P. S. and J. W. Wells (2004). "Oligomeric potential of the M2 muscarinic cholinergic receptor." *J Neurochem* 90(3): 537–48.

61. Damian, M., S. Mary, et al. (2008). "G protein activation by the leukotriene B4 receptor dimer. Evidence for an absence of trans-activation." *J Biol Chem* 283(30): 21084–92.

62. Gurevich, V. V. and E. V. Gurevich (2008). "How and why do GPCRs dimerize?" *Trends Pharmacol Sci* 29(5): 234–40.

63. Farrens, D. L., C. Altenbach, et al. (1996). "Requirement of rigid-body motion of transmembrane helices for light activation of rhodopsin." *Science* 274(5288): 768–70.

64. Sheikh, S. P., J. P. Vilardarga, et al. (1999). "Similar structures and shared switch mechanisms of the beta2-adrenoceptor and the parathyroid hormone receptor. Zn(II) bridges between helices III and VI block activation." *J Biol Chem* 274(24): 17033–41.

65. Ballesteros, J. A., A. D. Jensen, et al. (2001). "Activation of the beta 2-adrenergic receptor involves disruption of an ionic lock between the cytoplasmic ends of transmembrane segments 3 and 6." *J Biol Chem* 276(31): 29171–7.

66. Greasley, P. J., F. Fanelli, et al. (2002). "Mutagenesis and modelling of the alpha(1b)-adrenergic receptor highlight the role of the helix 3/helix 6 interface in receptor activation." *Mol Pharmacol* 61(5): 1025–32.

67. Shapiro, D. A., K. Kristiansen, et al. (2002). "Evidence for a model of agonist-induced activation of 5-hydroxytryptamine 2A serotonin receptors that involves the disruption of a strong ionic interaction between helices 3 and 6." *J Biol Chem* 277(13): 11441–9.

68. Lodowski, D. T., T. E. Angel, et al. (2009). "Comparative analysis of GPCR crystal structures." *Photochem Photobiol* 85(2): 425–30.

69. Gether, U., S. Lin, et al. (1997). "Agonists induce conformational changes in transmembrane domains III and VI of the beta2 adrenoceptor." *Embo J* 16(22): 6737–47.

70. Yao, X., C. Parnot, et al. (2006). "Coupling ligand structure to specific conformational switches in the beta2-adrenoceptor." *Nat Chem Biol* 2(8): 417–22.

71. Swaminath, G., Y. Xiang, et al. (2004). "Sequential binding of agonists to the beta2 adrenoceptor. Kinetic evidence for intermediate conformational states." *J Biol Chem* 279(1): 686–91.

72. Swaminath, G., X. Deupi, et al. (2005). "Probing the beta2 adrenoceptor binding site with catechol reveals differences in binding and activation by agonists and partial agonists." *J Biol Chem* 280(23): 22165–71.

73. Ghanouni, P., Z. Gryczynski, et al. (2001). "Functionally different agonists induce distinct conformations in the G protein coupling domain of the beta 2 adrenergic receptor." *J Biol Chem* 276(27): 24433–6.

74. Ghanouni, P., J. J. Steenhuis, et al. (2001). "Agonist-induced conformational changes in the G-protein-coupling domain of the beta 2 adrenergic receptor." *Proc Natl Acad Sci U S A* 98(11): 5997–6002.

75. Nobles, M., A. Benians, et al. (2005). "Heterotrimeric G proteins precouple with G protein-coupled receptors in living cells." *Proc Natl Acad Sci U S A* 102(51): 18706–11.

76. Gales, C., J. J. Van Durm, et al. (2006). "Probing the activation-promoted structural rearrangements in preassembled receptor-G protein complexes." *Nat Struct Mol Biol* 13(9): 778–86.

77. Audet, N., C. Gales, et al. (2008). "Bioluminescence resonance energy transfer assays reveal ligand-specific conformational changes within preformed signaling complexes containing delta-opioid receptors and heterotrimeric G proteins." *J Biol Chem* 283(22): 15078–88.

PART II: OLIGOMERIZATION OF GPCRS

3 GPCR-G protein fusions:

Use in functional dimerization analysis

Graeme Milligan

GPCRS AND DRUG DISCOVERY

It is frequently noted that G protein-coupled receptors (GPCRs) are the most tractable family of proteins for small molecule drug discovery, and small molecules that alter the activity of GPCRs are the most commonly used group of medicines in clinical practice.[1] Furthermore, GPCRs remain attractive targets for drug discovery, with close to 50% of high throughput screens directed at such targets. Despite this, although the human genome encodes on the order of 500 non-olfactory GPCRs, currently approved medicines target only a small subset. Furthermore, the majority of the successfully targeted GPCRs are regulated by aminergic transmitters and hormones. This may reflect the early evolution and important roles played by these groups of ligands and GPCRs in the control of key physiological processes with other systems playing more modulatory roles. Until recent times, drug discovery at GPCRs had focused almost exclusively on the identification and optimization as drugs of orthosteric ligands, that is, molecules that share at least part of the binding site of the endogenous agonist(s) and therefore act in a competitive manner. However, recent efforts have begun to consider other modes of action, based on a better understanding of the three-dimensional structure,[2–4] allosteric nature,[5] and super-organizational architecture[6–8] of GPCRs. Each of these features, as well as the potential to identify ligands that modulate only a subset of the functions of GPCRs,[9,10] offers opportunities for drug development and a means to break away from the strictures of orthodox thinking in this area, as long as appropriate assays can be developed.

GPCR DIMERIZATION

All current medicines that target GPCRs have been developed based on the assumption that GPCRs are monomeric. In large part, this stemmed from early cDNA cloning studies, in which expression in heterologous cells of a single cDNA of interest resulted in the production of a functional receptor displaying pharmacology predicted from earlier physiological and tissue pharmacology studies. As such, a single polypeptide was usually sufficient to generate the functional GPCR. This was clearly untrue in the case of the GABA$_B$ receptor, where a single cDNA was able to generate a binding site for selective antagonists with affinities in reasonable accordance with native tissues but with affinity for agonist ligands displaced some hundredfold to higher concentrations.[11] Subsequent identification of a related GPCR-like sequence, encoded by a distinct gene, resulted in appreciation that this was a second subunit of the GABA$_B$ receptor, required for effective function of the receptor but not for binding the transmitter,[12-14] and that co-expression of the two cDNAs and interactions between the encoded polypeptides was required to produce a pharmacologically consistent and fully functional receptor. This remains the most fully characterized GPCR heteromer, and indeed it has been suggested that the GABA$_B$ receptor may exist as a 'dimer of dimers' and hence as a tetramer.[15] Although a number of commentators consider that much of the evidence that supports dimeric or oligomeric organization of GPCRs can be understood without recourse to such an explanation or requirement[16] or is simply overinterpreted,[17] there is no doubt that the idea of GPCRs being able to display quaternary structure, incorporating either multiple copies of the same GPCR (homomers) or different GPCRs (heteromers),[18,19] has become widely accepted, although there remain caveats over the physiological relevance of certain reported interactions.[18,20] Approaches including co-immunoprecipitation, various forms of resonance energy transfer, and functional complementation have driven the widespread belief in the existence of oligomerization.[21,22] Differential pharmacology and/or modes of signal transduction in cells expressing pairs of receptors[23-27] has also added impetus to the case for the importance and relevance of heterodimerization/oligomerization. This general topic[28,29-32] as well as the basis for protein-protein interactions in at least GPCR homomers,[6] has been reviewed extensively in recent times and will not be revisited here.

A number of groups have suggested that GPCR heteromers represent attractive drug targets as they may display novel pharmacological characteristics and limit side effects of prospective medicines if such heteromers display more limited tissue expression patterns that the corresponding homomers.[28,33,34] Molecular modeling studies are also beginning to be employed[35] to understand the basis and selectivity of heteromeric pairs. To date, only a limited number of studies have reported on the pharmacology of heteromer-selective ligands. This may reflect that such ligands will be unusual, but may also reflect a lack of approaches that are appropriate to identify such ligands in random, high-throughput screens. Considerable effort is currently being expended on efforts to develop approaches suitable to detect the pharmacological distinctiveness of GPCR heteromers. This chapter will concentrate on one such approach that employs the co-expression of pairs of GPCR-Gα subunit fusion proteins.

GPCR-Gα PROTEIN FUSIONS

GPCRs have been linked in tandem to a variety of other proteins to generate single open reading frames containing the functionalities of both the GPCR and the partner protein. These include both a range of fluorescent proteins to allow visual identification of a location of the GPCR[36] and enzymes, including both firefly (*Photinus*) and sea pansy (*Renilla*) forms of luciferase, to measure receptor levels and regulation of receptor amount in response to treatment of cells expressing such constructs with either known receptor ligands[37] or in random compound screens.[38] More recently, combinations of GPCRs C-terminally tagged with fluorescent proteins and/or forms of *Renilla* luciferase have been used most widely to identify GPCR homomers and heteromers via either fluorescence or bioluminescence resonance energy transfer,[21,28] whereas the addition of complementary fragments of either fluorescent proteins[39] or enzymes[40] that can restore fluorescence or function when brought together by interacting GPCRs have also been employed. Other proteins that have been linked directly to the C-terminal tail of one or more GPCRs include Regulators of G protein Signaling, such as RGS4,[41,42] β-arrestins,[43,44] and a variety of G protein α subunits.[45–47]

Initial studies that linked together a GPCR and a corresponding G protein α subunit to generate a β₂-adrenoceptor-Gα_s fusion protein demonstrated that such constructs could be successfully expressed and respond to agonist ligands to generate intracellular signals.[48] Although a number of reports have noted differences in the function and regulation of such fusions when compared with the isolated receptor,[49] they can be used to explore a range of aspects of receptor pharmacology that are difficult to analyze quantitatively by other means.[46] The generation and use of such fusions was rapidly expanded, initially using a fusion of the α_{2A}-adrenoceptor and a $Cys^{351}Gly$, Pertussis toxin-resistant, mutant of the G protein $G\alpha_{i1}$.[50] Direct measurement of V_{max} of the GTPase activity of this construct in response to an agonist, combined with saturation binding studies to define expression levels, allowed assessment of the catalytic constant of the guanine nucleotide exchange rate, and co-expression of excess β/γ complex confirmed both a key role for this complex in the guanine nucleotide exchange and hydrolysis cycle and that the fusion of GPCR and G protein α subunit did not abrogate interaction with β/γ.[50] In related studies, such constructs were used to demonstrate that they were able to effectively report on ligand partial agonism/efficacy[51] and, following a detailed co-expression study in which the effectiveness of the α_{2A}-adrenoceptor to activate Pertussis toxin-insensitive forms of $G\alpha_{i1}$ in which the Pertussis toxin-sensitive Cys^{351} was converted to all other possible amino acids,[52] how agonist function at the α_{2A}-adrenoceptor was modulated by mutation of this residue.[53] Central to the use of such constructs was the demonstration that due to linkage of a GPCR to a G protein it was unable to activate in co-expression studies, and it therefore did not result in nonspecific activation following fusion.[54] This allowed direct comparisons of the capacity and potency of different ligands to activate different G proteins via the same receptor. For example, Moon and colleagues demonstrated that the DOP opioid receptor was more efficient in promoting activation of $G\alpha_{i1}$ than $G\alpha_{o1}$ when measures of the catalytic effectiveness of guanine nucleotide exchange and hydrolysis were obtained,[55] whereas in studies on dopamine D_2 receptor-G protein fusions, Lane and colleagues demonstrated both that (S)-3-(3-hydroxyphenyl)-N-propylpiperidine and p-tyramine acted as 'biased ligands', being agonists for activation of $G\alpha_{o1}$ but antagonist/inverse agonists for the activation of each of $G\alpha_{i1}$, $G\alpha_{i2}$, and $G\alpha_{i3}$, and that dopamine and certain (but not all) synthetic agonists displayed distinct potency and efficacy for the activation of different G proteins.[56] In a similar vein, in another study, Lane and colleagues also demonstrated the higher potency and efficacy of a number of ligands to activate the dopamine D_3 receptor than the D_2 receptor when measuring activation of $G\alpha_{o1}$, and confirmed selective coupling of the D_3 receptor to $G\alpha_{o1}$.[57] Importantly, the dopamine D_2 receptor-G protein fusions displayed clear high and low affinity 'two-site' binding characteristics for agonist but not antagonist ligands,[56] a feature frequently considered to reflect G protein-coupled and -uncoupled states of the receptor. The ability of GPCRs to be linked to a range of G proteins, including both promiscuous and engineered, chimaeric G proteins, has also been employed usefully in ligand screening studies, including efforts to identify both endogenous and surrogate ligands for orphan GPCRs.[46,47]

APPLICATIONS TO GPCR DIMERIZATION

Functional GPCRs can be generated by the co-expression of split fragments of the receptor. Such studies provide support for the concept that different elements of a GPCR can, at least in part, fold independently and combine to produce an organizational structure akin to the native protein. Such studies do not, however, address directly issues of the potential for quaternary structure in GPCRs. By contrast, studies employing distinct mutants in the third and fifth transmembrane helix of the angiotensin AT1 receptor that were each unable to bind angiotensin II demonstrated that their co-expression resulted in restoration of a binding site for the endogenous peptide.[58] This 'trans-complementation' was compatible with AT1 receptor dimerization. However, this was inefficient because the number of binding sites generated was only a small fraction of that produced by expression of wild type cDNA. Similarly, co-expression of two distinct mutants of the histamine H1 receptor, each unable to bind the antagonist [³H]mepyramine when expressed alone, was able to generate a limited number of specific [³H]mepyramine binding sites.[59] This study was also consistent with GPCR dimerization but, by using the small molecule [³H]mepyramine believed to bind within the polar cavity generated by the tertiary organization of the transmembrane domains, appears to require a dimeric interaction that involves a sharing of transmembrane helixes between the co-expressed mutants and a mechanism of interaction that has been described as 'domain swapping.'[60] Domain swapping inherently appears energetically unfavorable, and the studies mentioned earlier in the chapter were unable to assess if dimerization was an uncommon event that required this mechanism, or if a more energetically favorable mechanism, such as direct contact interfaces between pre-folded 7-transmembrane domain monomers, generated a substantial number of 'contact dimers' that could not be detected in the ligand binding 'trans-complementation' assays. More usefully, analysis of pairs of luteinizing hormone receptor mutants, in which one lacked ligand binding whereas the second was able to bind ligand but was deficient in signal generation, allowed complementation of substantial levels of function when they were co-expressed.[61] Because the luteinizing hormone receptor, like other glycoprotein hormone receptors, is a 'dual-domain' protein consisting of an exo-domain that binds the ligand and the 7-transmembrane domain element that conveys the signal to the G protein and hence into the cell, I reasoned that it should be possible to generate pairs of inactive GPCR-G protein fusions that would also have the capacity to complement function when co-expressed. Mutation of the G protein α subunit to convert a key Gly residue, required for guanine nucleotide exchange and conserved in all mammalian Gα subunits, provided one generic means to generate fusion proteins that bind ligands but are unable to cause G protein activation. Similarly, mutation of one or more amino acids in the intracellular loops of GPCRs provided a distinct means to generate fusion proteins able to bind ligands but unable to activate the linked Gα subunit. In studies with a series of such α_{1B}-adrenoceptor-$G\alpha_{11}$ fusion proteins, maintenance of the orthosteric binding pocket meant that [^{3}H]prazosin binding assays could be employed to define levels of expression of each construct and consequently allow addition of equal and defined amounts of each to the functional assays.[62] As anticipated, agonists promoted the binding of [^{35}S]GTPγS to the wild type α_{1B}-adrenoceptor-$G\alpha_{11}$ fusion and this was eliminated, both when either mutant was expressed alone and when membranes expressing each mutant individually were combined. By contrast, with co-expression of the two distinct, inactive mutants, agonist stimulation of [^{35}S]GTPγS binding was restored.[62] With the addition to the assay of twice as many [^{3}H]prazosin fusion protein binding sites in membranes co-expressing the two mutants as in those expressing the wild type fusion, agonist-stimulated [^{35}S]GTPγS binding was some 60% of that produced by the wild type construct.[62] Theoretically, if the constructs produced strict dimers, 50% of these should be complementary and functional. This would predict that a twofold higher level of binding site number should result in the reconstitution of 100% of the signal generated by membranes expressing the wild-type fusion. As there was no direct means to assess if an equal amount of each inactive fusion protein was present in the membranes co-expressing the two inactive forms, such quantitative analysis of the data is restricted. However, as the individual mutants were expressed to similar levels when expressed alone, these results are certainly consistent with a substantial proportion of active dimers being generated. Equivalent studies with histamine H1 receptor-$G\alpha_{11}$ fusion constructs resulted in an even higher proportion of reconstitution of function[62] and markedly greater predictions of the proportion of dimers than obtained from the reconstitution of 'domain swap' [^{3}H]mepyramine binding sites reported by Bakker and colleagues.[59] Because essentially all cells express endogenously $G\alpha_q/G\alpha_{11}$ family G proteins that might have been activated by the α_{1B}-adrenoceptor and histamine H1 receptor constructs, Carrillo and colleagues also employed mouse embryo fibroblasts lacking both $G\alpha_q$ and $G\alpha_{11}$.[62] In this model system, agonist addition to cells transfected to express either wild-type fusion protein resulted in transient but robust elevation of intracellular [Ca^{2+}], and this was recapitulated with co-expression of the pairs of inactive fusions. However, although this could also be achieved by co-expression of an inactive α_{1B}-adrenoceptor and the potentially complimentary inactive histamine H1 receptor-$G\alpha_{11}$ fusion, quantitative analysis demonstrated this to be an inefficient process, unlikely to reflect the presence of substantial amounts of such a heteromeric α_{1B}-adrenoceptor-histamine H1 receptor complex.[62] These results suggested that significant selectivity in hetero-dimer interactions should result in selectivity of functional re-constitution. Equivalent functional reconstitution studies that measured agonist stimulation of [^{35}S]GTPγS binding have also provided evidence for quaternary organization of the α_{1A}-adrenoceptor.[63]

As anticipated from the conservation across G protein α subunits of the Gly residue that was mutated to produce the inactive $G\alpha_{11}$ subunit, mutation of the equivalent residue in a Pertussis toxin-insensitive mutant of $G\alpha_{i1}$ allowed the production of inactive fusion proteins of 'G_i'-coupled receptors such as the DOP opioid receptor.[64] Using the same strategy of mutating key amino acids in the second intracellular loop of the receptor[21] to produce a second distinct, ligand binding but functionally inactive DOP receptor-$G\alpha_{i1}$ fusion, co-expression of these two forms again resulted in complementation of function. This occurred in membranes of cells that had been pre-treated with Pertussis-toxin to ensure that ligand stimulation of [^{35}S]GTPγS binding must be to the Pertussis toxin-resistant form of $G\alpha_{i1}$ linked to the DOP receptor rather than to endogenously expressed G_i-family G proteins.[64] Equivalent studies with fusions of the KOP and MOP opioid receptors produced generally similar results,[64] indicating and confirming[65] that each of the well-studied opioid receptor subtypes is able to form dimers/oligomers. Based on binding site expression levels, the effectiveness of reconstitution of DOP receptor activation of $G\alpha_{i1}$ was greater than for either the equivalent KOP or MOP receptor constructs,[64] but the implications of this in terms of quaternary structure of the opioid receptor subtypes remain unclear because of the inability, as noted earlier in the chapter, to measure separately the proportion of each inactive fusion protein in co-expression studies. Importantly, for the use of this approach in the detection of potential dimer/heteromer-selective ligands, reconstitution of such pairs of inactive GPCR-Gα protein fusions generates pharmacology akin to that observed at the equivalent wildtype fusion protein (Figure 3.1).

The fusion protein approach, now termed DimerScreen™ (http://www.caratherapeutics.com/dimerScreen-technology.php), has already been applied to studies of the pharmacology and function of GPCR heterodimerization. Based

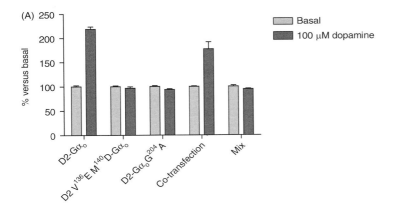

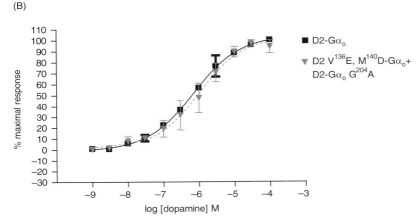

Figure 3-1: Reconstitution of function by co-expression of a pair of inactive dopamine D2-$G\alpha_{o1}$ fusion proteins: unaltered potency of dopamine.

Fusion proteins between the long isoform of the human dopamine D2 receptor and a Pertussis toxin resistant $C^{351}I$ variant of $G\alpha_{o1}$ were generated by Lane and colleagues. [56,57] These were otherwise either wildtype in both receptor and G protein, had the mutations $V^{136}E$ + $M^{140}D$ within the dopamine D2 receptor or also carried a $G^{204}A$ mutation in the G protein. Each of these was expressed individually in HEK cells or $V^{136}E$ + $M^{140}D$ D2- $C^{351}I$-$G\alpha_{o1}$ and D2-$G^{204}A$, $C^{351}I$-$G\alpha_{o1}$ were co-expressed and the cells were treated with Pertussis toxin to cause ADP-ribosylation of endogenously expressed G_i-family G proteins. Membranes were prepared and saturation [³H]spiperone binding studies performed to determine expression levels. Membranes containing equal numbers of [³H]spiperone binding sites (except for the membranes co-expressing $V^{136}E$ + $M^{140}D$ D2- $C^{351}I$-$G\alpha_{o1}$ and D2-$G^{204}A$, $C^{351}I$-$G\alpha_{o1}$ where 2 x as many [³H]spiperone binding sites were added) were used in [³⁵S]GTPγS binding studies. In a final set of experiments membranes expressing either $V^{136}E$ + $M^{140}D$ D2- $C^{351}I$-$G\alpha_{o1}$ or D2-$G^{204}A$, $C^{351}I$-$G\alpha_{o1}$ were mixed.

Basal [³⁵S]GTPγS binding (green) and the effect of addition of 10 μM dopamine (pink) were assessed.

The effect of various concentrations of dopamine was assessed in membranes expressing D2-$C^{351}I$-$G\alpha_{o1}$ (black) and $V^{136}E$ + $M^{140}D$ D2- $C^{351}I$-$G\alpha_{o1}$ plus D2-$G^{204}A$, $C^{351}I$-$G\alpha_{o1}$ (red) was assessed and plotted as % maximal effect.

on published interactions between various opioid and chemokine receptors[66,67] and the known cross-regulation of chemokine and opioid ligands in the control of chemotaxis in white cells, Parenty and colleagues established cell lines expressing constitutively a nonfunctional, Pertussis toxin-resistant DOP-$G\alpha_{i2}$

fusion protein containing a mutated G protein.[68] These cells were also able to express, in an antibiotic-dependent fashion, a potentially complementary, non-functional CXCR2-Pertussis toxin-resistant $G\alpha_{i2}$ fusion protein in which the mutation was located in the second intracellular loop of the CXCR2 receptor. Only after expression of the CXCR2-Pertussis toxin-resistant $G\alpha_{i2}$ fusion protein was D-Ala-D-Leu enkephalin and other opioid agonists with affinity for the DOP receptor able to stimulate binding of $[^{35}S]GTP\gamma S$ in membranes from these cells.[68] Furthermore, the presence of the CXCR2 blocker SB225002 enhanced the action of DOP agonists, although it had no direct affinity for the DOP receptor and did not regulate directly opioid function in the absence of expression of the CXCR2 receptor fusion protein.[68] Moreover, control of the timing of expression of the CXCR2 receptor fusion protein correlated with when the effect of the CXCR2 blocker could be observed.[68] Such studies are consistent with heteromeric interactions between CXCR2 and DOP receptors, and such allosteric properties of GPCR heteromers and their implications for drug design have recently been reviewed.[29,69] The mechanism of action of SB225002 remains to be established clearly. Compounds related to SB225002 have recently been reported to bind to an intracellular site on the CXCR2 receptor[70] rather than competing with inter-leukin 8 at the orthosteric binding site and may do so by interfering directly with G protein activation. This might make more G protein available to the DOP receptor in the experimental system employed by Parenty and colleagues.[68] Clearly further studies are required to test this possibility. Although yet to be assessed, the DimerScreen™ fusion protein strategy has the potential to reexamine and explore details of some of the more fascinating observations in the GPCR heteromer literature. For example, although neither the dopamine D_1 nor D_2 receptor couples selectively to the mobilization of intracellular Ca^{2+} via G_q/G_{11} G proteins when expressed separately, there is substantial evidence that this is the case in cells co-expressing the two receptors, that this occurs in the striatum and such interactions generate distinct pharmacology.[26] It would be expected from the foregoing that fusions of either D_1 or D_2 receptor to G_q/G_{11} would not result in agonist stimulation of $[^{35}S]GTP\gamma S$ binding but might well do so in cells in which the two (even wildtype) fusions are co-expressed. This and other examples of differential G protein coupling associated with receptor co-expression and potential heteromerization are well suited for analysis via the use of GPCR-G protein α subunit fusion proteins.

HETEROMER-SELECTIVE LIGANDS

A key goal of assays developed to study GPCR heteromers is to identify heteromer-selective ligands that may be used to explore the function of GPCR heteromers in physiological settings and to gain greater insight into the potential of such heteromers as therapeutic targets. There has been great interest in the report that 6'-guanidinonaltrindole acts as DOP-KOP opioid heteromer-selective agonist,[24] not least because this ligand is reported to be an analgesic only when administered into the spinal cord but not into the brain. Such results suggest that the organization of heteromers may be tissue-specific. Opioid receptor pharmacology has long been considered an area in which heteromer formation may account for the complexities observed,[65] and the capacity to observe such effect may also reflect the rich pharmacology available for these receptors due to the search for non-addictive opioid analgesics. Quite how this ligand might interact selectively with the DOP-KOP heteromer remains to be unravelled. Furthermore, the concept of widespread opioid receptor heteromerization has recently been challenged by the observation that DOP and MOP receptors are expressed in distinct subsets of primary afferents in mouse, rather than being widely co-expressed.[71] There is also evidence that the dopaminergic ligand SKF 83959 may act as a dopamine D1-D2 heteromer selective agonist[26] because it robustly stimulates calcium signals but not adenylyl cyclase activity in cells co-expressing these two receptors and lacks this activity in cells from mice lacking expression of either receptor. Random screening is likely to be required to identify and understand the molecular basis of function of heteromer-selective ligands because for many potential receptor pairs, available pharmacology is currently too limited. Such screens have been initiated by Cara Therapeutics using the DimerScreen™ approach.

As well as simple, small molecule ligands such as 6'-guanidinonaltrindole, there has been considerable emphasis in recent times on the synthesis of bivalent and multivalent ligands, with the hope that these might provide tool compounds to explore the functional relevance of GPCR heteromers. As well as modified opioid congeners[72] and bivalent ligands containing both opioid and other pharmacophores[73] separated by a series of spacer arms, a number of other 'bivalent' ligands have been synthesized and studied.[74],[75] Although not inherently 'drug-like,' such compounds will be of considerable value for 'proof of concept' studies and, once more, the DimerScreen™ approach offers an excellent means to characterize such ligands in detail prior to their use in native cells and animal models of health and disease.

ACKNOWLEDGMENTS

I thank the Medical Research Council U.K. (grants G9811527 and G0900050) for supporting my research in this area.

SUGGESTED READING

1. Jacoby, E., Bouhelal, R., Gerspacher, M., and Seuwen, K. (2006) The 7 TM G-protein-coupled receptor target family. ChemMedChem. *1*, 761–782.
2. Huber, T., Menon, S., and Sakmar, T.P. (2008) Structural basis for ligand binding and specificity in adrenergic receptors: implications for GPCR-targeted drug discovery. Biochemistry *47*, 11013–11023.
3. Topiol, S., and Sabio, M. (2009). X-ray structure breakthroughs in the GPCR trans-membrane region. Biochem. Pharmacol. *78*, 11–20.

4. Rosenbaum, D.M., Rasmussen, S.G., and Kobilka, B.K. (2009) The structure and function of G-protein-coupled receptors. Nature *459*, 356–363.

5. Conn, P.J., Christopoulos, A., Lindsley, C.W. (2009) Allosteric modulators of GPCRs: a novel approach for the treatment of CNS disorders. Nat. Rev. Drug Discov. *8*, 41–54.

6. Milligan, G. (2007) G protein-coupled receptor dimerisation: molecular basis and relevance to function. Biochim. Biophys. Acta *1768*, 825–835.

7. Milligan, G. (2008) A day in the life of a G protein-coupled receptor: the contribution to function of G protein-coupled receptor dimerization. Br. J. Pharmacol. *153* Suppl 1:S216–229.

8. Milligan, G. (2009) G protein-coupled receptor hetero-dimerization: contribution to pharmacology and function. Br. J. Pharmacol. [Epub ahead of print] PMID: 19309353

9. Violin, J.D., and Lefkowitz, R.J. (2007) Beta-arrestin-biased ligands at seven-transmembrane receptors. Trends Pharmacol. Sci. *28*, 416–422.

10. Verkaar, F., van Rosmalen, J.W., Blomenröhr, M., van Koppen, C.J., Blankesteijn, W.M., Smits, J.F., and Zaman, G.J. (2008) G protein-independent cell-based assays for drug discovery on seven-transmembrane receptors. Biotechnol. Annu. Rev. *14*, 253–274.

11. Kaupmann, K., Huggel, K., Heid, J., Flor, P.J., Bischoff, S., Mickel, S.J., McMaster, G., Angst, C., Bittiger, H., Froestl, W., and Bettler, B. (1997) Expression cloning of GABA(B) receptors uncovers similarity to metabotropic glutamate receptors. Nature *386*, 239–246.

12. Kaupmann, K., Malitschek, B., Schuler, V., Heid, J., Froestl, W., Beck, P., Mosbacher, J., Bischoff, S., Kulik, A., Shigemoto, R., Karschin, A., and Bettler, B. (1998) GABA(B)-receptor subtypes assemble into functional heteromeric complexes. Nature *396*, 683–687

13. Jones, K.A., Borowsky, B., Tamm, J.A., Craig, D.A., Durkin, M.M., Dai, M., Yao, W.J., Johnson, M., Gunwaldsen, C., Huang, L.Y., Tang, C., Shen, Q., Salon, J.A., Morse, K., Laz, T., Smith, K.E., Nagarathnam, D., Noble, S.A., Branchek, T.A., and Gerald, C. (1998) GABA(B) receptors function as a heteromeric assembly of the subunits GABA(B) R1 and GABA(B)R2. Nature *396*, 674–679.

14. White, J.H., Wise, A., Main, M.J., Green, A., Fraser, N.J., Disney, G.H., Barnes, A.A., Emson, P., Foord, S.M., and Marshall, F.H. (1998) Heterodimerization is required for the formation of a functional GABA(B) receptor. Nature *396*, 679–682.

15. Maurel, D., Comps-Agrar, L., Brock, C., Rives, M.L., Bourrier, E., Ayoub, M.A., Bazin, H., Tinel, N., Durroux, T., Prézeau, L., Trinquet, E., and Pin, J.P. (2008) Cell-surface protein-protein interaction analysis with time-resolved FRET and snap-tag technologies: application to GPCR oligomerization. Nat. Methods *5*, 561–567.

16. Chabre, M., Deterre, P., and Antonny, B. (2009) The apparent cooperativity of some GPCRs does not necessarily imply dimerization. Trends Pharmacol. Sci. *30*, 182–187.

17. James, J.R., Oliveira, M.I., Carmo, A.M., Iaboni, A., and Davis, S.J. (2006) A rigorous experimental framework for detecting protein oligomerization using bioluminescence resonance energy transfer. Nat. Methods *3*, 1001–1006.

18. Pin, J.P., Neubig, R., Bouvier, M., Devi, L., Filizola, M., Javitch, J.A., Lohse, M.J., Milligan, G., Palczewski, K., Parmentier, M., and Spedding, M. (2007) International Union of Basic and Clinical Pharmacology. LXVII. Recommendations for the recognition and nomenclature of G protein-coupled receptor heteromultimers. Pharmacol. Rev. *59*, 5–13.

19. Ferré, S., Baler, R., Bouvier, M., Caron, M.G., Devi, L.A., Durroux, T., Fuxe, K., George, S.R., Javitch, J.A., Lohse, M.J., Mackie, K., Milligan, G., Pfleger, K.D., Pin, J.P., Volkow, N.D., Waldhoer, M., Woods, A.S., and Franco, R. (2009) Building a new conceptual framework for receptor heteromers. Nat. Chem. Biol. *5*, 131–134.

20. Gurevich, V.V., and Gurevich, E.V. (2008) How and why do GPCRs dimerize? Trends Pharmacol. Sci. *29*, 234–240.

21. Milligan, G., and Bouvier, M. (2005) Methods to monitor the quaternary structure of G protein-coupled receptors. FEBS J. *272*, 2914–2925.

22. Harrison, C., and van der Graaf, P.H. (2006) Current methods used to investigate G protein coupled receptor oligomerisation. J. Pharmacol. Toxicol. Methods *54*, 26–35.

23. Kearn, C.S., Blake-Palmer, K., Daniel, E., Mackie, K., and Glass, M. (2005) Concurrent stimulation of cannabinoid CB1 and dopamine D2 receptors enhances heterodimer formation: a mechanism for receptor cross-talk? Mol. Pharmacol. *67*, 1697–1704.

24. Waldhoer, M., Fong, J., Jones, R.M., Lunzer, M.M., Sharma, S.K., Kostenis, E., Portoghese, P.S., and Whistler, J.L. (2005) A heterodimer-selective agonist shows in vivo relevance of G protein-coupled receptor dimers. Proc. Natl. Acad. Sci. U S A *102*, 9050–9055.

25. Ellis, J., Pediani, J.D., Canals, M., Milasta, S., and Milligan, G. (2006) Orexin-1 receptor-cannabinoid CB1 receptor heterodimerization results in both ligand-dependent and -independent coordinated alterations of receptor localization and function. J. Biol. Chem. *281*, 8812–8824.

26. George, S.R., and O ' Dowd, B.F. (2007) A novel dopamine receptor signaling unit in brain: heterooligomers of D1 and D2 dopamine receptors. Scientific World Journal. *7*, 58–63.

27. González-Maeso, J., Ang, R.L., Yuen, T., Chan, P., Weisstaub, N.V., López-Giménez, J.F., Zhou, M., Okawa, Y., Callado, L.F., Milligan, G., Gingrich, J.A., Filizola, M., Meana, J.J., and Sealfon, S.C. (2008) Identification of a serotonin/glutamate receptor complex implicated in psychosis. Nature *452*, 93–97.

28. Milligan, G. (2006) G-protein-coupled receptor heterodimers: pharmacology, function and relevance to drug discovery. Drug Discov. Today *11*, 541–549.

29. Milligan, G., and Smith, N.J. (2007) Allosteric modulation of heterodimeric G-protein-coupled receptors. Trends Pharmacol. Sci. *28*, 615–620.

30. Kent, T., McAlpine, C., Sabetnia, S., and Presland, J. (2007) G-protein-coupled receptor heterodimerization: assay technologies to clinical significance. Curr. Opin. Drug Discov. Devel. *10*, 580–589.

31. Dalrymple, M.B., Pfleger, K.D., and Eidne, K.A. (2008) G protein-coupled receptor dimers: functional consequences, disease states and drug targets. Pharmacol. Ther. *118*, 359–371.

32. Satake, H., and Sakai, T. (2008) Recent advances and perceptions in studies of heterodimerization between G protein-coupled receptors. Protein Pept. Lett. *15*, 300–308.

33. Panetta, R. and Greenwood, M.T. (2008) Physiological relevance of GPCR oligomerization and its impact on drug discovery. Drug Discov. Today *13*, 1059–1066.

34. Franco, R., Casadó, V., Cortés, A., Pérez-Capote, K., Mallol, J., Canela, E., Ferré, S., and Lluis, C. (2008) Novel pharmacological targets based on receptor heteromers. Brain Res. Rev. *58*, 475–482.

35. Filizola, M. (2009) Increasingly accurate dynamic molecular models of G-protein coupled receptor oligomers: Panacea or Pandora's box for novel drug discovery? Life Sci. [Epub ahead of print] PMID: 19465029

36. Groarke, D.A., Wilson, S., Krasel, C., and Milligan, G. (1999) Visualization of agonist-induced association and trafficking of green fluorescent protein-tagged forms of both beta-arrestin-1 and the thyrotropin-releasing hormone receptor-1. J. Biol. Chem. *274*, 23263–23269.

37. Ramsay, D., Bevan, N., Rees, S., and Milligan, G. (2001) Detection of receptor ligands by monitoring selective stabilization of a Renilla luciferase-tagged, constitutively active mutant, G-protein-coupled receptor. Br. J. Pharmacol. *133*, 315–323.

38. Zeng, F.Y., McLean, A.J., Milligan, G., Lerner, M., Chalmers, D.T., and Behan, D.P. (2003) Ligand specific up-regulation of a Renilla reniformis luciferase-tagged, structurally unstable muscarinic M3 chimeric G protein-coupled receptor. Mol. Pharmacol. *64*, 1474–1484

39. Vidi, P.A., and Watts, V.J. (2009) Fluorescent and bioluminescent protein-fragment complementation assays in the study of G protein-coupled receptor oligomerization and signaling. Mol. Pharmacol. *75*, 733–739.

40. Luker, K.E., Gupta, M., and Luker, G.D. (2009) Imaging chemokine receptor dimerization with firefly luciferase complementation. FASEB J. *23*, 823–834.

41. Bahia, D.S., Sartania, N., Ward, R.J., Cavalli, A., Jones, T.L., Druey, K.M., and Milligan, G. (2003) Concerted stimulation and deactivation of pertussis toxin-sensitive G proteins by chimeric G protein-coupled receptor-regulator of G protein signaling 4 fusion proteins: analysis of the contribution of palmitoylated cysteine residues to the GAP activity of RGS4. J. Neurochem. *85*, 1289–1298

42. Schneider, E.H., and Seifert, R. (2009) Histamine H(4) receptor-RGS fusion proteins expressed in Sf9 insect cells: A sensitive and reliable approach for the functional characterization of histamine H(4) receptor ligands. Biochem. Pharmacol. May 21 PMID: 19464266.

43. Martini, L., Hastrup, H., Holst, B., Fraile-Ramos, A., Marsh, M., and Schwartz, T.W. (2002) NK1 receptor fused to beta-arrestin displays a single-component, high-affinity molecular phenotype. Mol. Pharmacol. *62*, 30–37.

44. Jafri, F., El-Shewy, H.M., Lee, M.H., Kelly, M., Luttrell, D.K., and Luttrell, L.M. (2006) Constitutive ERK1/2 activation by a chimeric neurokinin 1 receptor-beta-arrestin1 fusion protein. Probing the composition and function of the G protein-coupled receptor "signalsome". J. Biol. Chem. *281*, 19346–19357.

45. Milligan, G., Parenty, G., Stoddart, L.A., and Lane, J.R. (2007) Novel pharmacological applications of G-protein-coupled receptor-G protein fusions. Curr. Opin. Pharmacol. *7*, 521–526.

46. Milligan, G., Feng, G.-J., Ward, R.J., Sartania, N., Ramsay, D., McLean, A.J., and Carrillo, J.J. (2004) G protein-coupled receptor fusion proteins in drug discovery. Current Pharmaceutical Design *10*, 1989–2001.

47. Suga, H., and Haga, T. (2007) Ligand screening system using fusion proteins of G protein-coupled receptors with G protein alpha subunits. Neurochem. Int. *51*, 140–164.

48. Bertin, B., Freissmuth, M., Jockers, R., Strosberg, A.D., and Marullo, S. (1994) Cellular signaling by an agonist-activated receptor/Gs alpha fusion protein. Proc. Natl. Acad. Sci. U S A *91*, 8827–8831.

49. Di Certo, M.G., Batassa, E.M., Casella, I., Serafino, A., Floridi, A., Passananti, C., Molinari, P., and Mattei, E. (2008) Delayed internalization and lack of recycling in a beta2-adrenergic receptor fused to the G protein alpha-subunit. BMC Cell Biol. *7*, 9:56.

50. Wise, A., Carr, I.C., and Milligan, G. (1997a) Measurement of agonist-induced guanine nucleotide turnover by the G-protein Gi1alpha when constrained within an alpha2A-adrenoceptor-Gi1alpha fusion protein. Biochem. J. *325*, 17–21.

51. Wise, A., Carr, I.C., Groarke, D.A., and Milligan, G. (1997b) Measurement of agonist efficacy using an alpha2A-adrenoceptor-Gi1alpha fusion protein. FEBS Lett. *419*, 141–146.

52. Bahia, D.S., Wise, A., Fanelli, F., Lee, M., Rees, S., and Milligan, G. (1998) Hydrophobicity of residue351 of the G protein Gi1 alpha determines the extent of activation by the alpha 2A-adrenoceptor. Biochemistry *37*, 11555–11562.

53. Jackson, V.N., Bahia, D.S., and Milligan, G. (1999) Modulation of relative intrinsic activity of agonists at the alpha-2A adrenoceptor by mutation of residue 351 of G protein Gi1alpha. Mol. Pharmacol. *55*, 195–201.

54. Fong, C.W., Bahia, D.S., Rees, S., and Milligan, G. (1998) Selective activation of a chimeric Gi1/Gs G protein alpha subunit by the human IP prostanoid receptor: analysis using agonist stimulation of high affinity GTPase activity and [35S]guanosine-5'-O-(3-thio)triphosphate binding. Mol. Pharmacol. *54*, 249–257.

55. Moon, H.E., Cavalli, A., Bahia, D.S., Hoffmann, M., Massotte, D., and Milligan, G. (2001) The human delta opioid receptor activates G(i1)alpha more efficiently than G(o1)alpha. J. Neurochem. *76*, 1805–1813.

56. Lane, J.R., Powney, B., Wise, A., Rees, S., and Milligan, G. (2007) Protean agonism at the dopamine D2 receptor: (S)-3-(3-hydroxyphenyl)-N-propylpiperidine is an agonist for activation of Go1 but an antagonist/inverse agonist for Gi1,Gi2, and Gi3. Mol. Pharmacol. *71*, 1349–1359.

57. Lane, J. R., Powney, B., Wise, A., Rees, S., and Milligan, G. (2008) G protein-coupling and ligand selectivity of the D2L and D3 dopamine receptors. J. Pharmacol. Exp. Therap. *325*, 319–330.

58. Monnot, C., Bihoreau, C., Conchon, S., Curnow, K.M., Corvol, P., and Clauser, E. (1996) Polar residues in the transmembrane domains of the type 1 angiotensin II receptor are required for binding and coupling. Reconstitution of the binding site by co-expression of two deficient mutants. J. Biol. Chem. *271*, 1507–1513.

59. Bakker, R.A., Dees, G., Carrillo, J.J., Booth, R.G., Lopez-Gimenez, J.F., Milligan, G., Strange, P.G. and Leurs, R. (2004) Domain swapping in the human histamine H1 receptor. J. Pharmacol. Exp. Ther. *311*, 131–318.

60. Gouldson, P.R., Higgs, C., Smith, R.E., Dean, M.K., Gkoutos, G.V., and Reynolds, C.A. (2000) Dimerization and domain swapping in G-protein-coupled receptors: a computational study. Neuropsychopharmacology *23*(4 Suppl), S60–77.

61. Lee, C., Ji, I., Ryu, K., Song, Y., Conn, P.M., and Ji, T.H. (2002) Two defective heterozygous luteinizing hormone receptors can rescue hormone action. J. Biol. Chem. *277*, 15795–15800.

62. Carrillo, J.J., Pediani, J. and Milligan, G. (2003) Dimers of class A G protein-coupled receptors function via agonist-mediated trans-activation of associated G proteins. J. Biol. Chem. *278*, 42578–42587.

63. Ramsay, D., Carr, I.C., Pediani, J., Lopez-Gimenez, J.F., Thurlow, R., Fidock, M., and Milligan, G. (2004) High-affinity interactions between human alpha1A-adrenoceptor C-terminal splice variants produce homo- and heterodimers but do not generate the alpha1L-adrenoceptor. Mol. Pharmacol. *66*, 228–239.

64. Pascal, G. and Milligan, G. (2005) Functional complementation and the analysis of opioid receptor homo-dimerization. Mol. Pharmacol. *68*, 905–915.

65. Alfaras-Melainis, K., Gomes, I., Rozenfeld, R., Zachariou, V., and Devi L. (2009) Modulation of opioid receptor function by protein-protein interactions. Front. Biosci. *14*, 3594–3607.

66. Pello, O.M., Martínez-Muñoz, L., Parrillas, V., Serrano, A., Rodríguez-Frade, J.M., Toro, M.J., Lucas, P., Monterrubio, M., Martínez-A, C., and Mellado, M. (2008) Ligand stabilization of CXCR4/delta-opioid receptor heterodimers reveals a mechanism for immune response regulation. Eur. J. Immunol. *38*, 537–549.

67. Hereld, D., and Jin, T. (2008) Slamming the DOR on chemokine receptor signaling: heterodimerization silences ligand-occupied CXCR4 and delta-opioid receptors. Eur. J. Immunol. *38*, 334–337.

68. Parenty, G., Appelbe, S., and Milligan, G. (2008) CXCR2 chemokine receptor antagonism enhances DOP opioid receptor function via allosteric regulation of the CXCR2-DOP receptor hetero-dimer. Biochem. J. *412*, 245–256.

69. Springael, J.Y., Urizar, E., Costagliola, S., Vassart, G., and Parmentier, M. (2007) Allosteric properties of G protein-coupled receptor oligomers. Pharmacol. Ther. *115*, 410–418.

70. Nicholls, D.J., Tomkinson, N.P., Wiley, K.E., Brammall, A., Bowers, L., Grahames, C., Gaw, A., Meghani, P., Shelton, P., Wright, T.J., and Mallinder, P.R. (2008) Identification of a putative intracellular allosteric antagonist binding-site in the CXC chemokine receptors 1 and 2. Mol. Pharmacol. *74*, 1193–1202.

71. Scherrer, G., Imamachi, N., Cao, Y.Q., Contet, C., Mennicken, F., O'Donnell, D., Kieffer, B.L., and Basbaum, A.I. (2009) Dissociation of the opioid receptor mechanisms that control mechanical and heat pain. Cell *37*, 1148–1159.

72. Xie, Z., Bhushan, R.G., Daniels, D.J., and Portoghese, P.S. (2005) Interaction of bivalent ligand KDN21 with heterodimeric delta-kappa opioid receptors in human embryonic kidney 293 cells. Mol. Pharmacol. *68*, 1079–1086.

73. Zheng, Y., Akgün, E., Harikumar, K.G., Hopson, J., Powers, M.D., Lunzer, M.M., Miller, L.J., and Portoghese, P.S. (2009) Induced association of mu opioid (MOP) and type 2 cholecystokinin (CCK2) receptors by novel bivalent ligands. J. Med. Chem. *52*, 247–258.

74. Berque-Bestel, I., Lezoualc ' h, F., Jockers, R. (2008) Bivalent ligands as specific pharmacological tools for G protein-coupled receptor dimers. Curr. Drug Discov. Technol. *5*, 312–318.

75. Liu, Z., Zhang, J., and Zhang, A. (2009) Design of multivalent ligand targeting G-protein-coupled receptors. Curr. Pharm. Des. *15*, 682–718.

76. Milligan, G., Carrillo, J.J., and Pascal, G. (2005) Functional complementation and the analysis of GPCR dimerization. In: The G protein-coupled receptors handbook (Ed. Devi, LA) pp 267–285. Humana press, Totowa NJ.

77. Pfleger, K.D., and Eidne, K.A. (2006) Illuminating insights into protein-protein interactions using bioluminescence resonance energy transfer (BRET). Nat. Methods *3*, 165–174.

4 Time-resolved FRET approaches to study GPCR complexes

Jean Philippe Pin, Damien Maurel, Laetitia Comps-Agrar, Carine Monnier, Marie-Laure Rives, Etienne Doumazane, Philippe Rondard, Thierry Durroux, Laurent Prézeau, and Erin Trinquet

INTRODUCTION

G protein-coupled receptors (GPCRs) are key players in intercellular communication, being the targets of both intercellular messengers, such as hormones and neurotransmitters, and cell adhesion membrane proteins.[1,2] GPCRs are encoded by the largest gene family in animal genomes, are expressed in every cell, and thus represent about 30% of all targets of the actual drugs on the market.[3] These cell surface proteins still represent the major target for drug development because their complexity offers new possibilities to identify drugs with fewer side effects. For many years, GPCRs were considered to function as monomeric entities. Indeed, monomeric GPCRs can activate G proteins.[4–8] However, a number of observations were more consistent with these receptors being able to assemble into dimers or even larger complexes.[9–12] Such quaternary structure is expected to offer even larger complexity to this transduction system resulting from allosteric interactions between the different receptors, and to allow a better control of their specific subcellular targeting and signaling properties. Best examples are the class C GPCRs, especially those activated by the neurotransmitters GABA and glutamate, or by the sweet and umami taste molecules.[13] Indeed, it is now well accepted that dimerization is mandatory for the functioning of these receptors possessing a large extracellular domain (ECD) where the agonist binding site is located.[14,15] For these receptors, a relative movement between each subunit at the level of their ECD is recognized as the necessary step for G protein activation, such that a monomer is not expected to be activated by agonists.[15–17]

However, despite a large number of studies analyzing the oligomerization and its consequence in class A, rhodopsin-like receptors, many questions remain unsolved. Indeed, a clear demonstration of class A GPCR dimers in native cells is still being awaited.[18] Moreover, the requirement of the physical assembly of these receptors for most of the functional interactions reported remains to be established because many of the reported effects may also be explained by indirect cross-talks.[18–20] Technologies to monitor GPCR dimers both in recombinant systems and in native cells are then needed to clarify the issue of GPCR dimerization.

Over the last ten years, energy transfer technologies were largely used to study the possible GPCR assembly in living cells.[21] These are based on the nonradiative transfer of energy from a donor fluorophore to an acceptor. Because this energy transfer occurs only when both partners are in close proximity (less than 100 Å), this is often considered as a good indication that both proteins are in very close proximity. Originally performed using purified proteins chemically labeled with organic fluorophores, the development of fluorescent proteins enabled the examination of protein proximity using recombinant fusion proteins in living cells. An alternative to this fluorescent energy transfer (FRET) approach makes use of luciferase as an energy donor (bioluminescence energy transfer [BRET]).

However, both GFP-based FRET and BRET technologies display a limited signal-to-noise ratio, thus requiring careful analysis of the signal to demonstrate that energy transfer occurs. This is mostly due to the overlap of the emission spectra of the donor and acceptor proteins, such that a large signal of the donor is being recorded at the optimal wavelength used to measure acceptor emission. Moreover, these approaches require the fusion of large proteins on the GPCR and the expression of the recombinant proteins in transfected cells where large amount of proteins commonly accumulate in intracellular organelles. This obviously can generate large nonspecific signals resulting from the high density of the fusion proteins in these organelles, and also limit the specific analysis of the cell surface fraction of the receptors.

In this chapter, we will report on the advantage of using lanthanide-based FRET (time-resolved FRET [TR-FRET])[22] to analyze protein-protein interaction. We will show that due to optimal energetic compatibilies and to the very long lifetime of the donor fluorophore, TR-FRET allows a drastic reduction in the background signal, leading to at least a one-thousand-fold increase in the signal-to-noise ratio.[22] Despite this major advantage, the TR-FRET approach, also called luminescence resonance energy transfer (LRET), was not commonly used because of the difficulty to label proteins in living cells with lanthanide-based fluorophores. First used with labeled antibodies to study GPCR dimerization, we will illustrate the recent new technologies that enable an easy specific labeling of cell surface proteins. This approach appears, therefore, as optimal to measure the proximity between cell surface proteins.

TIME-RESOLVED FRET PRINCIPLE AND ADVANTAGES

Basic principle of FRET and main parameters: distance and orientation

Described in the late 1940s, energy transfer became used in biology in the late 1950s to study protein-protein interaction, mostly in reconstituted system.[23]-[25] Indeed, this nonradiative transfer of energy from one fluorophore to another occurs at very short distances, often lower than 80 Å, such that, due to the mean size diameter of a protein around 30 Å, measurement of such transfer is a good indication that both proteins are in close proximity, interacting either directly or via an additional partner.

Several strict properties of the fluorophores are required for energy transfer to occur[26] (Figure 4.1). Firstly, there must be a good energetic compatibility in the FRET pair between the donor and the acceptor. Such energetic compatibility is assessed by calculating the integral overlap between the emission spectra of the donor and the absorption spectra of the acceptor (Figure 4.1A). The better the integral overlap, the more efficient the transfer. Secondly, the dipoles of the donor and acceptor should be correctly oriented, optimal FRET being obtained if the dipoles transition moments of the donor and the acceptor are parallel, but no transfer occurs if their dipoles transition moments are perpendicular[22] (Figure 4.1C). Thirdly, the distance between the donor and the acceptor should be compatible, and very close, lower than the 1.8x R_0, which is defined as the distance giving rise to a 50% FRET efficacy. As illustrated in Figure 4.1B, FRET efficacy indeed

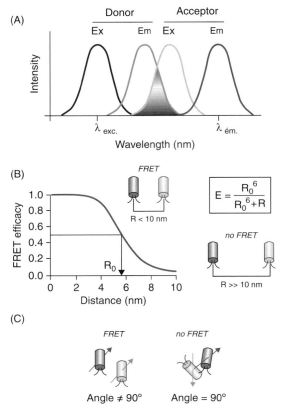

Figure 4-1: Conditions for energy transfer to occur between two fluorophores.
Energetic compatibilities between the fluorophores is illustrated by the overlap between the emission spectra of the donor (green curve), and the absorption spectra of the acceptor (yellow).

The FRET efficacy (E) is a function of the distance between the fluorophores (R), as indicated. The Förster radius (R_0) is defined as the distance giving rise to a 50% FRET efficacy and is commonly around 5–6 nm. FRET can be measured if R is lower than 1.8 x R_0.

Dipole orientation is important for FRET, because no FRET is observed when the dipoles are perpendicular. This parameter has to be taken into consideration only if the fluorophores are constraints and polarized. *See colour plate 5.*

varies as a function of the inverse of the distance to the sixth power, such that no FRET can be measured for distances 1.8 times greater than R_0.

Limitations of FRET using conventional fluorophores

When using organic fluorophores, or GFP-derived fluorescent proteins, the peak absorption and emission are not well separated such that in a FRET experiment, the peak emission of the acceptor is at wavelength where a large emission of the donor still occurs (Figure 4.2). Moreover, direct excitation of the acceptor also occurs at the wavelength used to excite the donor. Different techniques have therefore been developed to isolate the pure light emission resulting from the FRET excitation of the acceptor, from the emission of the donor, or the emission of the acceptor resulting from its direct excitation. To verify that FRET occurs, it

(A) - FRET

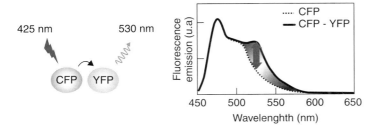

(B) - BRET

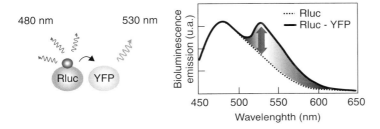

(C) - TR-FRET

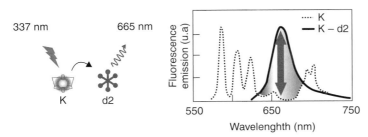

Figure 4-2: Emission spectra for CFP-YFP FRET (A), luciferase-YFP BRET (B), and Eu³⁺ cryptate d2 TR-FRET (C). Note that the peak emission of the acceptor (mostly resulting from FRET, indicated by the red arrow) occurs at a wavelength at which the donor still emits light in FRET (A) (not taking into account the background emission from living cells) and BRET (B), while the donor emission is very low in TR-FRET (C) giving rise to a larger signal to noise ratio.

is now common to represent the FRET as the ratio of the light emission at the acceptor wavelength, over that of the donor, and to make sure that this ratio is largely decreased upon photobleaching of the acceptor due both to an increase in light emission of the donor and to a decrease in the emission of the acceptor.

However, even with such careful analysis, the signal resulting from FRET is still low compare to the large background signal coming from (i) the emission of the donor and (ii) the natural fluorescence of many cell components.

BRET helps solve part of these limitations since the donor used is a bioluminescent enzyme (luciferase) emitting light in the presence of a substrate (either coelenterazine for BRET1 or deep-blue C for BRET2), then avoiding direct excitation of the acceptor or other natural cell fluorophores by excitation light, since no such light is used.

The orientation constraint in FRET can be a problem, especially when the fluorophores are oriented on the targeted protein. This implies that the FRET efficacy cannot be a simple indication of the distance between the fluorophores, nor can the absence of FRET be used as an indication of the absence of interaction between the partners.

TR-FRET principles

Lanthanide-based fluorophores have specific biophysical properties that give them specific advantages over classical organic fluorophores to perform FRET experiments in living cells.

The most important one is that they display a very long emission lifetime (this is why there are also often referred to luminescent molecules) (Figure 4.3). Indeed, upon excitation, the emission lasts for a few milliseconds (half-life in the range of one millisecond) while the lifetime of other fluorophores lifetime is in the range of a few nanoseconds (Figure 4.3C). Accordingly, by applying a short delay (around 50 μsec) between the excitation of the donor and the measurement of acceptor emission, it is possible to almost suppress any background emission resulting either from a direct excitation of the acceptor or from the native cell fluorescence.

Second, the stoke shift is much larger than that of any other fluorophores (more than 200 nm). Indeed, lanthanides – europium or terbium – are trapped either in an organic cage or a chelate that serves as a receiving chromophore that concentrates the excitation energy and transfers it to the ion through a FRET-based principle. This results in a large separation of the absorption spectra, which corresponds to that of the cage, and the emission of the lanthanide, then largely limiting any contamination of the excitation light in the emission window (Figures 4.3 and 4.4). The cage has two other advantages: One is to allow the complex to be linked to the protein of interest, and the second to protect the lanthanide from the quenching by water molecules.

Third, there is no orientation constraint in TR-FRET because the europium or terbium-based fluorescent probes are unpolarized.[22] Then, when using energy donors, FRET efficacy is a good indication of the distance between the fluorophores, and orientation constraints cannot prevent FRET to occur when the partners physically interact.

DONOR AND ACCEPTOR FLUOROPHORES FOR TR-FRET

Lanthanide-based fluorophores

As mentioned earlier in this chapter, lanthanides are trapped into either a chelate or a cage that serves for receiving chromophores, protects the ion from

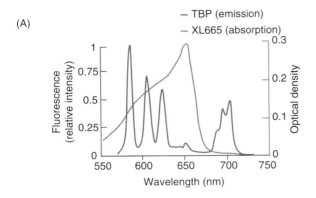

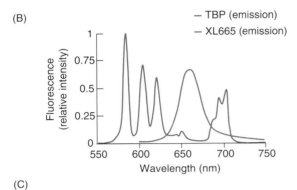

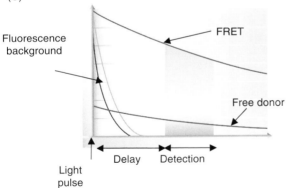

Figure 4-3: Biophysical properties of the TR-FRET fluorophore pairs.
(A). Superposition of the Eu^{3+} cryptate trisbipyridine emission spectra (red) and the absorption spectra of XL665 (green), indicating the large overlap required for efficient FRET. (B). Superposition of the Eu^{3+} cryptate trisbipyridine emission spectra (red) and the emission spectra of XL665 (blue), revealing the large window where acceptor emission can be detected without much contamination by the donor. (C). TR-FRET measurement take advantage of the very long lifetime emission of Eu^{3+} cryptate (t1/2 around 1ms), while that of common fluorophores is very short. Measuring emission after a short delay (50–100 µs, within the time window indicated in yellow) led to the removal of any contaminating light emission from natural fluorophores or directly activated acceptors.

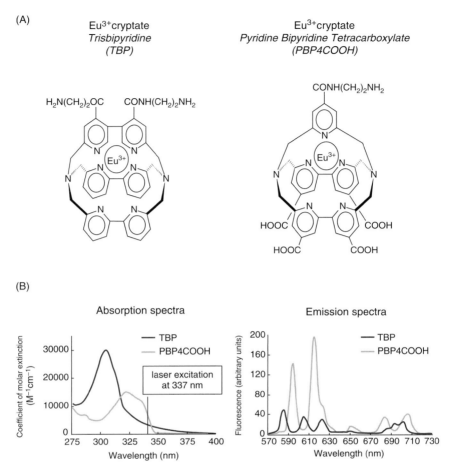

Figure 4-4: Structure and properties of two different Eu³⁺ cryptates. (A). Structure of original Eu³⁺ cryptates TBP and more recently developed PBP4COOH that does not accept water molecules quenching fluorescence. (B). Comparison of the absorption and emission spectra of the Eu³⁺ cryptates TBP and PBP4COOH. Note that both are compatible with an excitation at 337 nm, and a FRET with red acceptors like XL665, Cy5, or d2.

the quenching water molecules, and allows the fluorophore to be linked to the protein of interest.

The first TR-FRET-based fluorophores used europium either complexed by a chelate or trapped in the trisbipyridine cage of an europium cryptate (Figure 4.4A). Although the chelate displays a high affinity for Eu³⁺, the ion can easily be displaced by competing ions (Mn²⁺, Mg²⁺ or Ca²⁺) or be trapped by EGTA, such that the proportion of intact fluorophores in the assay can always be questioned, then making more difficult the interpretation of the data, especially in FRET. In contrast, Eu³⁺ cannot escape the trisbipyridine cage (TBP), making this fluorophore very stable. However, this first generation of cage accepts water molecules together with Eu³⁺, resulting in oxidation into Eu³⁺, then largely decreasing the emission. To overcome this problem, a high concentration of fluoride ions (potassium fluoride (KF) 100–400 mM) is used to displace the water molecules from the cage. This is clearly a difficulty when TR-FRET measurements must be done on

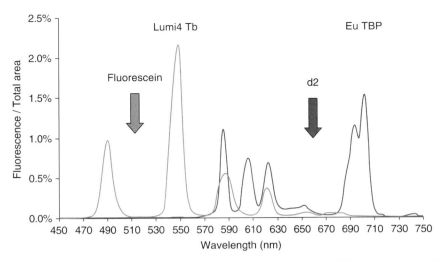

Figure 4-5: Comparison of the emission spectra of Eu^{3+} cryptates TBP (black) and Lumi4 Tb (gray). Note that while Eu^{3+} cryptates TBP is energetically compatible with red acceptors, lumi4 Tb is well compatible with both red and green acceptors.

living cells. Even though KF can be added at the last moment, just before reading the plate, such a high increase in osmolarity may well influence the biological event being studied.

To avoid the problem of europium oxidation, new generations of europium cryptates were generated with a smaller cage, pyridinebipyridine tetracarboxylate (PBP) that only accepts Eu^{3+} without water molecules (Figure 4.4A). Such crypates are then better adapted for experiments conducted with living cells.

Such Eu^{3+}-based fluorescent probes are best excited between 300 and 350 nm (Figure 4.4C) and have a complex emission spectra that depends on the cage. For example, the TBP has four major peaks at 585, 605, 620, and 700 nm, whereas the PBP has two major peaks at 595 and 615 and two minor peaks at 680 and 705 (Figure 4.4C). Of note, the emission of both probes is very low over a large window between 640 and 670 nm, allowing the measurement of acceptor emission with a minimal contamination with the donor (Figure 4.3B and 4.4).

More recently, a new lanthanide cryptate was generated, called Lumi4 Tb, which contains a Tb^{3+} ion instead of Eu^{3+}. This lanthanide cryptate displays interesting biophysical properties compared to the Eu^{3+}-based fluorophore. It has a better extinction coefficient when excited at 337 nm, as well as four emission peaks that allow transfer to both red (like the Eu^{3+}-cryptates) and green fluorophores due to a large emission peak at 490 nm (Figure 4.5).

Acceptor fluorophores for TR-FRET

The conditions, explained earlier in the chapter, to obtain efficient FRET also apply to TR-FRET. Originally, a chemically cross-linked allophycocyanin protein (XL665) was used as an acceptor due to its large absorption window between 580 and 650 nm, which overlaps the main emissions peaks of TBP. In addition, its peak of fluorescence emission at 660 nm allows a specific measurement of FRET at 665 nm, a wavelength at which the donor emission is limited. However, XL665 is a large protein (104 kDa) not necessarily easy to use to precisely label proteins. Then other small organic fluorophores have been validated for TR-FRET, including most of the "near infra-red" fluorophores such as Alexa647, Dy647, Cy5, and d2, for which the emission can also be measured at 665 nm.

Compared to the Eu^{3+}-cryptates, the lumi4 Tb has an additional peak emission at 490 nm, then largely overlapping the absorption spectra of many green fluorophores, including fluorescein and Cy3, but also GFP (Figure 4.5). In addition, lumi4 Tb has a very limited emission at 520 nm, allowing a specific measurement of fluorescein emission with very limited contamination from donor emission (Figure 4.5).

Analyzing GPCR oligomers using TR-FRET and conjugated antibodies

Antibodies to label cell surface receptors with TR-FRET compatible fluorophores

Although the biophysical properties of the TR-FRET fluorophores appear optimal to measure protein-protein interaction, one needs to specifically label the proteins of interest with such fluorophores. Antibodies conjugated either with TR-FRET donor or acceptor fluorophores were first used to examine the possible association between GPCRs. The use of receptors tagged at their N-terminal extracellular end, as an approach allowing the specific labeling of cell surface proteins, enables a specific analysis of their association at the cell surface.

It should be mentioned that FRET is largely dependent on the efficacy of labeling of the proteins. Indeed, if only 50% of the proteins are labeled, then only 25% of any possible dimers will be labeled on both partners. The concentration of antibodies used, as well as the incubation time for the labeling, need therefore to be optimized for a maximal labeling of the receptors.[27]

This approach can obviously be used to analyze the possible association between two distinct proteins carrying different epitopes, each being specifically labeled with an antibody, one carrying the donor, the other the acceptor.[27] It can also be used to analyze homo-oligomers formation using an equivalent expression of the protein carrying either one tag or another. Alternatively, oligomer formation of the same epitope-tagged receptor can be examined using an equimolar concentration of donor- and acceptor-labeled antibodies. Under such condition, in case of dimer formation, only 50% of the dimers will be detected since only 50% will be simultaneously labeled with both a donor and acceptor antibody.

Antibody-based TR-FRET approach validates GPCR dimers in transfected cells

This approach was first used to monitor the formation of hetero-oligomeric complexes of delta-opioid and β_2-adrenoceptors in transfected cells, using

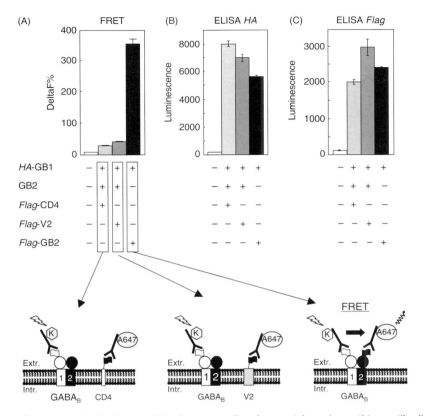

Figure 4-6: Testing the proximity between cell surface proteins using anti-tag antibodies conjugated with TR-FRET compatible fluorophores.
(A). Acceptor emission resulting from FRET measured on cells expressing the indicated tagged proteins. Note that a large FRET can be measured between both subunits of the GABA$_B$ receptor, but not between GABA$_B$ and either CD4 or the V2 vasopressin receptor. Note that the difference in FRET emission is not due to a difference in the cell surface expression of the different partners, either carrying the HA epitope (B) or the flag epitope (C).

Flag- and myc-tagged receptors labeled with either Eu^{3+} chelate or APC conjugated antibodies.[28] Although both delta-opioid and β_2-adrenoceptor homomers were nicely identified at the cell surface, no TR-FRET could be detected between these two receptors. This was in contrast to the co-immunoprecipitation and BRET data that detect both cell surface and intracellular protein complexes. It was therefore proposed that most delta-opioid β_2-AR complexes were intracellular, thus nicely illustrating the importance of specifically analyzing cell surface receptors.

Since then, a few studies have used this approach to monitor possible formation of GPCR complexes at the surface of living cells. This in turn has confirmed that several GPCRs can form homomers, including CXCR1 and CXCR2,[29] histamine H1 and H4,[30,31] alpha1A and 1B-adrenoceptors,[32,33] alpha2A and adenosine-A1,[34] and various types of mGlu receptors.[16,35–38]

Using differentially tagged receptors, several studies also validated the existence of heteromeric complexes using this antibody-based approach, including the $GABA_{B1}$-$GABA_{B2}$[27] (Figure 4.6), alpha2A-adenosine A1,[34] and CXCR1-CXCR2[29] heteromers.

As a proof of the selectivity of this approach, several GPCRs were found not to apparently associate at the cell surface. Indeed with such receptor pairs, very low TR-FRET signals not different from that measured with negative controls (and much lower than positive controls) were measured despite similar expression levels of the partners at the cell surface. We already mentioned the absence of TR-FRET measured between delta-opioid and β_2-adrenoceptors, but negative results were also reported for $GABA_B$-mGlu1,[20] $GABA_B$-V2 vasopressin or $GABA_B$-CD4[31] (Figure 4.6). It is important to note that the background signal measured remains very low, as expected due to the optimal biophysical properties of the TR-FRET fluorophore pairs (Figure 4.6).

Analyzing TR-FRET data

Compared to the commonly used GFP-based FRET or BRET approaches, the TR-FRET signals are simple to analyze. Simple measurements at the emission wavelength of the acceptor with and without the acceptor, or with and without the donor, lead to a simple measurement of the FRET emission in 96 or 384 well plates. As indicated in Figure 4.7, such signal is directly proportional to receptor expression at the cell surface as quantified using either fluorescence emission of the bound antibodies, an ELISA assay, or binding experiments. Such representations of TR-FRET data are different from the commonly used ratiometric analysis of GFP-based FRET or BRET data that better represents transfer efficacy. A good measure of TR-FRET efficacy can be obtained by calculating the ratio of the FRET emission over the amount of FRET donor at the cell surface (Figure 4.7). When this was examined in the case of $GABA_B$ or mGlu receptors, this ratio was found constant over a range of receptor density, further demonstrating that in these cases, the receptor assembly does not depend on receptor density at the cell surface.[16,20,39]

Possibilities to detect GPCR dimers in native cells

When analyzing the TR-FRET signal measured at various mGlu and $GABA_B$ receptor densities, it was found that large and significant signals can be measured with receptor expression at the cell surface as low as 100 fmol/mg (Figure 4.7). Such data were obtained with Eu^{3+}-cryptates as donor, and an almost tenfold increase in sensitivity is expected when using Tb^{3+} cryptate. Such a sensitivity is then sufficient to detect receptor oligomers in native tissues after labeling with specific monoclonal antibodies recognizing external epitopes of these receptors. Our preliminary data indeed prove that this is possible when targeting the native $GABA_B$ receptor in neurons (Comps-Agrar, Kniazeff, Trinquet, Pin, unpublished work).

Limitation on the use of antibodies-based TR-FRET approach

Despite the obvious quality of the data reported using antibodies-based TR-FRET to detect GPCR oligomers at the cell surface, this approach still suffers some

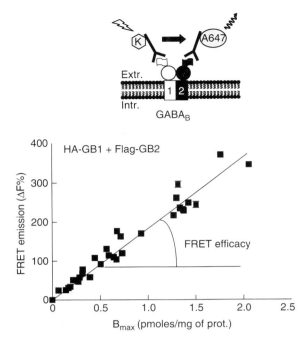

Figure 4-7: TR-FRET between antibody-labeled GABA$_B$ subunits measured at various receptor expression levels. TR-FRET is represented as the acceptor emission and is largely significant at receptor densities as low as 0.1 pmol/mg of protein. FRET is commonly represented as the ratio between the acceptor emission resulting from FRET and the donor emission intensity. Such FRET efficacy is represented by the slope of the present curve, illustrating that TR-FRET efficacy is constant over the tested range of receptor expression.

limitations. First, antibodies are large proteins (around 160 Å in length) that can lead to TR-FRET between distant proteins, depending on the position adopted by the antibodies relative to its target proteins. Second, chemical labeling of the antibodies with the fluorophore is commonly set up to a yield of three to six fluorophores per antibody, such that when considering the size of the antibodies, only a fraction of the fluorophores will be at a distance compatible with TR-FRET. Third, commonly used monoclonal antibodies are bivalent, such that a single antibody can possibly label both epitopes in a GPCR dimer, thus possibly limiting the double labeling of GPCR complexes or alternatively promoting clustering of monomeric proteins.

Other reagents were therefore developed to allow a more precise, specific labeling of GPCRs with TR-FRET compatible fluorophores.

COVALENT LABELING OF PROTEINS WITH TR-FRET COMPATIBLE FLUOROPHORES

Over the recent years, several approaches were developed to allow the specific labeling of recombinant proteins. We will concentrate here on the two that have been validated for the labeling of GPCRs with TR-FRET compatible fluorophores to avoid the use of antibodies and allow i) a more precise positioning of the fluorophores, and ii) a stoichiometric labeling of the proteins.

(A)

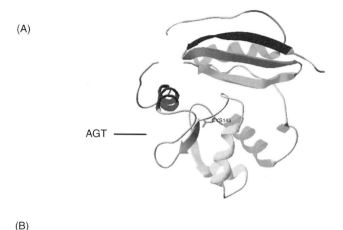

AGT ———

(B)

Figure 4-8: Using the snap-tag to label cell surface proteins.
(A). Structure of the O^6-alkylguanine-DNA alkyltransferase (AGT) (code pdb: 1EH6), the origin of the snap-tag. (B). Principle of the protein labeling using snap-tag fusion proteins. The fusion protein is exposed to the snap-tag substrate, a benzylguanine carrying the fluorophore of interest. After the reaction is completed, the benzyl group carrying the fluorophore is covalently linked to a serine residue of the active site of the snap-tag, leading to a covalent labeling of the protein with the fluorophore. *See colour plate 6.*

The snap-tag technology

The snap-tag (23 kDa) is derived from the enzyme O^6-alkylguanine-DNA alkyl-transferase (or AGT) (Figure 4.8), an enzyme involved in DNA repair, removing any alkyl-group inserted in the guanine bases in the DNA.[40,41] Interestingly, the reaction leads to a covalent linkage of the alkyl group on a cystein residue of the active site of AGT. As such, benzyl-guanine (BG) derivatives have been developed to inhibit AGT activity and improve the toxic action of anticancer drugs.[42] Snap-tag was then developed by the group of Kai Johnsson (EPFL, Lausanne, Switzerland) by mutating the DNA interaction site of AGT and optimizing its enzymatic activity to covalently label proteins with any chemical entities carried by the BG substrate.[43–47] Then, by inserting a snap-tag in the extracellular part of a receptor and using nonpermeant BG substrates conjugated with the fluorophore of interest, it is possible to specifically label cell surface proteins only[39] (Figure 4.8). By combining this approach and the use of TR-FRET compatible fluorophores conjugated to BG, it is then possible to circumvent most of the problems associated with the use of GFP fusion proteins to monitor GPCR proximity specifically at the cell surface by FRET measurement.

A critical aspect of such strategies to label protein for FRET measurements is to verify that most, and if possible all, proteins are indeed labeled, since a low labeling efficacy will have larger consequences of the simultaneous labeling of both proteins in a dimer (10% of the dimers will be doubly labeled if only 30% of the receptors are labeled). We verified that using an optimal concentration of substrate, all proteins were indeed labeled after one-hour incubation.[39]

By using in vitro evolution strategies, it has been possible to isolate mutants of the snap-tag that no longer react with BG derivatives, but instead with benzylcytosine (BC).[48] Such a snap-tag derivative, called clip-tag, can be used to specifically label two distinct proteins carrying either a snap or a clip with specific fluorophores, using a combination of BG and BC conjugated with two distinct fluorophores.

The ACP-tag technology

Another covalent labeling methodology makes use of the acyl-carrier protein (ACP), a 77 amino acid long protein found in various bacteria strains, such as *E. coli*, that can be covalently labeled by the phosphopantetheinyl group of coenzyme-A thanks to a phosphopantetheinyl transferase (PPTase) called acyl carrier protein synthase (AcpS) (Figure 4.9). Then, by inserting ACP into the extracellular side of a membrane protein, it is possible to label it with a fluorophore conjugated on the phosphopantetheinyl moiety of CoA, in the presence of AcpS.[49] Because soluble enzymes have to be used for these labeling, the reaction can only occur outside the cell, thus enabling a specific labeling of the cell surface proteins.

The same CoA derivatives can also be used to label proteins carrying the peptidyl carrier protein (PCP) from *B. subtilis*, and the PPTase Sfp.[50] Most interestingly, whereas the PPTase AcpS can label ACP exclusively, Sfp reacts with both ACP and PCP, thus allowing a specific labeling of two distinct proteins carrying the PCP- and ACP-tags, with two distinct fluorophores. To that aim, a first reaction using AcpS will exclusively transfer the phosphopantetheinyl group of CoA carrying a first fluorophore on the ACP; then the addition of Sfp with a second CoA derivative carrying a second fluorophore will allow the labeling of the PCP fusion proteins with this second fluorophore.

This approach, combined with TR-FRET, has recently been validated to demonstrate GPCR dimerization at the surface of living cells, as illustrated with the GABA$_B$ receptor used as a model system.[51] It was also used to examine NK1 receptor dimerization using a classical FRET fluorophore pair.[52] It was then proposed that NK1 receptors only assemble into complexes at high density, but the sensitivity of this approach based on the use of common fluorophores, then associated

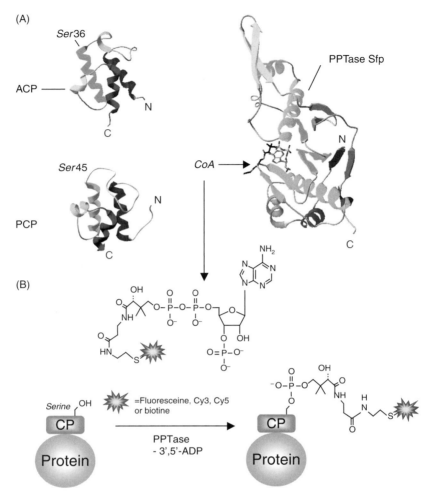

Figure 4-9: Using the ACP-tag to label cell surface proteins.
(A). Structures of the Acyl Carrier Protein (ACP) (code pdb: 1T8K), the Peptidyl Carrier Protein (PCP) (code pdb: 1DNY) and the phosphopantetheine transferase (PPTase) in the presence of coenzyme A (CoA) (code pdb: 1QR0). (B). Labeling reaction of ACP fusion proteins upon PPTase treatment. The enzyme transfers a 4'-phosphopantetheinyl (Ppant) group from CoA to a conserved serine residue of ACP. The derivatization of CoA with a fluorophore allows to label a fusion protein. *See colour plate 7.*

with a high background, may not have been sufficient to detect significant FRET at such low receptor density.

ANALYZING GPCR COMPLEXES USING SNAP-TAG AND TR-FRET

Validation using class C GPCRs

Using nonpermeant BG substrates carrying either a Eu^{3+}-cryptate as donor or a d2 as acceptor, we have been able to specifically label cell surface GPCRs carrying a snap-tag at their N-termini. Using first the GABA$_B$ or the mGlu receptors as

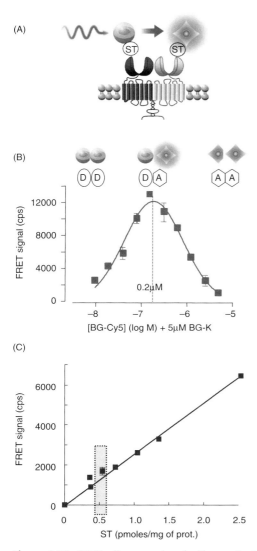

Figure 4-10: $GABA_B$ dimers analyzed with snap-tag-TR-FRET. (A). Schematic representation of the $GABA_B$ heterodimer with both subunits carrying a snap tag (ST). (B). Determination of the correct ratio of both donor and acceptor BG derivatives for an optimal labeling for efficient FRET; note that too low amount of acceptor, or too high amount of acceptor lead to a low FRET. (C). TR-FRET signal as a function of $GABA_B$ receptor expression level.

model GPCRs, we have been able to detect significant TR-FRET signals that can be easily measured in 96 or 384 well plates, thus allowing an easy quantitative analysis of the signal (Figure 4.10).

To analyze the association of such oligomeric entities in which all subunits carry a snap-tag, it was necessary to optimize the relative concentration of both substrates, one conjugated with the donor and the other conjugated with the acceptor. Thus, the concentration ratio of both substrates giving rise to the maximal TR-FRET signal has to be determined (Figure 4.10B). Indeed, this ratio depends on the relative reactivity of the two substrates with the snap-tag that can well depend on the environment of the snap-tag and on the fluorophore fused to the BG.

Using the first generation of BG derivatives compatible with TR-FRET, which were based on Eu^{3+} cryptate and d2, it was possible to detect $GABA_B$ receptor dimers at a density of 0.1 pmol/mg of total cell protein (Figure 4.10C), a density two- to five fold lower than that of the $GABA_B$ receptor in the brain and tenfold lower than that measured in cultured neurons.[39] As for the experiments conducted using conjugated antibodies, the TR-FRET emission was found to be directly proportional to the amount of receptors at the cell surface, indicating that the FRET efficacy is independent of receptor density, consistent with the $GABA_B$ receptor being a constitutive dimer (Figure 4.10C).

The specificity of the TR-FRET measured was validated by the absence of significant signal between $GABA_B$ receptor labeled with antibodies and various class A GPCRs carrying a snap-tag,[39] or between snap-tagged labeled $GABA_B$ receptors and either antibody-labeled mGlu1 or snap-tagged mGlu1 receptors.[20]

Class A GPCR dimers can be detected using snap-tag and TR-FRET

The snap-tag labeling approach was also used to compare the TR-FRET efficacy measured for class A dimers with that of class C GPCRs recognized as constitutive dimers. As for the $GABA_B$ receptor, a large and significant signal was measured, with a constant FRET efficacy at various receptor densities, consistent with a stable dimer formation of various class A GPCRs, including the V2 vasopressin, β_2-adrenoceptors, A1 adenosine, or PAR1 thrombin receptors.[39] However, the FRET efficacy was always slightly lower than that measured with class C GPCRs, an observation that can be explained either by a lower proportion of class A receptors in a homodimeric form at the cell surface (some being either monomeric or associated with another endogenous GPCR), or by a larger distance between the fluorophores.

Class C GPCR oligomerization determined using snap-tag labeling

Data obtained with atomic force microscopy suggested an oligomeric assembly of rhodopsin in native retinal disc membranes.[53] Association of more than two GPCRs in a complex was also reported using a combination of GFP complementation and BRET assay, three color FRET, or sequential BRET-FRET,[54–56] further suggesting that GPCRs can form larger oligomers.

We then examined whether class C GPCRs could also assemble into larger complexes by using an optimized $GABA_B$ receptor quality control system. This system is based on the retention of the $GABA_{B1}$ C-terminal region in the endoplasmic reticulum, and its targeting to the cell surface after proper interaction with the $GABA_{B2}$ C-terminal region. To prevent $GABA_{B2}$ to reach the cell surface alone, a retention signal that could be masked by the $GABA_{B1}$ C-terminal region was introduced. As such, any of these proteins could not reach the cell surface alone, but did so when assembled into an heterodimeric complex.[38,57]

(A)

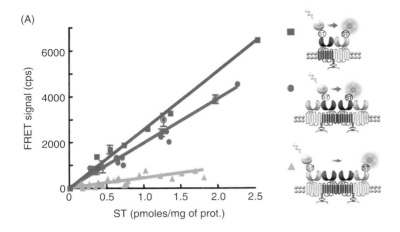

(B)

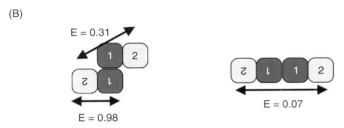

Figure 4-11: $GABA_B$ oligomers revealed with snap-tag-TR-FRET. (A). TR-FRET measured between $GABA_B$ subunits (blue squares), between $GABA_{B1}$ subunits (red circles), or between $GABA_{B2}$ subunits (green triangles). Because $GABA_{B1}$ is always associated with $GABA_{B2}$, the TR-FRET measured between $GABA_{B1}$ subunits indicates the proximity between $GABA_B$ heterodimers, via their $GABA_{B1}$ subunits. (B). The TR-FRET measured under these three conditions is consistent with the expected distances between the subunits in such organized $GABA_B$ oligomers (indicated are the FRET efficacy, assuming a R_0 value of 6.5 nm). Note that the FRET efficacy (slopes of the curves in A) is independent of receptor density and can be observed at a physiological density of the $GABA_B$ receptor.

Using this system, we have been able to label a single subunit within either a $GABA_B$ receptor heterodimer or a mGlu receptor dimer, and then examined the possible interaction between dimers at the cell surface. These studies revealed a closed proximity between $GABA_B$ receptor dimers through an interaction of their $GABA_{B1}$ subunit (Figure 4.11), but no close proximity between mGlu receptor dimers was detected even when expressed at a high density.[39] The possible heteromeric assembly between mGlu or $GABA_B$ receptors and some class A GPCRs was also examined and did not reveal significant TR-FRET signals.[20]

CONCLUSION AND PERSPECTIVES

Although still a matter of intense debate, the ability of GPCRs to form oligomers was then further validated by specific labeling at the cell surface with TR-FRET compatible fluorophores, either using conjugated antibodies or specific substrates that can be covalently linked to specific tags attached to the extracellular domain of the receptor. It is important to note that using either approach, the TR-FRET efficacy was much larger than that of negative controls and was constant over a range of cell surface densities, illustrating well the ability of these receptors to form oligomers even when expressed at low density in transfected cells. Such an approach was found sensitive enough using conjugated antibodies to detect GABA$_B$ oligomers in native brain membrane.

With the development of a number of labeling technologies, such as snap-, clip-, ACP- or PCP-tags, and many others described every year, it will then become possible to analyze in a quantitative way the association between various GPCRs in larger complexes.

Through the recent development of new TR-FRET compatible fluorophore pairs, such as Lumi4-Tb that can transfer to both green and red acceptors, with better biophysical properties, an even higher sensitivity of this system is expected, offering the possibility to detect oligomers expressed at even lower densities. Such an approach may then be very well adapted to detect GPCR oligomers using fluorescent high affinity ligands, then not requiring any modification of the receptor. Our preliminary experiments validate this approach and provide a way to detect GPCR oligomers in native tissues. This will certainly bring important new information for the validation of GPCR oligomers, but will also provide the necessary tools to elucidate the possible functional significance of this phenomenon in native cells.

ACKNOWLEDGMENTS

This work was supported by the Centre National de la Recherche Scientifique (CNRS), the Institut National de la Santé et de la Recherche Médicale (INSERM), CisBio International, and by grants from the French Ministry of Research, Action Concertée Incitative "Biologie Cellulaire Moléculaire et Structurale" (ACI-BCMS 328), the Agence Nationale de la Recherche (ANR-05-PRIB-02502, ANR-BLAN06–3_135092, ANR-05-NEUR-035).

REFERENCES

1. Bockaert J, Pin J-P (1999) Molecular tinkering of G-protein coupled receptors: an evolutionary success. EMBO J 18:1723–1729.
2. Fredriksson R, Lagerstrom MC, Lundin LG, Schioth HB (2003) The G-protein-coupled receptors in the human genome form five main families. Phylogenetic analysis, paralogon groups, and fingerprints. Mol Pharmacol 63:1256–1272.
3. Overington JP, Al-Lazikani B, Hopkins AL (2006) How many drug targets are there? Nat Rev Drug Discov 5:993–996.
4. Chabre M, le Maire M (2005) Monomeric G-protein-coupled receptor as a functional unit. Biochemistry 44:9395–9403.
5. Bayburt TH Leitz AJ, Xie G, Oprian DD, Sligar SG (2007) Transducin activation by nanoscale lipid bilayers containing one and two rhodopsins. J Biol Chem 282:14875–14881.

6. Ernst OP, Gramse V, Kolbe M, Hofmann KP, Heck M (2007) Monomeric G protein-coupled receptor rhodopsin in solution activates its G protein transducin at the diffusion limit. Proc Natl Acad Sci U S A 104:10859–10864.

7. Whorton MR, Bokoch MP, Rasmussen SG, Huang B, Zare RN, Kobilka B, Sunahara RK (2007) A monomeric G protein-coupled receptor isolated in a high-density lipoprotein particle efficiently activates its G protein. Proc Natl Acad Sci U S A 104: 7682–7687.

8. Whorton MR, Jastrzebska B, Park PS, Fotiadis D, Engel A, Palczewski K, Sunahara RK (2008) Efficient coupling of transducin to monomeric rhodopsin in a phospholipid bilayer. J Biol Chem 283:4387–4394.

9. Bouvier M (2001) Oligomerization of G-protein-coupled transmitter receptors. Nat Rev Neurosci 2:274–286.

10. Milligan G (2004) G protein-coupled receptor dimerization: function and ligand pharmacology. Mol Pharmacol 66:1–7.

11. Gurevich VV, Gurevich EV (2008a) GPCR monomers and oligomers: it takes all kinds. Trends Neurosci 31:74–81.

12. Gurevich VV, Gurevich EV (2008b) How and why do GPCRs dimerize? Trends Pharmacol Sci 29:234–240.

13. Pin J-P, Galvez T, Prézeau L (2003) Evolution, structure and activation mechanism of family 3/C G-protein coupled receptors. Pharmacol Ther 98:325–354.

14. Pin J-P, Kniazeff J, Liu J, Binet V, Goudet C, Rondard P, Prézeau L (2005) Allosteric functioning of dimeric Class C G-protein coupled receptors. FEBS J 272:2947–2955.

15. Rondard P, Huang S, Monnier C, Tu H, Blanchard B, Oueslati N, Malhaire F, Li Y, Maurel D, Trinquet E, Labesse G, Pin J-P, Liu J (2008) Functioning of the dimeric GABAB receptor extracellular domain revealed by glycan wedge scanning. EMBO J 27:1321–1332.

16. Kniazeff J, Bessis A-S, Maurel D, Ansanay H, Prezeau L, Pin J-P (2004) Closed state of both binding domains of homodimeric mGlu receptors is required for full activity. Nat Str Mol Biol 11:706–713.

17. Tateyama M, Abe H, Nakata H, Saito O, Kubo Y (2004) Ligand-induced rearrangement of the dimeric metabotropic glutamate receptor 1alpha. Nat Struct Mol Biol 11:637–642.

18. Ferre S, Baler R, Bouvier M, Caron MG, Devi LA, Durroux T, Fuxe K, George SR, Javitch JA, Lohse MJ, Mackie K, Milligan G, Pfleger KD, Pin JP, Volkow ND, Waldhoer M, Woods AS, Franco R (2009) Building a new conceptual framework for receptor heteromers. Nat Chem Biol 5:131–134.

19. Pin J-P, Neubig RR, Bouvier M, Devi L, Filizola M, Javitch JA, Lohse MJ, Milligan G, Palczewski K, Parmentier M, Spedding M (2007) International Union of Basic and Clinical Pharmacology. LXVII. Recommendations for the recognition and nomenclature of G protein-coupled receptor heteromultimers. Pharmacol Rev 59:5–13.

20. Rives M-L, Vol C, Tinel N, Trinquet E, Ayoub MA, Pin J-P, Prézeau L (2009) Cross talk between GABAB and mGlu1a receptors reveals new insights on GPCRs signal integration. EMBO J 28:2195–2208.

21. Milligan G, Bouvier M (2005) Methods to monitor the quaternary structure of G protein-coupled receptors. FEBS J 272:2914–2925.

22. Selvin PR (2002) Principles and biophysical applications of lanthanide-based probes. Annu Rev Biophys Biomol Struct 31:275–302.

23. Förster T (1948) Intermolecular energy migration and fluorescence. Ann Phys 2:55–75.

24. Stryer L, Haugland RP (1967) Energy transfer: a spectroscopic ruler. Proc Natl Acad Sci U S A 58:719–726.

25. Stryer L (1978) Fluorescence energy transfer as a spectroscopic ruler. Annu Rev Biochem 47:819–846.

26. Vogel SS, Thaler C, Koushik SV (2006) Fanciful FRET. Sci STKE 2006:re2.

27. Maurel D, Kniazeff J, Mathis G, Trinquet E, Pin JP, Ansanay H (2004) Cell surface detection of membrane protein interaction with homogeneous time-resolved fluorescence resonance energy transfer technology. Anal Biochem 329:253–262.

28. McVey M, Ramsay D, Kellett E, Rees S, Wilson S, Pope AJ, Milligan G (2001) Monitoring receptor oligomerization using time-resolved fluorescence resonance energy transfer and bioluminescence resonance energy transfer. The human delta -opioid receptor displays constitutive oligomerization at the cell surface, which is not regulated by receptor occupancy. J Biol Chem 276:14092–14099.

29. Wilson S, Wilkinson G, Milligan G (2005) The CXCR1 and CXCR2 receptors form constitutive homo- and heterodimers selectively and with equal apparent affinities. J Biol Chem 280:28663–28674.

30. Bakker RA, Dees G, Carrillo JJ, Booth RG, Lopez-Gimenez JF, Milligan G, Strange PG, Leurs R (2004) Domain swapping in the human histamine H1 receptor. J Pharmacol Exp Ther 311:131–138.

31. van Rijn RM, Chazot PL, Shenton FC, Sansuk K, Bakker RA, Leurs R (2006) Oligomerization of recombinant and endogenously expressed human histamine H(4) receptors. Mol Pharmacol 70:604–615.

32. Carrillo JJ, Lopez-Gimenez JF, Milligan G (2004) Multiple interactions between transmembrane helices generate the oligomeric alpha1b-adrenoceptor. Mol Pharmacol 66:1123–1137.

33. Ramsay D, Carr IC, Pediani J, Lopez-Gimenez JF, Thurlow R, Fidock M, Milligan G (2004) High-affinity interactions between human alpha1A-adrenoceptor C-terminal splice variants produce homo- and heterodimers but do not generate the alpha1L-adrenoceptor. Mol Pharmacol 66:228–239.

34. Ciruela F, Casado V, Rodrigues RJ, Lujan R, Burgueno J, Canals M, Borycz J, Rebola N, Goldberg SR, Mallol J, Cortes A, Canela EI, Lopez-Gimenez JF, Milligan G, Lluis C, Cunha RA, Ferre S, Franco R (2006) Presynaptic control of striatal glutamatergic neurotransmission by adenosine A1-A2A receptor heteromers. J Neurosci 26:2080–2087.

35. Goudet C, Kniazeff J, Hlavackova V, Malhaire F, Maurel D, Acher F, Blahos J, Prézeau L, Pin J-P (2005) Asymmetric functioning of dimeric metabotropic glutamate receptors disclosed by positive allosteric modulators. J Biol Chem 280:24380–24385.

36. Hlavackova V, Goudet C, Kniazeff J, Zikova A, Maurel D, Vol C, Trojanova J, Prézeau L, Pin J-P, Blahos J (2005) Evidence for a single heptahelical domain being turned on upon activation of a dimeric GPCR. EMBO J 24:499–509.

37. Rondard P, Liu J, Huang S, Malhaire F, Vol C, Pinault A, Labesse G, Pin J-P (2006) Coupling of agonist binding to effector domain activation in metabotropic glutamate-like receptors. J Biol Chem 281:24653–24661.

38. Brock C, Oueslati N, Soler S, Boudier L, Rondard P, Pin J-P (2007) Activation of a Dimeric Metabotropic Glutamate Receptor by Inter-Subunit Rearrangement. J Biol Chem 282:33000–33008.

39. Maurel D, Comps-Agrar L, Brock C, Rives ML, Bourrier E, Ayoub MA, Bazin H, Tinel N, Durroux T, Prezeau L, Trinquet E, Pin JP (2008) Cell-surface protein-protein interaction analysis with time-resolved FRET and snap-tag technologies: application to GPCR oligomerization. Nat Methods 5:561–567.

40. Pegg AE, Dolan ME (1987) Properties and assay of mammalian O6-alkylguanine-DNA alkyltransferase. Pharmacol Ther 34:167–179.

41. Daniels DS, Mol CD, Arvai AS, Kanugula S, Pegg AE, Tainer JA (2000) Active and alkylated human AGT structures: a novel zinc site, inhibitor and extrahelical base binding. EMBO J 19:1719–1730.

42. Pegg AE, Swenn K, Chae MY, Dolan ME, Moschel RC (1995) Increased killing of prostate, breast, colon, and lung tumor cells by the combination of inactivators of O6-alkylguanine-DNA alkyltransferase and N,N'-bis(2-chloroethyl)-N-nitrosourea. Biochem Pharmacol 50:1141–1148.

43. Juillerat A, Gronemeyer T, Keppler A, Gendreizig S, Pick H, Vogel H, Johnsson K (2003) Directed evolution of O6-alkylguanine-DNA alkyltransferase for efficient labeling of fusion proteins with small molecules in vivo. Chem Biol 10:313–317.

44. Keppler A, Gendreizig S, Gronemeyer T, Pick H, Vogel H, Johnsson K (2003) A general method for the covalent labeling of fusion proteins with small molecules in vivo. Nat Biotechnol 21:86–89.

45. Keppler A, Pick H, Arrivoli C, Vogel H, Johnsson K (2004) Labeling of fusion proteins with synthetic fluorophores in live cells. Proc Natl Acad Sci U S A 101:9955–9959.

46. Juillerat A, Heinis C, Sielaff I, Barnikow J, Jaccard H, Kunz B, Terskikh A, Johnsson K (2005) Engineering substrate specificity of O6-alkylguanine-DNA alkyltransferase for specific protein labeling in living cells. Chembiochem 6:1263–1269.

47. Gronemeyer T, Chidley C, Juillerat A, Heinis C, Johnsson K (2006) Directed evolution of O6-alkylguanine-DNA alkyltransferase for applications in protein labeling. Protein Eng Des Sel 19:309–316.

48. Gautier A, Juillerat A, Heinis C, Correa IR Jr., Kindermann M, Beaufils F, Johnsson K (2008) An engineered protein tag for multiprotein labeling in living cells. Chem Biol 15:128–136.

49. George N, Pick H, Vogel H, Johnsson N, Johnsson K (2004) Specific labeling of cell surface proteins with chemically diverse compounds. J Am Chem Soc 126:8896–8897.

50. Yin J, Liu F, Li X, Walsh CT (2004) Labeling proteins with small molecules by site-specific posttranslational modification. J Am Chem Soc 126:7754–7755.

51. Monnier C, Bourrier E, Vol C, Lamarque L, Trinquet E, Pin J, Rondard P (2010) Transactivation between two 7TM domains: a key step in heterodimeric GABAB receptor activation under revision.

52. Meyer BH, Segura JM, Martinez KL, Hovius R, George N, Johnsson K, Vogel H (2006) FRET imaging reveals that functional neurokinin-1 receptors are monomeric and reside in membrane microdomains of live cells. Proc Natl Acad Sci U S A 103:2138–2143.

53. Fotiadis D, Liang Y, Filipek S, Saperstein DA, Engel A, Palczewski K (2003) Atomic-force microscopy: Rhodopsin dimers in native disc membranes. Nature 421:127–128.

54. Lopez-Gimenez JF, Canals M, Pediani JD, Milligan G (2007) The alpha1b-adrenoceptor exists as a higher-order oligomer: effective oligomerization is required for receptor maturation, surface delivery, and function. Mol Pharmacol 71:1015–1029.

55. Carriba P, Navarro G, Ciruela F, Ferre S, Casado V, Agnati L, Cortes A, Mallol J, Fuxe K, Canela EI, Lluis C, Franco R (2008) Detection of heteromerization of more than two proteins by sequential BRET-FRET. Nat Methods 5:727–733.

56. Guo W, Urizar E, Kralikova M, Mobarec JC, Shi L, Filizola M, Javitch JA (2008) Dopamine D2 receptors form higher order oligomers at physiological expression levels. EMBO J 27:2293–2304.

57. Brock C, Boudier L, Maurel D, Blahos J, Pin J-P (2005) Assembly-dependent surface targeting of the heterodimeric GABAB receptor is controlled by COPI, but not 14–3–3. Mol Biol Cell 16:5572–5578.

5 Signaling of dopamine receptor homo- and heterooligomers

Ahmed Hasbi, Brian F. O'Dowd, and Susan R. George

INTRODUCTION

Dopamine is involved in a wide range of physiological functions, including reward mechanisms, attention, motivation, locomotion, reinforcing behavior, hormone regulation, memory, and learning.[1,2] Dysfunctional dopaminergic system signaling has been linked to many important neurological and neuropsychiatric disorders, such as schizophrenia, obsessive-compulsive disorder (OCD), depression, attention deficit hyperactivity disorder (ADHD), Parkinson's disease, Huntington disease, and drug addiction.[1-4]

The striatum represents the major target of dopamine (DA) neurons and is the structure with the highest concentration of dopamine receptors in the brain.[5,6] Medium spiny neurons (MSNs) represent the major constituent of the striatum (90–95%) whereas interneurons are a minor component (1–3%).[7] The mesolimbic dopamine neurons that target the striatum originate from substantia nigra (SN) and ventral tegmental area (VTA)[8,9] and interact through synapses with the MSNs and interneurons to release dopamine in the striatum. The SN dopaminergic neurons project to the corpus striatum forming the nigrostriatal pathway, important in locomotor function. The dopaminergic neurons originating from VTA project their axons to different parts of the limbic system, including nucleus accumbens and different brain structures responsible for the modulation of emotional and behavioral responses.

DOPAMINE RECEPTORS

Dopamine produces its physiological effects through the activation of specific receptors belonging to the G protein-coupled receptor (GPCR) superfamily. The five dopaminergic receptors (D1-D5) cloned were divided into two subclasses: D1-like (D1 and D5) and D2-like (D2, D3, and D4) receptors.[2] This subdivision was based on amino acid homology, pharmacology, and the ability of the receptors to activate (D1-like) or inhibit (D2-like) adenylyl cyclase (AC), thus modulating the accumulation of cyclic adenosine monophosphate (cAMP) through a mechanism involving either $G_{s/olf}$ or $G_{i/o}$ proteins, respectively.[2,10] The modulation of this signaling pathway is the most studied in attempts to understand the molecular events mediating dopaminergic effects. There are, however, other signaling cascades that have been reported to be involved in the effects exerted by dopamine, such as the reported D1-like receptor stimulation of phospholipase C and the regulation of ion channels and their signaling pathways.[2,10,11] D2-like receptors also regulate the PAR4 signaling pathway[12] and the activation of Akt/glycogen synthase kinase 3 (GSK3) cascade.[13] We have also described a novel signaling cascade associated with activation of the D1-D2 heteromer linking to intracellular calcium release.[14-16]

OLIGOMERIZATION OF DOPAMINE RECEPTORS

Oligomerization, whether between the monomeric units of the same receptor (homooligomerization) or between different receptors (heterooligomerization),[17] represents one of the fascinating aspects of GPCR biology documented at least a decade-and-a-half ago.[18-20] Dopamine receptors have been shown to form both homooligomers and heterooligomers, either within their subfamily or with other GPCRs or receptor-channels. Dopamine receptors also interact with a number of different proteins called DA receptor-interacting proteins (DRIPs).[6,21] These are proteins of different functions, including scaffolding, cytoskeletal, and signaling proteins. We will describe briefly some of the interactions of DA receptors with each other and with other receptors, either GPCRs or ion-channels, and we will concentrate more on the D1-D2 receptor heterooligomerization as a model for generating a novel signaling pathway.

DOPAMINE RECEPTOR HOMOOLIGOMERIZATION

Similar to other GPCRs,[18-20] dopamine receptors form homodimers or higher molecular entities (homooligomers). The first evidence for the presence of dopamine receptor homomers came from SDS-PAGE experiments, which were identified based on the resistance of these receptor complexes to denaturation by SDS. We and others have shown that dopamine D1 as well as D2 receptors (D1R and D2R, respectively) exist as higher molecular weight forms, constituting homodimers or homooligomers when expressed in heterologous systems.[22-26] Physical interactions were also shown by other methods including coimmunoprecipitation, substituted cystein accessibility method, or cysteine crosslinking.[26-28] A clear demonstration of the formation of physically interacting complexes for D1R homooligomers was shown using a novel experimental methodology developed in our laboratory, the nuclear localization sequence (NLS) strategy.[29,30] NLSs are specific motifs that, when contained in a protein sequence, are recognized by nuclear transport proteins, which then mediate the translocation of the NLS-containing protein to the nucleus.[31] In our strategy, we have added an NLS sequence to a GPCR, forcing its trafficking to the nucleus, and examined if another GPCR partner would be cotranslocated to the nucleus, in which case a physical interaction, namely oligomerization, had occurred.[29,30] The contact interfaces of dopamine receptor homomers were also examined by different methods, and results have shown that the homomerization of D2R involved the symmetrical interface of the fourth transmembrane domain (TM 4) and a patch of residues at the extracellular end of TM 1.[14,26,27,32] We also showed that D1R homomers are formed in the endoplasmic reticulum (ER) before being targeted to the plasma membrane after being subjected to a stringent quality control mechanism,[33] and that D1R homomers interacted with caveolae, rather than clathrin, at the plasma membrane.[34] Dopamine receptor homomers have also been shown for D3R,[35] D4R, and D5R (unpublished data).

HETEROOLIGOMERIZATION BETWEEN DOPAMINE RECEPTORS AND OTHER RECEPTORS

Heterooligomerization with other GPCRs

Heterooligomerization between adenosine A1 receptor and dopamine D1 receptor

The antagonism between adenosine and dopamine in the CNS is well documented by biochemical as well as behavioral studies.[36–38] Possible cross-talk between these systems has been suggested to involve physical interactions between adenosine and dopamine receptors.[39–41] The presence of D1R-A1R complexes has been shown in mouse fibroblast Ltk⁻ cells coexpressing both receptors and in cultured cortical neurons.[42] The oligomerization between these receptors had a direct impact on receptor trafficking. Three hours of exposure to A1 agonist led to a coaggregation of D1R and A1R, whereas exposure to D1 agonist for the same period of time led to an aggregation solely of D1R, in agreement with the fact that D1 agonist was able to disrupt the D1R-A1R heteromers. Functionally, D1R signaling was not desensitized by the formation of D1R-A1R aggregates (clusters). However, costimulation of the receptors did not result in the formation of receptor clusters and D1R desensitization was then preserved.[42]

These observations suggest that the heteromerization of D1R-A1R may be relevant not only for acute antagonism of D1R signaling by A1R, but also for a persistent long-term antagonism of D1R signaling.[37] This may have some clinical relevance for Parkinson's disease treatment as reviewed.[37] Levodopa (L-DOPA), a dopamine precursor, is commonly used for the treatment of the symptoms of Parkinson's disease, such as tremor, rigidity, and bradykinesia. After several years of treatment (5–10) and beneficial effects, the condition of the majority of patients decline, with motor and behavioral alterations. The design of other clinical approaches that capitalize on the antagonistic effects between dopamine and adenosine and the heteromer formation between their receptors (A1R-D1R and also A2AR-D2R)[11,38] is under investigation.[37] Thus an antagonist of adenosine receptors currently in clinical trials,[37,43,44] which would revert the antagonism exerted by endogenous adenosine on the striatal dopaminergic system. An antagonist in this case may have anti-Parkinsonian actions by blocking the adenosine receptors engaged in heteromeric complexes with dopamine receptors to preserve dopamine function.[11,37,38]

Heterooligomerization between adenosine A2A receptor and dopamine D2 receptor

Adenosine and dopamine systems involve another specific interaction between adenosine A2A receptor (A2AR) and dopamine D2 receptor (D2R). Reports of adenosine A2AR reducing the affinity of D2R for its agonists were previously shown,[45,46] and a molecular mechanism in which heteromerization between D2R and A2AR occurs was proposed.[47] This was demonstrated by coexpressing both receptors in different cell lines,[48–51] and was also revealed in A2R and D2R knockout mice.[52] Co-immunoprecipitation studies showed interaction between A2AR and D2R in the SH-SY5Y neuroblastoma cell line,[51] which was confirmed by quantitative and qualitative bioluminescence and fluorescence resonance energy transfer (BRET and FRET, respectively) studies,[53] as well as by mass spectrometry and pull-down techniques.[54] This interaction was shown to be specific to the A2AR-D2R pair, since D1R was not co-immunoprecipitated by A2AR. As a consequence of the A2AR-D2R heteromerization, both the affinities of D2R for its agonists and for G_i proteins were reduced, suggesting reduced functionality of D2R in the heteromeric complex.[49-51] Another consequence is the co-aggregation, co-internalization and co-desensitization of both receptors after prolonged (three hours) treatment with either agonist.[51]

The physiological relevance of A2AR-D2R heteromerization is based on the antagonism between dopamine and adenosine, which may serve as a basis of new clinical approaches in diseases such as Parkinson's disease and schizophrenia.[11,37] After degeneration of the nigrostriatal dopamine pathway in Parkinson's disease, the striatopallidal pathway remains. This pathway contains A2AR-D2R heteromers, and the D2R is strongly antagonized by the effect of A2AR due to the low dopamine tone. Antagonists of A2AR may block exaggerated A2AR signaling and would therefore increase the therapeutic effects of L-DOPA and other D2 agonists in Parkinson's disease.[37,38] Another area of relevance for D2R-A2AR heteromeric complexes may be in schizophrenia. In this case, the ventral striatopallidal pathway is known to be the target of antipsychotic drugs, which are D2R antagonists. Interestingly, A2AR agonists are effective in reducing D2-like agonist binding in this pathway, notably in the nucleus accumbens.[55,56] One new approach for schizophrenia treatment may then be the use of A2AR agonists that reduce D2R-A2AR heteromer signaling.

Heterooligomerization between μ-opioid receptor and dopamine D1 receptor

We investigated a possible interaction between opioidergic and dopaminergic systems and focused on the dopamine D1R and μ-opioid receptor (μOR), since reduced μ-opioid receptor expression was reported in striatal patches from dopamine D1 receptor null mice (D1R[-/-]).[57] This reduction was specific to μOR since the expression of the other opioid receptors, delta- or kappa-opioid receptors was intact. We postulated that heterooligomerization between μOR and D1R may occur at the cellular level. We first showed that these two receptors were co-localized within certain neurons in rat striatum.[58] Physical interaction between D1R and μOR was first shown using the NLS strategy,[58] whereby we added a NLS to D1R (D1R-NLS). When D1-NLS was coexpressed with μOR, both receptors were targeted to the nucleus, indicating that D1R and μOR formed a heteromer.[58] This interaction was shown to be specific, because when the same strategy was applied to D1-NLS and delta opioid receptor (δOR), we observed that D1-NLS was targeted to the nucleus, whereas δOR remained at the cell surface, suggesting an interaction between D1R and μOR, but not with δOR.[58] These results were confirmed by BRET analysis, which showed that the oligomerization was specific to the D1R and μOR pair, and not between D1R and δOR. It also indicated that the D1R-μOR complexes were formed constitutively, suggesting an interaction in the endoplasmic reticulum as well as at the plasma membrane.

Interestingly, we demonstrated that the D1R-μOR heterooligomerization resulted in a quantitative increase in the cell surface expression of μOR. We also showed that the C-tail of D1R was involved, at least in part, in promoting μOR cell surface expression, with probably the involvement of other domains as well. We postulated that heterooligomerization may mask an ER retention motif present on the μOR, or conversely may expose an ER export motif present on the μOR allowing receptor heteromer complexes to traffick to the cell surface through interactions with molecular chaperones. We have previously shown another example of this possibility with the enhancement of cell surface expression of D2R coexpressed with D1R.[59]

It has been reported that D1R-deficient mice (D1$^{-/-}$) were unable to generate a normal locomotor response to morphine.[60] This suggests that D1R-μOR heteromer complexes may have a role in morphine-induced effects in locomotion. D1R-μOR heteromerization may also be relevant in the behavioral effects resulting from the interaction between the dopaminergic and opioidergic systems. It was reported that administering dopaminergic agonists can alter morphine-induced place preference through changes in the sensitivity of dopamine D1 and D2 receptors.[61] Thus D1R-μOR heteromerization may have a role in reward perception and drug addiction, as well as locomotion.

Heterooligomerization with receptor-channels

In addition to the interaction with other GPCRs, dopamine receptors can also interact directly with ionotropic receptors/ligand-gated ion channels. Two major examples of these physical interactions are the interactions of D5R with γ-aminobutyric-acid (GABA) receptor (GABAR) and of dopamine D1R with N-methyl-D-aspartate (NMDA) receptor (NMDAR).

Heterooligomerization of GABAergic receptor with dopamine D5 receptor

The GABAA receptor is a heteropentameric structure constituted by the combination of five different subunits[62] and a member of the ligand-gated receptor family, which mediates fast inhibitory synaptic transmission in the brain. The modulation of GABAA receptor-mediated synaptic activity by dopamine D1-like receptor stimulation was shown to occur after the activation of receptor cAMP-protein kinase-dependent pathways and phosphatases.[63,64] The characterization of the interaction between the dopamine D5 receptor (D5R) and GABAA receptor was provided by co-immunoprecipitation, GST pull down, and in vitro binding in hippocampal neurons and co-transfected cells.[65] The GABAAR was specifically precipitated with D5R, but not with the related receptor D1R, and conversely, D5R, but not D1R, was precipitated by GABAAR γ$_2$ second intracellular (IL2) loop. The interaction of D5R-GABAAR was shown to occur through the direct binding of the D5R C-tail domain with IL2 of the GABAA receptor γ2 subunit.[65] Agonist-dependent dopamine D5 and GABAA γ2 receptor complex formation was found to be obligatory for the functional expression and maintenance of reciprocal receptor cross-talk. For example, the activation of D5R attenuated GABAAα1β2γ2 receptor-mediated inhibitory postsynaptic currents, and the stimulation of GABAA receptor reduced D5R-mediated cAMP accumulation. These results suggest that the physical interaction may represent a mechanism controling synaptic efficacy, by which a mutual functional regulation between a GPCR and a ligand-gated channel can occur. It also provides a mechanistic basis for the functional differentiation of dopamine D5R and D1R.[21] This is interesting because aberrations in cortical or hippocampal GABAergic and dopaminergic D1-like receptor activity[66] have been postulated to play a role in the pathophysiology of schizophrenia.[21,67,68]

Heterooligomerization of NMDA and D1 receptors

NMDA receptors are ligand-gated ion channels that, through their ability to permit ion fluxes, facilitate fast excitatory synaptic transmission. Previous studies have shown that the cross-talk in signal-transduction cascades activated by D1 and NMDA receptors occurred through downstream signaling pathways. However, direct interaction between both receptors was also shown to occur, affecting both NMDAR and D1R functions.[69–71] Two interaction sites have been found between the D1R and the NMDA receptor.[21,70] Two different regions in the C-tail of D1R were shown to interact with either the C-tail region of NR1 NMDA receptor subunit or the C-tail region of the NR2A subunit.[14,70] These interactions were specific, because different regions of D1R interacted with NR1 and NR2A subunits of NMDAR, but no direct physical interaction was observed between D1R and the NR2B subunit.

The oligomerization seems to have direct effects on both receptors. It modulates NMDAR-mediated currents in cultured neurons by allowing D1 receptors to inhibit these currents in hippocampal neurons and in HEK293 cells coexpressing D1R and NR1/NR2A NMDA receptor subunits. In addition, the hetrooligomerization exhibited functional effects on D1R activity by regulating D1 receptor targeting to the plasma membrane. Hence, the coexpression of D1R with NR1 was shown to induce the retention of both in the cytoplasmic compartments. This retention was abolished, however, when D1R-NR1 were coexpressed with NR2B, which suggests that D1R-NR1 complexes are formed in the ER before being trafficked to the plasma membrane in the presence of NR2B. Stimulation of NMDA receptors leads to increased D1R insertion into the plasma membrane and subsequently to an enhanced D1R-mediated cAMP accumulation.[72] Another effect of the D1R-NMDAR oligomerization on D1R consists in the abolishment of the D1R internalization-sequestration, usually observed after agonist treatment. This suggests that the oligomerization of D1R with NMDAR may affect the desensitization and also the regulation of D1R.[21,70]

A peptide corresponding to the region of D1R that interacts with NR1 led to suppression of NMDA-receptor-mediated cell death, revealing the physiological relevance to the D1-NMDAR interaction. This protective effect may be due to the dissociation of the D1R-NMDAR heteromer and also increased association of the NR1 subunit with calmodulin (CaM) and post-synaptic density (PSD)-95. Hence, D1R activation resulted in a decrease in the D1R-NR1 complex but also in an increased association of the NR1 subunit with CaM, which directly interacted with NR1 subunit as was shown by coimmunoprecipitation assays.[21,70] Recently, PSD-95, a major component of the PSD, was also shown to play a role in the assembly of the D1R-NMDAR heterooligomers,[73] with a possible role of preventing excessive D1R density at the synapse. Hence, PSD-95 was found to be critical for D1R-modulated NR1A/NR2B receptor function. The activation of D1R fails to modulate NR1A/NR2B receptor-mediated influx unless PSD-95 is present.[74] Since NMDAR activation was shown to enhance D1R-mediated cAMP accumulation by recruiting more D1R to the plasma membrane,[72] and that the D1R-mediated AC activation led to the NMDA receptor potentiation via the protein kinase A (PKA)/DARPP-32/PP1 cascade,[63,64] a positive feedback loop is thus created.[14,21] If this loop is not controlled, it may result in the overactivation of both dopaminergic and NMDAergic systems, leading to NMDA receptor- and D1-dependent neurotoxicity and cell death.[75] This D1R/PSD-95 interaction may therefore serve as a mechanism to antagonize the NMDA-dependent excessive insertion of D1R to the synapse.

Heterooligomerization of somatostatin SSTR5 and dopamine D2 receptor

The functional interaction between somatostatin (SST) and dopamine has long been known,[36] and their respective receptors share ~30% sequence homology and seem to be structurally related. Evidence for a direct physical interaction was shown using different techniques.[76] First it was shown by functional rescue of a partially active C-tail deleted mutant of human SSTR5 (Δ318 SSTR5). This mutant, while retaining full affinity for its agonist, displayed a complete loss of its ability to couple to AC. However, when coexpressed in CHO-K1 cells with D2R, SST induced a dose-response inhibition of AC, indicating a restoration of the mutant's functionality and suggesting a heteromerization between this mutant and D2R.[76] The heteromerization of the wild-type SSTR5 and D2R was also investigated and shown to occur. The coexpressed receptors showed a high (thirtyfold) increase in binding affinity of SST 14 upon treatment with quinpirole, a D2 agonist. In contrast, the affinity for SST 14 decreased by 80% when a D2 antagonist, sulpiride, was added. A role of receptor conformation states in promoting receptor association was suggested.[76] Photobleaching FRET (pbFRET) analysis showed that the SSTR5-D2R heteromers were induced by treatment with either agonist. As a consequence, the heteromerization induced a synergy in the binding affinity of each receptor such that the binding of the agonist to one receptor led to enhanced affinity of the second receptor for its agonist, with also enhanced coupling to G protein and AC. These findings suggest that the formation of SSTR5-D2R heteromers may explain the enhancement of dopaminergic and somatostatinergic transmission induced by in vivo administration of SST or dopamine agonists.

HETEROOLIGOMERIZATION OF DOPAMINE RECEPTORS WITH EACH OTHER

Interaction between D1 and D3 receptors

D1R and D3R were shown to colocalize in certain neurons in rat brain.[16,77,78] This colocalization was highest in neurons of the island of Calleja major and also present in nucleus accumbens.[16] This emphasized the potential for functional cross-talk between the two receptors at a cellular level, with behavioral and biochemical consequences.[78–80] Recently, using FRET and BRET techniques, two different teams reported that D1R and D3R form heterooligomers.[81,82] As a consequence of the D1R-D3R heteromerization, a synergistic D1-D3 interaction was detected, in which stimulation of D3R leads to an enhancement of the affinity of D1R in HEK293 T cells coexpressing D1R and D3R, which, according to the authors, represents a characteristic of D1-D3 receptor heterooligomers observed also in striatal membrane preparations.[82] Along the same lines, dopamine was shown to displace [^{3}H] (SCH23390) binding from D1R with higher affinity in HEK293 cells expressing D1-D3 heteromers than in HEK293 cells expressing the D1R alone.[81] As a functional consequence of this heteromerization, it was shown that dopamine stimulated AC with higher potency in HEK293 cells coexpressing D1R and D3R, than in cells expressing individual D1R.[81] A regulatory role for this heteromerization was suggested since agonist-induced D1R cytoplasmic sequestration induced by selective D1R agonists was abolished in cells coexpressing D1R and D3R. At the same time, the D1R-D3R heteromerization led to the internalization of the D1R/ D3R complex in response to co-stimulation of both D1R and D3R through a mechanism involving β-arrestin.[81]

Interaction between D2 and D5 receptors

The oligomerization of D2 and D5 receptors was shown by different techniques.[83] Interestingly, coactivation of D2R and D5R within D2-D5 heterooligomers resulted in the generation of a calcium signal. This intracellular calcium release was sensitive to chelation of extracellular calcium, inhibition of receptor-dependent, store-operated calcium channels, as well as inhibition of $G_{q/11}$ and phospholipase C (PLC). Notably, we showed that this mechanism used by D2-D5 receptor heteromers was different from that used by either D5R alone[83] or the D1-D2 receptor heteromers.[14] In fact, the agonist-activated D5R, expressed alone, showed a robust calcium mobilization, requiring influx of extracellular calcium and which was somewhat attenuated when D5R was coexpressed with D2R. This effect was reversed by costimulation of D5-D2 heteromers.[83] We also showed that D2-D5 heteromers differed from D1-D2 heteromers by its dependence on extracellular calcium.[14,83,84] Thus we showed that the D2R was able to modulate the D5R-induced calcium elevations resulting from calcium influx. This, and the fact that anatomical specificity of D5 and D1 receptor localization in the brain exists,[85] emphasizes that dopamine can regulate rapid calcium-mediated effects in the brain through distinct mechanisms represented by D5-D5 homomers, the D1-D2 heteromers, and D2-D5 heteromers.

Interaction between D2 and D3 receptors

Using co-immunoprecipitation studies, it was shown that D2R and D3R, when coexpressed in HEK293 cells, were able to form heteromers.[86] The presence of such heteromers was confirmed using functional complementation, a technique based on the ability of two nonfunctional chimera to rescue their functional response by interaction. This methodology was previously used to demonstrate physical interactions between other receptors.[87] For the D2R-D3R interaction, the coexpression of chimera of D2R and D3R resulted in functional dopamine receptors, with a pharmacological profile different, however, from those of D2R or D3R expressed individually.[86] Interestingly, the D3R, which cannot inhibit AC VI under certain conditions,[88] was able to exert its inhibition once coexpressed with D2R.[86] Although the existence of D2-D3 heteromers as physical entities has not yet been demonstrated in the brain, the possibility of heteromerization between D2R and D3R is substantiated by their colocalization in some brain regions.[77]

HETEROOLIGOMERIZATION OF D1-D2 RECEPTORS

Studies reported the modulation of phosphatidyl-inositol (PI) turnover through a D1-like receptor-activated PLC signaling pathway in rat striatal preparations.[89–93] However, the expression of the D1 receptor in different heterologous systems was unable to activate PLC-induced calcium mobilization.[15] In contrast, the injection of striatal mRNA in *Xenopus* oocytes induced a PLC-dependent PI turnover,[94] indicating that the heterologous cell systems are devoid of an important component that, together with D1R, could activate the PI signaling pathway, which in turn was present in the brain tissue.[15]

We considered the possibility of a physical complex between D1R and D2R, which may be responsible for the enhanced PI turnover and calcium mobilization,[15] since synergism between D1 and D2 receptors has been shown by biochemical,[95] electrophysiological,[28,96] as well as behavioral[97] studies. For example, this synergism was reported by many authors in the induction of c-fos production,[98] and the co-activation of D1 and D2 was shown to be necessary for physiological phenomena such as behavioral sensitization to cocaine[97] and striatal long-term synaptic depression.[96] Co-immunoprecipitation studies in cells coexpressing both D1R and D2R showed that these receptors physically interacted with each other.[99] This interaction was confirmed by FRET studies from our laboratory as well as from others,[59,100] which revealed that D1R and D2R do interact at the cell surface as well as in the ER.[59] A recent report indicated that D1-D2 heterooligomerization seems to involve, at least in part, two acidic residues in the C-tail of D1R, as well as the Arg-rich region of the third intracellular loop (IL3) of D2R.[101] GTPγS incorporation analysis revealed that D1-D2 heterooligomers were able to activate $G_{q/11}$, in contrast to either D1 or D2R, which activate G_s and G_i, respectively.[84] Furthermore, the activation of these D1-D2 heterooligomers, but not each receptor expressed individually, led to a rise in intracellular calcium in absence of any extracellular calcium influx.[84,99] The components of this D1-D2 heteromer-induced mobilization of intracellular calcium were further analyzed using specific inhibitors, demonstrating that this signaling pathway involved G_q protein, PLC and IP_3 receptors.[84,99]

Radioligand binding and intracellular calcium mobilization studies showed differences in the pharmacological profiles of some SKF compounds for the D1 receptor in D1-D2 and D1-D1 complexes. Hence, SKF 83959 and SKF 83822 showed very specific functional effects, with SKF 83959 robustly stimulating a calcium signal and not AC, whereas SKF 83822 robustly stimulated AC with no effect on calcium release.[84,102] We found that these agonists have selective actions, with SKF 83959 activating specifically the D1-D2 heterooligomer by acting as a full agonist at the D1 receptor and a partial agonist at the D2 receptor, whereas SKF 83822 stimulated the D1-D1 homooligomer exclusively.[15,84,102] In addition, we showed that the calcium signal rapidly desensitized following agonist occupancy of the D1 receptor, and that both receptors cointernalized after stimulation of either D1 or D2 receptors.[59]

THE EXISTENCE OF D1-D2 RECEPTOR HETEROOLIGOMERS IN THE BRAIN

An important point of discordance between investigators working in the dopamine receptor field relates to the neuronal localization of D1 and D2 receptors in the brain, which is still a matter of controversy. Generally, it is believed that D1 and D2 receptors are largely localized in two anatomically segregated sets of neurons, forming the "direct" (striatonigral) and the "indirect" (striatopallidal) pathways, respectively,[9,103,104] with the striatonigral neurons enriched in D1, whereas striatopallidal neurons express the D2 receptor. Results from bacterial artificial chromosome (BAC) transgenic mice analysis, using promoters for D1R or D2R tagged with enhanced green fluorescent protein (EGFP) (Drd1-EGFP and Drd2-EGFP, respectively), were consistent with this conclusion,[104–108] although the analysis of the number of GFP-positive neurons in NAcc from Drd1-EGFP and Drd2-EGFP mice showed that 50–60% of MSNs express only D1 receptors, that 35–40% express only D2 receptors, and that at least 10–15% co-express both D1 and D2 receptors.[104 108] Evidence for some neuronal colocalization of D1R and D2R has also been indicated by electrophysiological studies[96] as well as from studies using immunohistochemistry,[4,109,110] electron microscopy,[111] retrograde labeling,[112] and single-cell reverse-transcription PCR.[113]

We showed by immunocytochemistry that the D1R and D2R were colocalized within a significant proportion of cultured neonatal striatal neurons using highly specific antibodies.[99,114] Our immunohistochemistry studies indicated that this colocalization exists also in adult rat brain, as well as some regions of human brain, but at a lower level than in striatal neurons in culture. Furthermore, our co-immunoprecipitation studies showed that these receptors physically interacted in the striatum.[99] Moreover, using confocal FRET methodology, we demonstrated the presence of the D1–D2 receptor heterooligomer as physical entities separated by 5–7 nm (50–70 Å) in a large proportion of cultured early postnatal striatal neurons and in the rat striatum, although with some differences in their localization, with 20% of neurons in the nucleus accumbens exhibiting D1–D2 heteromers and only 6% in the caudate putamen.[114] The specific activation of the D1–D2 receptor heteromer induced a concentration-dependent increase in intracellular calcium in the striatal neurons, which was mediated through the activation of G_q and phospholipase C from IP_3 receptor-sensitive calcium stores.[114]

Analysis of direct $GTP\gamma^{35}S$ incorporation into specific G proteins showed that the stimulation of the D1-D2 heterooligomer by dopamine agonists in rat and mouse striatum induced rapid activation of $G_{q/11}$.[84] This effect was absent in mice, where either the D1 or D2 receptor gene was deleted, indicating the necessity for activation of both receptors to activate G_q. Moreover, pretreatment with antagonists specific to either D1 or D2 receptors abolished the agonist-induced effect, underscoring the necessary participation of both receptors within the heterooligomeric complex for dopamine agonist-induced activation of G_q in striatum. In rat striatal neurons in culture, the specific activation of the D1-D2 heteromer for 5 min triggered an enrichment of G_q at the plasma membrane.[114] This also led to an increase in the activated form of CaMKIIα, both in the cytoplasm and in the nucleus compartments,[114] and the activation of the heterooligomer by concurrent administration of a D1 and D2 agonist led to an increase in total and activated forms of CaMKIIα in the nucleus accumbens of mice and rats.[84,115] This effect was absent in D1 or D2 receptor gene-deleted mice and could be blocked by either D1- or D2-specific antagonists.[84,114]

Longer activation of the D1–D2 receptor heterooligomer (more than one hour) led to an increase in BDNF levels in striatal neurons and in rat nucleus accumbens, which was accompanied by an accelerated morphological maturation of neonatal striatal neurons, marked by an increased neuronal arborization and microtubule-associated protein 2 (MAP2) production.[114] These effects were absent in neurons from D1$^{-/-}$ mice but present in striatal neurons from D5$^{-/-}$ mice, indicating a specificity for the D1-D2 heteromer in these events. Thus, the specific activation of the D1-D2 receptor heteromer triggered a signaling pathway cascade of four steps that links dopamine signaling to BDNF production and neuronal growth: (i) activation of G_q, PLC, and IP_3 leading to a rapid mobilization of intracellular calcium, (ii) a rapid activation of cytosolic and nuclear CaMKIIα, (iii) followed by an enhanced BDNF expression, and (iv) accelerated morphological maturation and/or differentiation of striatal neurons.[114] This signaling cascade was also activated in adult rat nucleus accumbens.[84,114,115]

PHYSIOLOGICAL RELEVANCE

Compelling data from different studies indicate that dopamine exerts its broad range of effects by using different signaling pathways, each of them leading to a particular physiological effect, with a possibility of convergence between some or all of these signaling pathways. The calcium signaling pathway that we showed to occur in the brain through the activation of D1-D2 heterooligomers clearly indicates that the oligomerization of dopamine receptors led to the discovery of a novel mechanism through which dopamine may modulate some brain functions, differently from the sole activation of one of the individual receptors involved in the heterooligomerization. Hence, it was reported that activation of the PI-linked dopamine receptor agonist SKF 83959 stimulated cdk5 and CaMKII activities in frontal cortex slices.[116] This activation was shown to be involved in SKF 83959-induced change in levels of phospho-DARPP-32. This activation, while involving a D1 receptor contribution, since it was blocked by SCH 23390, was not seen in PC-12 cells transfected with DNA encoding for D1R. This suggests that this activation may involve the D1-D2 heteromers. It is important to underline that in regard to the CaMKII activation that we showed occurring in response to D1-D2 heteromer stimulation, and the results showing changes in the levels of phospho-DARPP-32 via PI-linked dopamine receptor, CaMKII and cdk5 activation may represent a novel mechanism for DARPP-32 regulation, different from the well-known cAMP-PKA-dependent one.

Intracellular calcium has key roles in many aspects of neuronal function including gene expression, protein activity, and regulation of synaptic transmission.[117] An aspect of the physiological relevance of the D1-D2-mediated calcium signaling cascade would be the possibility that this pathway leads to rapid and direct effects on synaptic plasticity with potential long-term changes, because the specific activation of D1-D2 dopamine receptor complexes increases the total and activated forms of CaMKIIα in the nucleus accumbens. Evidence has indicated that this form of CaMKIIα is a critical regulator of synaptic plasticity in neurons.[118–120] In this line of evidence, it has been shown that 50% of CaMKIIα-deficient mice present changes in behavior and learning.[121] This may suggest that alterations in D1-D2 receptor heteromer-induced calcium-CaMKIIα signaling may result in alterations in synaptic plasticity leading to the alterations of cognition, learning, and memory that contribute to the pathophysiology of dopamine-related disorders such as schizophrenia. All the components involved in this calcium pathway, including G_q, PLC, and CaMKII, have been shown to be affected in the brains of schizophrenia patients.[122] Another physiological aspect is the finding that activation of CaMKIIα was necessary for the induction of behavioral sensitization to cocaine, a physiological phenomenon that also requires the coactivation of D1 and D2 dopamine receptors,[97] and which may help explain the important role of dopamine in drug addiction. A recent report showed that CaMKIIα binds to and inhibits D3R in nucleus accumbens, leading to the promotion of cocaine-induced locomotion.[123] One can speculate about a mechanism by which the activation of D1-D2 heteromer signaling leads to CaMKII activation, which in turn binds to D3R, inhibiting D3R signaling, and leads to the promotion of cocaine-induced locomotion. Another physiological relevance for the D1-D2 receptor heterooligomer-induced CaMKIIα activation may be related to the involvement of CaMKIIα in neuronal growth and maturation.[120,124–126] The molecular mechanisms by which these diverse effects may be mediated through D1-D2 heteromers are still not completely delineated, although our results showing an enhanced BDNF production, and neuronal growth and arborisation represent a significant link between the D1–D2 receptor heteromer signaling pathway activation and neuronal growth and/or differentiation through BDNF expression. The role of this D1–D2 heteromer-triggered signaling pathway may represent an important new direction to pursue in unraveling the pathophysiology of disorders such as schizophrenia, drug addiction, and depression. It is interesting to observe the growing reports of different receptor heteromer combinations by dopaminergic receptors within their subfamily and with other GPCRs, receptor-channels, and proteins. While these findings provide insight into physiological mechanisms and represent interesting possibilities for novel drug design, it is undoubtedly adding more complexity by complicating the understanding of the functional roles for the different heterooligomers formed. The next exciting challenge before us may be to understand how different combinations of these different dopamine receptor heteromers and homooligomers act in concert to produce a physiological effect.

ACKNOWLEDGMENTS

The authors work on dopamine receptors is supported by a grant from the U.S. National Institute on Drug Abuse. Susan R. George holds a Canada Research Chair in Molecular Neuroscience.

SUGGESTED READING

1. Gainetdinov RR and Caron MG. (2001). Genetics of childhood disorders: XXIV. ADHD, part 8: hyperdopaminergic mice as an animal model of ADHD. *J Am Acad Child Adolesc Psychiatry* 40:380–382.
2. Nestler EJ (2001). Molecular basis of neural plasticity underlying addiction. *Nat Rev Neurosci* 2:119–128.
3. Neve KA, et al. (2004). Dopamine receptor signaling. *J Recept Signal Transduct Res* 24:165–205.
4. Aizman O, Brismar H, Uhlen P, Zettergren E, Levey AI, et al. (2000) Anatomical and physiological evidence for D1 and D2 dopamine receptor colocalization in neostriatal neurons. *Nat Neurosci* 3:226–230.
5. Agnati LF, Franzen O, Ferre S, Franco R, and Fuxe K (2003). Possible role of intramembrane receptor/receptor interactions in memory and learning via formation of long-lived heteromeric complexes: focus on motor learning in basal ganglia. *J Neural Transm* 65:195–222.
6. Zarrindast MR, Azami B, Rostami P, Rezayof A (2006). Repeated administration of dopaminergic agents in the nucleus accumbens and morphine-induced place preference. *Behav Brain Res* 169:248–255.
7. Zoli M, Agnati LF, Hedlund PB, Li XM, Ferre S, and Fuxe K (1993). Receptor/receptor interactions as an integrative mechanism in nerve cells. *Mol Neurobiol* 7:293–334.
8. Gerfen CR (1992). The neostriatal mosaic: multiple levels of compartmental organization in the basal ganglia. *Annu Rev Neurosci* 15:285–320.
9. Gerfen CR (2004). The basal ganglia in the rat nervous system, Ed 3 (Paxinos G, ed), pp 455–508. New York: Academic.

10. Ng GY, George SR, Zastawny RL, Caron M, Bouvier M, Dennis M, and O ' Dowd BF (1993). Human serotonin1B receptor expression in Sf9 cells: phosphorylation, palmitoylation, and adenylyl cyclase inhibition. *Biochemistry* 32: 11727–11733.

11. Ferre S, Ciruela F, Canals M, Marcellino D, Burgueno J, Casado V, Hillion J, Torvinen M, Fanelli F, Benedetti P, Goldberg SR, Bouvier M, Fuxe K, Agnati LF, Lluis C, Franco R, and Woods A (2004). Adenosine A2A-dopamine D2 receptor–receptor heteromers. Targets for neuropsychiatric disorders. *Parkinsonism Relat Disord* 10: 265–271.

12. Pei L, Lee FJ, Moszczynska A, Vukusic B, and Liu F (2004). Regulation of dopamine D1 receptor function by physical interaction with the NMDA receptors. *J Neurosci* 24:1149–1158.

13. Beaulieu JM, Sotnikova TD, Marion S, Lefkowitz RJ, Gainetdinov RR, and Caron MG (2005). An Akt/beta-arrestin 2/PP2A signaling complex mediates dopaminergic neurotransmission and behavior. *Cell* 122:261–273.

14. Lee FJ and Liu F (2004). Direct interactions between NMDA and D1 receptors: a tale of tails. *Biochem Soc Trans* 32:1032–1036.

15. George, SR and O'Dowd BF (2007). Novel dopamine receptor signaling unit in brain: heterooligomers of D1 and D2 dopamine receptors. *Scient World J* 7:58–63.

16. Rocheville M, Lange DC, Kumar U, Patel SC, Patel RC, and Patel YC (2000). Receptors for dopamine and somatostatin: formation of hetero-oligomers with enhanced functional activity. *Science (Wash DC)* 288:154–157.

17. Ferre S, Baler R, Bouvier M, Caron MG, Devi LA, Durroux T, Fuxe K, George SR, Javitch JA, Lohse MJ, Mackie K, Milligan G, Pfleger KD, Pin JP, Volkow ND, Waldhoer M, Woods AS, and Franco R (2009). Building a new conceptual framework for receptor heteromers. *Nat Chem Biol* 5:131–134.

18. Bouvier M (2001). Oligomerization of G-protein-coupled transmitter receptors. *Nat Neurosci* 2:274–286.

19. George, SR, O'Dowd, BF, and Lee, SP (2002). G-protein-coupled receptor oligomerization and its potential for drug discovery. *Nat. Rev Drug Discov* 1:808–820.

20. Missale C, Nash SR, Robinson SW, Jaber M, and Caron MG (1998). Dopamine receptors: from structure to function. *Physiol Rev* 78:189–225.

21. Wayman GA, Lee YS, Tokumitsu H, Silva A, and Soderling TR (2008). Calmodulin-kinases: modulators of neuronal development and plasticity. *Neuron* 59:914–931.

22. Ng GY, O'Dowd BF, Lee SP, Chung HT, Brann MR, Seeman P, and George SR (1996). Dopamine D2 receptor dimers and receptor-blocking peptides. *Biochem Biophys Res Commun* 227:200–204.

23. Nimchinsky EA, Hof PR, Janssen WG, Morrison JH, and Schmauss C (1997). Expression of dopamine D3 receptor dimers and tetramers in brain and in transfected cells. *J Biol Chem* 272:29229–29237.

24. George SR, Lee SP, Varghese G, Zeman PR, Seeman P, Ng GY, and O'Dowd BF (1998). A transmembrane domain-derived peptide inhibits D1 dopamine receptor function without affecting receptor oligomerization. *J Biol Chem* 273:30244–30248.

25. Zhang J, Vinuela A, Neely MH, Hallett PJ, Grant SG, Miller GM, Isacson O, Caron MG, and Yao WD (2007). Inhibition of the dopamine D1 receptor signaling by PSD-95. *J Biol Chem* 282:15778–15789.

26. Lee SP, Xie Z, Varghese G, Nguyen T, O'Dowd BF, and George SR (2000). Oligomerization of dopamine and serotonin receptors. *Neuropsychopharmacology* 23:S32–S40.

27. Guo W, Urizar E, Kralikova M, Mobarec JC, Shi, L, Filizola M, and Javitch JA (2008). Dopamine D2 receptors form higher order oligomers at physiological expression levels. *EMBO Journal* 27:2293–2304.

28. Harsing Jr.LG and Zigmond MJ (1997). Influence of dopamine on GABA release in striatum: Evidence for D1-D2 interactions and non-synaptic influences. *Neuroscience* 77:419–429.

29. Pacheco MA and Jope RS (1997). Comparison of [3H]phosphatidylinositol and [3H]phosphatidylinositol 4,5-bisphosphate hydrolysis in postmortem human brain membranes and characterization of stimulation by dopamine D1 receptors. *J Neurochem* 69:639–644.

30. O'Dowd BF, Ji X, Alijaniaram M, Rajaram RD, Kong MM, Rashid A, Nguyen T, and George SR (2005). Dopamine receptor oligomerization visualized in living cells. *J Biol Chem* 280:37225–37235.

31. Jans DA, Xiao CY, and Lam MH (2000). Nuclear targeting signal recognition: a key control point in nuclear transport? *Bioessays* 22, 532–544.

32. Guo W, Shi L, and Javitch JA (2003). The fourth transmembrane segment forms the interface of the dopamine D2 receptor homodimer. *J Biol Chem* 278: 4385–4388.

33. Kong MM, Fan T, Varghese G, O'Dowd BF, and George SR (2006). Agonist-induced cell surface trafficking of an intracellularly sequestered D1 dopamine receptor homooligomer. *Mol Pharmacol* 70, 78–89.

34. Kong MM, Hasbi A, Mattocks M, Fan T, O'Dowd BF, and George SR (2007). Regulation of D1 dopamine receptor trafficking and signaling by caveolin-1. *Mol Pharmacol* 72:1157–1170.

35. O'Dowd BF, Alijaniaram M, Ji X, Nguyen T, Eglen RM, George SR (2007). Using ligand-induced conformational change to screen for compounds targeting G-protein-coupled receptors. *J Biomol Screen* 12:175–185.

36. Agnati LF, Ferre S, Lluis C, Franco R, and Fuxe K (2003). Molecular mechanisms and therapeutic implications of intramembrane receptor/receptor interactions among heptahelical receptors with examples from the striatopallidal GABA neurons. *Pharmacol. Rev* 55:509–550.

37. Franco R, Lluis C, Canela EI, Mallol J, Agnati L, Casado V, Ciruela F, Ferre S, and Fuxe K (2007). Receptor–receptor interactions involving adenosine A1 or dopamine D1 receptors and accessory proteins. *J Neural Transm* 114: 93–104

38. Fuxe K, Stromberg I, Popoli P, Rimondini-Giorgini R, Torvinen M, Ogren SO, FrancoR, Agnati LF, and Ferre S (2001). Adenosine receptors and Parkinson's disease. Relevance of antagonistic adenosine and dopamine receptor interactions in the striatum. *Adv Neurol* 86:345–353.

39. Ferre S, Popoli P, Gimenez-Llort L, Finnman U-B, Martinez E, Scotti de Carolis A, and Fuxe K (1994). Postsynaptic antagonistic interaction between adenosine A1 and dopamine D1 receptors. *Neuroreport* 6:73–76.

40. Fuxe K, Ferre S, Zoli M, and Agnati LF (1998). Integrated events in central dopamine transmission as analyzed at multiple levels. Evidence for intramembrane adenosine A2A/dopamine D2 and adenosine A1/dopamine D1 receptor interactions in the basal ganglia. *Brain Res Rev* 26:258–273.

41. Franco R, Ferre S, Agnati L, Torvinen M, Gines S, Hillion J, Casado V, Lledo PM, Zoli M, Lluis C, et al. (2000). Evidence for adenosine/dopamine receptor interactions: indications for heteromerization. *Neuropsychopharmacology* 23:S50–S59.

42. Gines S, Hillion J, Torvinen M, LeCrom S, Casado V, Canela E, Rondin S, Lew J, Watson S, Zoli M, et al. (2000). Dopamine D1 and adenosine A1 receptors assemble into functionally interacting heteromeric complexes. *Proc Natl Acad Sci USA* 97: 8606–8611.

43. Kanda T, Jackson MJ, Smith LA, Pearce RK, Nakamura J, Kase H, Kuwana Y, and Jenner P (2000). Combined use of the adenosine A(2A) antagonist KW-6002 with L-DOPA or with selective D1 or D2 dopamine agonists increases antiparkinsonian activity but not dyskinesia in MPTP-treated monkeys. *Exp Neurol* 162:321–327.

44. Koga K, Kurokawa M, Ochi M, Nakamura J, and Kuwana Y (2000). Adenosine A(2A) receptor antagonists KF17837 and KW-6002 potentiate rotation induced by dopaminergic drugs in hemi-Parkinsonian rats. *Eur J Pharmacol* 408:249–255.

45. Ferre S, Fuxe K, von Euler G, Johansson B, and Fredholm BB (1992). Adenosine-dopamine interactions in the brain. *Neuroscience* 51: 501–512.

46. Mayford M, Bach ME, Huang YY, Wang L, Hawkins RD, Kandel ER (1996). Control of memory formation through regulated expression of a CaMKII transgene. *Science* 274:1678–1683.

47. Becker A, Grecksch G, Kraus J, Peters B, Schroeder H, Schulz S, Höllt V. (2001). Loss of locomotor sensitisation in response to morphine in D1 receptor deficient mice. *Naunyn Schmiedebergs Arch Pharmacol* 363: 562–568.

48. Dasgupta S, Ferre S, Kull B, Hedlund PB, Finnman U-B, Ahlberg S, Arenas E, Fredholm BB, and Fuxe K (1996a). Adenosine A2A receptors modulate the binding characteristics of dopamine D2 receptors in stably cotransfected fibroblast cells. *Eur J Pharmacol* 316:325–331.

49. Kull B, Ferre S, Arslan G, Svenningsson P, Fuxe K, Owman C, and Fredholm BB (1999). Reciprocal interactions between adenosine A2A and dopamine D2 receptors in Chinese hamster ovary cells co-transfected with the two receptors. *Biochem Pharmacol* 58:1035–1045.

50. Scarselli M, Novi F, Schallmach E, Lin R, Baragli A, Colzi A, Griffon N, Corsini GU, Sokoloff P, Levenson R, et al. (2001). D2/D3 dopamine receptor heterodimers exhibit unique functional properties. *J Biol Chem* 276:30308–30314.

51. Hillion J, Canals M, Torvinen M, Casado V, Scott R, Terasmaa A, Hansson A, Watson S, Olah ME, Mallol J, et al. (2002). Coaggregation, cointernalization and codesensitization of adenosine A2A receptors and dopamine D2 receptors. *J Biol Chem* 277:18091–18097.

52. Chen JF, Moratalla R, Impagnatiello F, Grandy DK, Cuellar B, Rubinstein M, Beilstein MA, Hackett E, Fink JS, Low MJ, et al. (2001). The role of the D(2) dopamine receptor (D(2)R) in A(2A) adenosine receptor (A(2A)R)-mediated behavioral and cellular responses as revealed by A(2A) and D(2) receptor knockout mice. *Proc Natl Acad Sci USA* 98:1970–1975.

53. Canals M, Marcellino M, Fanelli F, Ciruela F, de Benedetti P, Goldberg S, Fuxe K, Agnati LF, Woods AS, Ferre S, Lluis C, Bouvier M, and Franco R (2003). Adenosine A2A-dopamine D2 receptor–receptor heteromerization. Qualitative and quantitative assessment by fluorescence and bioluminescence energy transfer. *J Biol Chem* 278:46741–46749.

54. Ciruela F, Burgueno J, Casado V, Canals M, Marcellino D, Goldberg SR, Bader M, Fuxe K, Agnati LF, Lluis C, Franco R, Ferre S, and Woods AS (2004). Combining mass spectrometry and pull-down techniques for the study of receptor heteromerization. Direct epitope-epitope electrostatic interactions between adenosine A2A and dopamine D2 receptors. *Anal Chem* 76:5354–5363.

55. Ferre S (1997). Adenosine-dopamine interactions in the ventral striatum. Implications for the treatment of schizophrenia. *Psychopharmacology* 133:107–120.

56. Diaz-Cabiale Z, Hurd Y, Guidolin D, Finnman UB, Zoli M, Agnati LF, Vanderhaeghen JJ, Fuxe K, and Ferre S (2001). Adenosine A2A agonist CGS 21680 decreases the affinity of dopamine D2 receptors for dopamine in human striatum. *Neuroreport* 12:1831–1834.

57. Gu WH, Yang S, Shi WX, Jin GZ, Zhen XC. (2007). Requirement of PSD-95 for dopamine D1 receptor modulating glutamate NR1a/NR2B receptor function. *Acta Pharmacol Sin* 28: 756–762.

58. Juhasz JR, Hasbi A, Rashid AJ, So CH, George SR, and O'Dowd BF (2008). Mu-opioid receptor heterooligomer formation with the dopamine D1 receptor as directly visualized in living cells. *Europ J Pharm* 581:235–243

59. So CH, Verma V, O'Dowd BF, and George SR (2007). Desensitization of the dopamine D1 and D2 receptor heterooligomer mediated calcium signal by agonist occupancy of either receptor. *Mol Pharmacol* 72:450–462.

60. Hnasko TS, Sotak B, and Palmiter R (2005). Morphine reward in dopamine deficient mice. *Nature* 438:854–857.

61. Zawarynski P, Tallerico T, Seeman P, Lee SP, O'Dowd BF, and George SR (1998). Dopamine D2 receptor dimers in human and rat brain. *FEBS Lett* 441:383–386.

62. Barnard EA, Skolnick P, Olsen RW, Mohler H, Sieghart W, Biggio G, Braestrup C, Bateson AN, and Langer SZ (1998). International Union of Pharmacology. XV. Subtypes of gamma-aminobutyric acid A receptors: classification on the basis of subunit structure and receptor function. *Pharmacol Rev* 50:291–313.

63. Greengard P, Nairn A, Girault JA, Ouimet CC, Snyder GL, Fisone G, Allen PB, Fienberg A, and Nishi A (1998). The DARPP-32/protein phosphate-1 cascade: a model for signal integration. *Brain Res Rev* 26:274–284.

64. Greengard P, Allen PB, and Nairn AC (1999). Beyond the dopamine receptor: the DARPP-32/protein pnosphatase-1 cascade. *Neuron* 23:435–447.

65. Liu F, Wan Q, Pristupa ZB, Yu XM, Wang YT, and Niznik HB (2000). Direct protein-protein coupling enables cross-talk between dopamine D5 and γ-aminobutyric acid A receptors. *Nature* 403:274–280.

66. Radnikow G and Misgeld U (1998). Dopamine D1 receptors facilitate GABAA synaptic currents in the rat substantia nigra pars reticulata. *J Neurosci* 18:2009–2016.

67. Rashid AJ, O'Dowd BF, Verma V, and George SR (2007b). Neuronal Gq/11-coupled dopamine receptors: an uncharted role for dopamine. *TRENDS Pharmacol Sci* 28:551–555.

68. So CH, Varghese G, Curley KJ, Kong MM, Alijaniaram M, Ji X, Nguyen T, O'Dowd BF, and George SR (2005). D1 and D2 dopamine receptors form heterooligomers and co-internalize after selective activation of either receptor. *Mol Pharmacol* 68:568–578.

69. Fiorentini C, Gardoni F, Spano P, Di Luca M, and Missale C (2003). Regulation of dopamine D1 receptor trafficking and desensitization by oligomerization with glutamate Nmethyl-D-aspartate receptors. *J Biol Chem* 278: 20196–20202.

70. Lee FJ, Xue S, Pei L, Vukusic B, Chery N, Wang Y, Wang YT, Niznik HB, Yu XM, and Liu F (2002). Dual regulation of NMDA receptor functions by direct protein protein interactions with the dopamine D1 receptor. *Cell* 111:219–230.

71. Seeman P, Niznik HB, Guan HC, Booth G, and Ulpian C (1989). Link between D1 and D2 dopamine receptors is reduced in schizophrenia and Huntington diseased brain. *Proc Natl Acad Sci USA* 86:10156–10160.

72. Piomelli D, et al. (1991). Dopamine activation of the arachidonic acid cascade as a basis for D1/D2 receptor synergism. *Nature* 353:164–167.

73. Zhen X, et al. (2004). Regulation of cyclin-dependent kinase 5 and calcium/calmodulin-dependent protein kinase II by phosphatidylinositol-linked dopamine receptor in rat brain. *Mol Pharmacol* 66:1500–1507.

74. Robinson SW, Caron MG. (1997). Selective inhibition of adenylyl cyclase type V by the dopamine D3 receptor. *Mol Pharmacol* 52: 508–514.

75. Bozzi Y and Borrelli E (2006). Dopamine in neurotoxicity and neuroprotection: what do D2 receptors have to do with it? *Trends Neurosci* 29:167–174.

76. Rongo C and Kaplan JM (1999). CaMKII regulates the density of central glutamatergic synapses in vivo. *Nature* 402:195–199.

77. Le Moine C and Bloch B (1996). Expression of the D3 dopamine receptor in peptideric neurons of the nucleus accumbens: comparison with the D1 and D2 dopamine receptors. *Neuroscience* 73:131–143.

78. Scott L, Kruse MS, Forssberg H, Brismar H, Greengard P, and Aperia A (2002). Selective up-regulation of dopamine D1 receptors in dendritic spines by NMDA receptor activation. *Proc Natl Acad Sci USA* 99:1661–1664.

79. Karasinska JM, George SR, Cheng R, and O'Dowd BF (2005). Deletion of dopamine D1 and D3 receptors differentially affects spontaneous behaviour and cocaine-induced locomotor activity, reward and CREB phosphorylation. *Eur J Neurosci* 22:1741–1750.

80. Schwartz JC, Ridray S, Bordet R, Diaz J, and Sokoloff P (1998a). D1/D3 receptor relationships in brain: coexpression, coactivation, and coregulation. *Adv Pharmacol* 42:408–411.

81. Fiorentini C, Busi C, Gorruso E, Gotti C, Spano P, Missale C (2008). Reciprocal regulation of dopamine D1 and D3 receptor function and trafficking by heterodimerization. *Mol Pharmacol* 74:59–69.

82. Mayfield R, Larson G, Orona RA, and Zahniser NR (1996). Opposing actions of adenosine A2a and dopamine D2 receptor activation on GABA release in the basal ganglia: evidence for an A2a/D2 receptor interaction in globus pallidus. *Synapse* 22:132–138.

83. Surmeier DJ, Reiner A, Levine MS, and Ariano MA (1993). Are neostriatal dopamine receptors co-localized? *Trends Neurosci* 16:299–305.

84. Ridray S, Griffon N, Mignon V, Souil E, Carboni S, Diaz J, Schwartz JC, and Sokoloff P (1998). Coexpression of dopamine D1 and D3 receptors in islands of Calleja and shell of nucleus accumbens of the rat: opposite and synergistic functional interactions. *Eur J Neurosci* 10:1676–1686.

85. Bergson C, Mrzljak L, Smiley JF, Pappy M, Levenson R, and Goldman-Rakic PS. (1995). Regional, cellular, and subcellular variations in the distribution of D1 and D5 dopamine receptors in primate brain. *J Neurosci* 15:7821–7836.

86. Schwartz JC, Diaz J, Bordet R, Griffon N, Perachon S, Pilon C, Ridray S, and Sokoloff P (1998b). Functional implications of multiple dopamine receptor subtypes: the D1/D3 receptor coexistence. *Brain Res Rev* 26:236–242.

87. Łukasiewicz S, Faron-Górecka A, Dobrucki J, Polit A, Dziedzicka-Wasylewska M. (2009). Studies on the role of the receptor protein motifs possibly involved in electrostatic interactions on the dopamine D1 and D2 receptor oligomerization. *FEBS J* 276: 760–775.

88. Maggio R, Vogel Z, Wess J. (1993). Reconstitution of functional muscarinic receptors by co-expression of amino- and carboxyl-terminal receptor fragments. *FEBS Lett.*;319: 195–200.

89. Wang HY, Undie AS, and Friedman E (1995). Evidence for the coupling of Gq protein to D1-like dopamine sites in rat striatum: possible role in dopamine-mediated inositol phosphate formation. *Mol. Pharmacol* 48:988–994.

90. Wang M, Lee FJS, and Liu F (2008). Dopamine receptor interacting proteins (DRIPs) of dopamine D1-like receptors in the central nervous system. *Mol Cells* 25:149–157.

91. Park SK, Nguyen MD, Fischer A, Luke MP, Affar el B, et al. (2005). Par-4 links dopamine signaling and depression. *Cell* 122:275–287.

92. Jin LQ, Goswami S, Cai G, Zhen X, and Friedman E (2003). SKF83959 selectively regulates phosphatidylinositol-linked D1 dopamine receptors in rat brain. *J Neurochem* 85:378–386.

93. Undie AS and Friedman E (1990). Stimulation of a dopamine D1 receptor enhances inositol phosphates formation in rat brain. *J Pharmacol Exp Ther* 253:987–992.

94. Marcellino D, Ferré S, Casadó V, Cortés A, Le Foll B, Mazzola C, Drago F, Saur O, Stark H, Soriano A, Barnes C, Goldberg SR, Lluis C, Fuxe K, and Franco R (2008). Identification of dopamine D1-D3 receptor heteromers. Indications for a role of synergistic D1-D3 receptor interactions in the striatum. *J Biol Chem* 283: 26016–26025.

95. Poisbeau P, Cheney MC, Browning MD, and Mody I (1999). Modulation of synaptic GABAA receptor function by PKA and PKC in adult hippocampal neurons. *J Neurosci* 19:674–683.

96. Calabresi P, Mercuri N, Stanzione P, Stefani A, and Bernardi G (1987). Intracellular studies on the dopamine induced firing inhibition of neostriatal neurons *in vitro*: evidence for D1 receptor involvement. *Neuroscience* 20:757–771.

97. Capper-Loup C, Canales JJ, Kadaba N, and Graybiel AM (2002). Concurrent activation of dopamine D1 and D2 receptors is required to evoke neural and behavioral phenotypes of cocaine sensitization. *J Neurosci* 22:6218–6227.

98. Tang TS and Bezprozvanny I (2004). Dopamine receptor-mediated Ca2+ signaling in striatal medium spiny neurons. *J Biol Chem* 279:42082–42094.

99. Lee SP, So CH, Rashid AJ, Varghese G, Cheng R, Lanca AJ, O'Dowd BF, and George SR (2004). Dopamine D1 and D2 receptor co-activation generates a novel phospholipase C-mediated calcium signal. *J Biol Chem* 279:35671–35678.

100. Hasbi A, Fan T, Alijaniaram M, Nguyen T, Perreault ML, O'Dowd BF, George SR.(2009). Calcium signaling cascade links dopamine D1-D2 receptor heteromer to striatal BDNF production and neuronal growth. *Proc Natl Acad Sci U S A* 106:21377–21382.

101. Dziedzicka-Wasylewska M, Faron-Górecka A, Andrecka J, Polit A, Kuśmider M, Wasylewski Z. (2006) Fluorescence studies reveal heterodimerization of dopamine D1 and D2 receptors in the plasma membrane. *Biochemistry* 45: 8751–8759.

102. Rashid AJ, So CH, Kong MM, Furtak T, El-Ghundi M, Cheng R, O'Dowd BF, and George SR (2007a). D1-D2 dopamine receptor heterooligomers with unique pharmacology are coupled to rapid activation of Gq/11 in the striatum. *Proc Natl Acad Sci USA* 104:654–659.

103. Le Moine C and Bloch B (1995). D1 and D2 dopamine receptor gene expression in rat striatum: sensitive cRNA probes demonstrate prominent segregation of D1 and D2 mRNAs in distinct neuronal populations of the dorsal and ventral striatum. *J Comp Neurol* 355:418–426.

104. Lee K-W, Kim Y, Kim AM, Helmin K, Nairn AC, and Greengard P (2006). Cocaine-induced dendritic spine formation in D1 and D2 dopamine receptor-containing medium spiny neurons in nucleus accumbens. *PNAS* 103:3399–3404.

105. Gong S, Zheng C, Doughty ML, Losos K, Didkovsky N, Schambra UB, Nowak NJ, Joyner A, Leblanc G, Hatten ME, and Heintz N (2003). A gene expression atlas of the central nervous system based on bacterial artificial chromosomes. *Nature* 425:917–925.

106. Mahan LC, Burch RM, Monsma Jr. FJ, and Sibley DR (1990). Expression of striatal D1 dopamine receptors coupled to inositol phosphate production and Ca2+ mobilization in Xenopus oocytes. *Proc Natl Acad Sci USA* 87:2196–2200.

107. Shetreat ME, Lin L, Wong AC, and Rayport S (1996). Visualization of D1 dopamine receptors on living nucleus accumbens neurons and their colocalization with D2 receptors. *J Neurochem* 66:1475–1482.

108. Bertran-Gonzalez J, Bosch C, Maroteaux M, Matamales M, Hervé D, et al. (2008). Opposing patterns of signaling activation in dopamine D1 and D2 receptor-expressing striatal neurons in response to cocaine and haloperidol. *J Neurosci* 28:5671–5685.

109. Smart TG (1997). Regulation of excitatory and inhibitory neurotransmitter-gated ion channels by protein phosphorylation. *Curr Opin Neurobiol* 7:358–367.

110. Levey AI, Hersch SM, Rye DB, Sunahara RK, Niznik HB, et al. (1993). Localization of D1 and D2 dopamine receptors in brain with subtype-specific antibodies. *Proc Nat Acad Sci USA* 90:8861–8865.

111. Hersch SM, Ciliax BJ, Gutekunst CA, Rees HD, Heilman CJ, Yung KK, et al. (1995). Electron microscopic analysis of D1 and D2 dopamine receptor proteins in the dorsal striatum and their synaptic relationships with motor corticostriatal afferents. *J Neurosci* 15:5222–5237.

112. Deng YP, Lei WL, and Reiner A (2006). Differential perikaryal localization in rats of D1 and D2 dopamine receptors on striatal projection neuron types identified by retrograde labeling. *J Chem Neuroanat* 32:101–116.

113. Svenningsson P, Fredholm BB, Bloch B, and Le Moine C (2000). Co-stimulation of D(1)/D(5) and D(2) dopamine receptors leads to an increase in c-fos messenger RNA in cholinergic interneurons and a redistribution of c-fos messenger RNA in striatal projection neurons. *Neuroscience* 98:749–757.

114. Ng J, Rashid AJ, So CH, O'Dowd BF, George SR. (2010). Activation of calcium/calmodulin-dependent protein kinase IIalpha in the striatum by the heteromeric D1-D2 dopamine receptor complex. *Neuroscience* 165: 535–541.

115. Zhou F-M, Wilson CJ, and. Dani JA (2002). Cholinergic interneuron characteristics and nicotinic properties in the striatum. *J Neurobiol* 53:590–605.

116. Berridge, MJ (1998). Neuronal calcium signaling. *Neuron* 21:13–26.

117. Blaeser F, Sanders MJ, Truong N, Ko S, Wu LJ, Wozniak DF, Fanselow MS, Zhuo M, and Chatila TA (2006). Long-term memory deficits in Pavlovian fear conditioning in Ca2+/calmodulin kinase kinase alpha-deficient mice. *Mol Cell Biol* 26:9105–9115.

118. Choi SS, Seo YJ, Shim EJ, Kwon MS, Lee JY, Ham YO, and Suh HW (2006). Involvement of phosphorylated Ca2+/calmodulin-dependent protein kinase II and phosphorylated extracellular signal-regulated protein in the mouse formalin pain model. *Brain Res* 1108:28–38.

119. Wong AC, Shetreat ME, Clarke JO, and Rayport S (1999). D1- and D2-like dopamine receptors are colocalized on the presynaptic varicosities of striatal and nucleus accumbens neurons in vitro. *Neuroscience* 89:221–233.

120. Milligan G and White JH (2001). Protein-protein interactions at G-protein-coupled receptors. *Trends Pharmacol Sci* 22:513–518.

121. Lidow MS (2003). Calcium signaling dysfunction in schizophrenia: a unifying approach. *Brain Res Rev* 43:70–84.

122. Liu XY, Mao LM, Zhang GC, Papasian CJ, Fibuch EE, Lan HX, Zhou HF, Xu M, Wang JQ (2009). Activity-dependent modulation of limbic dopamine D3 receptors by CaMKII. *Neuron* 61:425–438.

123. Borodinsky LN, Coso OA, and Fiszman ML (2002). Contribution of Ca2+ calmodulin-dependent protein kinase II and mitogen-activated protein kinase kinase to neural activity-induced neurite outgrowth and survival of cerebellar granule cells. *J Neurochem* 80:1062–1070.

124. Salim H, Ferre S, Dalal A, Peterfreund RA, Fuxe K, Vincent JD, and Lledo PM (2000). Activation of adenosine A1 and A2A receptors modulates dopamine D2 receptor induced responses in stably transfected human neuroblastoma cells. *J Neurochem* 74:432–439.

125. Yao W-D, Spealman RD, and Zhang J (2008). Dopaminergic signaling in dendritic spines. *Biochem pharmacol* 75 (11): 2055–2069.

126. Agnati LF, Benfenati F, Solfrini V, Biagini G, Fuxe K, Guidolin D, Carani C, and Zini I (1993). Intramembrane receptor/receptor interactions: integration of signal transduction pathways in the nervous system. *Neurochem Int* 22:213–222.

127. Gouldson PR, Snell CR, Bywater RP, Higgs C, and Reynolds CA (1998). Domain swapping in G-protein coupled receptor dimers. *Protein Eng* 11:1181–1193.

128. Guo W, Shi L, Filizola M, Weinstein H, and Javitch JA (2005). From the cover: crosstalk in G protein-coupled receptors: changes at the transmembrane homodimer interface determine activation. *Proc Natl Acad Sci USA* 102:17495–17500.

129. Lee SP, O'Dowd BF, Rajaram RD, Nguyen T, and George SR (2003). D2 Dopamine receptor homodimerization is mediated by multiple sites of interaction, including an intermolecular interaction involving transmembrane domain 4. *Biochemistry* 42:11023–11031.

130. Liu XY, Chu XP, Mao LM, Wang M, Lan HX, Li MH, et al. (2006). Modulation of D2R–NR2B interactions in response to cocaine. *Neuron* 52:897–909.

131. Lobo MK, Karsten SL, Gray M, Geschwind DH, and Yang XW (2006). FACS-array profiling of striatal projection neuron subtypes in juvenile and adult mouse brains. *Nat Neurosci* 9:443–452.

132. Ng GY, Mouillac B, George SR, Caron M, Dennis M, Bouvier M, and O'Dowd BF (1994b). Desensitization, phosphorylation and palmitoylation of the human dopamine D1 receptor. *Eur J Pharmacol* 267: 7–19

133. Ng GY, O'Dowd BF, Caron M, Dennis M, Brann MR, and George SR (1994a). Phosphorylation and palmitoylation of the human D2L dopamine receptor in Sf9 cells. *J Neurochem* 63:1589–1595.

134. Shen W, Tian X, Day M, Ulrich S, Tkatch T, Nathanson NM, Surmeier DJ (2007). Cholinergic modulation of Kir2 channels selectively elevates dendritic excitability in striatopallidal neurons. *Nat Neurosci* 10:1458–1466.

135. Wu GY and Cline HT (1998). Stabilization of dendritic arbor structure in vivo by CaMKII. *Science* 279:222–226.

6 Functional consequences of chemokine receptor dimerization

Mario Mellado, Carlos Martínez-A., and José Miguel Rodríguez-Frade

INTRODUCTION

Chemokines are low molecular weight proinflammatory cytokines[1] that bind to members of the class A rhodopsin-like family of G protein-coupled receptors (GPCRs).[2] Over the last decade, the chemokines have been studied extensively as critical mediators of cell movement. Their role was initially linked to leukocyte traffic, but there is now considerable evidence that they also have roles in many other aspects of normal physiology and in pathological states. These proinflammatory cytokines participate in homeostatic processes and in inflammatory and autoimmune diseases such as rheumatoid arthritis, multiple sclerosis, psoriasis, nephritis, and atherosclerosis, as well as in cancer and HIV–1 infection.[3–5] Long before science realized the importance of this family of proinflammatory cytokines, viruses were using the chemokine system by expressing chemokines, receptors, binding proteins, and antagonists.[6]

Considerable effort has been made to establish the role of specific chemokine/chemokine receptor pairs in inflammation, autoimmunity, viral infection, and development. The use of KO and transgenic mice, siRNA, genomics, and proteomics techniques have helped define the role of these proteins, justifying the substantial investment by pharmaceutical companies in large-scale compound screening, drug development, and clinical testing. Despite these efforts, only two chemokine-related drugs have been approved for use in AIDS treatment and in stem cell mobilization, and none have yet reached patients with inflammatory or autoimmune disease.[7] This discrepancy between efforts and results has been attributed to the poor pharmacodynamic properties of the drugs tested, and to the promiscuity and redundancy between many chemokine ligands and receptors.[7]

Several factors should be taken into account. There are more than fifty chemokines and twenty receptors, divided into four main families.[8] Some of these chemokine/chemokine receptor pairs are highly specific, whereas others show considerable promiscuity. To function efficiently, chemokines secreted in response to signals, such as proinflammatory cytokines, are thought to interact with glycosaminoglycans (GAG) on the blood vessel endothelial surface.[9] These interactions are critical for in vivo function, although not essential when chemokine activity is evaluated in vitro.[10]

Another level of complexity in the study of the chemokine derives from their ability to aggregate in vitro and in vivo. GAG binding, which could increase ligand binding affinity, can also influence formation of ligand homo- and heterodimeric complexes.[2,10] The functional consequence of these complexes, which are also observed in solution in specific experimental conditions, might be important but remain to be clarified. The nature of chemokine interaction with its receptor is also debated. Although oligomerization of chemokine receptors is now generally accepted, reports describing their role in chemokine function are few and contradictory.[11]

Here we discuss chemokine-activated signaling pathways, chemokine oligomers and their importance in GAG and receptor interaction, chemokine receptor complexes, preferential partners, the role of ligands, interaction with and activation of G proteins, and crosstalk with receptors or other signaling pathways, as all these concepts must be reconsidered in analyzing chemokine function.

CHEMOKINES

The structure of many chemokines has been resolved and shows that, regardless of whether they have greater or lesser sequence similarity, these proteins adopt similar tertiary structures.[12] In addition, many chemokines can form dimers or higher order oligomers, in solution or after binding to GAG.[13,14] Several lines of evidence nonetheless indicate that in vitro, a chemokine interacts as a monomer with its receptor, because dimerization-deficient mutants of several chemokines are still able to promote cell migration. Mutant CXCL8,[15] CCL5,[10] CCL2,[16] and CCL4[17] interact with their respective receptors as efficiently as the native protein. The in vivo function of these mutant chemokines is nonetheless impaired, indicating that dimerization is necessary for chemokine function. Although it is not directly related with receptor binding,[10,18] chemokine oligomerization can facilitate the interactions with GAG that are needed for in vivo function.[10] Some reports indicate that chemokine oligomerization can influence the signaling pathways activated by monomeric chemokines, acting as agonist[19] or antagonist[20] for the corresponding receptor.

Chemokines do not only form homodimers, as heterodimerization between chemokines has also been shown. These complexes, which are affected by pH and buffer conditions,[13] form not only between chemokines of the same family (CC or CXC), but also between families. Although evidence is only indirect, a correlation has been established between heterodimer formation and specific function, including protection from proteolysis, cooperativity, and synergy.[21,22] The overall effect of a chemokine might thus depend on the dominant homo- or heterodimer formed in a given microenvironment. In addition, each ligand can trigger specific qualitative responses, that is, different anti-CCR5 monoclonal antibodies (mAb) stabilize distinct receptor conformations that activate specific responses.[23]

Our understanding of chemokine biology should extend to functional monomeric chemokines,[10] homo- and heterodimers between specific chemokine pairs,[13] as well as oligomers.[19] It is important to study the affinity of these complexes for GAG, to determine the receptor affinity and specificity for each complex, to analyze their pharmacokinetic properties, measure chemokine:receptor stoichiometry, and confirm complex-specific functions. These concepts will remain largely speculative until specific tools are developed that allow us to test the contribution of each type of complex to chemokine function.

CHEMOKINE RECEPTORS

Chemokine receptors were originally thought to act as monomeric entities in interactions with monomeric ligands. After chemokine binding, conformational changes in the receptor's extracellular regions propagate the information to the intracytoplasmic regions, allowing coupling with and activation of G proteins and other signaling molecules. The concept of "ligand/chemokine receptor promiscuity" is widely accepted; a given ligand can interact with several different receptors, with no major differences in the function elicited, and a receptor might mediate the same activity through interaction with distinct ligands.[24]

Chemokine receptors are members of the class A GPCR family. They are single polypeptide sequences of ~350 amino acids.[24,25] There are currently no data available on the crystal structure of these receptors. In comparison with the established 3D model for rhodopsin, bacteriorhodopsin, and β-adrenoceptor receptor,[26–28] chemokine receptors are composed of 7-transmembrane (TM) helical structures connected by intra- and extracellular loops. They are thought to be organized like a barrel, with an N-terminal domain and three extracellular loops that participate in ligand binding and three intracellular loops and a C-terminal end involved in coupling to signaling pathways.[2] Ligand binding induces conformational changes in the receptor that are transmitted to intracellular regions involved in coupling of signaling molecules. Among the intracellular domains, the DRYLAIV motif in the second intracellular loop has a key role in chemokine receptor function. DRY-mutant receptors are functionally deficient and show dominant negative effects on the wildtype (wt) counterpart.[29]

The chemokine receptors are classified in four families according to the ligands they bind; there is an additional group of nonfunctional decoy receptors (D6, DARC, CCX-CKR).[30,31] As their name indicates (GPCR), and with a few exceptions,[25] chemokine receptor signaling involves binding to and activation of G proteins. Classically considered monomeric entities, the chemokine receptor model initially fits a 1:1:1 stoichiometry model, with one receptor binding a single ligand to initiate a signal through coupling to a single heterotrimeric G protein. Data indicate that many GPCRs have functional properties that require direct or indirect interactions between clustered receptors; such interactions are needed for modulation of T cell responses[32] and activation of signaling molecules such as JAK/STAT, PI3K and MAPK.[32,33]

Methods to determine receptor dimerization

Most authors consider GPCR dimerization a universal phenomenon with a functionally important role, although data that confirm its importance in living organisms are still extremely limited.[34] A large body of evidence indicates dimerization among identical receptors, close GPCR family members, or GPCR of distinct families.[35] As the effects of dimerization on ligand binding, receptor activation, signaling, desensitization, and receptor trafficking are still being documented, models that describe GPCR function require constant revision.[36]

The first evidence for chemokine receptor dimerization came from biochemical studies such as immunoprecipitation and Western blotting, with the aid of specific antibodies, tagged receptors, and crosslinking agents.[29,37,38] These methods require cell lysis and solubilization and therefore cannot be used to study interactions in living cells. Certain buffers and detergents used in these processes might cause receptor aggregation as a result of incomplete solubilization of membrane or might disrupt existing interactions. When protein- or peptide-tagged receptors are used, the experimental conditions for transfection and overexpression might induce interactions that do not reflect the physiological environment. These initial studies were thus criticized, because the limitations of the techniques could give rise to complexes due to receptor aggregation, rather than reflect a true interaction between receptors. This criticism was not trivial, because chemokine receptors aggregate at the leading edge of migrating cells[39] and internalize in clathrin-coated vesicles.[40,41] Our lab and others established that chemokine receptor dimerization was a ligand-induced process, as in our experimental conditions bands corresponding to dimers and high molecular weight complexes, were detected only after ligand addition.[29,37,38] Later use of more sensitive techniques, such as those based on fluorescence resonance energy transfer (FRET), has shown that receptor dimers are preformed in the absence of ligands.[23,42,43] It is thus possible that ligands stabilize these conformations in a way that resists detergent lysis.

The principal advantage of biochemical techniques is that they can be used with primary cells and cell lines when specific antibodies are available, but they lack specificity and are highly dependent on assay conditions. These methods were used to show the existence of CCR2,[29] CCR5,[23,44] CXCR2,[43] and CXCR4[37,45] homodimers, as well as of heterodimers between chemokine receptors[46] and even between chemokine receptors and GPCR of other families.[47,48] These studies showed the potential interaction between receptors but did not add much information about the functional relevance of such complexes; although they represent a starting point in the analysis of chemokine receptor oligomerization, additional methods are needed to establish the existence and functional relevance of such complexes.

Data from functional assays do not formally demonstrate physical interactions between chemokine receptors but support the existence of receptor complexes. One of the initial experiments to demonstrate the in vitro relevance of chemokine receptor dimerization used a nonfunctional CCR2 mutant that acts as dominant negative receptor for wt CCR2.[29] Chemokine receptor dimers explain negative binding cooperativity between CCR2 and CCR5[49] or CCR2 and CXCR4, specific signaling pathway activation by CCR2/CCR5 dimers,[38] intracellular retention of CCR5 in CCR5Δ32 heterozygous individuals,[50] or delayed HIV-1 progression in CCR2V64I individuals.[51] These assays also suggest that targeting of chemokine receptor dimers can efficiently modify chemokine receptor function. Peptides that block dimerization of CCR5[52] or CXCR4[53] have been used successfully to control lymphocyte and tumor cell migration, both in vitro and in vivo.

RET is a new biophysical technique that allow the study of interactions between an acceptor and a donor protein pair within a distance of <10 nm. There are two main types of RET depending on the nature of the donor protein, luminescent (bioluminescent RET, BRET) or fluorescent (FRET).[54,55] In addition to the numerous and painstaking controls required for these techniques, there are several limitations. These methods use heterologous expression systems with cloned and appropriately tagged receptors; a control of cell location and ligand-mediated function must therefore be performed prior to RET analysis. Expression of the receptors tested should be within the physiological range and in a 1:1 acceptor:donor ratio. Overexpression of fluorescently-labeled receptors could easily give rise to nonspecific, random interactions. Furthermore, absence of energy transfer does not confirm lack of interaction between two proteins, as it could also indicate that the distance between fluorochromes or their conformation does not allow energy transfer, even if the receptors of interest interact. These data are of particular interest when analyzing the role of ligands, because changes in RET might reflect conformational changes in the receptor rather than disruption of interacting proteins. Their main advantage is that these techniques allow determination not only of receptor oligomerization, but also of receptor dynamics, the role of the ligand or of receptor levels, and definition of the cell sites in which oligomerization takes place.[56,57]

Several methods are in wide use to analyze chemokine receptor interactions.[56,57] BRET and FRET techniques have been used successfully to determine homooligomers between CCR2,[23] CCR5[42,58] CXCR2,[43,59] and CXCR4[60] and to analyze heterooligomers between CXCR1-CXCR2,[59] CCR2-CCR5,[49,58] CCR2-CXCR4,[58,60] CCR5-CXCR4,[32,61] and CXCR4-CXCR7.[62,63]

There is no perfect method to study chemokine receptor dimerization, and data should be reviewed in the near future with more accurate techniques. Nonetheless, there is now sufficient evidence that oligomers form during chemokine receptor synthesis and maturation, exist at the cell surface, and function as dimers or higher order complexes. The nature of the chemokine receptor dimer interface, the existence of allosteric interactions, and the role of the ligand in promoting, disrupting, and stabilizing dimers and in regulating receptor interaction with G proteins are some of the critical points to be analyzed for a better understanding of chemokine biology.

Dynamics of chemokine receptor dimers

Chemokine receptors were originally thought to act entirely through G_i-mediated processes.[25] It is now known that chemokine signaling is more versatile and that they can also activate pertussis toxin-insensitive G proteins[64] or signal through G protein-independent pathways.[65–67] Signaling elicited by chemokines depends on the presence of appropriate receptors on the target cell surface and their distribution in specialized domains. Disruption of membrane lipid rafts can affect responses to a chemokine, because cholesterol depletion inhibits CXCL12-induced Ca^{2+} mobilization, chemotaxis, and cell polarization,[68] as well as CCR5 functions such as integrin activation or PI3K redistribution.[69,70] CCR5 oligomerization is nonetheless unaffected by methyl-β-cyclodextrin, a cholesterol chelator that disrupts membrane rafts.[23] Differences in receptor expression,[71] changes in G protein coupling,[38] and cross-desensitization and regulation of the functional properties of other GPCRs[72,73] are indications of the diversity of chemokine responses that can be achieved through receptor oligomerization (see Table 6.1).

While some authors prefer the concept that GPCRs are born, function, and die as oligomers, others suggest that GPCR dimers might be transient or form only in response to agonist binding, and even that GPCRs can function as monomers.[74] There is strong evidence supporting both views,[75,76] which could simply reflect the structural and functional diversity of the GPCR family. The strict ligand-dependent view of chemokine receptor dimerization, which promoted a model with monomeric receptors at the cell surface,[29,37] probably reflected the limitations of available techniques. Although newer data indicate that chemokine receptors associate even in the absence of ligand,[43,52,60] the role of ligands is still hotly debated. Some reports claim that homo- and heterodimers are unaffected by ligand binding,[59] whereas others indicate an active role for the ligand in stabilizing the most appropriate conformation; CXCL12 induces changes in its receptor (CXCR4), and synthetic peptides of CXCR4 TM domains inhibit receptor activation by blocking ligand-induced transition to the dimeric state.[60]

The role of chemokine dimers in facilitating receptor dimerization, if any, remains to be determined. The fact that chemokine dimerization is driven or at least facilitated by interaction with GAG makes it difficult to interpret in vitro data obtained with chemokines in suspension. Nevertheless, the in vivo relevance of chemokine dimers was recently established, because disruption of

Table 6.1. Summary of Chemokine Receptor Dimers

Homodimers	
CCR2	
CCR5	
CXCR1	
CXCR2	
CXCR4	
CXCR7	
DARC	

Heterodimers	Effect
CCR2/CCR5	Blocks HIV-1 entry; Specific function; Negative binding cooperativity[38,49,58]
CCR5/CXCR4	Allosteric transinhibition; Blocks HIV-1 entry[58,60]
CCR5/CXCR4	Blocks HIV-1 entry; Altered endocytosis; Specific signaling[32,44,58]
CXCR1/CXCR2	None
CXCR4/CXCR7	Regulation of signaling; Synergy; Integrin activation [62,63]
DARC/CCR5	Trans-inhibition[101]

Dimers with other GPCR	Effect
CCR5/opioid receptors	Desensitization; Reduce susceptibility to HIV-1 entry[48,73]
CXCR4/opioid receptors	Suppression of signaling; Desensitization[47,82]
CXCR2/GluR1	Impaired migration; Altered ligand binding[72]

CXCL4/CCL5 interactions attenuates monocyte recruitment to atheromatous plaques and reduces atherosclerosis.[77] CCR5/CXCR4 heterodimers are recruited to the immunological synapse, where they trigger activation of G_q rather than of G_i.[32] FRET and BRET studies showed the existence of CXCR4/CCR7 heterodimers, which correlates with modulation of CXCL12-mediated signaling in cells that coexpress these receptors.[62,63]

Therapeutic opportunities of chemokine receptor dimerization

GPCRs are major pharmacological targets, and their existence as oligomers increases the number of potential phenotypes; this should have important implications in the development and design of new therapeutic agents. For some GPCRs, dimer formation is clearly linked to specific functions. Appropriate membrane targeting,[78,79] alteration of pharmacological profiles,[80,81] and changes in desensitization/internalization rates[47,82] have been shown for several GPCRs, including chemokine receptors.

Although the functional relevance of chemokine receptor homo- and heterodimer complexes is being analyzed in detail, the existence of chemokine and chemokine receptor oligomers suggests considerable complexity in the regulation of functions associated with this proinflammatory cytokine family. From a therapeutic point of view, several targets are thought to modify chemokine function and have been analyzed in in vitro and in vivo animal models of disease or in clinical trials. HIV-1 infection, tumor metastasis, asthma, rheumatoid arthritis, multiple sclerosis, psoriasis, and bone marrow transplant[83] are among the conditions for which chemokine/chemokine receptor targeting is being explored for therapeutic use.

Strategies include the use of N-terminal modified chemokines such as AOP or Met-RANTES,[84,85] small molecules,[86–89] antibodies,[90,91] pan-chemokine inhibitors,[92,93] and viral chemokines/binding proteins,[6,94] all of which alter chemokine binding to the receptor. Targeting the chemokine-GAG interaction was attempted using heparin-based therapeutics[95] and truncated chemokines or chemokine fragments that interact with GAG and block their chemokine-binding domains.[95,96] Although an effect on chemokine receptor dimerization could contribute to the effects of these drugs,[41] dimerization has not yet been considered an appropriate target for modifying chemokine function.

A synthetic peptide that interferes with CXCL4/CCL5 dimerization was shown to attenuate monocyte recruitment in vivo,[77] indicating the feasibility of targeting dimerization interfaces for therapeutic use. CXCR4/CCR7 heterodimers underlie the modulation of CXCR4 signaling observed in cells cotransfected with both receptors.[63] Cell surface receptor expression has been blocked using modified chemokines that induce receptor internalization and inhibit recycling[84] and by the use of intracellular chemokines that retain receptors in the endoplasmic reticulum.[97,98] Interference with receptor dimerization[52] will probably emerge as a promising target for altering chemokine function. Synthetic peptides of CCR5 TM domains specifically block CCL5-mediated functions[52] and the use of synthetic peptides that block CXCR4 dimerization controls malignant cell migration.[53] More information is needed to determine whether homo- and heterodimers trigger specific signaling cascades that justify their stabilization or inhibition of a specific step in their associated function.

CONCLUDING REMARKS

Regulation of chemokine function involves several mechanisms, including interaction with GAG, proteolysis, natural antagonism, chemokine clearance, chemokine receptor oligomerization, receptor internalization and recycling, receptor signaling convergence, and chemokine receptor uncoupling. Different chemokines that interact in vitro with the same receptor could induce distinct functions, because in vivo, only a single chemokine might be recruited to cell surfaces and tissues by GAG; this would limit receptor interaction and reduce the apparent redundancy of the chemokine system. It seems clear that chemokine receptors can assume a variety of conformations rather than acting as simple on/off switches; each potentially triggers specific effector functions, which would allow selective targeting. In addition, chemokine function must be considered in the context of the signaling networks that link chemokine receptors with other cell mediators, such as growth factors and cytokines.[99,100] Identification of specific receptor sets in different physiological and pathological situations and of the functional consequences of their interaction could form the basis for novel, more effective chemokine therapies.

Although our knowledge of chemokine receptor function has increased greatly in recent years, the development of therapeutically effective chemokine-based drugs in humans has been relatively unsuccessful. Targeting chemokine receptors must consider a number of emerging questions: How is the chemokine receptor system organized? Do these receptors form signaling complexes, and what is their composition? Are there functional microdomains? What is the contribution of the membrane environment? What happens at the physiological level of receptors? Understanding all these concepts will change our view of chemokine/chemokine receptor biology and have considerable influence on drug development and screening programs.

REFERENCES

1. Baggiolini, M. Chemokines and leukocyte traffic. *Nature* 392, 565–568 (1998).
2. Allen, S.J., Crown, S.E. & Handel, T.M. Chemokine: receptor structure, interactions, and antagonism. *Annu Rev Immunol* 25, 787–820 (2007).
3. Charo, I.F. & Ransohoff, R.M. The many roles of chemokines and chemokine receptors in inflammation. *N Engl J Med* 354, 610–621 (2006).
4. Gerard, C. & Rollins, B.J. Chemokines and disease. *Nat Immunol* 2, 108–115 (2001).
5. Viola, A. & Luster, A.D. Chemokines and their receptors: drug targets in immunity and inflammation. *Annu Rev Pharmacol Toxicol* 48, 171–197 (2008).
6. Alcami, A. Viral mimicry of cytokines, chemokines and their receptors. *Nat Rev Immunol* 3, 36–50 (2003).
7. Horuk, R. Chemokine receptor antagonists: overcoming developmental hurdles. *Nat Rev Drug Discov* 8, 23–33 (2009).
8. Zlotnik, A. & Yoshie, O. Chemokines: a new classification system and their role in immunity. *Immunity* 12, 121–127 (2000).
9. Witt, D.P. & Lander, A.D. Differential binding of chemokines to glycosaminoglycan subpopulations. *Curr Biol* 4, 394–400 (1994).
10. Proudfoot, A.E. *et al.* Glycosaminoglycan binding and oligomerization are essential for the in vivo activity of certain chemokines. *Proc Natl Acad Sci U S A* 100, 1885–1890 (2003).
11. Mellado, M., Serrano, A., Martinez, C. & Rodriguez-Frade, J.M. G protein-coupled receptor dimerization and signaling. *Methods Mol Biol* 332, 141–157 (2006).
12. Clore, G.M. & Gronenborn, A.M. Three-dimensional structures of alpha and beta chemokines. *FASEB J* 9, 57–62 (1995).
13. Nesmelova, I.V., Sham, Y., Gao, J. & Mayo, K.H. CXC and CC chemokines form mixed heterodimers: association free energies from molecular dynamics simulations and experimental correlations. *J Biol Chem* 283, 24155–24166 (2008).
14. Clark-Lewis, I. *et al.* Structure-activity relationships of chemokines. *J Leukoc Biol* 57, 703–711 (1995).
15. Rajarathnam, K. *et al.* Neutrophil activation by monomeric interleukin-8. *Science* 264, 90–92 (1994).
16. Paavola, C.D. *et al.* Monomeric monocyte chemoattractant protein-1 (MCP-1) binds and activates the MCP-1 receptor CCR2B. *J Biol Chem* 273, 33157–33165 (1998).

17. Laurence, J.S., Blanpain, C., Burgner, J.W., Parmentier, M. & LiWang, P.J. CC chemokine MIP-1 beta can function as a monomer and depends on Phe13 for receptor binding. *Biochemistry* 39, 3401–3409 (2000).

18. Campanella, G.S. *et al.* Oligomerization of CXCL10 is necessary for endothelial cell presentation and in vivo activity. *J Immunol* 177, 6991–6998 (2006).

19. Czaplewski, L.G. *et al.* Identification of amino acid residues critical for aggregation of human CC chemokines macrophage inflammatory protein (MIP)-1alpha, MIP-1beta, and RANTES. Characterization of active disaggregated chemokine variants. *J Biol Chem* 274, 16077–16084 (1999).

20. Veldkamp, C.T. *et al.* Structural basis of CXCR4 sulfotyrosine recognition by the chemokine SDF-1/CXCL12. *Sci Signal* 1, ra4 (2008).

21. Nesmelova, I.V. *et al.* Platelet factor 4 and interleukin-8 CXC chemokine heterodimer formation modulates function at the quaternary structural level. *J Biol Chem* 280, 4948–4958 (2005).

22. Paoletti, S. *et al.* A rich chemokine environment strongly enhances leukocyte migration and activities. *Blood* 105, 3405–3412 (2005).

23. Issafras, H. *et al.* Constitutive agonist-independent CCR5 oligomerization and antibody-mediated clustering occurring at physiological levels of receptors. *J Biol Chem* 277, 34666–34673 (2002).

24. Graham, G.J. and Nibbs, R. J. Chemokine receptors: A structural overview, in *The chemokine receptors*. (ed. J.K. Harrison and N.W. Lukacs) 31–54 (Humana, Totowa, N.J.; 2007).

25. Bokoch, G.M. Chemoattractant signaling and leukocyte activation. *Blood* 86, 1649–1660 (1995).

26. Palczewski, K. *et al.* Crystal structure of rhodopsin: A G protein-coupled receptor. *Science* 289, 739–745 (2000).

27. Pebay-Peyroula, E., Rummel, G., Rosenbusch, J.P. & Landau, E.M. X-ray structure of bacteriorhodopsin at 2.5 angstroms from microcrystals grown in lipidic cubic phases. *Science* 277, 1676–1681 (1997).

28. Rasmussen, S.G. *et al.* Crystal structure of the human beta2 adrenergic G-protein-coupled receptor. *Nature* 450, 383–387 (2007).

29. Rodriguez-Frade, J.M. *et al.* The chemokine monocyte chemoattractant protein-1 induces functional responses through dimerization of its receptor CCR2. *Proc Natl Acad Sci U S A* 96, 3628–3633 (1999).

30. Zlotnik, A., Yoshie, O. & Nomiyama, H. The chemokine and chemokine receptor superfamilies and their molecular evolution. *Genome Biol* 7, 243 (2006).

31. Comerford, I. & Nibbs, R.J. Post-translational control of chemokines: a role for decoy receptors? *Immunol Lett* 96, 163–174 (2005).

32. Contento, R.L. *et al.* CXCR4-CCR5: a couple modulating T cell functions. *Proc Natl Acad Sci USA* 105, 10101–10106 (2008).

33. Wong, M.M. & Fish, E.N. Chemokines: attractive mediators of the immune response. *Semin Immunol* 15, 5–14 (2003).

34. Milligan, G. G protein-coupled receptor dimerization: function and ligand pharmacology. *Mol Pharmacol* 66, 1–7 (2004).

35. Barnes, P.J. Receptor heterodimerization: a new level of cross-talk. *J Clin Invest* 116, 1210–1212 (2006).

36. Terrillon, S. & Bouvier, M. Roles of G-protein-coupled receptor dimerization. *EMBO Rep* 5, 30–34 (2004).

37. Vila-Coro, A.J. *et al.* The chemokine SDF-1alpha triggers CXCR4 receptor dimerization and activates the JAK/STAT pathway. *Faseb J* 13, 1699–1710 (1999).

38. Mellado, M. *et al.* Chemokine receptor homo- or heterodimerization activates distinct signaling pathways. *Embo J* 20, 2497–2507 (2001).

39. Nieto, M. *et al.* Polarization of chemokine receptors to the leading edge during lymphocyte chemotaxis. *J Exp Med* 186, 153–158 (1997).

40. Neel, N.F., Schutyser, E., Sai, J., Fan, G.H. & Richmond, A. Chemokine receptor internalization and intracellular trafficking. *Cytokine Growth Factor Rev* 16, 637–658 (2005).

41. Vila-Coro, A.J., Mellado, M., Martin de Ana, A., Martinez, A.C. & Rodriguez-Frade, J.M. Characterization of RANTES- and aminooxypentane-RANTES-triggered desensitization signals reveals differences in recruitment of the G protein-coupled receptor complex. *J Immunol* 163, 3037–3044 (1999).

42. Babcock, G.J., Farzan, M. & Sodroski, J. Ligand-independent dimerization of CXCR4, a principal HIV-1 coreceptor. *J Biol Chem* 278, 3378–3385 (2003).

43. Trettel, F. *et al*. Ligand-independent CXCR2 dimerization. *J Biol Chem* 278, 40980–40988 (2003).

44. Vila-Coro, A.J. *et al*. HIV-1 infection through the CCR5 receptor is blocked by receptor dimerization. *Proc Natl Acad Sci USA* 97, 3388–3393 (2000).

45. Toth, P.T., Ren, D. & Miller, R.J. Regulation of CXCR4 receptor dimerization by the chemokine SDF-1alpha and the HIV-1 coat protein gp120: a fluorescence resonance energy transfer (FRET) study. *J Pharmacol Exp Ther* 310, 8–17 (2004).

46. Rodriguez-Frade, J.M., Mellado, M. & Martinez-A., C. Chemokine receptor dimerization: two are better than one. *Trends Immunol* 22, 612–617 (2001).

47. Pello, O.M. *et al*. Ligand stabilization of CXCR4/delta-opioid receptor heterodimers reveals a mechanism for immune response regulation. *Eur J Immunol* 38, 537–549 (2008).

48. Szabo, I. *et al*. Selective inactivation of CCR5 and decreased infectivity of R5 HIV-1 strains mediated by opioid-induced heterologous desensitization. *J Leukoc Biol* 74, 1074–1082 (2003).

49. El-Asmar, L. *et al*. Evidence for negative binding cooperativity within CCR5-CCR2b heterodimers. *Mol Pharmacol* 67, 460–469 (2005).

50. Benkirane, M., Jin, D.Y., Chun, R.F., Koup, R.A. & Jeang, K.T. Mechanism of transdominant inhibition of CCR5-mediated HIV-1 infection by ccr5delta32. *J Biol Chem* 272, 30603–30606 (1997).

51. Lee, B. *et al*. Influence of the CCR2-V64I polymorphism on human immunodeficiency virus type 1 coreceptor activity and on chemokine receptor function of CCR2b, CCR3, CCR5, and CXCR4. *J Virol* 72, 7450–7458 (1998).

52. Hernanz-Falcon, P. *et al*. Identification of amino acid residues crucial for chemokine receptor dimerization. *Nat Immunol* 5, 216–223 (2004).

53. Wang, J., He, L., Combs, C.A., Roderiquez, G. & Norcross, M.A. Dimerization of CXCR4 in living malignant cells: control of cell migration by a synthetic peptide that reduces homologous CXCR4 interactions. *Mol Cancer Ther* 5, 2474–2483 (2006).

54. Jares-Erijman, E.A. & Jovin, T.M. FRET imaging. *Nat Biotechnol* 21, 1387–1395 (2003).

55. Pfleger, K.D.G., Seeber, R.M.a. & Eidne, K.A. Bioluminescence resonance energy transfer (BRET) for the real-time detection of protein-protein interactions. *Nature protocols* 1, 337–345 (2006).

56. Appelbe, S. & Milligan, G. Chapter 10. Hetero-oligomerization of chemokine receptors. *Methods Enzymol* 461, 207–225 (2009).

57. Rodriguez-Frade, J., Munoz, L.M. & Mellado, M. Chapter 5. Multiple approaches to the study of chemokine receptor homo- and heterodimerization. *Methods Enzymol* 461, 105–122 (2009).

58. Rodriguez-Frade, J.M. *et al*. Blocking HIV-1 infection via CCR5 and CXCR4 receptors by acting in trans on the CCR2 chemokine receptor. *Embo J* 23, 66–76 (2004).

59. Wilson, S., Wilkinson, G. & Milligan, G. The CXCR1 and CXCR2 receptors form constitutive homo- and heterodimers selectively and with equal apparent affinities. *J Biol Chem* 280, 28663–28674 (2005).

60. Percherancier, Y. *et al*. Bioluminescence resonance energy transfer reveals ligand-induced conformational changes in CXCR4 homo- and heterodimers. *J Biol Chem* 280, 9895–9903 (2005).

61. Guan, E., Wang, J. & Norcross, M.A. Amino-terminal processing of MIP-1beta/CCL4 by CD26/dipeptidyl-peptidase IV. *J Cell Biochem* 92, 53–64 (2004).

62. Sierro, F. *et al.* Disrupted cardiac development but normal hematopoiesis in mice deficient in the second CXCL12/SDF-1 receptor, CXCR7. *Proc Natl Acad Sci USA* 104, 14759–14764 (2007).

63. Levoye, A., Balabanian, K., Baleux, F., Bachelerie, F. & Lagane, B. CXCR7 heterodimerizes with CXCR4 and regulates CXCL12-mediated G protein signaling. *Blood* 113, 6085–6093 (2009).

64. Soede, R.D., Zeelenberg, I.S., Wijnands, Y.M., Kamp, M. & Roos, E. Stromal cell-derived factor-1-induced LFA-1 activation during in vivo migration of T cell hybridoma cells requires Gq/11, RhoA, and myosin, as well as Gi and Cdc42. *J Immunol* 166, 4293–4301 (2001).

65. Mellado, M. *et al.* The chemokine monocyte chemotactic protein 1 triggers Janus kinase 2 activation and tyrosine phosphorylation of the CCR2B receptor. *J Immunol* 161, 805–813 (1998).

66. Vlahakis, S.R. *et al.* G protein-coupled chemokine receptors induce both survival and apoptotic signaling pathways. *J Immunol* 169, 5546–5554 (2002).

67. Prado, G.N. *et al.* Chemokine signaling specificity: essential role for the N-terminal domain of chemokine receptors. *Biochemistry* 46, 8961–8968 (2007).

68. Nguyen, D.H. & Taub, D. CXCR4 function requires membrane cholesterol: implications for HIV infection. *J Immunol* 168, 4121–4126 (2002).

69. Nguyen, D.H. & Taub, D. Cholesterol is essential for macrophage inflammatory protein 1 beta binding and conformational integrity of CC chemokine receptor 5. *Blood* 99, 4298–4306 (2002).

70. Mañes, S., Lacalle, R.A., Gómez-Mouton, C. and Martínez-A., C. From rafts to crafts: membrane asymmetry in living cells. *Trends Immunol* 24, 319–325 (2003).

71. Chelli, M. & Alizon, M. Determinants of the trans-dominant negative effect of truncated forms of the CCR5 chemokine receptor. *J Biol Chem* 276, 46975–46982 (2001).

72. Limatola, C. *et al.* Expression of AMPA-type glutamate receptors in HEK cells and cerebellar granule neurons impairs CXCL2-mediated chemotaxis. *J Neuroimmunol* 134, 61–71 (2003).

73. Szabo, I. *et al.* Heterologous desensitization of opioid receptors by chemokines inhibits chemotaxis and enhances the perception of pain. *Proc Natl Acad Sci USA* 99, 10276–10281 (2002).

74. Ernst, O.P., Gramse, V., Kolbe, M., Hofmann, K.P. & Heck, M. Monomeric G protein-coupled receptor rhodopsin in solution activates its G protein transducin at the diffusion limit. *Proc Natl Acad Sci USA* 104, 10859–10864 (2007).

75. Kroeger, K.M., Hanyaloglu, A.C., Seeber, R.M., Miles, L.E. & Eidne, K.A. Constitutive and agonist-dependent homo-oligomerization of the thyrotropin-releasing hormone receptor. Detection in living cells using bioluminescence resonance energy transfer. *J Biol Chem* 276, 12736–12743 (2001).

76. Ayoub, M.A. *et al.* Monitoring of ligand-independent dimerization and ligand-induced conformational changes of melatonin receptors in living cells by bioluminescence resonance energy transfer. *J Biol Chem* 277, 21522–21528 (2002).

77. Koenen, R.R. *et al.* Disrupting functional interactions between platelet chemokines inhibits atherosclerosis in hyperlipidemic mice. *Nat Med* 15, 97–103 (2009).

78. Minneman, K.P. Heterodimerization and surface localization of G protein coupled receptors. *Biochem Pharmacol* 73, 1043–1050 (2007).

79. Salahpour, A. *et al.* Homodimerization of the beta2-adrenergic receptor as a prerequisite for cell surface targeting. *J Biol Chem* 279, 33390–33397 (2004).

80. Springael, J.Y . *et al.* Allosteric modulation of binding properties between units of chemokine receptor homo- and hetero-oligomers. *Mol Pharmacol* 69, 1652–1661 (2006).

81. Franco, R. *et al.* G-protein-coupled receptor heteromers: function and ligand pharmacology. *Br J Pharmacol* 153 Suppl 1, S90–98 (2008).

82. Finley, M.J. *et al.* Bi-directional heterologous desensitization between the major HIV-1 co-receptor CXCR4 and the kappa-opioid receptor. *J Neuroimmunol* 197, 114–123 (2008).

83. Wells, T.N., Power, C.A., Shaw, J.P. & Proudfoot, A.E. Chemokine blockers – therapeutics in the making? *Trends Pharmacol Sci* 27, 41–47 (2006).

84. Elsner, J. *et al.* Differential activation of CC chemokine receptors by AOP-RANTES. *J Biol Chem* 275, 7787–7794 (2000).

85. Elsner, J. *et al.* The CC chemokine antagonist Met-RANTES inhibits eosinophil effector functions through the chemokine receptors CCR1 and CCR3. *Eur J Immunol* 27, 2892–2898 (1997).

86. Liang, Z. *et al.* Inhibition of breast cancer metastasis by selective synthetic polypeptide against CXCR4. *Cancer Res* 64, 4302–4308 (2004).

87. Dragic, T. *et al.* A binding pocket for a small molecule inhibitor of HIV-1 entry within the transmembrane helices of CCR5. *Proc Natl Acad Sci USA* 97, 5639–5644 (2000).

88. Muller, A. *et al.* Involvement of chemokine receptors in breast cancer metastasis. *Nature* 410, 50–56 (2001).

89. Hendrix, C.W. *et al.* Pharmacokinetics and safety of AMD-3100, a novel antagonist of the CXCR-4 chemokine receptor, in human volunteers. *Antimicrob Agents Chemother* 44, 1667–1673 (2000).

90. Huang, S. *et al.* Fully humanized neutralizing antibodies to interleukin-8 (ABX-IL8) inhibit angiogenesis, tumor growth, and metastasis of human melanoma. *Am J Pathol* 161, 125–134 (2002).

91. Mellado, M., Martin de Ana, A., Gomez, L., Martinez, C. & Rodriguez-Frade, J.M. Chemokine receptor 2 blockade prevents asthma in a cynomolgus monkey model. *J Pharmacol Exp Ther* 324, 769–775 (2008).

92. Deruaz, M. *et al.* Ticks produce highly selective chemokine binding proteins with antiinflammatory activity. *J Exp Med* 205, 2019–2031 (2008).

93. Reckless, J. & Grainger, D.J. Identification of oligopeptide sequences which inhibit migration induced by a wide range of chemokines. *Biochem J* 340 (Pt 3), 803–811 (1999).

94. Kledal, T.N. *et al.* A broad-spectrum chemokine antagonist encoded by Kaposi's sarcoma-associated herpesvirus. *Science* 277, 1656–1659 (1997).

95. Johnson, Z., Proudfoot, A.E. & Handel, T.M. Interaction of chemokines and glycosaminoglycans: a new twist in the regulation of chemokine function with opportunities for therapeutic intervention. *Cytokine Growth Factor Rev* 16, 625–636 (2005).

96. Johnson, Z. *et al.* Interference with heparin binding and oligomerization creates a novel anti-inflammatory strategy targeting the chemokine system. *J Immunol* 173, 5776–5785 (2004).

97. Chen, J.D., Bai, X., Yang, A.G., Cong, Y. & Chen, S.Y. Inactivation of HIV-1 chemokine co-receptor CXCR-4 by a novel intrakine strategy. *Nat Med* 3, 1110–1116 (1997).

98. Yang, A.G., Zhang, X., Torti, F. & Chen, S.Y. Anti-HIV type 1 activity of wild-type and functional defective RANTES intrakine in primary human lymphocytes. *Hum Gene Ther* 9, 2005–2018 (1998).

99. Pello, O.M. *et al.* SOCS up-regulation mobilizes autologous stem cells through CXCR4 blockade. *Blood* 108, 3928–3937 (2006).

100. Soriano, S.F. *et al.* Functional inactivation of CXC chemokine receptor 4-mediated responses through SOCS3 up-regulation. *J Exp Med* 196, 311–321 (2002).

101. Chakera, A., Seeber, R.M., John, A.Ee, Eidne, K.A. & Greaves, D.R. The duffy antigen/receptor for chemokines exists in an oligomeric form in living cells and functionally antagonizes CCR5 signaling through hetero-oligomerization. *Mol Pharmacol.* 73:1362–1370 (2008).

7 G protein functions identified using genetic mouse models

Stefan Offermanns

INTRODUCTION

Heterotrimeric G proteins couple heptahelical receptors to various effectors, and these three components constitute a transmembrane signaling system that operates in all cells of the mammalian organism to regulate many physiological and pathological processes.[1] While G proteins consist of an α-, β-, and γ-subunit, of which the β- and γ-subunit form an undissociable complex, the main properties of individual G proteins are determined by the identity of their α-subunits. More than twenty G protein α-subunits have been described in the mammalian system, and they can be divided into four subfamilies based on structural and functional homologies.

The G proteins of the G_i/G_o family are widely expressed and especially the α-subunits of G_{i1}, G_{i2}, and G_{i3} have been shown to mediate receptor-dependent inhibition of various types of adenylyl cyclases.[2] Since the expression levels of G_i and G_o are relatively high, their receptor-dependent activation results in the release of relatively high amounts of $\beta\gamma$-complexes. Activation of G_i/G_o is therefore believed to be the major coupling mechanism that results in the activation of $\beta\gamma$-mediated signaling processes.[3] The structural similarity between the three $G\alpha_i$-subforms suggests that they may have partially overlapping functions. In contrast to other G proteins the effects of G_o, which is particularly abundant in the nervous system, appears to be primarily mediated by its $\beta\gamma$-complex. A less widely expressed member of the $G\alpha_i/G\alpha_o$ family is $G\alpha_z$,[4] which, in contrast to G_i and G_o, is not a substrate for pertussis toxin. Several α-subunits like gustducin and transducins belong to the $G\alpha_i/G\alpha_o$ family and are involved in specific sensory functions.[5,6]

The G_q/G_{11} family of G proteins couples receptors to β-isoforms of phospholipase C.[7] The α-subunits of G_q and G_{11} are almost ubiquitously expressed, whereas the other members of this family, like $G\alpha_{14}$ and $G\alpha_{15/16}$ ($G\alpha_{15}$ being the murine, $G\alpha_{16}$ the human orthologue), show a rather restricted expression pattern. Receptors that are able to couple to the G_q/G_{11} family do not appear to discriminate between G_q and G_{11}.

The G proteins G_{12} and G_{13}, which are often activated by receptors coupling to G_q/G_{11}, constitute the G_{12}/G_{13} family and are expressed ubiquitously.[8–10] The analysis of cellular signaling processes regulated through G_{12} and G_{13} has been difficult since there are no specific inhibitors of these G proteins available. In addition, G_{12}/G_{13}-coupled receptors usually activate other G proteins. An important cellular function of G_{12}/G_{13} is their ability to regulate the actomyosin-based cellular contractility by increasing the activity of the small GTPase RhoA. Activation of RhoA by $G\alpha_{12}$ and $G\alpha_{13}$ is mediated by a subgroup of guanine nucleotide exchange factors (GEFs) for Rho, which include p115-RhoGEF, PDZ-RhoGEF and LARG.[11]

The ubiquitously expressed G protein G_s couples many receptors to adenylyl cyclase and mediates receptor-dependent adenylyl cyclase activation resulting in increases in the intracellular cAMP concentration. The α-subunit of G_s, $G\alpha_s$, is encoded by *GNAS*, a complex imprinted gene that gives rise to several gene products due to the presence of various promoters and splice variants. In addition to $G\alpha_s$, another transcripts encoding $XL\alpha_s$ is generated by promoters upstream of the $G\alpha_s$ promoter. $XL\alpha_s$ is structurally identical to $G\alpha_s$ but has an extra long amino-terminal extension encoded by a specific first exon and shares with $G\alpha_s$ the ability to bind to $\beta\gamma$-subunits and to mediate receptor-dependent stimulation of cAMP production.[12,13]

To elucidate the role of G protein-mediated signaling processes in the intact mammalian organism, almost all known genes encoding G protein α-subunits have been inactivated by gene targeting in mice (Table 7.1). This chapter summarizes the main phenotypical changes observed in mice lacking G protein α-subunits.

DEVELOPMENT

In various $G\alpha$-deficient mouse models, defects – in particular developmental processes – have been observed. For example, genetic ablation of the genes encoding $G\alpha_{12}$ and $G\alpha_{13}$ in mice has shown that mice lacking $G\alpha_{13}$ die at embryonic day 9.5 due to a defect in angiogenesis.[14] Whereas $G\alpha_{12}$-deficient mice are normal, $G\alpha_{12}/G\alpha_{13}$ double-deficient mice die at embryonic day 8.5. Thus both G proteins play critical roles during development, but their functions are not fully redundant. The angiogenic defect in constitutive $G\alpha_{13}$-deficient mice is due to the absence

Table 7.1. Phenotypical Changes in Mice Lacking α-subunits of Heterotrimeric G proteins

Class/Type	Gene	Expression	Effector(s)	Phenotype of Global or Tissue-restricted Knock-out	Reference
_Gα$_s$ class_					
Gα$_s$[1]	_Gnas_	ubiquitous	AC (all types) ↑	global: embryonic lethal[4]	[22]
				osteoblast: reduced bone turnover	[88]
				chondrocyte: epiphyseal and growth plate abnormalities	[89]
				hepatocyte: increased glucose tolerance	[73]
				juxtaglomerular cell: reduced renin formation	[90]
				pancreatic β-cell: reduced β-cell mass, diabetes	[72]
				haematopoietic system: defective engraftment of haematopoietic stem cells in bone marrow	[40]
				nervous system: obesity, insulin resistance, diabetes	[74]
				adipocytes: adipogenesis ↓, impaired cold tolerance responsiveness of brown adipose tissue to sympathetic stimulation ↓	[75]
Gα$_{sXL}$	(_GnasXL_)	neuroendocrine	AC ↑	global: perinatal lethal	[91,92]
Gα$_{olf}$	_Gnal_	olfact. epithelium, brain	AC ↑	global: anosmia, hyperactivity	[31]
_Gα$_{i/o}$ class_					
Gα$_{i1}$	_Gnai1_	widely distributed	AC ↓[5]	global: impaired memory formation	[93]
Gα$_{i2}$	_Gnai2_	ubiquitous		global: inflammatory bowel disease, impaired platelet activation, impaired lymphocyte trafficking	[42–45,84,85,94]
Gα$_{i3}$	_Gnai3_	widely distributed		global: lack of hepatic antiautophagic action of insulin	[24]
Gα$_o$[2]	_Gnao_	neuronal, neuroendocrine	VDCC↓, GIRK↑[6]	global: various CNS defects	[25–28]
Gα$_z$	_Gnaz_	neuronal, platelets	AC (e.g. V,VI) ↓	global: viable, increased bleeding time	[29,30]
Gα$_{gust}$	_Gnat3_	taste cells, brush cells	PDE ↑?	global: impaired bitter and sweet sensation	[78]

(continued)

127

Table 7.1. (continued)

Class/Type	Gene	Expression	Effector(s)	Phenotype of Global or Tissue-restricted Knock-out	Reference
Gα$_{tr}$	Gnat1	retinal rods, taste cells	PDE 6 (rod) ↑	global: mild retinal degeneration	[77]
Gα$_{tc}$	Gnat2	retinal cones	PDE 6 (cone) ↑	global: achromatopsia	[95]
Gα$_{i1}$ + Gα$_{i3}$				global: viable, immunological defects	[96]
Gα$_{i2}$ + Gα$_{i3}$				global: embryonic lethal before E10	[24]
Gα$_{q11}$ <u>class</u>					
Gα$_q$	Gnaq	ubiquitous	PLC-β1–4 ↑	global: ataxia, defective platelet activation	[35,82]
Gα$_{11}$	Gna11	almost ubiquitous	PLC-β1–4 ↑	global: no obvious phenotype seen so far	[19]
Gα$_{14}$	Gna14	kidney, lung, spleen	PLC-β1–4 ↑	global: no obvious phenotype seen so far[9]	
Gα$_{15}$	Gna15[3]	hematopoietic cells	PLC-β1–4 ↑	global: no obvious phenotype seen so far	[97]
Gα$_q$ + Gα$_{11}$				global: myocardial hypoplasia (lethal e11)	[19]
				cardiomyocyte: pressure overload induced hypertrophy ↓	[56]
				nervous system: perinatal lethal	[1]
				neural crest: craniofacial defects	[20]
				forebrain principal neurons: abnormal mothering behaviours; epilepsy, impaired endocannabinoid formation	[39,98]
				parathyroid: hypercalcemia, hyperparathyroidism	[99]
				thyrocyte: impaired thyroid function and goiter development	[68]
				smooth muscle: hypotension, salt-dependent hypertension ↓	[61]
				endothelial cell: protection against anaphylactic shock	[62]
				pancreatic β-cell: reduced insulin secretion	[69]

Gα[12/13] class

	Gene	Expression	Effectors	Phenotype	Ref.
Gα[12]	Gna12	ubiquitous	RhoGEF[7]	global: no obvious phenotype seen so far	[100]
Gα[13]	Gna13	ubiquitous	RhoGEF[8]	global: defective angiogenesis (lethal e9.5)	[14]
				platelet: activation defect	[86]
				platelet: severe platelet defect	[87]
Gα[q] + Gα[13]				global: embryonic lethal (e8.5)	[100]
Gα[12] + Gα[13]				platelet: like Gna13[-/-]	[86]
				neural crest: cardiac defects	[20]
				B cell: lack of marginal zone B-cells	[46,47]
				T cell: lymphadenopathy	[48]
				smooth muscle: salt-dependent hypertension ↓	[101]

[1] several splice variants; [2] two splice variants; [3] human orthologue: GNA16; [4] parent of origin specific defects in heterozygotes; [5] types I,III,VI,VIII,IX; [6] via Gβγ; [7] PDZ-RhoGEF/LARG + Btk, Gap1m, Cadherin; [8] p115RhoGEF, PDZ-RhoGEF/LARG + radixin; [9] H. Jiang and M.I. Simon (personal communication); AC, adenylyl cyclase; PDE, phosphodiesterase; PLC, phospholipase C; GIRK, G protein regulated inward rectifier potassium channel; VDCC, voltage-dependent Ca²⁺-channel; RhoGEF, Rho guanine nucleotide exchange factor

of $G\alpha_{13}$ in endothelial cells, since it is also observed in mice in which $G\alpha_{13}$ deficiency is restricted to endothelial cells, and since it can be rescued by a transgene expressing $G\alpha_{13}$ under the control of an endothelial cell-specific promoter.[15]

The absence of $G\alpha_{12}/G\alpha_{13}$ causes developmental defects in the nervous system as well. A conditional knockout of the genes encoding $G\alpha_{12}$ and $G\alpha_{13}$ in all glial and neuronal cells results in the overmigration of cortical neurons in particular regions of the cerebral and cerebellar cortices, resulting in neuronal ectopia of the cerebral cortex and a severe malformation of the rostral cerebellar cortex.[16] This phenotype is mimicked by a neuronal cell-specific deletion of $G\alpha_{12}/G\alpha_{13}$ indicating that radially migrating cortical neurons require intact signaling through G_{12} and G_{13} to stop migrating at the appropriate time and place. Since G_{12} and G_{13} can induce the retraction of cellular protrusions like neurites,[17,18] it has been speculated that radially migrating cortical neurons lacking G_{12}/G_{13} are unable to receive a stop signal acting via G_{12}/G_{13}-coupled receptors.

Mice deficient in both $G\alpha_q$ and $G\alpha_{11}$ suffer from a defect in heart development and die in utero (see further in this chapter). These mice die at embryonic day 11 due to a severe thinning of the myocardial layer of the heart.[19] Both the trabecular ventricular myocardium and the subepicardial layer appeared to be underdeveloped. In addition, signaling through G_q class members has also been implicated in the proliferation and/or migration of craniofacial neural crest cells.[19-21] The endothelin/endothelin receptor system plays an especially critical role in the differentiation and terminal migration of particular neural crest cell subpopulations. Studies in neural crest cell-specific $G\alpha_q/G\alpha_{11}$- and $G\alpha_{12}/G\alpha_{13}$-deficient mice have shown that G_q/G_{11}- but not G_{12}/G_{13}-mediated signaling processes mediate endothelin-1/ET_A-receptor-dependent development of the cephalic neural crest. In contrast, neither G_q/G_{11}- nor G_{12}/G_{13}-mediated signaling appears to be involved in endothelin-3/ET_B-receptor-mediated development of neural crest-derived melanocytes or enteric neurons.[20] Neural crest cell-specific $G\alpha_{12}/G\alpha_{13}$ deficiency results in characteristic cardiac malformations.[20]

The complete loss of $G\alpha_s$ in mice homozygous for an inactivating $G\alpha_s$ mutation leads to embryonic lethality before embryonic day 10.[22] It is interesting that heterozygotes show varying phenotypes depending on the paternal origin of the intact allele; these are probably caused by genetic haploinsufficiency and/or tissue-specific imprinting of the maternal $G\alpha_s$ allele.[13,23] Similarly, lack of both $G\alpha_{i2}$ and $G\alpha_{i3}$ results in embryonic lethality before embryonic day 10.[24] The exact developmental defects underlying embryonic lethality of $G\alpha_s$ and $G\alpha_{i2}/G\alpha_{i3}$-deficient mice are not known.

CENTRAL NERVOUS SYSTEM

In the central nervous system, many mediators and neurotransmitters function through G protein-coupled receptors to modulate neuronal activity or morphology. Neurotransmitters that induce an inhibitory modulation typically act on receptors that are coupled to members of the $G_{i/o}$ family, whereas G_q- and G_s-family members are primarily involved in excitatory responses.

The G protein G_o is highly abundant in the mammalian nervous system and has been shown to mediate inhibition of neuronal (N-, P/Q-, R-type) voltage-dependent Ca^{2+} channels via its $\beta\gamma$-complex, thereby reducing the excitability of the cell. $G\alpha_o$-deficient mice suffer from tremors and have occasional seizures.[25,26] In addition, $G\alpha_o$-deficient mice appear to be hyperalgesic when tested in the hot plate assay.[25] The latter finding is consistent with the observation that opioid receptor-mediated inhibition of Ca^{2+} currents in dorsal root ganglia (DRG) from $G\alpha_o$-deficient animals was reduced by about 30% compared to those in wild-type DRGs.[25] The inhibitory glutamatergic synapse between photoreceptor cells and ON bipolar cells in the retina has been studied in more detail using various $G\alpha$-deficient mouse lines. In the absence of light, ON bipolar cells are inhibited by glutamate released from rods and cones, an effect mediated by the metabotropic glutamate receptor mGluR6. In mice lacking the $G\alpha_o$ splice variant $G\alpha_{o1}$, light-induced modulation of ON bipolar cells is abolished,[27,28] indicating that G_o is the main mediator of glutamate-induced inhibition of ON bipolar cells in the absence of light.

G_z, a member of the $G_{i/o}$-family of G proteins, shares with G_{i1}, G_{i2}, and G_{i3} the ability to inhibit adenylyl cyclases but has a rather limited pattern of expression, being found in brain, adrenal medulla, and platelets. $G\alpha_z$-deficient mice exhibit altered responses to a variety of psychoactive drugs. Cocaine-induced increases in locomotor activity were more pronounced, and short-term antinociceptive effects of morphine were altered.[29,30] In addition, behavioral effects of catecholamine reuptake inhibitors were abolished in $G\alpha_z$-deficient mice,[30] indicating that G_z is involved in signaling processes regulated by various neurotransmitters.

$G\alpha_{olf}$ is expressed in various regions of the CNS, including olfactory sensory neurons and basal ganglia. $G\alpha_{olf}$-deficient mice exhibit clear motoric abnormalities such as hypermotoric behavior.[31] Subsequent studies indicate that G_{olf} is critically involved in dopamin(D_1)- and adenosine(A_{2A})-receptor-mediated effects in the striatum.[32,33]

The two main members of the G_q family, G_q and G_{11}, are widely expressed in the central nervous system. Mice lacking $G\alpha_q$ develop an ataxia with clear signs of motor coordination deficits, and functional defects could be observed in the cerebellar cortex of $G\alpha_q$-deficient mice.[34,35] These abnormalities were mainly due to defects in the parallel fibre (PF)-Purkinje cell (PC) synapse in which G_q is the predominant G protein mediating the effects of metabotropic glutamate group 1 receptors (mGluR1). The G_q-mediated increase in the dendritic Ca^{2+} concentration in response to glutamate is important for the induction of long-term depression (LTD) of the PF-PC synapse, and LTD is believed to be one of the cellular mechanisms underlying cerebellar motor learning.[36] Interestingly, Ca^{2+} responses to mGluR1 activation of PCs were absent in $G\alpha_q$-deficient mice, whereas they are indistinguishable between wild-type and $G\alpha_{11}$-deficient animals.[34] At the same time, synaptically evoked LTD was decreased in $G\alpha_{11}$-deficient animals, albeit to a lesser extent than in mice lacking $G\alpha_q$. The most likely explanation for the predominant role of G_q in Purkinje cells is the relatively high expression of $G\alpha_q$ when compared with $G\alpha_{11}$.[34,35] The defect in the modulation of the PF-PC synapse in mice lacking $G\alpha_q$ also results in a defect in the regression of supernumerary climbing fibres innervating Purkinje cells postnatally, which most likely contributes to the observed atactic phenotype in $G\alpha_q$-deficient mice.[35]

Long-term potentiation in the CA1 region of the hippocampus is equally impaired in mice lacking $G\alpha_q$ and $G\alpha_{11}$,[37] while the mGluR-mediated long-term depression in the hippocampal CA1-region was absent in mice lacking $G\alpha_q$ but was unaffected in $G\alpha_{11}$-deficient mice.[38]

Interestingly, the G_q/G_{11}-mediated signaling pathway appears to play an important role in the regulation of endocannabinoid formation.[39] In mice lacking $G\alpha_q$ and $G\alpha_{11}$ in principal neurons of the forebrain, the on-demand formation of endocannabinoids was reduced, a defect accompanied by an increased seizure susceptability and an impaired activation of neuroprotective mechanisms. This suggests that G_q/G_{11} negatively regulate neuronal excitability in vivo through their role in modulation of endocannabinoid formation.[39]

HEMATOPOIESIS AND IMMUNE SYSTEM

Hematopoietic stem and progenitor cells (HSPCs) change their location during development from the fetal liver to the bone marrow. Also later in life, HSPCs circulate and home in on their specific bone marrow niches. The ability of HSPCs to specifically engraft bone marrow is used clinically in bone marrow transplantation. In mice lacking $G\alpha_s$, it has been observed that HSPC engraftment of bone marrow in fetal development but also in the adult organism depends on G_s-mediated signaling.[40] $G\alpha_s$ was not required for chemotaxis of HSPCs or for maintaining HSPCs in the bone marrow. Interestingly, activation of $G\alpha_s$ using cholera toxin enhanced homing and engraftment of HCSPs in vivo. Thus G_s-mediated signaling plays a critical role in HSPC bone marrow engraftment. The mediators and receptors involved in this process are unknown, but prostaglandin E_2 appears to be a candidate.[41]

Mice lacking $G\alpha_{12}$ develop a lethal, diffuse inflammatory bowel disease that resembles in many aspects ulcerative colitis in humans.[42] In subsequent studies, dramatic changes in the phenotype and function of intestinal lymphocytes and epithelial cells have been described, likely due to defective lymphocyte homing in enteric epithelia.[43] On a cellular level, G_{12} may be involved in the regulation of T cell function and trafficking. In fact, studies in $G\alpha_{12}$-deficient mice clearly showed that G_{12} is required for trafficking of B lymphocytes into and within lymph nodes, as well as for the movement of T lymphocytes.[44,45] These processes are regulated through chemoattractant and chemokine receptors that show a predominant coupling to G_i-type G proteins. In addition to the colitis, many $G\alpha_{12}$-deficient mice develop colonic adenocarcinomas, which are probably secondary to colonic inflammation.[42]

Recently, G_{12}/G_{13} also have been involved in defined immunological functions. Inactivation of $G\alpha_{12}/G\alpha_{13}$, specifically in murine B cells, for example, resulted in the reduction of a specialized splenic B cell population, the marginal zone B (MZB)-cells.[46] The mechanisms underlying the reduction of MZB-cells are not completely understood; however, an impaired integrin-mediated adhesiveness of $G\alpha_{12}/G\alpha_{13}$-deficient MZB-cells and a disinhibition of migration in response to serum most likely contribute to the phenotype.[47] Lack of $G\alpha_{12}/G\alpha_{13}$ in T lymphocytes results in increased activity of the integrin leukocyte-funtion-antigen-1 (LFA-1) leading to defects in the fine-tuning of T cell trafficking, proliferation and increased susceptibility toward immune disease.[48] G_{12}/G_{13} has also repeatedly been implicated in cellular polarization and migration. Expression of dominant negative $G\alpha_{12}$ and $G\alpha_{13}$, for example, has been shown to disturb ordered polarization in a neutrophil cell line, resulting in cells containing multiple leading edges.[49] Abnormal polarization and migration has also been observed in $G\alpha_{12}/G\alpha_{13}$-deficient MZB-cells.[47] Thus G_{12}/G_{13} appears to be involved in establishing correct frontness and backness during leukocyte migration.

CARDIOVASCULAR SYSTEM

The $G\alpha_q/G\alpha_{11}$-mediated signaling pathway appears to play a pivotal role in the regulation of physiological myocardial growth during embryogenesis. This is demonstrated by the phenotype of $G\alpha_q/G\alpha_{11}$-double deficient mice that die at embryonic day 11 due to a severe thinning of the myocardial layer of the heart.[19] Adult cardiomyocytes are terminally differentiated postmitotic cells that respond to stimulatory signals with cell growth rather than proliferation. Evidence has been provided that the initial phase of myocardial hypertrophy involves the formation of cardiac paracrine/autocrine factors like endothelin-1 or angiotensin II, which act on G_q/G_{11}-coupled receptors.[50] A potential role for G_q/G_{11} in myocardial hypertrophy was indicated by the expression of wild-type $G\alpha_q$ or of a constitutively active mutant of $G\alpha_q$ in the mouse heart which resulted in cardiac hypertrophy.[51,52] While these approaches used the conventional αMHC promoter that drives transgene expression in the atrium starting perinatally and in the ventricle right after birth, two mouse lines have been established that allow for inducible expression of $G\alpha_q$ in adult cardiomyocytes.[53,54] The induction of cardiac $G\alpha_q$ expression in the neonatal period recapitulated the phenotype seen in conventional αMHC transgenic $G\alpha_q$ mice,[54] whereas the overexpression of $G\alpha_q$ in adult animals failed to induce myocardial hypertrophy,[53,54] suggesting that myocardial hypertrophy in response to transgenic overexpression of $G\alpha_q$ requires a growing heart. Two additional genetic approaches were employed to analyze the role of G_q/G_{11} in the induction of cardiac hypertrophy. Signaling through G_q/G_{11} was inhibited by the transgenic expression of a short fragment of the C-terminus of $G\alpha_q$,[55] which reduced myocardial hypertrophy. In an alternative approach, the G_q/G_{11}-mediated signaling pathway was abrogated by conditional cardiomyocyte-specific inactivation of the genes encoding $G\alpha_q/G\alpha_{11}$. In mice with cardiomyocyte-specific $G\alpha_q/G\alpha_{11}$ deficiency, no ventricular hypertrophy could be seen in response to pressure overload.[56] These genetic mouse models strongly support the concept that the G_q/G_{11}-mediated signaling pathway plays a pivotal role in the induction of myocardial hypertrophy in response to mechanical stress by coupling various receptors to the induction of a genetic program resulting in myocardial growth.

The heart rate regulation via the sympathetic and the parasympathetic nervous system involves G_s- as well as G_i/G_o-mediated signaling pathways. The involvement of particular G_i/G_o family members in the inhibitory regulation of the heart through the parasympathetic system has recently been analyzed.[57] Mice lacking either $G\alpha_{i2}$ or $G\alpha_o$ show a tachycardia under resting conditions in contrast to wild-type and $G\alpha_{i1}/G\alpha_{i3}$ double deficient mice. The diurnal heart rate variation was only affected in $G\alpha_o$-deficient mice. In contrast, cardioinhibitory effects induced by the muscarinic M_2 receptor agonist carbachol were only affected in mice lacking $G\alpha_{i2}$. Thus, G_{i2} appears to be the major mediator of parasympathetic heart rate modulation in mice, whereas G_o plays a role in the short-term heart rate regulation, and G_{i1} and G_{i3} are involved in neither of the two processes.[57] The mechanisms underlying the specific functions of G_{i2} and G_o in heart rate regulation are not known but are likely to involve differential regulation of G protein-gated inwardly rectifying K^+ channels (GIRKs) through G protein $\beta\gamma$-subunits, as well as different inhibitory effects on the adenylyl cyclase system. Interestingly, the inhibition of L-type Ca^{2+} channels in the heart through muscarinic M_2 receptors was found to be abrogated in hearts lacking $G\alpha_o$ as well as $G\alpha_{i2}$.[26,58] This unexpected finding suggests that both G proteins may regulate this downstream signaling event in a complex fashion.

G_q/G_{11} and G_{12}/G_{13} mediate the effects of most vasopressors, such as angiotensin A_2, endothelin-1, or thromboxane A_2, which induce contraction of blood vessels by acting through particular G protein-coupled receptors.[59] Both G-protein-mediated signaling pathways eventually induce myosin light chain phosphorylation and actomyosin-based contractility via Ca^{2+}-dependent activation of myosin light chain kinase and Ca^{2+}-independent Rho/Rho-kinase-mediated inhibition of the myosin phosphatase.[60] Studies in mice with smooth muscle-specific $G\alpha_q/G\alpha_{11}$ and $G\alpha_{12}/G\alpha_{13}$ deficiency have shown that G_q/G_{11} but not G_{12}/G_{13}-mediated signaling is required for the maintenance of the basal vascular tone underlying basal blood pressure.[61] However, both G_q/G_{11} and G_{12}/G_{13}, as well as the RhoGEF protein LARG, are critically involved in the development of salt-dependent hypertension, indicating that the increased vascular tone underlying salt-dependent hypertension requires signaling through G protein-coupled receptors and the parallel activation of G_q/G_{11}- and G_{12}/G_{13}-mediated signaling.[61]

Many inflammatory and anaphylactic mediators activate endothelial cells through receptors coupled to G_q/G_{11} and G_{12}/G_{13}, resulting in an opening of the endothelial barrier, the formation of nitric oxide (NO), or the exocytosis of Weibel-Palade bodies. Using endothelial-specific inducible $G\alpha_q/G\alpha_{11}$- or $G\alpha_{12}/G\alpha_{13}$-deficient mice, it could be shown that G_q/G_{11}-mediated signaling in endothelial cells is required for the opening of the endothelial barrier and the stimulation of NO formation by various inflammatory mediators as well as by local anaphylaxis. Interestingly, lack of G_{12}/G_{13} in endothelial cells blocked G protein-coupled receptor-mediated activation of RhoA, but did not affect regulation of the endothelial barrier under in vivo conditions.[62] Further analysis of these genetic mouse models showed that the systemic effects of anaphylactic mediators like histamine and platelet activating factor (PAF), but not of bacterial lipopolysaccharide, are blunted in mice with endothelial $G\alpha_q/G\alpha_{11}$ deficiency. These mice, but not those with endothelial $G\alpha_{12}/G\alpha_{13}$ deficiency, are protected against the fatal consequences of passive and active systemic anaphylaxis.[62]

The recruitment of leukocytes from the blood to sites of inflammation is a multistep process involving tethering, rolling, adhesion, and eventual transmigration of cells through the vessel wall. Some of these processes are regulated by chemokines and chemotactic factors acting via G protein-coupled receptors. Studies in mice lacking the α-subunit of G_{i2} combined with bone marrow transplantation have shown that efficient leukocyte extravasation requires intact G_{i2}-mediated signaling in endothelial cells.[63] This is consistent with earlier observations that inactivation of G_i-family G proteins by pertussis toxin treatment causes defects in leukocyte extravasation and homing to different organs.[64]

ENDOCRINE SYSTEM AND METABOLISM

The function of the thyroid gland is primarily regulated by thyroid-stimulating hormone (TSH) that acts on a receptor coupled to various G proteins including G_s and G_q/G_{11}.[65,66] The activation of the G_s/adenylyl cyclase-mediated pathway has been suggested for most of the biological effects of TSH on thyroid cells.[67] Interestingly, studies in thyroid-specific $G\alpha_q/G\alpha_{11}$-deficient mice have shown that lack of G_q/G_{11} blocks phospholipase C activation in response to thyroid-stimulating hormone and other mediators, which leads to a severely reduced iodine organification and thyroid hormone secretion in response to thyroid-stimulating hormone. These animals develop hypothyroidism when adult. Interestingly, the thyroid-specific deficiency of $G\alpha_q/G\alpha_{11}$ results in a strongly reduced proliferative thyroid response to thyroid-stimulating hormone or to goiterogenic diet.[68] Thus G_q/G_{11}-mediated signaling is essential for the regulation of thyroid function and growth.

Recent data based on the analysis of mice lacking $G\alpha_q/G\alpha_{11}$ specifically in pancreatic β-cells indicated that this pathway plays an important role in the stimulatory regulation of insulin secretion and that $G_{q/11}$-mediated signaling potentiates insulin secretion in an autocrine manner by mediating the effects of nucleotides and Ca^{2+} ions released together with insulin from β-cells.[69]

G_s-mediated signaling is involved in multiple endocrine and metabolic functions, and mice lacking $G\alpha_s$ constitutively have multiple endocrinological and metabolic phenotypes.[70] However, it has been difficult to evaluate the effects of Gα_s deficiency in specific tissues due to tissue-specific imprinting and the disruption of alternative GNAS gene products that are generated by alternative promoters and splicing events.[71] These problems have been overcome by the generation of conditional *Gnas* alleles that allow tissue-specific deletion of the Gα_s subunit alone.[12,13] These studies revealed, for example, that the G$_s$-mediated signaling pathway in pancreatic β-cells is not only necessary to mediate effects of hormones like glucagon-like peptide-1 to increase glucose-stimulated insulin secretion, but is also required for growth and survival of pancreatic β-cells.[72] A selective deletion of the GNAS gene in hepatocytes resulted in increased glucose tolerance, increased liver weight and glycogen content, and reduced adiposity. These effects were accompanied by strongly increased glucagon and glucagon-like peptide-1 plasma levels, as well as pancreatic α-cell hyperplasia, as a result of hepatic glucagon resistance and a tendency toward hypoglycemic states.[73] Mutations on the maternal allele of the GNAS gene in humans as well as in mice cause obesity and multihormone resistance, a phenomenon that is due to the fact that Gα_s is preferentially expressed from the maternal allele in various tissues.[13] Data based on central nervous system-specific Gα_s-deficient mice indicate that the imprinting of the Gα_s gene in the nervous system is responsible for these effects, and that Gα_s is imprinted in the paraventricular nucleus.[74] The recent analysis of mice lacking Gα_s in the adipose tissue revealed that adipocyte Gα_s is required for adipogenesis. In addition, the lack of Gα_s in adipocytes resulted in an impaired cold tolerance and reduced responsiveness of brown adipose tissue to sympathetic stimuli.[75]

SENSORY SYSTEMS

Odors, light, and many tastants act directly on G protein-coupled receptors. The G protein G$_{olf}$ is centrally involved in the transduction of odorant stimuli in olfactory cilia, and Gα_{olf}-deficient mice exhibit dramatically reduced electrophysiological responses to all odors tested.[31] Since nursing and mothering behavior in rodents is mediated a great deal by the olfactory system, most Gα_{olf}-deficient pups die a few days after birth due to insufficient feeding, and rare surviving mothers exhibit inadequate maternal behavior. In contrast to the olfactory epithelium, the vomeronasal organ, which detects pheromones, expresses receptors that are coupled to G$_{i/o}$. Absence of Gα_o results in apoptotic death of receptor cells that usually express Gα_o.[76]

Rod-transducin (G$_{t-r}$) and cone-transducin (G$_{t-c}$) play well-established roles in the phototransduction cascade in the outer segments of retinal rods and cones, where they couple light receptors to cGMP-phosphodiesterase. In mice lacking Gα_{t-r}, the majority of retinal rods does not respond to light anymore, and these animals develop mild retinal degeneration with age.[77] The light response is transferred from the receptor cell to bipolar cells of the retina. In mice lacking Gα_o, modulation of ON bipolar cells in response to light is abrogated, indicating that G_o is critically involved in the tonic inhibition of these cells mediated by metabotropic glutamate (mGluR6) receptors.[28]

Among the four taste qualities – sweet, bitter, sour, and salty – bitter and sweet tastes appear to signal through heterotrimeric G proteins. Gustducin is a G protein mainly expressed in taste cells, and $G\alpha_{gust}$-deficient mice show impaired electrophysiological and behavioral responses to bitter and sweet agents.[78] The residual bitter and sweet taste responsiveness of $G\alpha_{gust}$-deficient mice could be further diminished by a dominant-negative mutant of gustducin-α, suggesting the involvement of other G proteins related to $G\alpha_{gust}$.[79] Interestingly, recent evidence indicates that gustducin-mediated signaling is also involved in the regulation of various cells of the enteric epithelium.[80,81]

HEMOSTASIS

Hemostasis is a complex process involving platelet adhesion and aggregation as well as formation of fibrin through the coagulation cascade. Platelet activation results in a rapid shape-change reaction immediately followed by secretion of granule contents, as well as inside-out activation of the fibrinogen receptor, integrin $\alpha_{IIb}\beta_3$, leading to platelet aggregation. Most physiological platelet activators act through G protein-coupled receptors, which in turn activate $G_{i2/3}$, G_q, G_{12}, and G_{13}. In platelets from $G\alpha_q$-deficient mice, the effect of various platelet stimuli on aggregation and degranulation was abrogated, demonstrating that $G\alpha_q$-mediated phospholipase C activation represents an essential event in platelet activation.[82] However, platelet shape change can still be induced in the absence of $G\alpha_q$, indicating that it is mediated by G proteins other than G_q, most likely G_{12}/G_{13}.[83] The defective activation of $G\alpha_q$-deficient platelets results in a primary hemostasis defect, and $G\alpha_q$ (−/−) mice are protected against platelet-dependent thromboembolism.

The role of G proteins of the $G_{i/o}$ family in platelet activation has recently been elucidated. Platelets contain at least three members of this class, G_{i2}, G_{i3}, and G_z. ADP, which is released from activated platelets and functions as a positive feedback mediator during platelet activation, induces platelet activation through the G_q-coupled $P2Y_1$ receptor as well as through the G_i-coupled $P2Y_{12}$ purinergic receptor. The general importance of the G_i-mediated pathway is indicated by the fact that responses to ADP but also to thrombin were markedly reduced in platelets lacking $G\alpha_{i2}$.[84,85] In contrast to ADP or thrombin, epinephrine is not a full platelet activator per se in murine platelets. However, it is able to potentiate the effect of other platelet stimuli. In platelets from $G\alpha_z$-deficient mice, epinephrine's potentiating effects were clearly impaired, whereas the effects of other platelet activators appeared to be unaffected by the lack of $G\alpha_z$.[30] Thus members of the G protein families G_q, G_{12}, and $G_{i/o}$ are involved in processes leading to platelet activation.

Interestingly, activation of platelets by various stimuli was severely inhibited in platelets lacking $G\alpha_{13}$ but not in $G\alpha_{12}$-deficient platelets.[86] These defects were accompanied by a reduced activation of the RhoA-mediated signaling pathway as well as by an inability to form stable platelet thrombi under high sheer stress conditions. In addition, mice carrying $G\alpha_{13}$-deficient platelets have an increased bleeding time and are protected against the formation of arterial thrombi.[86] Various studies with $G\alpha$-deficient platelets have clearly shown that three G proteins are the major mediators of platelet activation: G_q, G_{i2}, and G_{13}. While in the absence of either G_q, G_{i2}, or G_{13}, some platelet activation can still be induced, in the absence of both $G\alpha_q$ and $G\alpha_{13}$, platelets are completely unresponsive to various stimuli.[87]

CONCLUSIONS

During the last fifteen years, basically all known genes encoding G protein α-subunits have been deleted in mice, and these mouse knock-out models have enabled researchers to gain new and in some cases first insights into the biological roles of particular G protein-mediated signaling pathways. Since in some cases the resulting phenotypes were rather complex or consisted of severe defects already during the development, many conditional alleles of genes encoding G protein α-subunits have been generated to allow for the conditional inactivation in a time- and tissue-specific manner. These conditional gene inactivation approaches and the combination of various constitutive or conditional mutant alleles have allowed to overcome embryonic lethality or complex defects resulting from some null mutations in $G\alpha$ genes. The analysis of these conditional and compound mutants in defined physiological and also pathophysiological contexts has already revealed unexpected new functions of particular G protein-mediated signaling pathways, and these approaches will in the future very likely lead to many more exciting new insights into the complex system of G protein-mediated signaling.

SUGGESTED READING

1. Wettschureck, N., Moers, A., Wallenwein, B., Parlow, A.F., Maser-Gluth, C., and Offermanns, S. (2005). Loss of Gq/11 family G proteins in the nervous system causes pituitary somatotroph hypoplasia and dwarfism in mice. Mol Cell Biol *25*, 1942–1948.
2. Sunahara, R.K., Dessauer, C.W., and Gilman, A.G. (1996). Complexity and diversity of mammalian adenylyl cyclases. Annu Rev Pharmacol Toxicol *36*, 461–480.
3. Clapham, D.E., and Neer, E.J. (1997). G protein beta gamma subunits. Annu Rev Pharmacol Toxicol *37*, 167–203.
4. Meng, J., and Casey, P.J. (2004). Signaling through Gz. In Handbook of cell signaling, R.A. Bradshaw, and E.A. Dennis, eds. (Amsterdam, Boston, Heidelberg), pp. 601–604.
5. Arshavsky, V.Y., Lamb, T.D., and Pugh, Jr., E.N. (2002). G proteins and phototransduction. Annu Rev Physiol *64*, 153–187.
6. Damak, S. (2004). G proteins mediating taste transduction. In Handbook of cell signaling, R.A. Bradshaw, and E.A. Dennis, eds. (Amsterdam, Boston, Heidelberg, Academic Press), pp. 657–661.

7. Rhee, S.G. (2001). Regulation of phosphoinositide-specific phospholipase C. Annu Rev Biochem *70*, 281–312.

8. Kelly, P., Casey, P.J., and Meigs, T.E. (2007). Biologic functions of the G12 subfamily of heterotrimeric g proteins: growth, migration, and metastasis. Biochemistry *46*, 6677–6687.

9. Kurose, H. (2003). Galpha12 and Galpha13 as key regulatory mediator in signal transduction. Life Sci *74*, 155–161.

10. Worzfeld, T., Wettschureck, N., and Offermanns, S. (2008). G(12)/G(13)-mediated signalling in mammalian physiology and disease. Trends Pharmacol Sci *29*, 582–589.

11. Fukuhara, S., Chikumi, H., and Gutkind, J.S. (2001). RGS-containing RhoGEFs: the missing link between transforming G proteins and Rho? Oncogene *20*, 1661–1668.

12. Plagge, A., Kelsey, G., and Germain-Lee, E.L. (2008). Physiological functions of the imprinted Gnas locus and its protein variants Galpha(s) and XLalpha(s) in human and mouse. J Endocrinol *196*, 193–214.

13. Weinstein, L.S., Xie, T., Zhang, Q.H., and Chen, M. (2007). Studies of the regulation and function of the G(s)alpha gene Gnas using gene targeting technology. Pharmacol Ther *115*, 271–291.

14. Offermanns, S., Mancino, V., Revel, J.P., and Simon, M.I. (1997b). Vascular system defects and impaired cell chemokinesis as a result of Galpha13 deficiency. Science *275*, 533–536.

15. Ruppel, K.M., Willison, D., Kataoka, H., Wang, A., Zheng, Y.W., Cornelissen, I., Yin, L., Xu, S.M., and Coughlin, S.R. (2005). Essential role for Galpha13 in endothelial cells during embryonic development. Proc Natl Acad Sci U S A *102*, 8281–8286.

16. Moers, A., Nurnberg, A., Goebbels, S., Wettschureck, N., and Offermanns, S. (2008). Galpha12/Galpha13 deficiency causes localized overmigration of neurons in the developing cerebral and cerebellar cortices. Mol Cell Biol *28*, 1480–1488.

17. Kranenburg, O., Poland, M., van Horck, F.P., Drechsel, D., Hall, A., and Moolenaar, W.H. (1999). Activation of RhoA by lysophosphatidic acid and Galpha12/13 subunits in neuronal cells: induction of neurite retraction. Mol Biol Cell *10*, 1851–1857.

18. Nürnberg, A., Bräuer, A.U., Wettschureck, N., and Offermanns, S. (2008). Antagonistic regulation of neurite morphology through Gq/G11 and G12/G13. J Biol Chem *283*, 35526–35531.

19. Offermanns, S., Zhao, L.P., Gohla, A., Sarosi, I., Simon, M.I., and Wilkie, T.M. (1998). Embryonic cardiomyocyte hypoplasia and craniofacial defects in G alpha q/G alpha 11-mutant mice. Embo J *17*, 4304–4312.

20. Dettlaff-Swiercz, D.A., Wettschureck, N., Moers, A., Huber, K., and Offermanns, S. (2005). Characteristic defects in neural crest cell-specific Galphaq/Galpha11- and Galpha12/Galpha13-deficient mice. Dev Biol *282*, 174–182.

21. Ivey, K., Tyson, B., Ukidwe, P., McFadden, D.G., Levi, G., Olson, E.N., Srivastava, D., and Wilkie, T.M. (2003). Galphaq and Galpha11 proteins mediate endothelin-1 signaling in neural crest-derived pharyngeal arch mesenchyme. Dev Biol *255*, 230–237.

22. Yu, S., Yu, D., Lee, E., Eckhaus, M., Lee, R., Corria, Z., Accili, D., Westphal, H., and Weinstein, L.S. (1998). Variable and tissue-specific hormone resistance in heterotrimeric Gs protein alpha-subunit (Gsalpha) knockout mice is due to tissue-specific imprinting of the gsalpha gene. Proc Natl Acad Sci U S A *95*, 8715–8720.

23. Weinstein, L.S., and Yu, S. (1999). The role of genomic imprinting of Galpha in the pathogenesis of Albright hereditary osteodystrophy. Trends Endocrinol Metab *10*, 81–85.

24. Gohla, A., Klement, K., Piekorz, R.P., Pexa, K., vom Dahl, S., Spicher, K., Dreval, V., Haussinger, D., Birnbaumer, L., and Nurnberg, B. (2007). An obligatory requirement for the heterotrimeric G protein Gi3 in the antiautophagic action of insulin in the liver. Proc Natl Acad Sci U S A *104*, 3003–3008.

25. Jiang, M., Gold, M.S., Boulay, G., Spicher, K., Peyton, M., Brabet, P., Srinivasan, Y., Rudolph, U., Ellison, G., and Birnbaumer, L. (1998). Multiple neurological abnormalities in mice deficient in the G protein Go. Proc Natl Acad Sci U S A *95*, 3269–3274.

26. Valenzuela, D., Han, X., Mende, U., Fankhauser, C., Mashimo, H., Huang, P., Pfeffer, J., Neer, E.J., and Fishman, M.C. (1997). G alpha(o) is necessary for muscarinic regulation of Ca2+ channels in mouse heart. Proc Natl Acad Sci U S A *94*, 1727–1732.

27. Dhingra, A., Jiang, M., Wang, T.L., Lyubarsky, A., Savchenko, A., Bar-Yehuda, T., Sterling, P., Birnbaumer, L., and Vardi, N. (2002). Light response of retinal ON bipolar cells requires a specific splice variant of Galpha(o). J Neurosci *22*, 4878–4884.

28. Dhingra, A., Lyubarsky, A., Jiang, M., Pugh, E.N., Jr., Birnbaumer, L., Sterling, P., and Vardi, N. (2000). The light response of ON bipolar neurons requires G[alpha]o. J Neurosci *20*, 9053–9058.

29. Hendry, I.A., Kelleher, K.L., Bartlett, S.E., Leck, K.J., Reynolds, A.J., Heydon, K., Mellick, A., Megirian, D., and Matthaei, K.I. (2000). Hypertolerance to morphine in G(z alpha)-deficient mice. Brain Res *870*, 10–19.

30. Yang, J., Wu, J., Kowalska, M.A., Dalvi, A., Prevost, N., O'Brien, P.J., Manning, D., Poncz, M., Lucki, I., Blendy, J.A., *et al.* (2000). Loss of signaling through the G protein, Gz, results in abnormal platelet activation and altered responses to psychoactive drugs. Proc Natl Acad Sci U S A *97*, 9984–9989.

31. Belluscio, L., Gold, G.H., Nemes, A., and Axel, R. (1998). Mice deficient in G(olf) are anosmic. Neuron *20*, 69–81.

32. Corvol, J.C., Studler, J.M., Schonn, J.S., Girault, J.A., and Herve, D. (2001). Galpha(olf) is necessary for coupling D1 and A2a receptors to adenylyl cyclase in the striatum. J Neurochem *76*, 1585–1588.

33. Zhuang, X., Belluscio, L., and Hen, R. (2000). G(olf)alpha mediates dopamine D1 receptor signaling. J Neurosci *20*, RC91.

34. Hartmann, J., Blum, R., Kovalchuk, Y., Adelsberger, H., Kuner, R., Durand, G.M., Miyata, M., Kano, M., Offermanns, S., and Konnerth, A. (2004). Distinct roles of Galpha(q) and Galpha11 for Purkinje cell signaling and motor behavior. J Neurosci *24*, 5119–5130.

35. Offermanns, S., Hashimoto, K., Watanabe, M., Sun, W., Kurihara, H., Thompson, R.F., Inoue, Y., Kano, M., and Simon, M.I. (1997a). Impaired motor coordination and persistent multiple climbing fiber innervation of cerebellar Purkinje cells in mice lacking Galphaq. Proc Natl Acad Sci U S A *94*, 14089–14094.

36. Boyden, E.S., Katoh, A., and Raymond, J.L. (2004). Cerebellum-dependent learning: the role of multiple plasticity mechanisms. Annu Rev Neurosci *27*, 581–609.

37. Miura, M., Watanabe, M., Offermanns, S., Simon, M.I., and Kano, M. (2002). Group I metabotropic glutamate receptor signaling via Galpha q/Galpha 11 secures the induction of long-term potentiation in the hippocampal area CA1. J Neurosci *22*, 8379–8390.

38. Kleppisch, T., Voigt, V., Allmann, R., and Offermanns, S. (2001). G(alpha)q-deficient mice lack metabotropic glutamate receptor-dependent long-term depression but show normal long-term potentiation in the hippocampal CA1 region. J Neurosci *21*, 4943–4948.

39. Wettschureck, N., van der Stelt, M., Tsubokawa, H., Krestel, H., Moers, A., Petrosino, S., Schutz, G., Di Marzo, V., and Offermanns, S. (2006). Forebrain-specific inactivation of Gq/G11 family G proteins results in age-dependent epilepsy and impaired endocannabinoid formation. Mol Cell Biol *26*, 5888–5894.

40. Adams, G.B., Alley, I.R., Chung, U.I., Chabner, K.T., Jeanson, N.T., Lo Celso, C., Marsters, E.S., Chen, M., Weinstein, L.S., Lin, C.P., *et al.* (2009). Haematopoietic stem cells depend on Galpha(s)-mediated signalling to engraft bone marrow. Nature *459*, 103–107.

41. North, T.E., Goessling, W., Walkley, C.R., Lengerke, C., Kopani, K.R., Lord, A.M., Weber, G.J., Bowman, T.V., Jang, I.H., Grosser, T., *et al.* (2007). Prostaglandin E2 regulates vertebrate haematopoietic stem cell homeostasis. Nature *447*, 1007–1011.

42. Rudolph, U., Finegold, M.J., Rich, S.S., Harriman, G.R., Srinivasan, Y., Brabet, P., Bradley, A., and Birnbaumer, L. (1995). Gi2 alpha protein deficiency: a model of inflammatory bowel disease. Journal of clinical immunology *15*, 101S–105S.

43. Hornquist, C.E., Lu, X., Rogers-Fani, P.M., Rudolph, U., Shappell, S., Birnbaumer, L., and Harriman, G.R. (1997). G(alpha)i2-deficient mice with colitis exhibit a local increase in memory CD4+ T cells and proinflammatory Th1-type cytokines. J Immunol *158*, 1068–1077.

44. Han, S.B., Moratz, C., Huang, N.N., Kelsall, B., Cho, H., Shi, C.S., Schwartz, O., and Kehrl, J.H. (2005). Rgs1 and Gnai2 regulate the entrance of B lymphocytes into lymph nodes and B cell motility within lymph node follicles. Immunity *22*, 343–354.

45. Hwang, I.Y., Park, C., and Kehrl, J.H. (2007). Impaired trafficking of Gnai2+/- and Gnai2-/- T lymphocytes: implications for T cell movement within lymph nodes. J Immunol *179*, 439–448.

46. Rieken, S., Sassmann, A., Herroeder, S., Wallenwein, B., Moers, A., Offermanns, S., and Wettschureck, N. (2006b). G12/G13 family G proteins regulate marginal zone B cell maturation, migration, and polarization. J Immunol *177*, 2985–2993.

47. Rieken, S., Herroeder, S., Sassmann, A., Wallenwein, B., Moers, A., Offermanns, S., and Wettschureck, N. (2006a). Lysophospholipids control integrin-dependent adhesion in splenic B cells through G(i) and G(12)/G(13) family G-proteins but not through G(q)/G(11). J Biol Chem *281*, 36985–36992.

48. Herroeder, S., Reichardt, P., Sassmann, A., Zimmermann, B., Jaeneke, D., Hoeckner, J., Hollmann, M.W., Fischer, K.D., Vogt, S., Grosse, R., *et al.* (2009). Guanine nucleotide-binding proteins of the G12 family shape immune functions by controlling CD4+ T cell adhesiveness and motility. Immunity *30*, 708–720.

49. Xu, J., Wang, F., Van Keymeulen, A., Herzmark, P., Straight, A., Kelly, K., Takuwa, Y., Sugimoto, N., Mitchison, T., and Bourne, H.R. (2003). Divergent signals and cytoskeletal assemblies regulate self-organizing polarity in neutrophils. Cell *114*, 201–214.

50. Frey, N., and Olson, E.N. (2003). Cardiac hypertrophy: the good, the bad, and the ugly. Annu Rev Physiol *65*, 45–79.

51. D'Angelo, D.D., Sakata, Y., Lorenz, J.N., Boivin, G.P., Walsh, R.A., Liggett, S.B., and Dorn, G.W. (1997). Transgenic Galphaq overexpression induces cardiac contractile failure in mice. Proc Natl Acad Sci U S A *94*, 8121–8126.

52. Mende, U., Kagen, A., Cohen, A., Aramburu, J., Schoen, F.J., and Neer, E.J. (1998). Transient cardiac expression of constitutively active Galphaq leads to hypertrophy and dilated cardiomyopathy by calcineurin-dependent and independent pathways. Proc Natl Acad Sci U S A *95*, 13893–13898.

53. Fan, G., Jiang, Y.P., Lu, Z., Martin, D.W., Kelly, D.J., Zuckerman, J.M., Ballou, L.M., Cohen, I.S., and Lin, R.Z. (2005a). A transgenic mouse model of heart failure using inducible Galpha q. J Biol Chem *280*, 40337–40346.

54. Syed, F., Odley, A., Hahn, H.S., Brunskill, E.W., Lynch, R.A., Marreez, Y., Sanbe, A., Robbins, J., and Dorn, G.W. (2004). Physiological growth synergizes with pathological genes in experimental cardiomyopathy. Circ Res *95*, 1200–1206.

55. Akhter, S.A., Luttrell, L.M., Rockman, H.A., Iaccarino, G., Lefkowitz, R.J., and Koch, W.J. (1998). Targeting the receptor-Gq interface to inhibit in vivo pressure overload myocardial hypertrophy. Science *280*, 574–577.

56. Wettschureck, N., Rutten, H., Zywietz, A., Gehring, D., Wilkie, T.M., Chen, J., Chien, K.R., and Offermanns, S. (2001). Absence of pressure overload induced myocardial hypertrophy after conditional inactivation of Galphaq/Galpha11 in cardiomyocytes. Nat Med *7*, 1236–1240.

57. Zuberi, Z., Birnbaumer, L., and Tinker, A. (2008). The role of inhibitory heterotrimeric G proteins in the control of in vivo heart rate dynamics. Am J Physiol Regul Integr Comp Physiol *295*, R1822–1830.

58. Chen, F., Spicher, K., Jiang, M., Birnbaumer, L., and Wetzel, G.T. (2001). Lack of muscarinic regulation of Ca(2+) channels in G(i2)alpha gene knockout mouse hearts. Am J Physiol Heart Circ Physiol *280*, H1989–1995.

59. Maguire, J.J., and Davenport, A.P. (2005). Regulation of vascular reactivity by established and emerging GPCRs. Trends Pharmacol Sci *26*, 448–454.

60. Somlyo, A.P., and Somlyo, A.V. (2003). Ca2+ sensitivity of smooth muscle and nonmuscle myosin II: modulated by G proteins, kinases, and myosin phosphatase. Physiol Rev *83*, 1325–1358.

61. Wirth, A., Benyo, Z., Lukasova, M., Leutgeb, B., Wettschureck, N., Gorbey, S., Orsy, P., Horvath, B., Maser-Gluth, C., Greiner, E., *et al*. (2008). G12-G13-LARG-mediated signaling in vascular smooth muscle is required for salt-induced hypertension. Nat Med *14*, 64–68.

62. Korhonen, H., Fisslthaler, B., Moers, A., Wirth, A., Habermehl, D., Wieland, T., Schutz, G., Wettschureck, N., Fleming, I., and Offermanns, S. (2009). Anaphylactic shock depends on endothelial Gq/G11. J Exp Med *206*, 411–420.

63. Pero, R.S., Borchers, M.T., Spicher, K., Ochkur, S.I., Sikora, L., Rao, S.P., Abdala-Valencia, H., O'Neill, K.R., Shen, H., McGarry, M.P., *et al*. (2007). Galphai2-mediated signaling events in the endothelium are involved in controlling leukocyte extravasation. Proc Natl Acad Sci U S A *104*, 4371–4376.

64. Warnock, R.A., Askari, S., Butcher, E.C., and von Andrian, U.H. (1998). Molecular mechanisms of lymphocyte homing to peripheral lymph nodes. J Exp Med *187*, 205–216.

65. Allgeier, A., Offermanns, S., Van Sande, J., Spicher, K., Schultz, G., and Dumont, J.E. (1994). The human thyrotropin receptor activates G-proteins Gs and Gq/11. J Biol Chem *269*, 13733–13735.

66. Laugwitz, K.L., Allgeier, A., Offermanns, S., Spicher, K., Van Sande, J., Dumont, J.E., and Schultz, G. (1996). The human thyrotropin receptor: a heptahelical receptor capable of stimulating members of all four G protein families. Proc Natl Acad Sci U S A *93*, 116–120.

67. Tonacchera, M., Van Sande, J., Parma, J., Duprez, L., Cetani, F., Costagliola, S., Dumont, J.E., and Vassart, G. (1996). TSH receptor and disease. Clin Endocrinol (Oxf) *44*, 621–633.

68. Kero, J., Ahmed, K., Wettschureck, N., Tunaru, S., Wintermantel, T., Greiner, E., Schutz, G., and Offermanns, S. (2007). Thyrocyte-specific G(q)/G(11) deficiency impairs thyroid function and prevents goiter development. J Clin Invest *117*, 2399–2407.

69. Sassmann A, Gier B, Grone HJ, Drews G, Offermanns S and Wettschureck N (2010) The Gq/G11-mediated signaling pathway is critical for autocrine potentiation of insulin secretion in mice. J Clin Invest doi:10.1172/JCI41541 [epub ahead of print].

70. Weinstein, L.S. (1999). Gs alpha knockouts in mice and man. Rinsho byori *47*, 425–429.

71. Weinstein, L.S., Liu, J., Sakamoto, A., Xie, T., and Chen, M. (2004). GNAS: Normal and abnormal functions. Endocrinology *145*, 5459–5464.

72. Xie, T., Chen, M., Zhang, Q.H., Ma, Z., and Weinstein, L.S. (2007). Beta cell-specific deficiency of the stimulatory G protein alpha-subunit Gsalpha leads to reduced beta cell mass and insulin-deficient diabetes. Proc Natl Acad Sci U S A *104*, 19601–19606.

73. Chen, M., Gavrilova, O., Zhao, W.-Q., Nguyen, A., Lorenzo, J., Shen, L., Nackers, L., Pack, S., Jou, W., and Weinstein, L.S. (2005). Increased glucose tolerance and reduced adiposity in the absence of fasting hypoglycemia in mice with liver-specific Gsalpha deficiency. J Clin Invest *115*, 3217–3227.

74. Chen, M., Wang, J., Dickerson, K.E., Kelleher, J., Xie, T., Gupta, D., Lai, E.W., Pacak, K., Gavrilova, O., and Weinstein, L.S. (2009). Central nervous system imprinting of the G protein G(s)alpha and its role in metabolic regulation. Cell Metab *9*, 548–555.

75. Chen M, Chen H, Nguyen A, Gupta D, Wang J, Lai EW, Pacak K, Gavrilova O, Quon MJ and Weinstein LS (2010) G(s)alpha deficiency in adipose tissue leads to a lean phenotype with divergent effects on cold tolerance and diet-induced thermogenesis. Cell Metab *11*, 320–330.

76. Tanaka, M., Treloar, H., Kalb, R.G., Greer, C.A., and Strittmatter, S.M. (1999). G(o) protein-dependent survival of primary accessory olfactory neurons. Proc Natl Acad Sci U S A *96*, 14106–14111.

77. Calvert, P.D., Krasnoperova, N.V., Lyubarsky, A.L., Isayama, T., Nicolo, M., Kosaras, B., Wong, G., Gannon, K.S., Margolskee, R.F., Sidman, R.L., *et al.* (2000). Phototransduction in transgenic mice after targeted deletion of the rod transducin alpha -subunit. Proc Natl Acad Sci U S A *97*, 13913–13918.

78. Wong, G.T., Gannon, K.S., and Margolskee, R.F. (1996). Transduction of bitter and sweet taste by gustducin. Nature *381*, 796–800.

79. Ruiz-Avila, L., Wong, G.T., Damak, S., and Margolskee, R.F. (2001). Dominant loss of responsiveness to sweet and bitter compounds caused by a single mutation in alpha-gustducin. Proc Natl Acad Sci U S A *98*, 8868–8873.

80. Jang, H.J., Kokrashvili, Z., Theodorakis, M.J., Carlson, O.D., Kim, B.J., Zhou, J., Kim, H.H., Xu, X., Chan, S.L., Juhaszova, M., *et al.* (2007). Gut-expressed gustducin and taste receptors regulate secretion of glucagon-like peptide-1. Proc Natl Acad Sci U S A *104*,15069–15074.

81. Margolskee, R.F., Dyer, J., Kokrashvili, Z., Salmon, K.S., Ilegems, E., Daly, K., Maillet, E.L., Ninomiya, Y., Mosinger, B., and Shirazi-Beechey, S.P. (2007). T1R3 and gustducin in gut sense sugars to regulate expression of Na+-glucose cotransporter 1. Proc Natl Acad Sci U S A *104*, 15075–15080.

82. Offermanns, S., Toombs, C.F., Hu, Y.H., and Simon, M.I. (1997c). Defective platelet activation in G alpha(q)-deficient mice. Nature *389*, 183–186.

83. Klages, B., Brandt, U., Simon, M.I., Schultz, G., and Offermanns, S. (1999). Activation of G12/G13 results in shape change and Rho/Rho-kinase-mediated myosin light chain phosphorylation in mouse platelets. J Cell Biol *144*, 745–754.

84. Jantzen, H.M., Milstone, D.S., Gousset, L., Conley, P.B., and Mortensen, R.M. (2001). Impaired activation of murine platelets lacking G alpha(i2). J Clin Invest *108*, 477–483.

85. Yang, J., Wu, J., Jiang, H., Mortensen, R., Austin, S., Manning, D.R., Woulfe, D., and Brass, L.F. (2002). Signaling through Gi family members in platelets. Redundancy and specificity in the regulation of adenylyl cyclase and other effectors. J Biol Chem *277*, 46035–46042.

86. Moers, A., Nieswandt, B., Massberg, S., Wettschureck, N., Gruner, S., Konrad, I., Schulte, V., Aktas, B., Gratacap, M.P., Simon, M.I., *et al.* (2003). G13 is an essential mediator of platelet activation in hemostasis and thrombosis. Nat Med *9*, 1418–1422.

87. Moers, A., Wettschureck, N., Gruner, S., Nieswandt, B., and Offermanns, S. (2004). Unresponsiveness of platelets lacking both G{alpha}q and G{alpha}13: implications for collagen-induced platelet activation. J Biol Chem *279*, 45354–45359.

88. Sakamoto, A., Chen, M., Nakamura, T., Xie, T., Karsenty, G., and Weinstein, L.S. (2005b). Deficiency of the G-protein alpha-subunit G(s)alpha in osteoblasts leads to differential effects on trabecular and cortical bone. J Biol Chem *280*, 21369–21375.

89. Sakamoto, A., Chen, M., Kobayashi, T., Kronenberg, H.M., and Weinstein, L.S. (2005a). Chondrocyte-specific knockout of the G protein G(s)alpha leads to epiphyseal and growth plate abnormalities and ectopic chondrocyte formation. J Bone Miner Res *20*, 663–671.

90. Chen, L., Kim, S.M., Oppermann, M., Faulhaber-Walter, R., Huang, Y., Mizel, D., Chen, M., Lopez, M.L., Weinstein, L.S., Gomez, R.A., *et al.* (2007). Regulation of renin in mice with Cre recombinase-mediated deletion of G protein Gsalpha in juxtaglomerular cells. Am J Physiol *292*, F27–37.

91. Plagge, A., Gordon, E., Dean, W., Boiani, R., Cinti, S., Peters, J., and Kelsey, G. (2004). The imprinted signaling protein XL alpha s is required for postnatal adaptation to feeding. Nat Genet *36*, 818–826.

92. Xie, T., Plagge, A., Gavrilova, O., Pack, S., Jou, W., Lai, E.W., Frontera, M., Kelsey, G., and Weinstein, L.S. (2006). The alternative stimulatory G protein alpha-subunit XLalphas is a critical regulator of energy and glucose metabolism and sympathetic nerve activity in adult mice. J Biol Chem *281*, 18989–18999.

93. Pineda, V.V., Athos, J.I., Wang, H., Celver, J., Ippolito, D., Boulay, G., Birnbaumer, L., and Storm, D.R. (2004). Removal of G(ialpha1) constraints on adenylyl cyclase in the hippocampus enhances LTP and impairs memory formation. Neuron *41*, 153–163.

94. He, J., Gurunathan, S., Iwasaki, A., Ash-Shaheed, B., and Kelsall, B.L. (2000). Primary role for Gi protein signaling in the regulation of interleukin 12 production and the induction of T helper cell type 1 responses. J Exp Med *191*, 1605–1610.

95. Chang, B., Dacey, M.S., Hawes, N.L., Hitchcock, P.F., Milam, A.H., Atmaca-Sonmez, P., Nusinowitz, S., and Heckenlively, J.R. (2006). Cone photoreceptor function loss-3, a novel mouse model of achromatopsia due to a mutation in Gnat2. Invest Ophthalmol Vis Sci *47*, 5017–5021.

96. Fan, H., Zingarelli, B., Peck, O.M., Teti, G., Tempel, G.E., Halushka, P.V., Spicher, K., Boulay, G., Birnbaumer, L., and Cook, J.A. (2005b). Lipopolysaccharide- and gram-positive bacteria-induced cellular inflammatory responses: role of heterotrimeric Galpha(i) proteins. Am J Physiol Cell Physiol *289*, C293–301.

97. Davignon, I., Catalina, M.D., Smith, D., Montgomery, J., Swantek, J., Croy, J., Siegelman, M., and Wilkie, T.M. (2000). Normal hematopoiesis and inflammatory responses despite discrete signaling defects in Galpha15 knockout mice. Mol Cell Biol *20*, 797–804.

98. Wettschureck, N., Moers, A., Hamalainen, T., Lemberger, T., Schutz, G., and Offermanns, S. (2004). Heterotrimeric G proteins of the Gq/11 family are crucial for the induction of maternal behavior in mice. Mol Cell Biol *24*, 8048–8054.

99. Wettschureck, N., Lee, E., Libutti, S.K., Offermanns, S., Robey, P.G., and Spiegel, A.M. (2007). Parathyroid-specific double knockout of Gq and G11 alpha-subunits leads to a phenotype resembling germline knockout of the extracellular Ca2+ -sensing receptor. Mol Endocrinol *21*, 274–280.

100. Gu, J.L., Muller, S., Mancino, V., Offermanns, S., and Simon, M.I. (2002). Interaction of G alpha(12) with G alpha(13) and G alpha(q) signaling pathways. Proc Natl Acad Sci U S A *99*, 9352–9357.

101. Wirth, A., Benyó, Z., Lukasova, M., Leutgeb, B., Wettschureck, N., Gorbey, S., Örsy, P., Horváth, B., Maser-Gluth, C., Greiner, E., *et al.* (2007). G12/G13-LARG-mediated signalling in vascular smooth muscle is required for salt-induced hypertension. Nat Med *in press*.

102. Wettschureck, N., and Offermanns, S. (2005). Mammalian G proteins and their cell type specific functions. Physiol Rev *85*, 1159–1204.

8 Kinetics of GPCR, G protein, and effector activation

Peter Hein

INTRODUCTION

G protein-coupled receptors (GPCRs) are cell surface receptors that can be activated by a plethora of stimuli – including peptide hormones, odors, small molecules, ions, and light – and GPCR-mediated signaling plays important physiological and pathophysiological roles. Because GPCRs are expressed in virtually every cell type, and because they possess a highly specific binding pocket, they have become a prime target for therapeutic drugs.[1] As detailed in Chapter 2 of this book, all GPCRs share a conserved structure consisting of seven transmembrane helixes connected by intra- and extracellular loops.[2–4] Upon activation by agonists, GPCRs interact with heterotrimeric G proteins initiating the exchange of basally associated GDP for GTP. This causes the G protein to become activated.[5,6] Activated G proteins regulate target enzymes, for example, ion channels or adenylyl cyclase, and these effectors in turn regulate intracellular second messenger pools.

Historically, signaling kinetics have been assessed mainly in two ways: first, in biochemical and cellular, and second, in physiological assays. Both of these approaches led to relatively slow kinetics. For example, isoprenaline-mediated heart rate increase is measurable only after several tens of seconds,[7] agonist binding in biochemical receptor binding experiments after several minutes,[8] and agonist-induced GTPγS binding in membranes is in the order of minutes as well.[9]

These data place GPCR signaling kinetics between two extremes of cell surface receptors: slower than ligand-gated ion channels like e.g. the ionotropic glutamate receptor, but faster than receptor tyrosine kinases receptors. However, it has also long been clear that at least certain GPCRs are capable of signaling significantly faster. Examples include the GPCR rhodopsin[10] or receptor-mediated ion channel activation,[11,12] where GPCR effectors are activated in less than one second. This being said, it should be kept in mind that the *speed of signaling* always depends on what effect is measured; proximal effects, like rhodopsin's transition to metarhodopsin II, can be observed in 1 ms, while distal effects, like changes in gene expression, naturally take much longer.

Kinetic measurements of physiological processes at a cellular level have been possible only in a few model systems until recently (in the rhodopsin system, for Ca^{2+} measurements using specific dyes or Ca^{2+} proteins and a kinetic reader, and for ion channel activation using patch clamp techniques). However, in the past couple of years, fluorescence-based assays have been developed allowing for direct measurement of some of these processes in living cells. These methods rely on resonance energy transfer (RET) measurements between a donor molecule (either a fluorescent or a bioluminescent protein for use in FRET- and BRET-based assays, respectively) and a suitable acceptor fluorophor.[13,14] The most widely used combinations are cyan fluorescent protein (CFP) and yellow fluorescent protein (YFP) for FRET, and *Renilla* luciferase and cyan fluorescent protein (GFP) for BRET as donor and acceptor molecules, respectively. Since those proteins can be genetically encoded and heterologously expressed, they permit measurements in living, intact cells. The energy transfer is limited by the distance between donor and acceptor and by their relative orientation; a signal is only measurable when the distance between these molecules is less than 10nm.[15] Fusion of both donor and acceptor to conformation-sensitive domains of one protein allows to measure its conformational change, whereas fusing donor and acceptor to different molecules allows to measure distance and/or relative orientation of these two molecules toward each other.

In this chapter, GPCR signaling kinetics will be discussed, starting at the most proximal step in the cascade, ligand binding, covering receptor activation, interaction with and activation of G proteins, effector activation, and finally describing kinetics of receptor desensitization. A special emphasis is put on studies in intact cells.

Ligand binding

Traditionally, ligand binding to receptors has been assessed in radioligand binding studies and done in membrane preparations from cells expressing the receptor of interest. In those kind of assays, the apparent binding kinetics are determined by two processes: ligand binding to, and ligand dissociation from, receptors. With the former usually being diffusion-controlled, the latter determines the apparent association kinetics.[16] Since usually only ligand binding to high affinity sites is measured, dissociation kinetics tend to be long, and the time to reach binding equilibrium is in the tens of minutes.[8] This clearly does not reflect the physiological situation.

Two reported studies measured FRET between labeled receptors and ligands during stimulation of intact cells using the neurokinin NK_2[17] and the parathyroid hormone receptor (PTHR).[18] Both GPCRs are peptide hormone receptors that have been N-terminally labeled with GFP; neurokinin A has been labeled with TexasRed, and parathyroid hormone (PTH) with tetramethyl-rhodamine. Upon stimulation of the receptors with their respective agonists, a FRET signal could be detected and was interpreted as ligand binding to the receptor. Interestingly, both studies determined biphasic agonist binding kinetics; the interpretations, however, varied. For the NK_2 receptor, a fast binding step with a time constant of ≈1s and an about tenfold slower secondary binding step were observed. Since the kinetic of the fast step was equal to the cAMP production kinetics, it was suggested to reflect ligand binding. The slow binding step was interpreted as binding of the agonist to a population of interconverted high-affinity receptors, and the interconversion limited binding to this population.[17] For ligand binding to the PTHR, the fast step was measured with a time constant of ≈140ms, and the slow step with a time constant of ≈1s. It was explained that the initial binding step may reflect the interaction of PTH with the N terminus of the PTHR, whereas the slow binding step may reflect binding of the hormone to the receptor core; this should coincide with the conformational event.[18] Thus, in both studies the receptor-activating binding of the peptide ligands to the receptor had a time constant of ≈1s, which may not be representative for other GPCRs activated by small ligands.

Receptor activation

Agonist binding to a receptor is intrinsically connected to a conformational change of that receptor establishing an active conformation. Whereas there may be multiple 'active' conformations – depending on which downstream signalling is evoked by receptor activation – the nature of the switch required to activate the classical GPCR mediator, a heterotrimeric G protein, has been fairly well established for adrenoceptors.[19] (Also see Chapter 2 of this book.) This conformational switch includes a movement of helices III and VI, making the ends of these helices a prime location for monitoring said conformational switch.

Kobilka and colleagues developed a method to monitor the activation switch of the β_2-adrenoceptor by introducing specific cysteine residues into a receptor mutant devoid of other accessible cysteines. After receptor purification and reconstitution in lipid vesicles, the introduced cysteine could be labeled with small fluorophores whose fluorescence depends on its local environment. This approach allowed to monitor agonist-induced changes of fluorescence, caused by a movement of the fluorophore.[20] Agonist stimulation led to a change in fluorescence with a biphasic kinetic; the fast component had a time constant of 2.6s, and the slow component of 147s.[21] The authors suggested that the fast component reflects the conformational switch that establishes a receptor state capable of G protein activation, and that the slow component indicates interaction of the receptor with G protein-coupled receptor kinases (GRKs) and/or arrestins.[21] Further studies established that binding of the agonist is not a one-step process in itself, but rather consists of sequential binding events of different functional groups of the agonist molecule to the receptor core.[21,22] Nevertheless, the proposed kinetics of several seconds for the fast, active conformation-establishing conformational change seemed rather slow, given that conformational switches in, for example, ion channels can occur on the time scale of at least μs.[23,24]

In the visual system, the switch from rhodopsin to metarhodopsin II, the form that activates G proteins, is established 1ms after light activation of rhodopsin.[25] This system is especially suited to measure conformational changes of the receptor, because different rhodopsin states possess different optical properties that can be relatively easy measured. However, this approach can not be applied to other GPCRs.

A more recent strategy to monitor receptor conformational switches has been to develop FRET-based biosensors by fusing fluorophores to the third intracellular loop and the C terminus of a GPCR, and to express these constructs heterologously. This approach has been taken for the α_{2A}-adrenoceptors,[26–28] the β_1-adrenoceptors,[29] the A_{2A} adenosine,[26] and the bradykinin B_2 receptor,[30] as well as the PTHR.[28,31] For these constructs, receptor activation leads to a change in distance and/or orientation of the fluorophores translating into a change in FRET. Stimulation of the α_{2A}- and β_1-adrenoceptors and the A_{2A} adenosine receptors with their respective agonists yielded signal changes that could be fitted with an exponential curve to calculate a time constant for receptor activation.[28] This time constant was ≈ 50ms for the aforementioned receptors, about 1s for the PTH receptor,[28] and even slower for the bradykinin B_2 receptor.[30] For the PTHR, the activation time constant is on the same time scale as the slower of the two agonist-binding steps observed at this receptor,[18] reinforcing the notion that this slower step reflects final ligand binding to the receptor. Hence, ligand binding to the PTHR seems time-limiting for receptor activation. This could also partially explain why there seem to be two groups of receptors – a 'fast' group consisting of the α_{2A}- and β_1-adrenoceptor and the A_{2A} adenosine receptor and a 'slow' group (PTH and bradykinin B_2 receptors) – considering that agonist binding could in each case limit the kinetic of the conformational switch, with smaller agonists having a faster binding kinetics and therefore enabling a faster receptor activation.

This is in line with a recent study on the metabotropic glutamate receptor type I ($mGluR_1$). This receptor belongs to the class C family of GPCRs and is a constitutive dimer. Glutamate is bound by the large N-terminal Venus Flytrap module and leads to a rearrangement of both receptors within the dimer, bringing together the bundles of the transmembrane helixes of both subunits. Insertion of either YFP or CFP to the second intracellular loop and coexpression of both constructs in PC12 cells led to a detectable change in FRET upon agonist stimulation, reflecting the said rearrangement.[32] Stimulation with the agonist glutamate at a saturating concentration led to a time constant for receptor activation of 10ms.

Together with glutamate, 30mM K$^+$ was applied and the potassium-evoked current measured simultaneously with the change in FRET to monitor the speed of solution exchange at the cell surface. Since the time constant of the K$^+$-induced current was 10ms as well, the authors concluded that the 10ms measured for receptor activation is the upper limit of the 'true' receptor activation kinetic, and is limited by the speed of agonist application.

Why are the receptor activation kinetics measured by FRET so slow, especially when compared to conformational switches of ligand-activated ion channels[33,34] or the GPCR rhodopsin?[25] In the light of the studies discussed earlier in this chapter, it seems most likely that at least for class A receptors, ligand access to receptors is time-limiting and not the activation of the receptor itself, because in whole-cell FRET experiments, thousands of receptors are examined simultaneously, and diffusible ligands may not all reach those receptors at the same time. (In the case of rhodopsin, receptor activation is triggered by light, which naturally can reach receptors much faster than any other ligand.) However, alternative explanations exist: Agonist binding to GPCRs may be a multistep process,[21,22] and the conformational switch measured in FRET experiments may coincide with the last binding step. Another possibility is that the conformational switch itself takes time – in this case, the receptor activation kinetic would be an intrinsic property of a specific receptor. The latter two possibilities are supported by the notion that different agonists elicit distinct receptor conformations with different kinetics on the α_{2A}-adrenoceptor.[27,35] These studies examined a number of ligands with varying efficacy and demonstrated that the agonist-stabilized and inverse agonist-stabilized receptor conformations reveal different kinetics, effectively ruling out ligand access as time-limiting and suggesting that either ligand binding or the conformational switch itself is time-limiting for receptor activation.

Receptor/G protein interaction

Following receptor activation but preceding G protein activation is the interaction of activated receptors with G proteins. This step is conceptually independent of the question of whether or not GPCRs and G proteins are precoupled[36]: receptor/G protein interaction and G protein activation can be separated experimentally.

Kinetics of the receptor/G protein interaction have been studied for the α_{2A}- and β_1-adrenoceptors, and the A$_{2A}$ adenosine receptors and their respective G proteins using FRET,[37,38] and for the α_{2A}- and β_2-adrenoceptors using BRET.[39,40] In both cases, the receptor and the G protein (on varying positions on the α and $\beta\gamma$ subunit) were labeled with GFP variants, and the constructs heterologously expressed to measure resonance energy transfer. While the FRET-based approaches yielded time constants of about 50ms for receptor/G protein interaction – and thus time constants not markedly different from those obtained for receptor activation itself – a time constant of about 300ms was determined in BRET experiments. The faster time constants measured with FRET approaches may either reflect measurement in a single cell (as opposed to measurements in a population of cells in the BRET-based experiments) or the possibility of higher sampling rates with FRET due to higher emission intensities.[14] Nevertheless, all studies showed a very fast interaction of activated receptors with G proteins, either because of precoupling or rapid collision coupling.[36]

Another set of studies again comes from the rhodopsin system. Light-activated rhodopsin activates its G protein, called transducin (G_t). It has been shown that monomeric rhodopsin in solution can activate G_t at the diffusion limit and at a rate of about 50 molecules per second,[41] indicating that the interaction must be at least that fast. Membrane preparations of rhodopsin even yielded activation rates between 600 and 1,300 G_t molecules per second, indicating a much faster interaction.[42] Moreover, a fast interaction of light-activated rhodopsin and transducin has also been shown using transgenic animals.[10]

G protein activation

Interaction of an activated GPCR with its G protein usually leads to activation of the G protein. Receptor-induced dissociation of the GDP is followed by GTP-binding to the $G\alpha$ subunit, a step that enables productive coupling of both $G\alpha$ and $G\beta\gamma$ to their respective effectors. GTP hydrolysis by a GTPase activity in $G\alpha$ then leads to reassembly of the inactive heterotrimer.[5,6] Hence, at least three distinct steps are required for a G protein to become activated: dissociation of GDP, binding of GTP, and at least one conformational change between and/or dissociation of $G\alpha$ and $G\beta\gamma$. (For the G_i system, subunit rearrangement rather than dissociation has been shown as a consequence of G_i activation.[43])

Traditionally, G protein activation has been measured quantifying receptor-stimulated [^{35}S]GTPγS binding in cell membranes. This assay reveals kinetics of several minutes for G protein activation,[44,45] and it is obvious that those times are slower than can be expected for G protein activation in physiological contexts. Moreover, this assay is difficult to perform for G_s,G_q, and G_{12} proteins due to low signal-to-noise ratios, if no $G\alpha$ subunit-specific antibodies are utilized. However, FRET- and BRET-based methods enabled to measure G protein activation kinetics, first in *Dictyostelium*,[46] then in yeast,[47] and in mammalian cells.[43] The basic principle here again is fusion of RET donor and acceptor molecules to the α and $\beta\gamma$ subunits of heterotrimeric G proteins; G protein activation then translates to a change in distance and/or orientation of the fluorophores that can be monitored as a change in RET.

Kinetic measurements of receptor-mediated G_s and G_i protein activation yielded activation time constants of $\approx$450ms[38] or several 100ms.[40] This is approximately ten times slower than the initial interaction of receptors and G proteins[38] and suggests that G protein activation is the first time-limiting step in the signaling cascade – at least in the heterologous expression systems suitable for direct measurements. Figure 8.1 shows sample traces for receptor activation, receptor/G protein interaction, and G protein activation.[38] Biochemical experiments suggest that GDP release is the step that is time-limiting for G protein activation.[48,49]

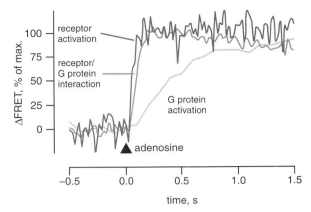

Figure 8-1: Sample FRET traces for early signaling processes. Original recordings of intramolecular FRET of the A_{2A} adenosine receptor, and FRET between the receptor and $G\alpha_s$, and between $G\alpha_s$ and $G\beta\gamma$. While receptor activation and receptor / G protein interaction are both fast processes (time constant 50ms), G protein activation is about tenfold slower (time constant 450ms) (Hein et al., 2006).

As discussed earlier in this chapter, in the visual system several hundreds (or even a thousand) G_t proteins become active per second and per light-activated rhodopsin molecule.[42,50] Thus this is the fastest G protein activation rate observed so far and points to a very efficient activation mechanism (which would include fast GDP release from inactive G_t) that may be unique to the rhodopsin/G_t system.

It should be noted that a regulator of G protein signaling (RGS) proteins can markedly speed up receptor-mediated G protein activation as well as G protein deactivation, thereby significantly modifying signaling kinetics.[51,52]

Effector molecules and second messengers

Downstream of the β_2-adrenoceptor are adenylyl cyclases (ACs), enzymes that produces cyclic AMP (cAMP) from ATP. cAMP is generally measured by antibody-based assays following cell lysis, making real-time measurements impossible. To overcome this limitation, FRET-based techniques to indirectly measure cAMP levels were developed. The first of those assays measured activation of protein kinase A (PKA), an effector of cAMP. cAMP leads to dissociation of the PKA regulatory subunits from the catalytical subunits, and PKA activation could directly be measured by attaching FRET donor and acceptor molecules to those different PKA subunits.[53] cAMP dynamics can be monitored in living cells using FRET-based biosensors based on cAMP effectors.[27,54–57] These studies revealed relatively slow kinetics for cAMP: Whole-cell FRET signals indicated relatively slow changes in cAMP concentrations after a lag of several seconds, and the increase was on a time-scale of tens of seconds or minutes.[38,55,57] These kinetics seem to be determined largely by phosphodiesterase (PDE)-mediated cAMP hydrolysis, since PDEs decrease cAMP faster than ACs can synthesize it.[58] These studies have been performed in whole cells, so it is possible that locally, cAMP may be regulated with faster kinetics.

A second, well-studied effector system is G protein-mediated activation of ion channels. Those can be analyzed directly, without any labeling, using the patch-clamp technique with a very high temporal resolution. For example, G protein-activated, inwardly rectifying potassium (GIRK) channels are activated by $G_{\beta g}$ subunits in less than 1s after activation of A_1 adenosine or α_{2A}-adrenoceptor, less than 2s after D_{2S} dopamine, and less than 4s after M_4 muscarinergic acetylcholine (M_4ACh) receptor activation;[59] activation kinetics mediated by stimulation of D_{2S} dopamine or M_4ACh receptors could be markedly accelerated by RGS proteins.[60] The time constant of μ-opioid receptor-mediated GIRK channel activation is less than 100ms and depends on amount of receptor cDNA transfected.[61] Since electrophysiological measurements are not dependent on fluorescence, it is even possible to simultaneously measure ion channel activation and FRET signals. This has been done for the α_{2A}-adrenoceptor/G_i interaction and the G_i-mediated activation of GIRK channels,[37] as well as for G_i activation itself .[43] The latter studies demonstrated that G protein activation measured by FRET follows exactly the same kinetics as GIRK channel currents stimulated by those G proteins. The same kinetics for channel activation were observed when wild-type G proteins were used, indicating that labeled G proteins do not activate slower than their non-labelled, wild-type counterparts.[43]

Interlude – where is the time-limiting step?

G protein activation seems to represent the time-limiting step in initiating signaling. However, the studies discussed in this chapter were carried out in heterologous expression systems because overexpression is necessary for bioluminescent RET readouts So, what step is time-limiting in physiological systems?

The first time-limiting step is agonist binding to the receptors, even at relatively high agonist concentrations.[32] Under conditions of moderately high or high receptor and G protein expression levels, coupling of receptors to G proteins can be as fast as receptor activation;[37,42] then, activation of G proteins is the next time-limiting step,[38] at least in the absence of RGS proteins. However, the law of mass action proposes that coupling kinetics depend on the concentration of receptors and G proteins, with lower concentrations of receptors and/or G proteins yielding slower interaction kinetics. This has been shown experimentally for α_{2A}-adrenoceptorsand G_i proteins.[37] Therefore, under conditions of low receptor expression, the receptor/G protein interaction may become another time-limiting step. This has, in fact, been demonstrated for channel activation in native systems: In supraventricular cardiomyocytes, A_1 adenosine receptors activate GIRK channels with a slow kinetic,[62] but channel activation kinetics increase upon adenoviral overexpression of the receptors.[63] Likewise, M_2AChR-mediated GIRK channel activation kinetics depend on the availability of M_2AChRs in ventricular myocytes.[12,64]

In conclusion, the physiological context will dictate what the time-limiting step in the signaling cascade is, and different experimental setups will likely show different time-limiting steps. Especially in non-overexpressing systems, the relative concentrations of signaling molecules, their subcellular localization, and the presence of accessory proteins like RGS proteins will influence kinetics of distinct steps. Moreover, there may be more than one time-limiting step within a single signaling cascade, making it even harder to generalize findings.

Noncanonical signaling kinetics

In recent years, it has become increasingly clear that GPCRs do not only signal via heterotrimeric G proteins, but also through other pathways (termed 'noncanonical' signaling). While there is a number of noncanonical signaling pathways,[65] the best-studied of these are the ones mediated by arrestins.[66] Perhaps the best-studied example of these is the activation of the extracellular signal-related kinase (ERK), which can be mediated by arrestins but, of course, also by G_i proteins. Those pathways are, in general, non-overlapping and follow different kinetics. Usually, G protein-mediated ERK activation kinetics are faster than arrestin-mediated ERK activation kinetics. After stimulation of AT_{1A} angiotensin receptors, G protein-mediated ERK activation peaks about 2min after agonist stimulation and decreases thereafter, whereas arrestin-mediated ERK activation follows a slower kinetic and is most pronounced 10min after agonist stimulation.[67] Both times are much longer for the PTHR: G protein-mediated ERK activation peaks at 10min, but arrestin-mediated at around 60min.[68] These studies employed disrupting the cells following agonist stimulation to determine ERK activities using phospho-specific antibodies. Although the time-resolution is on a minute scale at best, it seems enough to analyze these relatively slow kinetics.

GPCR phosphorylation and arrestin binding

After agonist stimulation, receptors get phosphorylated by PKA or members of the G protein-coupled receptor kinase (GRK) family. Phosphorylation is believed to uncouple the receptor from G proteins and to enable arrestin binding to activated phosphorylated receptors. These processes have been monitored using FRET-based approaches. While direct measurement of the receptor / GRK interaction has not been measured so far, FRET has been measured between β_2-adrenoceptors and βarrestins. These studies demonstrated that the initial recruitment of βarrestin to activated receptors has a half-time of about ≈20s.[69,70] However, after agonist withdrawal followed by a second agonist stimulation, the interaction kinetics significantly increased to a half-time around 2s.[69] This suggests that GRK phosphorylation of receptors is limiting – the first receptor/βarrestin interaction depends on prior receptor phosphorylation, whereas βarrestins can bind much faster to the still phosphorylated receptors during the second agonist stimulation.[69] In HEK293 cells as well, measurements of β_2 adrenoceptor phosphorylation by GRK and PKA using specific antibodies yield much slower kinetics: Here, receptor phosphorylation can be detected as most pronounced a few minutes after stimulation with maximal epinephrine concentration.[71]

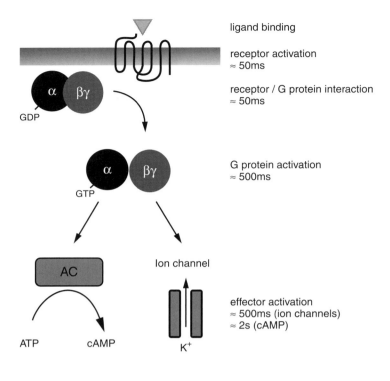

ligand binding

receptor activation
≈ 50ms

receptor / G protein interaction
≈ 50ms

G protein activation
≈ 500ms

effector activation
≈ 500ms (ion channels)
≈ 2s (cAMP)

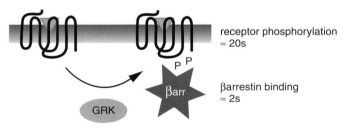

receptor phosphorylation
≈ 20s

βarrestin binding
≈ 2s

Figure 8-2: Kinetics of different steps of signal transduction. Given are approximate time constants / half times for different steps in a GPCR-originating signaling cascade. Stimulation with a saturating agonist concentration is followed by fast receptor activation and receptor / G protein interaction. G protein activation is time-limiting, but subsequent signaling steps can be as fast.

EPILOGUE

The last five years brought unprecedented insights in cellular signaling, made possible largely by novel optical techniques. FRET- and BRET-based methods showed that proximal signaling kinetics – up to the receptor/G protein interaction – can have time constants of 50ms or less. The initial steps, receptor activation and receptor/G protein interaction, seem to depend on ligand binding to the receptors – a technical obstacle that may be overcome by using, for example, caged ligands that can be set free in the immediate vicinity of a receptor. In high-expressing systems, G protein activation is time-limiting by at least an order of magnitude. Subsequent steps may be as fast (like ion channel activation) or slower (like cAMP accumulation). Deactivation of receptors is again slower by an order of magnitude when compared to G protein activation. Figure 8.2 gives approximate times for different steps within a GPCR-originating signaling cascade.

However, with the high temporal resolution necessary for these kinds of experiments, spatial resolution is often low or missing, and it seems plausible that, for example, second messenger pools get regulated faster near their site of synthesis than if looked at on a whole-cell basis. This is one major challenge for future experiments. A second challenge is to examine signaling kinetics not in over-expression, but in native systems. For this, either different labeling methods of native cells are needed, or label-free approaches. Ultimately, the goal is to be able to monitor signaling not only in intact cells, but in intact tissues or even animals, to assess kinetics of physiological signaling processes.

SUGGESTED READING

1. Overington, J.P., Al-Lazikani, B., and Hopkins, A.L. (2006). How many drug targets are there? Nat Rev Drug Discov 5, 993–996.
2. Cherezov, V., Rosenbaum, D.M., Hanson, M.A., Rasmussen, S.G., Thian, F.S., Kobilka, T.S., Choi, H.J., Kuhn, P., Weis, W.I., Kobilka, B.K., *et al.* (2007). High-resolution crystal structure of an engineered human beta2-adrenergic G protein-coupled receptor. Science 318, 1258–1265.
3. Palczewski, K., Kumasaka, T., Hori, T., Behnke, C.A., Motoshima, H., Fox, B.A., Le Trong, I., Teller, D.C., Okada, T., Stenkamp, R.E., *et al.* (2000). Crystal structure of rhodopsin: A G protein-coupled receptor. Science 289, 739–745.
4. Rosenbaum, D.M., Cherezov, V., Hanson, M.A., Rasmussen, S.G., Thian, F.S., Kobilka, T.S., Choi, H.J., Yao, X.J., Weis, W.I., Stevens, R.C., *et al.* (2007). GPCR engineering yields high-resolution structural insights into beta2-adrenergic receptor function. Science 318, 1266–1273.
5. Bourne, H.R. (1997). How receptors talk to trimeric G proteins. Curr Opin Cell Biol 9, 134–142.
6. Gilman, A.G. (1987). G proteins: transducers of receptor-generated signals. Annu Rev Biochem 56, 615–649.
7. Tyagi, A., Sethi, A.K., and Chatterji, C. (2002). Comparison of isoprenaline with adrenaline as components of epidural test dose solutions for halothane anaesthetized children. Anaesthesia and intensive care 30, 29–35.
8. Gantzos, R.D., and Neubig, R.R. (1988). Temperature effects on a2-adrenergic receptor-Gi interactions. Biochem Pharmacol 37, 2815–2821.
9. Biddlecome, G.H., Berstein, G., and Ross, E.M. (1996). Regulation of phospholipase C-b1 by Gq and m1 muscarinic cholinergic receptor. Steady-state balance of receptor-mediated activation and GTPase-activating protein-promoted deactivation. J Biol Chem 271, 7999–8007.
10. Makino, C.L., Wen, X.H., and Lem, J. (2003). Piecing together the timetable for visual transduction with transgenic animals. Curr Opin Neurobiol 13, 404–412.
11. Bünemann, M., Bücheler, M.M., Philipp, M., and Hein, L. (2001). Activation and deactivation kinetics of alpha 2A- and alpha 2C-adrenergic receptor-activated G protein-activated inwardly rectifying K+ channel currents. J Biol Chem 276, 47512–47517.
12. Bünemann, M., Brandts, B., and Pott, L. (1997). In vivo downregulation of M2 receptors revealed by measurement of muscarinic K+ current in cultured guinea-pig atrial myocytes. J Physiol 501, 549–554.
13. Förster, T. (1948). Zwischenmolekulare Energiewanderung und Fluoreszenz. Ann Phys (Leipzig) 2, 55–75.

14. Marullo, S., and Bouvier, M. (2007). Resonance energy transfer approaches in molecular pharmacology and beyond. Trends Pharmacol Sci.

15. Vogel, S.S., Thaler, C., and Koushik, S.V. (2006). Fanciful FRET. Sci STKE *2006*, re2.

16. Motulsky, H.J., and Mahan, L.C. (1984). The kinetics of competitive radioligand binding predicted by the law of mass action. Mol Pharmacol *25*, 1–9.

17. Palanche, T., Ilien, B., Zoffmann, S., Reck, M.P., Bucher, B., Edelstein, S.J., and Galzi, J.L. (2001). The neurokinin A receptor activates calcium and cAMP responses through distinct conformational states. J Biol Chem *276*, 34853–34861.

18. Castro, M., Nikolaev, V.O., Palm, D., Lohse, M.J., and Vilardaga, J.P. (2005). Turn-on switch in parathyroid hormone receptor by a two-step parathyroid hormone binding mechanism. Proc Natl Acad Sci USA *102*, 16084–16089.

19. Gether, U. (2000). Uncovering molecular mechanisms involved in activation of G protein-coupled receptors. Endocr Rev *21*, 90–113.

20. Gether, U., Lin, S., and Kobilka, B.K. (1995). Fluorescent labeling of purified beta 2 adrenergic receptor. Evidence for ligand-specific conformational changes. J Biol Chem *270*, 28268–28275.

21. Swaminath, G., Xiang, Y., Lee, T.W., Steenhuis, J.J., Parnot, C., and Kobilka, B.K. (2004). Sequential binding of agonists to the b2 adrenoceptor. Kinetic evidence for intermediate conformational states. J Biol Chem *279*, 686–691.

22. Yao, X., Parnot, C., Deupi, X., Ratnala, V.R., Swaminath, G., Farrens, D.L., and Kobilka, B.K. (2006). Coupling ligand structure to specific conformational switches in the β2-adrenoceptor. Nat Chem Bio *2*, 417–422.

23. Colquhoun, D., and Sakmann, B. (1981). Fluctuations in the microsecond time range of the current through single acetylcholine receptor ion channels. Nature *294*, 464–466.

24. Robert, A., and Howe, J.R. (2003). How AMPA receptor desensitization depends on receptor occupancy. J Neurosci *23*, 847–858.

25. Okada, T., Ernst, O.P., Palczewski, K., and Hofmann, K.P. (2001). Activation of rhodopsin: new insights from structural and biochemical studies. Trends Biochem Sci *26*, 318–324.

26. Hoffmann, C., Gaietta, G., Bünemann, M., Adams, S.R., Oberdorf-Maass, S., Behr, B., Vilardaga, J.P., Tsien, R.Y., Ellisman, M.H., and Lohse, M.J. (2005). A FlAsH-based FRET approach to determine G protein-coupled receptor activation in living cells. Nat Methods *2*, 171–176.

27. Nikolaev, V.O., Hoffmann, C., Bünemann, M., Lohse, M.J., and Vilardaga, J.P. (2006). Molecular basis of partial agonism at the neurotransmitter α2A-adrenergic receptor and Gi-protein heterotrimer. J Biol Chem *281*, 24506–24511.

28. Vilardaga, J.P., Bünemann, M., Krasel, C., Castro, M., and Lohse, M.J. (2003). Measurement of the millisecond activation switch of G protein-coupled receptors in living cells. Nat Biotechnol *21*, 807–812.

29. Rochais, F., Vilardaga, J.P., Nikolaev, V.O., Bünemann, M., Lohse, M.J., and Engelhardt, S. (2007). Real-time optical recording of beta1-adrenergic receptor activation reveals supersensitivity of the Arg389 variant to carvedilol. J Clin Invest *117*, 229–235.

30. Chachisvilis, M., Zhang, Y.L., and Frangos, J.A. (2006). G protein-coupled receptors sense fluid shear stress in endothelial cells. Proc Natl Acad Sci USA *103*, 15463–15468.

31. Zhang, Y., Frangos, J.A., and Chachisvilis, M. (2009). Mechanical stimulus alters conformation of type 1 parathyroid hormone receptor in bone cells. Am J Physiol, Cell Physiol.

32. Marcaggi, P., Mutoh, H., Dimitrov, D., Beato, M., and Knöpfel, T. (2009). Optical measurement of mGluR1 conformational changes reveals fast activation, slow deactivation, and sensitization. Proc Natl Acad Sci USA.

33. Auerbach, A. (2005). Gating of acetylcholine receptor channels: brownian motion across a broad transition state. Proc Natl Acad Sci USA *102*, 1408–1412.

34. Macdonald, R.L., and Olsen, R.W. (1994). GABAA receptor channels. Annu Rev Neurosci *17*, 569–602.

35. Vilardaga, J.P., Steinmeyer, R., Harms, G.S., and Lohse, M.J. (2005). Molecular basis of inverse agonism in a G protein-coupled receptor. Nat Chem Bio *1*, 25.

36. Hein, P., and Bünemann, M. (2009). Coupling mode of receptors and G proteins. Naunyn Schmiedebergs Arch Pharmacol *379*, 435–443.

37. Hein, P., Frank, M., Hoffmann, C., Lohse, M.J., and Bünemann, M. (2005). Dynamics of receptor/G protein coupling in living cells. EMBO J *24*, 4106–4114.

38. Hein, P., Rochais, F., Hoffmann, C., Dorsch, S., Nikolaev, V.O., Engelhardt, S., Berlot, C.H., Lohse, M.J., and Bünemann, M. (2006). Gs activation is time-limiting in initiating receptor-mediated signaling. J Biol Chem *281*, 33345–33351.

39. Gales, C., Rebois, R.V., Hogue, M., Trieu, P., Breit, A., Hebert, T.E., and Bouvier, M. (2005). Real-time monitoring of receptor and G-protein interactions in living cells. Nat Methods *2*, 177–184.

40. Gales, C., Van Durm, J.J., Schaak, S., Pontier, S., Percherancier, Y., Audet, M., Paris, H., and Bouvier, M. (2006). Probing the activation-promoted structural rearrangements in preassembled receptor-G protein complexes. Nat Struct Biol *13*, 778–786.

41. Ernst, O.P., Gramse, V., Kolbe, M., Hofmann, K.P., and Heck, M. (2007). Monomeric G protein-coupled receptor rhodopsin in solution activates its G protein transducin at the diffusion limit. Proc Natl Acad Sci USA *104*, 10859–10864.

42. Heck, M., and Hofmann, K.P. (2001). Maximal rate and nucleotide dependence of rhodopsin-catalyzed transducin activation: initial rate analysis based on a double displacement mechanism. J Biol Chem *276*, 10000–10009.

43. Bünemann, M., Frank, M., and Lohse, M.J. (2003). Gi protein activation in intact cells involves subunit rearrangement rather than dissociation. Proc Natl Acad Sci USA *100*, 16077–16082.

44. Akam, E.C., Carruthers, A.M., Nahorski, S.R., and Challiss, R.A. (1997). Pharmacological characterization of type 1alpha metabotropic glutamate receptor-stimulated [35S]-GT-PgammaS binding. Br J Pharmacol *121*, 1203–1209.

45. Cowburn, R.F., Wiehager, B., Ravid, R., and Winblad, B. (1996). Acetylcholine muscarinic M2 receptor stimulated [35S]GTP gamma S binding shows regional selective changes in Alzheimer's disease postmortem brain. Neurodegeneration : a journal for neurodegenerative disorders, neuroprotection, and neuroregeneration *5*, 19–26.

46. Janetopoulos, C., Jin, T., and Devreotes, P.N. (2001). Receptor-mediated activation of heterotrimeric G-proteins in living cells. Science *291*, 2408–2411.

47. Yi, T.M., Kitano, H., and Simon, M.I. (2003). A quantitative characterization of the yeast heterotrimeric G protein cycle. Proc Natl Acad Sci USA *100*, 10764–10769.

48. Higashijima, T., Ferguson, K.M., Sternweis, P.C., Smigel, M.D., and Gilman, A.G. (1987). Effects of Mg2+ and the bg-subunit complex on the interactions of guanine nucleotides with G proteins. J Biol Chem *262*, 762–766.

49. Mukhopadhyay, S., and Ross, E.M. (1999). Rapid GTP binding and hydrolysis by Gq promoted by receptor and GTPase-activating proteins. Proc Natl Acad Sci USA *96*, 9539–9544.

50. Fung, B., and Stryer, L. (1980). Photolyzed rhodopsin catalyzes the exchange of GTP for bound GDP in retinal rod outer segments. Proc Natl Acad Sci USA *77*, 2500–2504.

51. Doupnik, C.A., Davidson, N., Lester, H.A., and Kofuji, P. (1997). RGS proteins reconstitute the rapid gating kinetics of Gbg-activated inwardly rectifying K+ channels. Proc Natl Acad Sci USA *94*, 10461–10466.

52. Ross, E.M. (2008). Coordinated Speed and Amplitude in G-Protein Signaling. Curr Biol *18*, R777–R783.

53. Adams, S.R., Harootunian, A.T., Buechler, Y.J., Taylor, S.S., and Tsien, R.Y. (1991). Fluorescence ratio imaging of cyclic AMP in single cells. Nature *349*, 694–697.

54. Dipilato, L.M., Cheng, X., and Zhang, J. (2004). Fluorescent indicators of cAMP and Epac activation reveal differential dynamics of cAMP signaling within discrete subcellular compartments. Proc Natl Acad Sci USA *101*, 16513–16518.

55. Nikolaev, V.O., Bünemann, M., Hein, L., Hannawacker, A., and Lohse, M.J. (2004). Novel Single Chain cAMP Sensors for Receptor-induced Signal Propagation. J Biol Chem *279*, 37215–37218.

56. Ponsioen, B., Zhao, J., Riedl, J., Zwartkruis, F., van der Krogt, G., Zaccolo, M., Moolenaar, W.H., Bos, J.L., and Jalink, K. (2004). Detecting cAMP-induced Epac activation by fluorescence resonance energy transfer: Epac as a novel cAMP indicator. EMBO Rep *5*, 1176–1180.

57. Zaccolo, M., De Giorgi, F., Cho, C.Y., Feng, L., Knapp, T., Negulescu, P.A., Taylor, S.S., Tsien, R.Y., and Pozzan, T. (2000). A genetically encoded, fluorescent indicator for cyclic AMP in living cells. Nat Cell Biol *2*, 25–29.

58. Nikolaev, V.O., Gambaryan, S., Engelhardt, S., Walter, U., and Lohse, M.J. (2005). Real-time Monitoring of the PDE2 Activity of Live Cells: hormone-stimulated cAMP hydrolysis is faster than hormone-stimulated cAMP synthesis. J Biol Chem *280*, 1716–1719.

59. Benians, A., Leaney, J.L., Milligan, G., and Tinker, A. (2003). The dynamics of formation and action of the ternary complex revealed in living cells using a G-protein-gated K+ channel as a biosensor. J Biol Chem *278*, 10851–10858.

60. Benians, A., Nobles, M., Hosny, S., and Tinker, A. (2005). Regulators of G-protein signaling form a quaternary complex with the agonist, receptor, and G-protein. A novel explanation for the acceleration of signaling activation kinetics. J Biol Chem *280*, 13383–13394.

61. Lober, R.M., Pereira, M.A., and Lambert, N.A. (2006). Rapid activation of inwardly rectifying potassium channels by immobile G-protein-coupled receptors. J Neurosci *26*, 12602–12608.

62. Bünemann, M., and Pott, L. (1995). Down-regulation of A1 adenosine receptors coupled to muscarinic K+ current in cultured guinea-pig atrial myocytes. J Physiol (Lond) *482 (Pt 1)*, 81–92.

63. Wellner-Kienitz, M.C., Bender, K., Meyer, T., Bünemann, M., and Pott, L. (2000). Overexpressed A(1) adenosine receptors reduce activation of acetylcholine-sensitive K(+) current by native muscarinic M(2) receptors in rat atrial myocytes. Circ Res *86*, 643–648.

64. Bünemann, M., Brandts, B., and Pott, L. (1996). Downregulation of muscarinic M2 receptors linked to K+ current in cultured guinea-pig atrial myocytes. J Physiol *494*, 351–362.

65. Hall, R.A., Premont, R.T., and Lefkowitz, R.J. (1999). Heptahelical receptor signaling: beyond the G protein paradigm. The Journal of cell biology *145*, 927–932.

66. DeWire, S.M., Ahn, S., Lefkowitz, R.J., and Shenoy, S.K. (2007). Beta-arrestins and cell signaling. Annu Rev Physiol *69*, 483–510.

67. Ahn, S., Shenoy, S.K., Wei, H., and Lefkowitz, R.J. (2004). Differential kinetic and spatial patterns of beta-arrestin and G protein-mediated ERK activation by the angiotensin II receptor. J Biol Chem *279*, 35518–35525.

68. Gesty-Palmer, D., Chen, M., Reiter, E., Ahn, S., Nelson, C.D., Wang, S., Eckhardt, A.E., Cowan, C.L., Spurney, R.F., Luttrell, L.M., *et al.* (2006). Distinct beta-arrestin- and G protein-dependent pathways for parathyroid hormone receptor-stimulated ERK1/2 activation. J Biol Chem *281*, 10856–10864.

69. Krasel, C., Bünemann, M., Lorenz, K., and Lohse, M.J. (2005). b-Arrestin Binding to the b2-Adrenergic Receptor Requires Both Receptor Phosphorylation and Receptor Activation. J Biol Chem *280*, 9528–9535.

70. Violin, J.D., Ren, X.R., and Lefkowitz, R.J. (2006). G-protein-coupled receptor kinase specificity for β-arrestin recruitment to the β2-adrenergic receptor revealed by fluorescence resonance energy transfer. J Biol Chem *281*, 20577–20588.

71. Tran, T.M., Friedman, J., Qunaibi, E., Baameur, F., Moore, R.H., and Clark, R.B. (2004). Characterization of agonist stimulation of cAMP-dependent protein kinase and G protein-coupled receptor kinase phosphorylation of the b2-adrenergic receptor using phosphoserine-specific antibodies. Mol Pharmacol *65*, 196–206.

72. Nikolaev, V.O., and Lohse, M.J. (2006). Monitoring of cAMP synthesis and degradation in living cells. Physiology *21*, 86–92.

9 RGS-RhoGEFs and other RGS multidomain proteins as effector molecules in GPCR-dependent and GPCR-independent cell signaling

José Vázquez-Prado and J. Silvio Gutkind

INTRODUCTION

Regulators of G protein signaling (RGS) proteins were originally discovered as small GTPase activating proteins that contribute to shutting down the signaling of Gα subunits by accelerating the hydrolysis of GTP bound to active GTPases.[1–3] Their activities helped explain the rapid regulation of Gα signaling in multiple biological systems. With the definition of a ~120 amino acid consensus sequence that constitutes an RGS domain, it was observed that RGS homology domains are also found in larger and complex proteins that also contain additional structural features and domains as part of their diverse architecture (Figure 9.1). These additional protein regions endow large RGS-containing proteins with the ability to interact with lipids and other molecules. In some cases, RGS-containing proteins are also equipped with catalytic domains that provide a link from Gα subunits to downstream effectors via direct interactions. These complex multidomain proteins, now considered bona fide effectors and scaffolds of heterotrimeric G protein α subunits, represent central nodes of complex intracellular networks that integrate the actions of GPCRs and other plasma membrane receptors. This group of multidomain RGS proteins includes members of the family of G protein-coupled receptor kinases (GRKs),[4] RGS-RhoGEFs,[5–7] RGS7 and RGS12,[1] and Axin.[8] In these cases, interaction with the cognate Gα subunit putatively causes a conformational change in the RGS protein that exposes other domains. In the case of RhoGEFs harboring an RGS homology domain (also known as rgRGS, RGL, or RH), they are specific for the small GTPase RhoA, previously known to be activated in response to GPCRs such as lysophosphatidic acid and thrombin receptors. The family of RGS-RhoGEFs, formed by p115-RhoGEF,[9] PDZ-RhoGEF,[10] and Leukemia Associated RhoGEF (LARG),[11] is activated by direct interactions with Gα$_{12/13}$ family of G proteins, constituting the molecular link between this family of Gα subunits and the small GTPase RhoA, in the signaling cascade regulating the actin cytoskeleton and nuclear events, independent of the generation

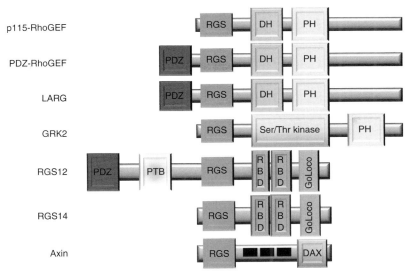

Figure 9-1: Structure of multidomain RGS proteins including $G\alpha_{12/13}$-regulated Rho guanine exchange factors (p115-RhoGEF, PDZ-RhoGEF and LARG), G protein-coupled receptor kinase 2 (GRK2), RGS12 and 14, and Axin. PDZ domains are primarily involved in protein-protein interactions; the best known involve the association with plasma membrane proteins containing a PDZ-interacting consensus sequence at their carboxyl terminus. PTB domain interacts with phospho-tyrosine containing proteins, RGS domain is mostly known to act as a GAP for $G\alpha$ subunits of heterotrimeric G proteins; it has evolved in multidomain RGS-containing proteins as an effector domain linking G proteins and their coupled receptors to downstream signaling pathways. Dbl-homology domain (DH) defines a family of guanine exchange factors specific for Rho GTPases; GRK2 serine-threonine kinase mainly phosphorylates agonist activated G protein-coupled receptors, although other non-GPCR targets have been identified; PH domain interacts primarily with membrane lipids. GoLoco domain interacts with GDP-bound $G\alpha_i$. The DAX domain of Axin is homologous to the DIX domain in Disheveled; both signaling proteins of the Wnt pathway; this domain is involved in protein-protein interactions, including self-association. *See colour plate 8.*

of second messengers. Enzymes such as PLC epsilon can be activated downstream of this pathway, thus generating second messengers.[12–14] Other plasma membrane receptors including plexins, which are receptors for semaphorins that contribute to the recognition of migratory patterns of the developing nervous and vascular systems, also interact with and activate these GEFs via the interaction with the PDZ domain present at the amino terminus of PDZ-RhoGEF and LARG, two members of this family of RGS-RhoGEFs.[15–17] RGS12 and RGS14, characterized by the presence of a couple of Ras-binding domains located in tandem, act as scaffolds in signaling downstream of growth factor receptors.[18,19] GRKs are versatile regulators of G protein signaling; the ubiquitous GRK2, which has served as a model to study this family of kinases, acts as an effector of $G\beta\gamma$ in the process of GPCR desensitization by phosphorylating the carboxyl terminus and third intracellular loop of agonist-activated receptors[20] and, at the same time, inhibit $G\alpha_q$ through a direct interaction with the RGS homology domain present at its amino terminus.[21] Axin has been recently recognized as a participant of an aberrant proliferative pathway in colon cancer cells stimulated by G_s-coupled prostaglandin E2 receptors and likely other G_{13}-coupled receptors, by a mechanism that depends on the interaction between Gα proteins and the RGS homology domain of Axin, leading to the accumulation of active β-catenin and the activation of a pathway relevant for the progression of colon cancer.[8] This chapter emphasizes the role of complex RGS-containing proteins that act as molecular scaffolds and as integral components of key signaling pathways initiated by both GPCR-dependent and GPCR-independent molecular mechanisms.

RGS-RHOGEFS AS EFFECTORS OF GPCR SIGNALING

Gα$_{12}$/Gα$_{13}$-regulated RGS-RhoGEFs (p115RhoGEF, PDZ-RhoGEF, LARG)

p115RhoGEF, PDZ-RhoGEF, and LARG

In the early 1990s, it was described that G protein-coupled receptors, such as those for lyosphosphatidic acid, promote remodeling of the actin cytoskeleton via the intervention of the small GTPase RhoA.[22] Similar effects were later observed when expressing constitutively active mutants of Gα$_{13}$, which caused Rho-dependent stress fiber formation and focal adhesion assembly.[23] Parallel efforts addressing the mechanisms by which GPCRs signal from the membrane to the nucleus thereby regulating the expression of growth promoting genes revealed that Rho link GPCRs to the activation of the serum response factor.[24] Of interest, a parallel pathway to the nucleus was found to be activated by the small GTPase Rac1[25–27] that also plays an important role in the regulation of actin-based cytoskeleton. These pioneering observations, aligned with the finding that mouse embryonic fibroblasts from Gα$_{13}$ knockout mice showed defective cell migration in response to thrombin,[28] evidenced the existence of a signaling pathway involved in the communication between "Big" (Gα$_{12/13}$, heterotrimeric) and "Small" (RhoA, monomeric) GTP-binding proteins. This biochemical route was elucidated at the molecular level with the discovery of a group of proteins that act as effectors linking G proteins directly to Rho activation. These effectors were identified as Gα$_{12/13}$-sensitive RhoGEFs, p115RhoGEF, PDZ-RhoGEF, and LARG,[6,9–11] which are regulated by direct interaction with active Gα$_{12/13}$ subunits. These RhoGEFs are characterized by the presence of an RGS-homology domain that serves as the site of physical interaction of GTP-bound Gα$_{12/13,}$ causing the activation of the guanine nucleotide exchange factor (GEF) activity of these RhoGEFs and the subsequent loading of GTP within the small GTPase RhoA (Figure 9.2).[6,9–11] Then, the active form of RhoA initiates the activation of multiple effectors, such as the serine/threonine kinase ROCK (also known as ROK), which promotes the assembly of actin stress fibers by regulating myosin-dependent contractile complexes via phosphorylation of myosin light chain (MLC) and MLC phosphatase (which is inhibited as a result of this phosphorylation).[29] A second mechanism depends on the phosphorylation of cofilin, an actin filament severing protein, by LIMK, which is itself phosphorylated by ROCK.[30] In addition, other Rho effectors, such as mDia, contribute to the assembly of stress fibers by directly promoting actin polymerization in response to its interaction with Rho, and to the acquisition

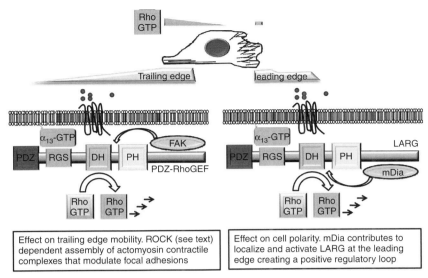

Figure 9-2: In migrating fibroblasts, Rho activation is important to promote the removal of focal adhesions at the trailing edge in response to lysophosphatidic acid, which elicits a $G\alpha_{13}$-dependent activation of PDZ-RhoGEF; this process is blocked by inhibiting focal adhesion kinase (FAK), which interacts with PDZ-RhoGEF. At the leading edge, LARG interacts with mDia in a Rho-GTP dependent manner and activates the GEF, creating a positive feedback loop that contributes to establish and maintain cell polarity. *See colour plate 9.*

of a polarized phenotype by contributing to the asymmetric localization of the microtubule organizing center.[31] In this case, Rho-GTP recognizes a region at the amino terminal domain of mDia, which causes the release of an auto-inhibitory interaction and uncovering two formin-homology domains present in this protein: FH1 that binds profilin, a known monomeric-actin sequestering protein, and FH2, which shows an actin nucleation and polymerization activity.[32] Whether mDia and other members of the formin-homology family are involved in GPCR signaling downstream of RGS-RhoGEFs is starting to be addressed. Current evidence indicates that, in the case of thrombin, it induces stress fibers in endothelial cells by the intervention of one of the formins. In this case, the auto-inhibitory interaction that keeps FHOD1 (the formin involved) is released in response to its phosphorylation by ROCK.[33] mDia1 also localizes, together with a fraction of active RhoA, to the front of migrating fibroblasts in a $G\alpha_{12/13}$-dependent manner, and this localization is essential to promote cell polarization and migration in a wound-healing assay.[31] Interestingly, as it will be described below, cell polarity is acquired in response to the $G\alpha_{12/13}$/LARG signaling pathway by adjusting the localization of the microtubule organizing center (mTOC), thus affecting microtubule dynamics.[34] In this event, it is important that LARG interacts with pericentrin, a protein that interacts with γ-tubulin and directly nucleates microtubules.[34] These results are coincident with earlier findings in Drosophila, in which, by genetic epistasis analysis, it was demonstrated that DRhoGEF2, a likely ancestral homolog of LARG and PDZ-RhoGEF, is activated downstream of concertina (the homolog of $G\alpha_{12/13}$) in a pathway that determines cell polarity during developmental cell movements occurring during gastrulation.[35] The effect of Rho in the actin cytoskeleton helps assemble a contractile machinery required for polarized cell motility. A wide array of fundamental biological events are based on this signaling pathway, which include cell movements during embryogenesis, wound healing, immune response, vascular development, and axonal growth.[7,36] It is also considered that alteration in these Rho-dependent events contribute to multiple pathological situations such as hypertension and uncontrolled cell proliferation.[6,36–38]

As discussed above, $G\alpha_{12/13}$ interact directly with structural elements of a family of RGS-RhoGEFs composed by three members: PDZ-RhoGEF, LARG, and p115Rho-GEF, all of them characterized by the presence of an RGS homology domain.[6] Interestingly, these RhoGEFs contain a series of additional structural elements that may enable a complex array of molecular interactions and thus confer specific regulatory properties.[5] Indeed, the multidomain nature of PDZ-RhoGEF and LARG provides elements for regulation by $G\alpha_{12/13}$ proteins, the plexin family of single-pass transmembrane receptors for semaphorins, and multiple serine/threonine and tyrosine kinases.[10,11,15,16,39–42] Each activity is clearly attributable to a different structural feature. Specifically, the amino terminal PDZ domain is recognized by plexins of the B family;[15–17,43,44] the RGS domain interacts with $G\alpha_{12/13}$ subunits in their active form;[10,11] the DH catalytic domain, followed by a PH domain, promotes the exchange of GDP for GTP on RhoA; while the carboxyl terminal domain determines the ability of the GEFs to dimerize and establishes direct interactions with PAK4,[39,40,42] a serine/threonine kinase identified as a Cdc42 effector.[45] The latter suggests the existence of a regulatory circuitry by which small GTPases of the Rho family can regulate the activation of each other (Figure 9.3). This may contribute to achieving the temporal regulation and spatial distribution of active Rho GTPases that is required for biological processes involving directional migration.

An additional member of the Dbl-homology group of RhoGEFs that has been linked to $G\alpha_{12/13}$ signaling to Rho is Lbc.[46,47] The overall structural characteristics of Lbc are different from those that define the family of RGS-RhoGEFs. Lbc does not contain an RGS-homology domain toward the amino terminus of the DH domain. Experimental evidence suggesting the inclusion of Lbc in the signaling pathway from $G\alpha_{12/13}$ to Rho includes the effect of a mutant Lbc lacking the DH domain, which inhibits rounding of 1321N1 astrocytoma cells in response to thrombin,[46] physical interactions between Lbc and active $G\alpha_{12/13}$ subunits,[48] and the presence at the C-terminal region of Lbc of a putative RGS-homology domain with a 34% identity at the amino acid level with the RGS-homology domain of PDZ-RhoGEF.[47] Further biochemical and structural studies are needed to elucidate the detailed molecular basis determining the reported regulation of Lbc by G proteins.

Cloning of $G\alpha_{12/13}$-regulated RGS-RhoGEFs

The identification of p115-RhoGEF[49] and LARG[50] preceded their characterization as $G\alpha_{12/13}$-effectors, whereas PDZ-RhoGEF was recognized as a $G\alpha_{12/13}$-sensitive

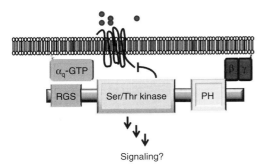

Signaling?

Figure 9-3: GRK2 is an effector of Gβγ that phosphorylates agonist stimulated G protein-coupled receptors initiating the process of desensitization. In addition, through its RGS domain, GRK2 interacts with Gα$_q$ in a similar way as an effector does, suggesting that GRK2 is an effector of Gα$_q$. *See colour plate 10.*

RhoGEF from its original discovery and characterization.[10] In the three cases, an RGS homology domain is located toward the amino terminus of a RhoGEF domain (also known as DH due to its homology with Dbl) followed by a PH domain. In addition, PDZ-RhoGEF and LARG contain an amino terminal PDZ domain that interacts with plexins of the B family, GPCRs, and even tyrosine kinase receptors; these interactions contribute to attract the GEFs to the plasma membrane.[15–17,43,51,52] The DH-PH cassette defines a family of RhoGEFs that in humans is composed by sixty-one members, most of them with a complex architecture putatively involved in the integration of different signals into the activation of Rho GTPases, a group of small GTPases whose representative members are RhoA, Rac1, and Cdc42.[37,53–55] These GTPases are linked to the control of cell shape by modulating the reorganization of actin cytoskeleton: RhoA promotes the assembly of stress fibers and contractile actomyosin complexes, whereas Cdc42 and Rac interact with effectors that lead to the formation of dynamic cellular extensions known as filopodia and lamellipodia, based also on the polymerization of actin.[29,53] Moreover, these GTPases are central nodes in the activation of signaling cascades that regulate gene expression.[25]

p115-RhoGEF. In order to understand the complexity of elements controlling the regulation of Rho, a nucleotide-depleted form of this GTPase was used as an affinity ligand in a protein purification protocol, using cell lysates of *src*-transformed NIH3T3 fibroblasts and Cos cells, leading to the identification of p115-RhoGEF. A partial peptide sequence of the isolated protein served to design oligonucleotides that helped to identify p115-RhoGEF cDNA clones from a human library. The sequence of p115-RhoGEF provided evidence of homology with Dbl.[49] The original description reported a particular abundance of p115-RhoGEF mRNA in peripheral blood leukocytes, spleen, and thymus.[49] Even before discovering the existence of an RGS homology domain at the amino terminus of p115-RhoGEF, its initial characterization demonstrated that a construct lacking this region exhibited transforming properties in transfected fibroblasts, indirectly suggesting that the amino terminus contained a negative regulatory element.[49] The identification of an RGS-homology domain in p115-RhoGEF resulted from multiple protein alignments that, beside this GEF, included bona fide RGS proteins such as RGS4, RGS2, and GAIP, and other RhoGEFs such as the drosophila DRhoGEF2, and the sequence of KIAA380 obtained from a database of long cDNAs coding for complex human proteins. The alignment was adjusted with the known structure of RGS4 and revealed the existence of an atypical RGS-homology region located at the amino terminus of p115-RhoGEF.[56] This RGS-homology domain was found to be a specific GAP for $G\alpha_{12}$ and $G\alpha_{13}$.[56] This specificity, in the context of the known capacity of $G\alpha_{12}$ and $G\alpha_{13}$ to activate Rho, suggested that p115-RhoGEF, besides acting as a GAP for $G\alpha_{12}$ and $G\alpha_{13}$, might also be an effector in the signaling pathway of this group of heterotrimeric G proteins to Rho. This hypothesis was demonstrated by in vitro reconstitution assays in parallel to the characterization of p115-RhoGEF as a GAP,[56] thus helping define a subset of RGS proteins with a dual activity as GAPs and effectors coupling RhoGEF activity to $G\alpha_{13}$ subunits.[9]

PDZ-RhoGEF. PDZ-RhoGEF was identified in protein databases as a multidomain RhoGEF hypothetically regulated by G proteins based on its structural properties.[10] The initial characterization validated this hypothesis. It was found that PDZ-RhoGEF directly interacts and forms a molecular complex with active $G\alpha_{12}$ and $G\alpha_{13}$ GTPases in vivo, thereby activating Rho, demonstrating that PDZ-RhoGEF represents a direct target of this family of heterotrimeric G proteins as part of the pathway that results in RhoA activation.[10] The structure of PDZ-RhoGEF, composed by an amino terminal PDZ domain followed by an RGS-like region homologous to p115-RhoGEF and RGS14, upstream of the characteristic DH-PH cassette and a long carboxyl terminal tail, further suggested that this RhoGEF might be regulated via multiple interactions. In fact, PDZ-RhoGEF mutants lacking the amino terminal region are highly active, indicating its important regulatory function. The RGS domain of PDZ-RhoGEF and its interaction with $G\alpha_{12/13}$ has been characterized at the structural level.[57] In addition, a systematic search for additional interactors demonstrated that the PDZ domain associates with the carboxyl terminal tail of plexins-B and other plasma membrane proteins, and that the carboxyl terminal tail of PDZ-RhoGEF associates with the kinase PAK4.[5] The PDZ domain is specialized in protein-protein recognition, in particular by interacting with PDZ-binding consensus sequences often located at the carboxyl terminus of plasma membrane proteins (S/TXV motif); this property is frequently used as part of the mechanism that controls the mobilization of PDZ containing proteins to the plasma membrane. As described throughout this chapter, the multiple structural features present in PDZ-RhoGEF and LARG enable these related RhoGEF to interact directly or indirectly with a variety of signaling systems initiated by GPCRs and other cell surface receptors.

LARG. LARG (for leukemia-associated RhoGEF) was originally found as a result of the characterization of the chromosomic anomaly of a patient with primary acute myeloid leukemia.[50] The gene located at the 3' extreme of MLL resulted homologous to the group of RhoGEFs characterized by the presence of a dbl-homology domain. The mRNA of LARG is expressed in all human tissues examined by Northern analysis, being particularly abundant in ovary and colon; and clearly detectable in peripheral blood leukocytes, spleen, prostate, testis, small intestine, and minimally in thymus. When the full-length LARG was cloned, it revealed the presence of PDZ domain and an RGS domain highly related to PDZ-RhoGEF.[11] This prompted its investigation as a potential G protein effector molecule, which led to the demonstration that LARG establishes a functional interaction with active $G\alpha_{12}$ and $G\alpha_{13}$, and that this interaction results in the activation of the RhoA pathway.[11] As described in more detail further in this chapter, emerging information supports multiple physiological and pathological functions for LARG, whose complexity and molecular basis have just begun to be elucidated.

Thrombin and lysophosphatidic acid receptors as paradigms of GPCR signaling to Rho through RGS-RhoGEFs

Thrombin and lysophosphatidic acid receptors promote the rapid modification of the cell shape by provoking changes in the actin cytoskeleton; these effects are caused by the action of RhoA-dependent effectors that also contribute to changes in gene expression. Thus, these receptors have served as paradigmatic elements in the characterization of GPCR signaling to RhoA via the $G\alpha_{12/13}$/RGS-RhoGEF/ RhoA signaling pathway. Although these $G_{12/13}$-coupled receptors are also coupled to G_i and $G_{q/11}$ families of heterotrimeric G proteins, genetic evidence has highlighted the importance of coupling to $G\alpha_{13}$ in the biochemical route leading to RhoA activation. Studies with $G\alpha_{13}$ knockout mouse embryo fibroblasts showed defective cell migration in response to thrombin, whereas the proliferative effect or the production of inositol phosphates were unaltered.[28] The morphological changes caused by thrombin and LPA are highly cell type dependent, such as cell rounding in 1321N1 astrocytoma cells; cell retraction in neurons, which stop the extension of neurites or growing axons in response to LPA; and cell contraction in the case of smooth muscle cells. Whereas a detailed description of the multiple biological functions of thrombin and LPA receptors is beyond the scope of this chapter, the actions of these GPCRs range from developmental events and stem cell function to hemostasis and endothelial permeability, and contribute to many pathological events linked to abnormal cell migration and proliferation, including cancerous growth and metastasis.

Commonly used strategies to study the molecular and cellular effects of thrombin or LPA-dependent RhoA activation include direct measurements of RhoA activation by an assay designed to isolate the active fraction of RhoA, using an affinity system based on the interaction of GTP-bound RhoA with an effector domain, such as the one from Rhotekin, which is expressed as a recombinant fusion protein coupled to an immobilized matrix. The use of SRE (serum response element) dependent reporter genes gives also an alternative approach to investigate RhoA signaling. At the cellular level, the most conspicuous effects can be revealed by staining the cells with fluorescent phalloidin, which specifically stains actin filaments, or alternatively by monitoring the effects on cell rounding.[46]

Even though thrombin and LPA receptors couple to $G_{12/13}$ heterotrimeric G proteins and activate RhoA downstream of the $G_{12/13}$-RGS-signaling pathway, they show, at least in a limited number of cell lines, a differential specificity for RGS-RhoGEFs. For example, a knockdown approach in PC-3 prostate cancer cell lines, using small interfering RNAs, demonstrated the requirement of LARG in thrombin signaling mediated by the PAR1 receptor, whereas LPA-dependent activation of Rho requires the expression of PDZ-RhoGEF.[58] These observations suggest that beside the availability of active $G\alpha_{12/13}$, additional elements are involved in defining the specificity of the RGS-RhoGEF utilized by specific receptors. These could involve events occurring in the vicinity of the membrane, including interactions between RhoGEFs and the receptors themselves or other plasma membrane proteins. In the case of LPA_1 and LPA_2 receptors, which have classical PDZ interacting motifs at their C-terminus, it has been documented that, in transfected HEK293 cells, they establish direct interactions with the PDZ domains of PDZ-RhoGEF and LARG. The functional significance of such interaction was revealed by the dominant negative effect of transfected PDZ domains of PDZ-RhoGEF and LARG on the activation of Rho by LPA receptors, and the inability of carboxyl terminal LPA receptors mutants to activate Rho.[51]

MOLECULAR MECHANISMS OF REGULATION OF RGS-RHOGEFS

p115, PDZ-RhoGEF, LARG

Phosphorylation

Phosphorylation is one of the most striking mechanisms by which guanine exchange factors are regulated. It promotes conformational changes that affect intramolecular interactions or the association of regulatory proteins, resulting in the acquisition of an active conformation by exposing the catalytic domain or, on the contrary, the inhibition of its activity. Considering the extensive possibilities of regulatory phosphorylations based on the presence of multiple potential sites putatively recognized by specific tyrosine kinases or serine/threonine kinases, this posttranslational modification offers endless possibilities of regulatory events that we are just beginning to understand. The role of phosphorylation in the activity of RGS-RhoGEFs is variable. PDZ-RhoGEF interacts with PAK4 and is negatively regulated by this kinase (Figure 9.3). Since PAK4 is a serine/threonine kinase considered an effector of Cdc42, its action on PDZ-RhoGEF is postulated to be part of a mechanism by which different RhoGTPases control the temporality and spatial distribution of their activation.[39] On the contrary, the interaction of PKCα with p115RhoGEF in endothelial cells stimulated with thrombin results in a positive effect. In this case, PKCα-dependent phosphorylation of p115RhoGEF activates the GEF via a mechanism sensitive to overexpression of p115RhoGEF-RGS domain, suggesting a joint action of $G\alpha_{12/13}$ and PKCα on the activation of p115RhoGEF in endothelial cells responding to thrombin.[59]

Functional assays used as readouts for the activation of effectors responding to GPCR/$G_{12/13}$/RGS-RhoGEFs/Rho-dependent pathways are sensitive to tyrosine kinase inhibitors, suggesting that signal transducers acting upstream of Rho are regulated by phosphorylation in tyrosine residues. For example, thrombin receptors cause a rapid peak of RhoA activation followed by a sustained phase that lasts for hours. The sustained phase is sensitive to tyrosine kinase inhibitors; as both PDZ-RhoGEF and LARG can be phosphorylated by PYK2 and FAK, it has been reported that the sustained activity of these GEFs is maintained by their phosphorylation in tyrosine residues.[41] Whereas the molecular basis for this increased activity of RGS-RhoGEFs upon tyrosine phosphorylation is still unclear, it is possible to speculate that phosphorylation may promote structural changes causing the displacement of an auto-inhibitory region exposing the DH catalytic domain. Alternatively, additional proteins that recognize phosphotyrosines may bind these RGS-RhoGEFs and modulate their stimulating activity on RhoA. These and additional possibilities warrant further investigations. RGS-RhoGEFs may not be directly phosphorylated by receptor tyrosine kinases, but it is interesting to notice that a direct interaction between this group of receptors and LARG has been reported. Insulin-like growth factor receptor interacts with the PDZ domain of LARG, and the overexpression of LARG-PDZ domain inhibits insulin-like growth factor-dependent activation of Rho and the consequent changes in the actin cytoskeleton, suggesting that insulin-like growth factor receptors activate Rho via a direct interaction with LARG.[52] Interestingly, the effect of insulin-like growth factor is also sensitive to LARG-RGS-overexpression, raising the interesting possibility of a cross-talk between insulin-like growth factor receptors and the G_{12} family of heterotrimeric G proteins mediated via LARG. The fact that LARG has to be phosphorylated in order to be activated in response to $G\alpha_{12}$[60] further supports these interesting possibilities.

Oligomerization

PDZ-RhoGEF and LARG form homooligomers and heterooligomers, whereas p115-RhoGEF just homooligomerizes (Figure 9.3).[42] The ability of these GEFs to form oligomers is determined by their carboxyl terminal tail, which is considered an important regulatory element based on experimental findings demonstrating that deletion mutants lacking this region are more active in cellular assays. Interestingly, two splicing variants of Lsc (the murine version of p115-RhoGEF) expressed in the spleen lack the carboxyl terminal coiled-coil dimerizing region and show increased Rho-dependent transcriptional activity.[61] In the particular case of Lsc, biochemical analysis suggested the existence of homotetramers in B lymphocytes.[62] In addition, as indicated, the carboxyl terminus of PDZ-RhoGEF constitutes the site for interaction with PAK4, which phosphorylates and inactivates this GEF. In the case of LARG, it has been suggested that dimerization contributes to regulate the spatial localization of this GEF, whereas LARG dimers are cytosolic, the monomers shuttle to the nucleus.[63] Whether the assembly of oligomers is a constitutive or signal transduction dependent dynamic event requires further investigation.

STRUCTURAL BASIS OF GPCR SIGNALING TO RGS-RHOGEFS

$G\alpha_{12/13}$ interact with the RGS-homology domain of RGS-RhoGEFs. This RGS-homology domain has been often referred to as rgRGS to differentiate it from canonical RGS proteins with which it shares close to 20% amino acid sequence identity and similar structural characteristics in the region known as the RGS box, but differs in the adjacent regions known to be important for the functional interaction with $G\alpha_{12/13}$. The overall structure of this rgRGS resembles RGS4 and other canonical RGS proteins; it is constituted by nine alpha-helixes characteristic of the RGS-box; in addition, it includes, toward the carboxyl extension of the domain, three additional alpha-helices that are tightly bound, via hydrophobic interactions, to the RGS-box and constitute part of the same globular structure.[64,65] The amino terminal extension of this rgRGS includes an acidic region at the amino terminus not present in RGS4. This acidic region is involved in the GAP activity of p115-RGS toward $G\alpha_{13}$.[57]

The binding of $G\alpha_{13}$ to the RGS of RGS-RhoGEFs occurs at two sites. $G\alpha_{13}$ interacts with the RGS box of this rgRGS in a similar way as other $G\alpha$ subunits interact with their effectors, thus suggesting a primary role of this interaction in the activation of the signaling pathway that links $G\alpha_{13}$ to RhoA, rather than an interaction that would shut down $G\alpha_{13}$ signaling. The second site of interaction within the amino terminal extension of rgRGS, is the acidic region critical for the GAP activity in p115-RhoGEF.[57] Our studies with chimeric $G\alpha_{12/13}$ demonstrated the importance of the switch regions of $G\alpha_{13}$ for signaling to RhoA, and also demonstrated that most of the GTPase domain of $G\alpha_{13}$ is required for a functional interaction with rgRGS-RhoGEFs, whereas the alpha helical domain can be swapped for the one of $G\alpha_{12}$.[66] On the contrary, for the GAP activity of p115-RGS toward $G\alpha_{13}$, the alpha helical domain is critical for the sensitivity of $G\alpha_{13}$ to the GAP activity of p115-RGS.[67] The switch regions of $G\alpha_{13}$ and other GTPases are the most dynamic regions, according to the structural differences evidenced when the GTPases are compared between the GDP-bound form, which interacts with $G\beta\gamma$, and the GTP-bound form that interacts with effectors. However, as mentioned earlier in this chapter, the RGS-containing family of RhoGEFs has a dual action, being at the same time effectors and GAPs for $G\alpha_{12/13}$. In particular, p115-RhoGEF and LARG are GAPs for $G\alpha_{13}$ whereas PDZ-RhoGEF is not. The main difference, when seen from the GAP perspective, resides precisely at the level of the acidic amino terminal region of the rgRGS domain, as demonstrated by crystallographic analysis of the complexes formed by $G\alpha_{13}$ and the RGS domain of PDZ-RhoGEF[57] and between p115-RGS domain with a chimeric $G\alpha_{13/i1}$ protein.[67] In particular, the amino terminal extension outside of the RGS box is fundamental for the GAP activity of p115-RhoGEF toward $G\alpha_{13}$. Mutagenesis experiments demonstrated that PDZ-RhoGEF can gain GAP activity by introducing the equivalent residues of p115 in the amino terminal acidic extension of the rgRGS domain.[57] This is expected to result in functional differences in terms of the timing and duration of the activation of p115-RhoGEF or PDZ-RhoGEF by $G\alpha_{13}$. It would be expected that PDZ-RhoGEF (and probably LARG) would maintain a persistent signaling to RhoA, whereas p115-RhoGEF would be subject to a rapid turnover.[57] Thus, the interaction between $G\alpha_{12/13}$ and rgRGS shows a recognizable difference when compared with the paradigmatic complex formed by $G\alpha_i$ and RGS4. Intriguingly, rgRGS-PDZ-RhoGEF could interact with the different activation states of $G\alpha_{13}$, including the GDP-bound form of the GTPase, suggesting a specialization of the canonical two-step model of GDP-off and GTP-on status of GTPases.[57] This finding would imply that $G\alpha_{13}$ might still be able to activate rgRGS-RhoGEFs in the GDP-bound form and would be switched to the inactive condition once that $G\beta\gamma$ interacts with it, displacing the GEF. It is important to keep in mind that structural studies have analyzed just the interaction of $G\alpha_{13}$, or a chimera that includes part of $G\alpha_{i1}$, and the isolated RGS domain. Since it is known that RGS-RhoGEFs are large multidomain proteins participating in macromolecular complexes, it is very likely that internal or external interactions and physical constrains may modify the availability of the rgRGS domain to interact with $G\alpha_{13}$. This may add an additional level of regulation that is yet to be fully characterized in order to completely understand the structural basis of RhoGEF activation by $G\alpha_{13}$, particularly in terms of the conformational changes in the GEF necessary to achieve an active conformation in response to $G\alpha_{13}$-coupled GPCRs.

RGS-RHOGEFS IN SIGNALING BY NON-GPCR CELL SURFACE RECEPTORS

Role in Plexin-B Signaling

In the developing nervous system, the correct wiring of synaptic connections depends on the extension of axons via a mechanism supported by polarized movement of axonal growth cones. Numerous plasma membrane receptors are expressed at the tip of axonal growth cones; some of them are specialized in recognizing either positive or negative signals that guide the path of the developing axon. Many of these receptors act by regulating the activity of small GTPases of the Rho family thus causing dynamic changes in the actin cytoskeleton, which result in the extension of filopodia and lamellipodia in response to chemoattractants, or the actomyosin-dependent retraction of the growth cone in response to repulsive signals. The semaphorins are repellents that limit the extension of growing axons by inducing the contraction of growth cones upon interaction with their receptors, known as plexins. In this case, the intracellular pathway triggered by plexins is linked to the local activation of RhoA. In 2002, we and others reported the existence of a functional interaction between plexins and the two members of $G\alpha_{12/13}$-sensitive RGS-RhoGEFs that contain a PDZ domain at their amino terminus: PDZ-RhoGEF and LARG.[15-17,43,44] Plexins, in particular plexin-B family members, through their PDZ-interacting motif present at their carboxyl terminus bind the PDZ domain of PDZ-RhoGEF and LARG, activating those GEFs in response to semaphorin 4D (Sema4D) (Figure 9.3). This interaction appears to be constitutive, but these GEFs are activated by plexins only in the presence of Sema4D in the extracellular milieu, either soluble or bound to the plasma membrane of neighboring cells. This suggests that the contact between Sema4D and plexin-B1 results in a conformational change in the intracellular domain of plexin-B1 that is transduced to PDZ-RhoGEF and LARG, leading to the activation of these GEFs and the consequent local activation of RhoA, causing the collapse of extending axonal growth cones. The functional interaction between plexin-B1 and PDZ-RhoGEF is further enhanced by Rnd, a small GTPase member of the Rho family, that interacts with the intracellular domain of plexin-B1.[68] Other plasma membrane receptors have been reported to participate in the pathway from plexin-B1 to Rho; in particular, ErbB-2 tyrosine kinase receptor interacts with plexin-B1 and is activated in response to the binding of Sema4D to plexin-B1; this results in the phosphorylation of plexin-B1 and contributes to Rho activation.[69] It is also known that GPCRs, such as those for lysophosphatidic acid, promote axonal cone collapse; thus it can be speculated that PDZ-RhoGEF and LARG may integrate signals from plexins and $G\alpha_{12/13}$-coupled receptors.

Accumulating evidence supports a parallelism in the molecular programs that mediate axon guidance in the central nervous system and angiogenic events during development and under pathological conditions. It has been reported that angioblasts and endothelial cells express different plexins that recognize the migratory cues defined by semaphorins. Both PDZ-RhoGEF and LARG are also expressed in endothelial cells, suggesting that the plexinB/PDZ-RhoGEF and plexinB/LARG signaling pathways are relevant in the vascular system. Sema4D acting on plexin-B1 induces endothelial cell chemotaxis and in vitro angiogenesis, and enhances blood vessel formation in an in vivo mouse model consisting of an implant of basement membrane proteins mixed with soluble Sema4D.[70] In this model, plexin-B1 dependent angiogenesis requires the interaction of plexin-B1 and the PDZ domain of PDZ-RhoGEF and LARG and the activation of RhoA.[70] Whether plexin-B signaling to RhoA is regulated via the intervention of the Met receptor tyrosine kinase is not clear, and is likely to be cell-specific.[70–72] Available data support a differential role of plexin-B1 signaling to RhoA depending on its interaction with tyrosine kinase receptors; the complex plexin-B1/ErbB-2 leads to RhoA activation, while the complex plexin-B1/Met prevents RhoA activation in transfected HEK-293 cells.[72] Endothelial cells of different origin lead to distinct results regarding the role of Met in Sema4D-dependent chemotactic migration,[70,71] further suggesting that the final outcome of plexin-B1 signaling in response to Sema4D depends on the balance conferred by tyrosine kinase receptors,[72] as well as on the activity of PI3K/AKT and other signaling pathways that can be activated in parallel by plexin-B via RhoA-dependent or independent mechanisms.[73] The proangiogenic role of plexin-B1 seems to be particularly relevant under pathological conditions but can be compensated by the action of other members of the plexin-B family under physiological conditions and during developmental angiogenesis, as suggested by the fact that a plexin-B1 knockout mouse is viable.[74] Independent evidence indicates that mice lacking plexin-B1 presents an abnormally large kidney and augmented ureteric branching;[75]

accordingly, the Sema4D/plexin-B1-dependent Rho/ROCK signaling pathway is relevant to restrict the branching morphogenesis of the ureteric collecting duct during early stages of kidney development.[75] In addition, the role of this signaling pathway in tumor-induced angiogenesis is variable in different tumors studied in mice models; in head and neck squamous cell carcinomas, knocking down Sema4D reduces the size and vascularity of tumor xenografts,[76] whereas in breast cancer, it has been reported that tumor-associated macrophages secrete Sema4D promoting tumor-induced angiogenesis and metastasis via this proangiogenic factor.[77] On the contrary, in the B16-melanoma model, which also expresses Sema4D, the lack of expression of plexin-B1 in endothelial cells does not affect tumor growth, suggesting that either this pathway is irrelevant for melanoma-induced angiogenesis or that other plexins of the B family, such as plexin-B2, might be playing a role.[74] RGS-RhoGEF-independent signaling pathways are also activated by other plexins in endothelial cells. For instance, plexin-D1, one of the plexins unable to bind RGS-RhoGEFs, is involved in the development of the vascular system, as indicated by developmental alterations of endothelial-restricted plexin-D1 knockout mice, which presents congenital heart disease due to vascular patterning defects,[78] and by the developmental vascular defects of zebrafish with plexin-D1 gene disruption.[79] Endothelial cells may represent an ideal system to investigate the possible cross-talk between plexins and $G\alpha_{13}$ mediated by RGS-RhoGEFs, considering the strong evidence supporting a proangiogenic role of $G\alpha_{13}$ and $G\alpha_{13}$-coupled receptors, such as the thrombin receptor PAR1. $G\alpha_{13}$ knockout in mice,[28] and most recently endothelial specific knockout of $G\alpha_{13}$,[80] fail to develop a fully functional vasculature. A similar phenotype is seen in half of PAR1 knockout mice[81] that die in utero.

Role in tyrosine kinase receptor signaling

The PDZ domains of PDZ-RhoGEF and LARG establish functional interactions with different plasma membrane receptors that activate Rho-dependent signaling pathways through these GEFs. As mentioned, plexin signaling to Rho involves its interaction with the PDZ domain of these GEFs and is positively regulated by the interaction of plexin-B with the receptor tyrosine kinase Erb-B2.[69] Similarly, in human head and neck squamous carcinoma cells, a hyaluronan receptor known as CD44 has a positive effect on LARG activity via direct interactions.[82] In this case, the complex formed by LARG/CD44 serves as a node for interaction with additional proteins including EGFR, thus promoting parallel activation of Rho and Ras signaling pathways, constituting an interesting example of signal integration consistent with a role in the migratory and proliferative capacity of these cancer cells. The tyrosine kinase receptor for insulin-like growth factor-1 also interacts through its carboxyl terminus with LARG.[52] This interaction leads to Rho kinase activation, which is sensitive to overexpression of either PDZ or RGS domains of LARG, suggesting that IGF-1 receptor and LARG form an active complex that somehow involves the G_{12} family of heterotrimeric GTPases in the pathway leading to Rho activation.

Role in Integrin signaling

Adhesion of fibroblasts to fibronectin elicits Rho-dependent events such as assembly of stress fibers and focal adhesions. LARG and p115-RhoGEF are activated in cells adhering to fibronectin, suggesting that these RhoGEFs are signal transducers of integrins, the adhesion receptors that interact with the extracellular matrix.[83] Based on adhesive and migratory properties of Lsc/p115-RhoGEF knockout cells, an equivalent role has been suggested for this GEF in migrating leukocytes.[84–86]

CELLULAR DYNAMICS AND EFFECTS OF RGS-RHOGEFS DOWNSTREAM OF GPCR SIGNALING

Translocation to the cell membrane

To be activated by $G\alpha_{12/13}$-coupled GPCRs or other receptors, RGS-RhoGEFs must reach the plasma membrane, either in response to intracellular signaling cascades generated by activated receptors or by interacting with plasma membrane proteins that keep them in a nearly constitutive association with the plasma membrane. In Drosophila S2 cells, it has been observed that DRhoGEF2 interacts with the microtubule plus-end tracking protein EB1 and uses this contact to reach the cell cortex by moving on the tips of growing microtubules.[87] The GEF remains bound to the microtubules until it interacts with Concertina, the $G\alpha_{13}$ homolog that activates it, causing a localized actomyosin contraction that provides the force that sustains the morphogenetic movements during development.[87] In NIH-3T3 fibroblasts, lysophosphatidic acid induces translocation of a fraction of p115-RhoGEF from the cytosol to the membrane.[88] Experiments assessing the translocation of RGS-RhoGEFs demonstrated that constitutively active $G\alpha_{13}$ promote the translocation of these RhoGEFs to the plasma membrane.[89] Interestingly, in the case of agonist-dependent activation of $G\alpha_{12/13}$, the association of p115-RhoGEF to the plasma membrane is transient and can be reversed by specific receptor antagonists; thus it has been proposed that this dynamic cellular event can be used as a readout to characterize small molecules with pharmacological properties acting on $G_{12/13}$-coupled receptors.[90]

Role in spatiotemporal control of RhoGTPases, actin and microtubule dynamics, cell migration and polarity

Migrating cells acquire a polarized morphology that involves rapid redistribution of proteins that recognize directionality and move the cells forward. Accumulating evidence sustains a role for $G_{12/13}$-coupled receptors in cell migration (Figure 9.2). RhoA activation at the trailing edge of moving cells is relevant for the assembly of contractile actomyosin complexes. In the model of starved fibroblasts stimulated with lysophosphatidic acid (LPA), it has been demonstrated that a focal adhesion kinase (FAK)/PDZ-RhoGEF signaling complex is involved in the local activation of Rho/ROCKII needed for the disassembly of focal adhesions at the trailing edge that cells must remove in order to move directionally.[91] In addition, LPA stimulates

RhoA at the leading edge of moving fibroblasts via a mechanism that depends on $G\alpha_{12/13}$.[31] The role of $G\alpha_{12/13}$-dependent pathways in the acquisition of cell polarity in moving cells has started to be elucidated; studies with $G\alpha_{12/13}$-deficient mouse embryonic fibroblasts revealed a defect in the polarized distribution of the microtubule organizing center affecting the orientation of microtubules.[34] The contribution of $G_{12/13}$-coupled receptors in cell polarity depends on the concerted actions of LARG and the actin nucleating formin mDia, by a mechanism that involves the direct association of LARG with pericentrin, a protein involved in the nucleation of microtubules that interacts with γ-tubulin at the centrosome, and the distribution of this GEF along microtubules.[34] Considering the fundamental role of cell polarity in processes such as asymmetric cell division, cell differentiation, and polarized migration, the observed effect could be potentially relevant in numerous cell populations whose function depends on a polarized phenotype, such as epithelial stem cells and neurons, to mention some examples. An interesting example in which the contribution of this signaling pathway has been demonstrated is during *Drosophila* gastrulation, where striking cell migratory events need to occur.[35,92–94] As mentioned, in *Drosophila melanogaster* S2 cells, the PDZ-RhoGEF/LARG ortholog dRhoGEF2 moves along the microtubules through interaction with EB1, a protein that associates to the growing end of microtubules. It is suggested that dRhoGEF2 uses this mechanism to reach restricted regions of the plasma membrane were the morphogenetic stimuli involving concertina (the $G\alpha_{13}$ ortholog) are being transduced, thus leading to localized activation of Rho and contractile events that depend on actomyosin complexes.[87] Whether this mechanism is conserved in mammalian cells remains to be determined. It is clear that EGFP-tagged LARG associates with pericentrin and microtubules in mammalian cells and LARG knockdown prevent polarization of microtubule-organizing centers and cell migration in a scratch wound assay; however, the interaction of LARG with EB1 was not detected, suggesting that an alternative protein could lead LARG to the cell periphery through its interaction with the microtubules; in this case mDia may contribute to localize LARG along the microtubules.[34] Interestingly, mDia, which is an effector of RhoA that promotes the polymerization of actin, interacts with LARG in response to RhoA-GTP; in addition, it has been reported that conformational changes occurring in mDia upon interaction with RhoA-GTP exposes a LARG interacting site that, in vitro, activates LARG,[95] suggesting the existence of a positive feedback that maintains LARG activity.

PHYSIOLOGICAL FUNCTIONS OF RGS-RHOGEFS

RGS-RhoGEFs in the nervous system

The nervous system coordinates all the actions of the human body. To do so, it depends on the establishment of a sophisticated network of precisely communicating cells. During development, neural progenitor cells migrate and developing neurons emit protrusions that follow environmental cues that define the paths by which the cell projections advance; then synapses are established when growing axons encounter the appropriate partner. This process depends on the identification of positive and negative signals that restrict the routes of migration. Thus, a growing axon retracts when it finds an undesirable partner, such as a glial cell. This retraction requires the activation of Rho and occurs in response to the activation of different plasma membrane receptors including GPCRs such as GPR56[96] and others such as plexin-B2,[97] which is critically required in vivo for proliferation, migration, and pattern formation in the mouse forebrain and the cerebellums; or the complex formed by Neogenin and Unc5B[98] In vitro, lyso-phosphatidic acid activates a $G\alpha_{12/13}$-coupled GPCRs that induce neurite retraction.[99] PDZ-RhoGEF and LARG are expressed in the central nervous system.[100] They are also known to interact with plexins, which recognize repulsive signals marked by the presence of semaphorins. The molecular basis of this process constitutes an interesting example of signal integration in which LARG is one of the main characters. Similarly, RGMa (repulsive guidance molecule A) exposed in the glial cell activates Neogenin, a receptor on the growing axon, which activates RhoA through its partner Unc5B that interacts with LARG. Interestingly, while Unc5B keeps LARG associated to the membrane by interacting with its PDZ domain, Neogenin promotes the phosphorylation of LARG and its consequent activation.[98]

RGS-RhoGEFs in the cardiovascular system

$G\alpha_{13}$ knockout mice die during development at embryonic day 9.5, due to angiogenic defects.[28] Since $G\alpha_{12}$ knockout mice are normal, these findings indicate that the functions of the two members of the G_{12} family of heterotrimeric G proteins are not redundant. The specific requirement of $G\alpha_{13}$ in endothelial cells during development is further confirmed by similar phenotype observed in endothelial restricted $G\alpha_{13}$ knockout mice[80] and by rescuing $G\alpha_{13}$ knockout mice from dying during development by expressing $G\alpha_{13}$ in endothelial cells. Interestingly, knockout of the thrombin receptor PAR1, known to be coupled to different G proteins including G_{13}, shows a similar, less penetrant, phenotype.[81] Beside its role during development, this signaling pathway is a fundamental regulator of physiological vascular function, keeping a tight control on vascular permeability and influencing the vascular tone.

RGS-RhoGEFs in the immune system

The member of the family of RGS-RhoGEF preferentially expressed in the immune system is p115-RhoGEF. The role of this RGS-RhoGEF in immune cells has been revealed by a knockout approach. Leukocytes require Lsc/p115-RhoGEF to migrate properly,[86] as demonstrated by defects of p115-RhoGEF knockout leucocytes in their ability to regulate the assembly and disassembly of focal adhesions needed to advance the trailing edge of migrating cells. Also, in differentiated HL60 cells that acquire neutrophil characteristics, a role has been revealed for PDZ-RhoGEF localizing RhoA activity and promoting the assembly of contractile cytoskeletal complexes to the back of polarizing neutrophil-like cells.[101]

PATHOLOGICAL IMPLICATIONS OF RGS-RHOGEF SIGNALING

Role in cancer

The G_{12} family of heterotrimeric G proteins has been associated with cell invasion in breast and prostate cancer cells.[102,103] Surprisingly, cytogenetic analysis of colorectal and breast carcinoma patients frequently revealed a deletion of 11q23–q24, the region in which the gene coding for LARG is present, suggesting that this RGS-RhoGEF has a tumor suppressor potential.[104] Coincidently, a fraction of cancer samples and colorectal and breast carcinoma cell lines express reduced amounts of LARG, which, when introduced in deficient cell lines, reduced their proliferative behavior and migration, further suggesting its role as a metastasis suppressor.[104] On the contrary, in a model of transfected NIH3T3 fibroblasts, LARG cooperates with activated Raf-1 to transform these cells.[105] Since LARG was originally identified as a gene fused in a primary acute myeloid leukemia translocation event,[50] its role in cancer progression deserves further investigations. Interestingly, since PDZ-RhoGEF and LARG interact with plexin-B and can be activated in response to semaphorin, the role of this signaling pathway in semaphorin-dependent tumor-induced angiogenesis could also contribute to cancer progression in different neoplasias.[76]

Role in hypertension

In recent studies, it was observed that LARG knockout mice are resistant to salt-induced hypertension, but their constitutive vascular tone is normal.[106] These results highlight the potential of LARG as a target for pharmacological intervention in hypertensive patients.

Potential role in diabetes

Studies of LARG polymorphisms resulted in the identification of Tyr1306Cys as a variant allele significantly associated with increased glucose uptake in nondiabetic individuals of an isolated population of Pima Indians known to present high incidence of diabetes.[107] In vitro analysis, using a reporter gene as readout, revealed that LARG-Tyr1306Cys was less active than the variant version. Since Rho activity is inversely correlated with insulin function, these data suggest that individuals carrying the variant allele could be less susceptible to develop the disease.[107] In addition, a LARG-R1467H variant was reported by the same group to be present in young Pima Indians with a higher risk to develop diabetes.[108] However, a similar analysis in a German Caucasian population did not reveal a correlation between the Tyr1306Cys allele and insulin action,[109] whereas the LARG-R1467H variant was independently found in a German Caucasian group and considered associated with a higher risk of type 2 diabetes,[110] indicating that additional factors influence the potential role of LARG variants in this disease.

RGS-RHOGEFS AS NOVEL PHARMACOLOGICAL TARGETS

Considering the multiple physiological and pathological processes influenced by the function of RGS-RhoGEFs, it has a clear therapeutic potential based on the modulation of their activities. For instance, it can be speculated that inhibiting PDZ-RhoGEF and LARG would promote nerve regeneration. LARG inhibition would be desirable to treat hypertensive patients, whereas inhibition of p115-RhoGEF would potentially have anti-inflammatory or immunosuppressive properties. Thus, the identification of small molecule inhibitors of RGS-RhoGEFs will open interesting possibilities to pursue these goals. Initial studies are already aimed at identifying these potential inhibitors. A high-throughput screen for inhibitors of LARG resulted in the identification of a group of five compounds that block LARG ability to stimulate RhoA in vitro.[111]

MULTIPLE RGS-CONTAINING PROTEINS ACT AS G PROTEIN EFFECTORS IN GPCR SIGNALING

G_q signaling to Rho

It is now clear that GPCRs coupled to G_q also stimulate RhoA signaling.[112] Interestingly, genetic evidence using either PTX-treated $G\alpha_{12/13}$- or $G\alpha_{q/11}$- knockout cells demonstrated that receptors coupled exclusively to $G\alpha_{q/11}$ activate RhoA via a second messenger independent mechanism.[113] Some G_q-coupled receptors such α1-adrenergic or muscarinic [M_1] depend on $G\alpha_q$ to activate RhoA. The mechanism by which $G\alpha_q$ activates RhoA is clearly different from the one used by $G\alpha_{12/13}$. Comparative analysis of interactions between constitutively active mutants of $G\alpha_{13}$ or $G\alpha_q$ demonstrated that the RGS-domain of RGS-RhoGEFs recognizes $G\alpha_{13}$ but does not interact with $G\alpha_q$.[112] The importance of G_q signaling to RhoA has been highlighted in neonatal rat ventricular myocytes that transit to a hypertrophic phenotype in response to agonists activating α_1-adrenoceptors or ET-1 receptors, among others, which induce this response with the participation of RhoA.[114,115] Emerging evidence indicates that an important mechanism by which G_q-coupled receptors activate RhoA depends on the direct interaction between active $G\alpha_q$ and p63-RhoGEF, a RhoA-specific GEF lacking any recognizable domain with homology to RGS proteins.[116,117] However, an alternative mechanism has been proposed in transfected and knockout cells in which G_q signaling to RhoA is sensitive to the expression of a dominant negative mutant of LARG in $G\alpha_{12/13}$-deficient mouse embryo fibroblasts,[113] G_q signaling to the nucleus is sensitive to overexpression of a construct corresponding to the RGS domain of LARG, suggesting that $G\alpha_q$ potentially activate RhoA through interactions with LARG.[118]

G_q/GRK2

G protein-coupled receptor kinases (GRKs) play a fundamental role in the homologous desensitization of GPCRs by phosphorylating serine and threonine residues at their carboxyl terminal tail and intracellular loops.[20] GRK-phosphorylated receptors bind arrestins that serve as adaptors for their internalization and also as novel links to additional signaling pathways.[119] The fate of the internalized receptors is influenced by their ability to interact with sorting proteins such as GPCR-associated sorting protein (GASP), which determine whether they are recycled or sorted to degradation.[120] The amino terminus of GRKs contain an RGS homology domain, whereas a $G\beta\gamma$-interacting region is present at the carboxyl terminus, which contains a PH domain. $G\alpha_q$ interacts with the RGS homology domain of GRK2 (Figure 9.4). The molecular aspects of this interaction have been studied by structural and mutational analysis and differ from those that determine the association between $G\alpha_q$ and RGS3 or RGS4. It is clear that $G\alpha_q$ recognizes the RGS domain of the kinase as an effector more than as a GAP.[21] Thus, the inhibition that GRK2 exerts on $G\alpha_q$ is due to a physical interference more than to a GAP activity. The current model postulates that GRK2 is an effector of $G\beta\gamma$ that is translocated to the membrane where a GPCR, upon the presence of its cognate ligand, promotes the dissociation of the heterotrimeric G protein coupled to the intracellular loops of the activated receptor. Then, GRK phosphorylates the receptor, attenuating its ability to interact with additional heterotrimeric G proteins, while the RGS domain interferes with the signaling of the $G\alpha$ subunit. Thus, GRK2 is a multifunctional inhibitor of G protein signaling (Figure 9.4).

Similar to other multidomain RGS proteins, GRK2 and other members of the GRK family might have an effector role in GPCR signaling based on their affinity for proteins involved in fundamental signaling cascades. This possibility has been explored by investigating the identity of additional GRK2-interacting partners. The group of novel GRK2-interacting partners includes classic signal transduction proteins such as caveolin, PI3K, AKT, and MEK.[4] It is tempting to speculate that GRK2 links G_q to PI3K/AKT and ERK signaling cascades; however, available evidence indicates a versatile effect of GRK2 on the activity of these signaling effectors.[4] GRK2 contributes to the translocation of PI3Kγ to the membrane[121] but inhibits the activation of AKT; thus, PI3Kγ translocated in response to GRK2 might contribute to the activation of other effectors, either by generating PIP_3 at the plasma membrane or, as described, by phosphorylating proteins, such as tropomyosin, important in the internalization and trafficking of GPCRs, as demonstrated for β-adrenoceptors.[122] Emerging evidence sustaining a role for GRK2 as a bona fide $G\alpha_q$ effector is the characterization of a ternary complex in which both RGS4 and GRK2 bind simultaneously to $G\alpha_q$; in this complex, GRK2

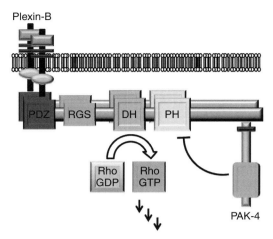

Figure 9-4: Additional mechanisms of regulation of PDZ-RhoGEF and LARG include activation in response to interaction of Plexin-B with Semaphorin, oligomerization and phosphorylation. Other plasma membrane receptors linked to these RhoGEFs include LPA receptors, and Insulin-like growth factor receptor, which also interact with the PDZ domain of PDZ-RhoGEF or LARG. *See colour plate 11.*

displays functional properties of a bound effector, in particular by stimulating the GAP activity of RGS4 present in the complex.[123]

G Proteins/Axin

Axin is a negative regulator of the canonical Wnt signaling pathway. This pathway has an essential role during development and it is altered in cancer.[124] Within the Wnt signaling pathway, Axin modulates GSK3β-dependent β-catenin phosphorylation, thus regulating the availability of β-catenin, a transcription factor that is either phosphorylated and degraded, or non-phosphorylated and accumulates in the nucleus, where it activates the expression of genes that promote cell proliferation.[124] Axin contains an RGS homology domain known to interact with the adenomatous polyposis coli protein (APC); this interaction is critical to assemble the multicomponent complex that includes GSK3β and β-catenin, maintaining organized elements involved in the phosphorylation of β-catenin needed to keep control of the levels of this transcription factor. Other domains of Axin interact with known components of the Wnt canonical pathway and other signaling proteins. The existence of an RGS-homology domain, located at the amino terminal region of Axin, suggest a link of this protein to heterotrimeric G protein signaling, either as a GAP as expected for classical RGS proteins, or as an effector or scaffold, as it has been confirmed for several multidomain RGS-homology proteins. The potential role of Axin-RGS domain in heterotrimeric G protein signaling is starting to be addressed. A role for Axin in pathological signaling of Gα$_s$ has been demonstrated. Based on the known potent proliferative action of G$_s$-coupled prostaglandin PGE$_2$ receptors in colon cancer cells, it was hypothesized that G$_s$ may be directly linked to aberrant proliferative pathways through β-catenin. Evidence was recently provided that a direct interaction between $G\alpha_s$ and Axin alters the tight influence that Axin exerts on β-catenin regulation, thus promoting the proliferating consequences of β-catenin deregulation.[8] Subsequently, other heterotrimeric G protein alpha subunits have been reported to interact with Axin. $G\alpha_{12}$ interacts in vitro with Axin-RGS which, in this system, does not show GAP activity toward $G\alpha_{12}$.[125] In addition, in transfected MDA-MB231 breast cancer cells, Axin-RGS prevents $G\alpha_{12}$-dependent cell rounding, suggesting that it competes with the interaction between p115-RhoGEF and $G\alpha_{12}$; whether the complex $G\alpha_{12}$/Axin alters the equilibrium of the Axin/APC complex, and thus the phosphorylation and degradation of β-catenin, remains to be explored.[125] Recently, the in vivo importance of the Axin-RGS domain has been reported[126] using a genetic knock in approach in mice in which endogenous Axin was substituted by a mutant lacking the RGS domain. Interestingly, these mice expressing ΔRGS-Axin were not viable, showing a lethal phenotype similar to that observed in Axin knockouts. Although the lethal phenotype could be explained by the inability of ΔRGS-Axin to interact with APC, the essential role of other potential functions of the Axin RGS domain remains a possibility that will be surely explored in the future.[127]

G_i/RGS12, RGS14

The structure of RGS12 and RGS14 suggests a complex array of possible regulatory interactions with heterotrimeric and monomeric Ras-related G proteins. RGS12 and RGS14 are characterized by the presence of two $G\alpha$-interacting domains: an RGS domain and a carboxyl terminal GoLoco domain. In addition, a tandem of two RBD domains with affinity for the small GTPases Ras and Rap is located between the RGS and GoLoco domains (Figure 9.5).[1] Beside these common structural characteristics that make RGS14 look like a short version of RGS12, RGS12 contains an amino terminal extension that includes a PDZ domain followed by a PTB domain. This amino terminal extension of RGS12 makes this multidomain protein a functional regulator of neuronal calcium channels by interactions of the PTB domain with tyrosine phosphorylated channels.[128] It also helps the scaffolding role of this RGS in the signaling pathway to the ERK cascade induced by nerve growth factor (NGF).[19] Current evidence points to a role for RGS12 and RGS14 as dual inhibitors of G_i family of heterotrimeric G proteins. RGS12 and RGS14 may inhibit G_i signaling by the additive action of both the RGS domain that acts as a GAP, and the GoLoco domain, which inhibits the dissociation of GDP, thus acting as a GDI (Figure 9.5).[129–131] However, an interesting turn to the role of GoLoco proteins as promoters of G protein-dependent downstream events emerged from the documented role of these proteins in asymmetric cell division, in which they contribute to the specification of spindle polarity,[132] which is fundamental for stem cell renewal and early development. Strikingly, RGS14 is a microtubule-associated protein that regulates the mitotic spindle and is needed for the first division of the mouse zygote, an event that is believed to be independent of GPCRs.[133] Since the binding of the GoLoco domain to $G\alpha_i$ keeps this subunit in the GDP-bound inactive

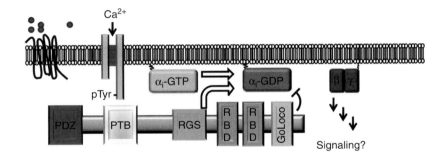

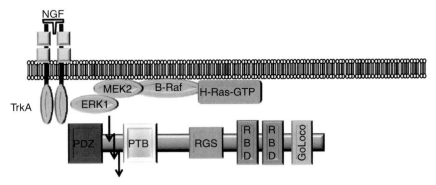

Figure 9-5: RGS12 and RGS14 regulate G protein signaling and growth factor receptor signaling. Upper panel, RGS12 (shown in the figure) and RGS14 contain two regulatory domains for Gα$_i$, the RGS domain that acts as a GAP and the GoLoco domain that functions as a GDI. This suggests the possibility that, besides regulating Gα$_i$, RGS12 and RGS14 may contribute to initiate Gβγ signaling. In addition, RGS12 contains a PDZ and a PTB domains at its amino terminal extension not present in RGS14. The PDZ domain can interact with the carboxyl terminal tail of plasma membrane receptors, and the PTB domain recognizes tyrosine phosphorylated neuronal calcium channels. Bottom panel, RGS12 serves as a scaffold in the signaling of NGF-TrkA receptors to sustained ERK activation, which contributes to neuronal differentiation. *See colour plate 12.*

state, thereby freeing Gβγ, GoLoco-domain containing proteins may potentially contribute to initiate Gβγ signaling even in the absence of GPCR activation. However, experiments oriented to test this hypothesis using potassium currents via the Kir3.1/3.2 channels as the readouts of Gβγ-dependent signaling have not confirmed this possibility.[134] Considering the increasing number of Gβγ effectors, perhaps the hypothesis of GoLoco proteins as activators of Gβγ effectors could still be valid for some Gβγ-dependent pathways but not for others.

RGS12 as a scaffold in NGF-mediated differentiation

RGS12 interacts with the NGF receptor tyrosine kinase TrkA and components of the ERK signaling cascade, including active H-Ras, B-Raf, and MEK2.[19] Together, these signaling proteins generate a prolonged signaling effect in response to nerve growth factor (Figure 9.5). The role attributed to RGS12 in this signaling pathway as a protein scaffold may be important for neurogenesis, as evidenced by a RNAi-mediated knockdown strategy that prevents the neuritogenic effect of NGF in PC12 cells and the axonal growth in embryonic neurons.[19] An equivalent functional effect has been reported for RGS14, although with differences in the molecular aspects of the relevant interactions.[18]

RGS12 and the control of ion channels

The proper function of the central nervous system depends on the regulation of diverse ion channels. Both voltage-dependent and agonist-dependent ion channels are tightly regulated. In cultured chicken root ganglion neurons, $GABA_B$ receptors, coupled to G_o, inhibit N-type calcium channels via a mechanism that involves tyrosine kinases, suggesting that they are regulated by phospho-tyrosine-dependent protein-protein interactions. A key component of the regulatory mechanism is RGS12, which interacts, through its PTB domain, with tyrosine phosphorylated calcium channels (Figure 9.5). This interaction occurs in response to neurotransmitter stimulation and attenuates the activity of the interacting N-type calcium channels.[128]

RGS12 as a regulator of transcription

A splice variant of RGS12 (RGS12TS-S) is a nuclear protein, showing a restricted punctuate subnuclear localization suggesting a specialized action in the nucleus not likely related to the control of heterotrimeric G proteins. Initial characterization of its role in the nucleus revealed that RGS12TS-S works as a transcriptional repressor that interferes with the progression of the cell cycle and promotes the emergence of multinucleated cells.[135] These nuclear activities of RGS12TS-S do not involve the RGS domain.

Role of RGS14 in cell division

A knockout approach in mouse revealed that RGS14 participates in the first division of the zygote.[136] RGS14 is ubiquitously expressed, mainly in the nucleus of cells in interphase, and associated with the mitotic spindle and centrioles of mitotic cells.[136] Whether the role of RGS14 in early division and further mitotic events is a $G\alpha_i$-dependent event remains to be determined. Available evidence shows the presence of $G\alpha_i$ associated to RGS14 in centrosomes; moreover, experimental strategies that reduce the expression of $G\alpha_i$ result in altered cytokinesis in Hela cells, suggesting a functional role for the interaction between these proteins during cell division.[137] Studies in *Drosophila* and *C. elegans* have evidenced a role for $G\alpha$ and GoLoco proteins in asymmetric cell division via a mechanism that, instead of depending on the action of a G protein-coupled receptor, requires the intervention of AGS3, a GoLoco protein that dissociates $G\alpha$ from $G\beta\gamma$, and a cytosolic guanine exchange factor known as Ric-8.[138]

ACKNOWLEDGMENTS

Work in JVP laboratory is supported by the National Council for Science and Technology of Mexico (CONACyT Grant # 61127). Work in JSG is supported by the Intramural Research Program of NIH, National Institute of Dental and Craniofacial Research.

REFERENCES

1. Siderovski, D. P., and Willard, F. S. (2005) *Int J Biol Sci* **1**, 51–66
2. Dohlman, H. G., and Thorner, J. (1997) *J Biol Chem* **272**, 3871–3874
3. De Vries, L., Zheng, B., Fischer, T., Elenko, E., and Farquhar, M. G. (2000) *Annu Rev Pharmacol Toxicol* **40**, 235–271
4. Ribas, C., Penela, P., Murga, C., Salcedo, A., Garcia-Hoz, C., Jurado-Pueyo, M., Aymerich, I., and Mayor, F., Jr. (2007) *Biochim Biophys Acta* **1768**, 913–922
5. Vazquez-Prado, J., Basile, J., and Gutkind, J. S. (2004) *Methods Enzymol* **390**, 259–285
6. Fukuhara, S., Chikumi, H., and Gutkind, J. S. (2001) *Oncogene* **20**, 1661–1668
7. Suzuki, N., Hajicek, N., and Kozasa, T. (2009) *Neurosignals* **17**, 55–70
8. Castellone, M. D., Teramoto, H., Williams, B. O., Druey, K. M., and Gutkind, J. S. (2005) *Science* **310**, 1504–1510
9. Hart, M. J., Jiang, X., Kozasa, T., Roscoe, W., Singer, W. D., Gilman, A. G., Sternweis, P. C., and Bollag, G. (1998) *Science* **280**, 2112–2114
10. Fukuhara, S., Murga, C., Zohar, M., Igishi, T., and Gutkind, J. S. (1999) *J Biol Chem* **274**, 5868–5879
11. Fukuhara, S., Chikumi, H., and Gutkind, J. S. (2000) *FEBS Lett* **485**, 183–188
12. Hains, M. D., Wing, M. R., Maddileti, S., Siderovski, D. P., and Harden, T. K. (2006) *Mol Pharmacol* **69**, 2068–2075
13. Citro, S., Malik, S., Oestreich, E. A., Radeff-Huang, J., Kelley, G. G., Smrcka, A. V., and Brown, J. H. (2007) *Proc Natl Acad Sci U S A* **104**, 15543–15548
14. Malcolm, K. C., Elliott, C. M., and Exton, J. H. (1996) *J Biol Chem* **271**, 13135–13139
15. Perrot, V., Vazquez-Prado, J., and Gutkind, J. S. (2002) *J Biol Chem* **277**, 43115–43120
16. Aurandt, J., Vikis, H. G., Gutkind, J. S., Ahn, N., and Guan, K. L. (2002) *Proc Natl Acad Sci U S A* **99**, 12085–12090
17. Swiercz, J. M., Kuner, R., Behrens, J., and Offermanns, S. (2002) *Neuron* **35**, 51–63
18. Willard, F. S., Willard, M. D., Kimple, A. J., Soundararajan, M., Oestreich, E. A., Li, X., Sowa, N. A., Kimple, R. J., Doyle, D. A., Der, C. J., Zylka, M. J., Snider, W. D., and Siderovski, D. P. (2009) *PLoS ONE* **4**, e4884
19. Willard, M. D., Willard, F. S., Li, X., Cappell, S. D., Snider, W. D., and Siderovski, D. P. (2007) *Embo J* **26**, 2029–2040
20. Pitcher, J. A., Freedman, N. J., and Lefkowitz, R. J. (1998) *Annu Rev Biochem* **67**, 653–692
21. Tesmer, V. M., Kawano, T., Shankaranarayanan, A., Kozasa, T., and Tesmer, J. J. (2005) *Science* **310**, 1686–1690
22. Ridley, A. J., and Hall, A. (1992) *Cell* **70**, 389–399
23. Buhl, A. M., Johnson, N. L., Dhanasekaran, N., and Johnson, G. L. (1995) *J Biol Chem* **270**, 24631–24634
24. Fromm, C., Coso, O. A., Montaner, S., Xu, N., and Gutkind, J. S. (1997) *Proc Natl Acad Sci U S A* **94**, 10098–10103
25. Coso, O. A., Chiariello, M., Yu, J. C., Teramoto, H., Crespo, P., Xu, N., Miki, T., and Gutkind, J. S. (1995) *Cell* **81**, 1137–1146
26. Minden, A., Lin, A., Claret, F. X., Abo, A., and Karin, M. (1995) *Cell* **81**, 1147–1157
27. Hill, C. S., Wynne, J., and Treisman, R. (1995) *Cell* **81**, 1159–1170
28. Offermanns, S., Mancino, V., Revel, J. P., and Simon, M. I. (1997) *Science* **275**, 533–536
29. Burridge, K., and Wennerberg, K. (2004) *Cell* **116**, 167–179
30. Maekawa, M., Ishizaki, T., Boku, S., Watanabe, N., Fujita, A., Iwamatsu, A., Obinata, T., Ohashi, K., Mizuno, K., and Narumiya, S. (1999) *Science* **285**, 895–898
31. Goulimari, P., Kitzing, T. M., Knieling, H., Brandt, D. T., Offermanns, S., and Grosse, R. (2005) *J Biol Chem* **280**, 42242–42251
32. Kovar, D. R. (2006) *Curr Opin Cell Biol* **18**, 11–17

33. Takeya, R., Taniguchi, K., Narumiya, S., and Sumimoto, H. (2008) *Embo J* **27**, 618–628

34. Goulimari, P., Knieling, H., Engel, U., and Grosse, R. (2008) *Mol Biol Cell* **19**, 30–40

35. Hacker, U., and Perrimon, N. (1998) *Genes Dev* **12**, 274–284

36. Worzfeld, T., Wettschureck, N., and Offermanns, S. (2008) *Trends Pharmacol Sci* **29**, 582–589

37. Etienne-Manneville, S., and Hall, A. (2002) *Nature* **420**, 629–635

38. Dorsam, R. T., and Gutkind, J. S. (2007) *Nat Rev Cancer* **7**, 79–94

39. Barac, A., Basile, J., Vazquez-Prado, J., Gao, Y., Zheng, Y., and Gutkind, J. S. (2004) *J Biol Chem* **279**, 6182–6189

40. Barac, A., Basile, J., Vazquez-Prado, J., Gao, Y., Zheng, Y., and Gutkind, J. S. (2003) *J Biol Chem*

41. Chikumi, H., Fukuhara, S., and Gutkind, J. S. (2002) *J Biol Chem* **277**, 12463–12473

42. Chikumi, H., Barac, A., Behbahani, B., Gao, Y., Teramoto, H., Zheng, Y., and Gutkind, J. S. (2004) *Oncogene* **23**, 233–240

43. Driessens, M. H., Olivo, C., Nagata, K., Inagaki, M., and Collard, J. G. (2002) *FEBS Lett* **529**, 168–172

44. Hirotani, M., Ohoka, Y., Yamamoto, T., Nirasawa, H., Furuyama, T., Kogo, M., Matsuya, T., and Inagaki, S. (2002) *Biochem Biophys Res Commun* **297**, 32–37

45. Abo, A., Qu, J., Cammarano, M. S., Dan, C., Fritsch, A., Baud, V., Belisle, B., and Minden, A. (1998) *Embo J* **17**, 6527–6540

46. Majumdar, M., Seasholtz, T. M., Buckmaster, C., Toksoz, D., and Brown, J. H. (1999) *J Biol Chem* **274**, 26815–26821

47. Dutt, P., Nguyen, N., and Toksoz, D. (2004) *Cell Signal* **16**, 201–209

48. Sagi, S. A., Seasholtz, T. M., Kobiashvili, M., Wilson, B. A., Toksoz, D., and Brown, J. H. (2001) *J Biol Chem* **276**, 15445–15452

49. Hart, M. J., Sharma, S., elMasry, N., Qiu, R. G., McCabe, P., Polakis, P., and Bollag, G. (1996) *J Biol Chem* **271**, 25452–25458

50. Kourlas, P. J., Strout, M. P., Becknell, B., Veronese, M. L., Croce, C. M., Theil, K. S., Krahe, R., Ruutu, T., Knuutila, S., Bloomfield, C. D., and Caligiuri, M. A. (2000) *Proc Natl Acad Sci U S A* **97**, 2145–2150

51. Yamada, T., Ohoka, Y., Kogo, M., and Inagaki, S. (2005) *J Biol Chem* **280**, 19358–19363

52. Taya, S., Inagaki, N., Sengiku, H., Makino, H., Iwamatsu, A., Urakawa, I., Nagao, K., Kataoka, S., and Kaibuchi, K. (2001) *J Cell Biol* **155**, 809–820

53. Jaffe, A. B., and Hall, A. (2005) *Annu Rev Cell Dev Biol* **21**, 247–269

54. Schmidt, A., and Hall, A. (2002) *Genes Dev* **16**, 1587–1609

55. Rossman, K. L., Der, C. J., and Sondek, J. (2005) *Nat Rev Mol Cell Biol* **6**, 167–180

56. Kozasa, T., Jiang, X., Hart, M. J., Sternweis, P. M., Singer, W. D., Gilman, A. G., Bollag, G., and Sternweis, P. C. (1998) *Science* **280**, 2109–2111

57. Chen, Z., Singer, W. D., Danesh, S. M., Sternweis, P. C., and Sprang, S. R. (2008) *Structure* **16**, 1532–1543

58. Wang, Q., Liu, M., Kozasa, T., Rothstein, J. D., Sternweis, P. C., and Neubig, R. R. (2004) *J Biol Chem* **279**, 28831–28834

59. Holinstat, M., Mehta, D., Kozasa, T., Minshall, R. D., and Malik, A. B. (2003) *J Biol Chem* **278**, 28793–28798

60. Suzuki, N., Nakamura, S., Mano, H., and Kozasa, T. (2003) *Proc Natl Acad Sci U S A* **100**, 733–738

61. Eisenhaure, T. M., Francis, S. A., Willison, L. D., Coughlin, S. R., and Lerner, D. J. (2003) *J Biol Chem* **278**, 30975–30984

62. Hu, J., Strauch, P., Rubtsov, A., Donovan, E. E., Pelanda, R., and Torres, R. M. (2008) *Mol Immunol* **45**, 1825–1836

63. Grabocka, E., and Wedegaertner, P. B. (2007) *Mol Pharmacol* **72**, 993–1002

64. Chen, Z., Wells, C. D., Sternweis, P. C., and Sprang, S. R. (2001) *Nat Struct Biol* **8**, 805–809

65. Longenecker, K. L., Lewis, M. E., Chikumi, H., Gutkind, J. S., and Derewenda, Z. S. (2001) *Structure* **9**, 559–569

66. Vazquez-Prado, J., Miyazaki, H., Castellone, M. D., Teramoto, H., and Gutkind, J. S. (2004) *J Biol Chem* **279**, 54283–54290

67. Chen, Z., Singer, W. D., Sternweis, P. C., and Sprang, S. R. (2005) *Nat Struct Mol Biol* **12**, 191–197

68. Oinuma, I., Katoh, H., Harada, A., and Negishi, M. (2003) *J Biol Chem* **278**, 25671–25677

69. Swiercz, J. M., Kuner, R., and Offermanns, S. (2004) *J Cell Biol* **165**, 869–880

70. Basile, J. R., Barac, A., Zhu, T., Guan, K. L., and Gutkind, J. S. (2004) *Cancer Res* **64**, 5212–5224

71. Conrotto, P., Valdembri, D., Corso, S., Serini, G., Tamagnone, L., Comoglio, P. M., Bussolino, F., and Giordano, S. (2005) *Blood* **105**, 4321–4329

72. Swiercz, J. M., Worzfeld, T., and Offermanns, S. (2008) *J Biol Chem* **283**, 1893–1901

73. Basile, J. R., Afkhami, T., and Gutkind, J. S. (2005) *Mol Cell Biol* **25**, 6889–6898

74. Fazzari, P., Penachioni, J., Gianola, S., Rossi, F., Eickholt, B. J., Maina, F., Alexopoulou, L., Sottile, A., Comoglio, P. M., Flavell, R. A., and Tamagnone, L. (2007) *BMC Dev Biol* **7**, 55

75. Korostylev, A., Worzfeld, T., Deng, S., Friedel, R. H., Swiercz, J. M., Vodrazka, P., Maier, V., Hirschberg, A., Ohoka, Y., Inagaki, S., Offermanns, S., and Kuner, R. (2008) *Development* **135**, 3333–3343

76. Basile, J. R., Castilho, R. M., Williams, V. P., and Gutkind, J. S. (2006) *Proc Natl Acad Sci U S A* **103**, 9017–9022

77. Sierra, J. R., Corso, S., Caione, L., Cepero, V., Conrotto, P., Cignetti, A., Piacibello, W., Kumanogoh, A., Kikutani, H., Comoglio, P. M., Tamagnone, L., and Giordano, S. (2008) *J Exp Med* **205**, 1673–1685

78. Gitler, A. D., Lu, M. M., and Epstein, J. A. (2004) *Dev Cell* **7**, 107–116

79. Torres-Vazquez, J., Gitler, A. D., Fraser, S. D., Berk, J. D., Van, N. P., Fishman, M. C., Childs, S., Epstein, J. A., and Weinstein, B. M. (2004) *Dev Cell* **7**, 117–123

80. Ruppel, K. M., Willison, D., Kataoka, H., Wang, A., Zheng, Y. W., Cornelissen, I., Yin, L., Xu, S. M., and Coughlin, S. R. (2005) *Proc Natl Acad Sci U S A* **102**, 8281–8286

81. Griffin, C. T., Srinivasan, Y., Zheng, Y. W., Huang, W., and Coughlin, S. R. (2001) *Science* **293**, 1666–1670

82. Bourguignon, L. Y., Gilad, E., Brightman, A., Diedrich, F., and Singleton, P. (2006) *J Biol Chem* **281**, 14026–14040

83. Dubash, A. D., Wennerberg, K., Garcia-Mata, R., Menold, M. M., Arthur, W. T., and Burridge, K. (2007) *J Cell Sci* **120**, 3989–3998

84. Girkontaite, I., Missy, K., Sakk, V., Harenberg, A., Tedford, K., Potzel, T., Pfeffer, K., and Fischer, K. D. (2001) *Nat Immunol* **2**, 855–862

85. Rubtsov, A., Strauch, P., Digiacomo, A., Hu, J., Pelanda, R., and Torres, R. M. (2005) *Immunity* **23**, 527–538

86. Francis, S. A., Shen, X., Young, J. B., Kaul, P., and Lerner, D. J. (2006) *Blood* **107**, 1627–1635

87. Rogers, S. L., Wiedemann, U., Hacker, U., Turck, C., and Vale, R. D. (2004) *Curr Biol* **14**, 1827–1833

88. Wells, C. D., Gutowski, S., Bollag, G., and Sternweis, P. C. (2001) *J Biol Chem* **276**, 28897–28905

89. Bhattacharyya, R., and Wedegaertner, P. B. (2003) *Biochem J* **371**, 709–720

90. Meyer, B. H., Freuler, F., Guerini, D., and Siehler, S. (2008) *J Cell Biochem* **104**, 1660–1670

91. Iwanicki, M. P., Vomastek, T., Tilghman, R. W., Martin, K. H., Banerjee, J., Wedegaertner, P. B., and Parsons, J. T. (2008) *J Cell Sci* **121**, 895–905

92. Parks, S., and Wieschaus, E. (1991) *Cell* **64**, 447–458

93. Barrett, K., Leptin, M., and Settleman, J. (1997) *Cell* **91**, 905–915

94. Nikolaidou, K. K., and Barrett, K. (2004) *Curr Biol* **14**, 1822–1826

95. Kitzing, T. M., Sahadevan, A. S., Brandt, D. T., Knieling, H., Hannemann, S., Fackler, O. T., Grosshans, J., and Grosse, R. (2007) *Genes Dev* **21**, 1478–1483

96. Iguchi, T., Sakata, K., Yoshizaki, K., Tago, K., Mizuno, N., and Itoh, H. (2008) *J Biol Chem* **283**, 14469–14478

97. Deng, S., Hirschberg, A., Worzfeld, T., Penachioni, J. Y., Korostylev, A., Swiercz, J. M., Vodrazka, P., Mauti, O., Stoeckli, E. T., Tamagnone, L., Offermanns, S., and Kuner, R. (2007) *J Neurosci* **27**, 6333–6347

98. Hata, K., Kaibuchi, K., Inagaki, S., and Yamashita, T. (2009) *J Cell Biol* **184**, 737–750

99. Kranenburg, O., Poland, M., van Horck, F. P., Drechsel, D., Hall, A., and Moolenaar, W. H. (1999) *Mol Biol Cell* **10**, 1851–1857

100. Kuner, R., Swiercz, J. M., Zywietz, A., Tappe, A., and Offermanns, S. (2002) *Eur J Neurosci* **16**, 2333–2341

101. Wong, K., Van Keymeulen, A., and Bourne, H. R. (2007) *J Cell Biol* **179**, 1141–1148

102. Kelly, P., Stemmle, L. N., Madden, J. F., Fields, T. A., Daaka, Y., and Casey, P. J. (2006) *J Biol Chem* **281**, 26483–26490

103. Kelly, P., Moeller, B. J., Juneja, J., Booden, M. A., Der, C. J., Daaka, Y., Dewhirst, M. W., Fields, T. A., and Casey, P. J. (2006) *Proc Natl Acad Sci U S A* **103**, 8173–8178

104. Ong, D. C., Ho, Y. M., Rudduck, C., Chin, K., Kuo, W. L., Lie, D. K., Chua, C. L., Tan, P. H., Eu, K. W., Seow-Choen, F., Wong, C. Y., Hong, G. S., Gray, J. W., and Lee, A. S. (2009) *Oncogene*

105. Reuther, G. W., Lambert, Q. T., Booden, M. A., Wennerberg, K., Becknell, B., Marcucci, G., Sondek, J., Caligiuri, M. A., and Der, C. J. (2001) *J Biol Chem* **276**, 27145–27151

106. Wirth, A., Benyo, Z., Lukasova, M., Leutgeb, B., Wettschureck, N., Gorbey, S., Orsy, P., Horvath, B., Maser-Gluth, C., Greiner, E., Lemmer, B., Schutz, G., Gutkind, J. S., and Offermanns, S. (2008) *Nat Med* **14**, 64–68

107. Kovacs, P., Stumvoll, M., Bogardus, C., Hanson, R. L., and Baier, L. J. (2006) *Diabetes* **55**, 1497–1503

108. Ma, L., Hanson, R. L., Que, L. N., Cali, A. M., Fu, M., Mack, J. L., Infante, A. M., Kobes, S., Bogardus, C., Shuldiner, A. R., and Baier, L. J. (2007) *Diabetes* **56**, 1454–1459

109. Holzapfel, C., Klopp, N., Grallert, H., Huth, C., Gieger, C., Meisinger, C., Strassburger, K., Giani, G., Wichmann, H. E., Laumen, H., Hauner, H., Herder, C., Rathmann, W., and Illig, T. (2007) *Eur J Endocrinol* **157**, R1–5

110. Bottcher, Y., Schleinitz, D., Tonjes, A., Bluher, M., Stumvoll, M., and Kovacs, P. (2008) *J Hum Genet* **53**, 365–367

111. Evelyn, C. R., Ferng, T., Rojas, R. J., Larsen, M. J., Sondek, J., and Neubig, R. R. (2009) *J Biomol Screen* **14**, 161–172

112. Chikumi, H., Vazquez- Prado, J., Servitja, J. M., Miyazaki, H., and Gutkind, J. S. (2002) *J Biol Chem* **277**, 27130–27134

113. Vogt, S., Grosse, R., Schultz, G., and Offermanns, S. (2003) *J Biol Chem* **278**, 28743–28749

114. Sah, V. P., Hoshijima, M., Chien, K. R., and Brown, J. H. (1996) *J Biol Chem* **271**, 31185–31190

115. Brown, J. H., Del Re, D. P., and Sussman, M. A. (2006) *Circ Res* **98**, 730–742

116. Lutz, S., Shankaranarayanan, A., Coco, C., Ridilla, M., Nance, M. R., Vettel, C., Baltus, D., Evelyn, C. R., Neubig, R. R., Wieland, T., and Tesmer, J. J. (2007) *Science* **318**, 1923–1927

117. Lutz, S., Freichel-Blomquist, A., Yang, Y., Rumenapp, U., Jakobs, K. H., Schmidt, M., and Wieland, T. (2005) *J Biol Chem* **280**, 11134–11139

118. Booden, M. A., Siderovski, D. P., and Der, C. J. (2002) *Mol Cell Biol* **22**, 4053–4061

119. DeWire, S. M., Ahn, S., Lefkowitz, R. J., and Shenoy, S. K. (2007) *Annu Rev Physiol* **69**, 483–510

120. Whistler, J. L., Enquist, J., Marley, A., Fong, J., Gladher, F., Tsuruda, P., Murray, S. R., and Von Zastrow, M. (2002) *Science* **297**, 615–620

121. Naga Prasad, S. V., Barak, L. S., Rapacciuolo, A., Caron, M. G., and Rockman, H. A. (2001) *J Biol Chem* **276**, 18953–18959

122. Naga Prasad, S. V., Jayatilleke, A., Madamanchi, A., and Rockman, H. A. (2005) *Nat Cell Biol* **7**, 785–796
123. Shankaranarayanan, A., Thal, D. M., Tesmer, V. M., Roman, D. L., Neubig, R. R., Kozasa, T., and Tesmer, J. J. (2008) *J Biol Chem* **283**, 34923–34934
124. Huang, H., and He, X. (2008) *Curr Opin Cell Biol* **20**, 119–125
125. Stemmle, L. N., Fields, T. A., and Casey, P. J. (2006) *Mol Pharmacol* **70**, 1461–1468
126. Chia, I. V., Kim, M. J., Itoh, K., Sokol, S. Y., and Costantini, F. (2009) *Genetics*
127. Chia, I. V., Kim, M. J., Itoh, K., Sokol, S. Y., and Costantini, F. (2009) *Genetics* **181**, 1359–1368
128. Schiff, M. L., Siderovski, D. P., Jordan, J. D., Brothers, G., Snow, B., De Vries, L., Ortiz, D. F., and Diverse-Pierluissi, M. (2000) *Nature* **408**, 723–727
129. Traver, S., Splingard, A., Gaudriault, G., and De Gunzburg, J. (2004) *Biochem J* **379**, 627–632
130. Snow, B. E., Hall, R. A., Krumins, A. M., Brothers, G. M., Bouchard, D., Brothers, C. A., Chung, S., Mangion, J., Gilman, A. G., Lefkowitz, R. J., and Siderovski, D. P. (1998) *J Biol Chem* **273**, 17749–17755
131. Kimple, R. J., De Vries, L., Tronchere, H., Behe, C. I., Morris, R. A., Gist Farquhar, M., and Siderovski, D. P. (2001) *J Biol Chem* **276**, 29275–29281
132. Willard, F. S., Kimple, R. J., and Siderovski, D. P. (2004) *Annu Rev Biochem* **73**, 925–951
133. Martin-McCaffrey, L., Hains, M. D., Pritchard, G. A., Pajak, A., Dagnino, L., Siderovski, D. P., and D'Souza, S. J. (2005) *Dev Dyn* **234**, 438–444
134. Webb, C. K., McCudden, C. R., Willard, F. S., Kimple, R. J., Siderovski, D. P., and Oxford, G. S. (2005) *J Neurochem* **92**, 1408–1418
135. Chatterjee, T. K., and Fisher, R. A. (2002) *Mol Cell Biol* **22**, 4334–4345
136. Martin-McCaffrey, L., Willard, F. S., Oliveira-dos-Santos, A. J., Natale, D. R., Snow, B. E., Kimple, R. J., Pajak, A., Watson, A. J., Dagnino, L., Penninger, J. M., Siderovski, D. P., and D'Souza S. J. (2004) *Dev Cell* **7**, 763–769
137. Cho, H., and Kehrl, J. H. (2007) *J Cell Biol* **178**, 245–255
138. Knoblich, J. A. (2008) *Cell* **132**, 583–597

10 Adenylyl cyclase isoform-specific signaling of GPCRs

Karin F. K. Ejendal, Julie A. Przybyla, and Val J. Watts

INTRODUCTION

Cyclic AMP signaling pathway

Adenylyl cyclase is a membrane-bound enzyme responsible for converting cellular ATP into cyclic AMP (cAMP), the first described intracellular second messenger.[1] The cAMP signaling pathway is highly conserved among species and is involved in a multitude of physiological functions including development, cardiovascular function, learning and memory, and aging.[2–4] The cAMP signaling system is comprised of a number of components such as G protein-coupled receptors (GPCRs), G protein subunits, adenylyl cyclase isoforms, phosphodiesterase isoforms, regulators of G protein signaling (RGS) proteins, and a growing list of novel modulators.[4–6] The ever-growing complexity of this system provides additional opportunities for modulating cAMP levels ultimately influencing diverse cellular signaling pathways associated with GPCRs.

The initial cloning of adenylyl cyclase type 1 (AC1) occurred in 1989, and eight additional membrane-bound isoforms have since been cloned and characterized.[4,7] A tenth, cytosolic adenylyl cyclase, is expressed exclusively in male testis and will not be discussed further.[8] The nine membrane-bound isoforms of adenylyl cyclase share an overall similar primary structure and topology that includes an intracellular N terminus followed by two membrane-spanning domains (M1, M2) alternating with two cytoplasmic loops, C1 and C2, that can be further divided into a and b regions (Figure 10.1). The C1a and C2a regions share approximately 40% identity and together form the catalytic domain of the enzyme. In contrast, the N terminus of adenylyl cyclase and C-terminal end of the C1b or the C2b (when present) are quite divergent and appear to represent unique sites for regulation. Each adenylyl cyclase isoform is regulated in a unique manner by Gα and Gβγ subunits, protein kinases, Ca^{2+}, post-translational modifications, subcellular localization, and small molecules.[4,9,10]

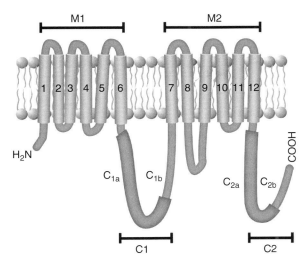

Figure 10-1: Adenylyl cyclase is a membrane-bound enzyme that contains an intracellular N terminus, followed by a membrane-bound region (M1). The third intracellular loop is much larger than the other intracellular loops and forms the C1 catalytic loop. The C1 loop is often divided into a conserved C1a region, important for the formation of the catalytic unit, and the more divergent C1b region. The second membrane-spanning region is called the M2 region. The C terminus (C2) forms the other half of the catalytic unit and is divided into C2a and C2b regions. The sequences are more divergent at important regulatory regions such as the N terminus, the C1b region, the fourth intracellular loop, and the C2b.

Reprinted from Beazely and Watts,[9] with permission from Elsevier.

Overview of adenylyl cyclases

The adenylyl cyclases are typically classified into four distinct groups based on their overall regulatory properties (Table 10.1, Figure 10.2). Group I adenylyl cyclases are also referred to as the Ca^{2+}-stimulated family and include AC1, AC3, and AC8. Each responds to Ca^{2+} in a calmodulin-dependent fashion.[10] AC2, AC4, and AC7 comprise the Group II adenylyl cyclase isoforms and are characterized by their ability to be conditionally activated by $G\beta\gamma$ subunits.[4,10] In addition, AC2 and AC7, but not AC4, are activated by phosphorylation by PKC. Group III adenylyl cyclases are the Ca^{2+}-inhibited adenylyl cyclase isoforms, AC5 and AC6. AC5 and AC6 share a very high amino acid sequence identity and several regulatory properties such as robust stimulation by $G\alpha_s$ and the diterpene forskolin (a small molecule activator of AC1–8), and inhibition by Ca^{2+} and PKA.[9, 10] Group IV contains only one member, AC9, the most divergent membrane-bound adenylyl cyclase isoform and the only isoform that is not robustly activated by forskolin.[4,11,12]

Expression of adenylyl cyclases

It appears that most tissues and cells typically express multiple adenylyl cyclase isoforms as measured by mRNA expression patterns.[13–15] The overall adenylyl cyclase expression patterns have been reviewed previously,[14,16,17] and studies examining the expression of adenylyl cyclase have largely been limited to examining mRNA as opposed to proteins.[4] Despite years of efforts, high quality and

Table 10-1. Regulatory Properties of Adenylyl Cyclase Isoforms

Group	Type	Response to Regulators					
		$G\alpha_s$	$Ca^{2}+$	$\beta\gamma$	$G\alpha_{i/o}$	FSK	PKA and PKC
	AC1	↑	↑ (Cam)	↓	↓ (α_i)	↑	↑ PKC (weak)
1	AC3	↑	↑ (Cam)	↓	↓ (weak)	↑	↑ PKC (weak)
	AC8	↑	↑ (Cam)	↓	↓ (weak)	↑	
	AC2	↑		↑*		↑	↑ PKC
2	AC4	↑		↑*		↑	↓ PKC
	AC7	↑		↑*		↑	↑ PKC
3	AC5	↑	↓	↑*	↓	↑	↓ PKA, ↑ PKC
	AC6	↑	↓;↑ (Cam)	↑*	↓	↑	↓ PKA; ↓ or ↑* PKC
4	AC9	↑	↓ Calcineurin ↑ (CamK)		↓ (weak)	↑(weak);↓*	↓ PKC

↑ = stimulatory; ↓ = inhibitory; *, conditional; Cam, Calmodulin
Reprinted from Watts and Neve,[59] with permission from Elsevier.

selective adenylyl cyclase antibodies have been too difficult to obtain by individual investigators, and those available commercially have limited utility.[4,18] In
general, the expression of the Group I adenylyl cyclases (AC1, AC3, and AC8)
appears to be primarily in neuronal tissue.[17,19] In contrast, the expression pattern
of Group II adenylyl cyclases is generally broader, with the highest expression
levels seen in brain and muscle tissue.[17] Group III adenylyl cyclases are highly
expressed in cardiac muscle and striatal neurons; however, AC6 is also highly
expressed in many other tissues.[17] AC9 is highly expressed in brain, lung, and
skeletal muscle.[17]

GPCR REGULATION OF ADENYLYL CYCLASES

Historical perspective of GPCR modulation of adenylyl cyclase

Early studies of adenylyl cyclase signal transduction were aimed at understanding
how hormones regulate carbohydrate metabolism. Through the course of more
than twenty-five years, a number of seminal experiments allowed scientists to
eventually determine that hormones were binding to a "receptor" that then
transduced its signal through a guanine nucleotide binding protein (G protein).
The G protein was capable of activating adenylyl cyclase, thereby increasing
cAMP accumulation. Thus, this G protein was designated $G\alpha_s$ for the stimulatory properties it exerts on all adenylyl cyclase isoforms.[4] Throughout this process, scientists discovered the receptor-G protein cycle. Agonist activation of the
GPCR activation causes a conformational change in heterotrimeric ($G\alpha\beta\gamma$) G
protein that promotes a GDP to GTP nucleotide exchange on the $G\alpha$ subunit of
the interacting trimeric G protein. The GTP-bound $G\alpha$ subunits then dissociate

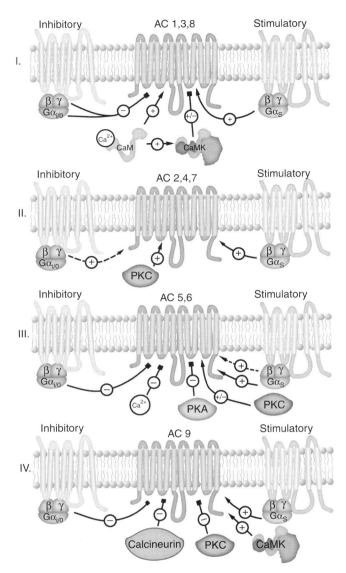

Figure 10-2: The nine membrane-bound isoforms of adenylyl cyclase are classified into four categories/groups based on their regulatory properties. Group I (top) includes AC1, AC3, and AC8. Group II is represented by AC2, 4, and 7. Group III is made up of AC5 and AC6. AC9 (bottom) is the sole member of Group IV (see text for details on their regulation).

from (or change conformation relative to) the Gβγ subunits, allowing interactions between Gα-GTP and Gβγ with effector proteins.[5, 20] The intrinsic GTPase activity of the Gα subunit returns Gα-GTP to Gα-GDP, allowing it to reassociate with Gβγ and its receptor.

To date, there are four families of Gα subunits, five Gβ subunits, and twelve Gγ subunits (Table 10.2). The Gα$_s$ family is comprised of a series of splice variants (i.e., Gα$_s$short, Gα$_s$long, and Gα$_s$XL) and Gα$_{olf}$ that is expressed in the olfactory neuroepithelium and the striatum.[4] All Gα$_s$ splice variants and Gα$_{olf}$ stimulate adenylyl cyclase. In contrast, there are several inhibitory Gα subunits (i.e., Gα$_i$,

Table 10-2. Subunits of the Heterotrimeric G proteins

Class	Members	Effect(s) on Adenylyl Cyclase	Comments
$G\alpha_s$	$G\alpha_s$ short $G\alpha_s$ long $G\alpha_s$ XL $G\alpha_{olf}$	Activation of AC	$G\alpha_s$ activated by cholera toxin. $G\alpha_s$ stimulates AC by enhancing C1-C2 interactions
$G\alpha_{i/o}$	$G\alpha_{i1-3}$ $G\alpha_o$ $G\alpha_t$ $G\alpha_g$ $G\alpha_z$	Inhibition of ACs in group 1, 3, 4	Pertussis toxin prevents receptor activation by ADP-ribosylation of $G\alpha_{i/o/t}$ subunits. $G\alpha_i$ inhibition occurs via binding to the C1 domain
$G\alpha_q$	$G\alpha_q$ $G\alpha_{11}$ $G\alpha_{14}$ $G\alpha_{15}$ $G\alpha_{16}$	Indirect, AC isoform dependent.	Activates PLC, which generates the second messengers DAG and IP_3 and subsequent downstream effectors (i.e. PKC and Ca^{2+}).
$G\alpha_{12/13}$	$G\alpha_{12}$ $G\alpha_{13}$	Selective activation of AC7	
$G\beta/\gamma$	$G\beta_{1-5}$ $G\gamma_{1-5, 7, 8, 10-14}$	Inhibition or conditional activation	AC dependent; may be dependent on $G\alpha$.

$G\alpha_o$, and $G\alpha_z$) that can inhibit select adenylyl cyclase isoforms (Tables 10.1 and 10.2). Very recently it was suggested that receptor activation of $G\alpha_{12/13}$ subunits can selectively and directly stimulate the activity of AC7.[21] In contrast, GPCR activation of $G\alpha_q$ leads to an indirect modulation of adenylyl cyclases that involves stimulation of phospholipase C (PLC) and subsequent modulation of intracellular Ca^{2+} levels, as well as the activation of members of the PKC family (see group discussions further in this chapter). The co-expression of stimulatory and inhibitory $G\alpha$ subunits and their respective receptors appears to provide balance for regulation of adenylyl cyclase activity and consequently for cAMP signaling in the cell. Many adenylyl cyclases are also regulated by G protein $\beta\gamma$ subunits following GPCR activation.[4] Group II adenylyl cyclases are characterized by their ability to be conditionally activated by $G\beta\gamma$. In contrast, several other adenylyl cyclases are negatively modulated by $G\beta\gamma$ (see group discussions further in this chapter). The ability of $G\alpha_s$, $G\alpha_{i/o}$, and $G\beta\gamma$ to regulate adenylyl cyclases is thought to involve direct interactions with adenylyl cyclases.

G protein signaling and thus regulation of adenylyl cyclase can also be modulated by RGS proteins.[4] RGS proteins were originally identified as GTPase activators for $G\alpha_{i/o}$ and $G\alpha_q$; however, they also have a variety of effects on other signaling molecules, including adenylyl cyclase. Studies to date reveal that RGS2 inhibits several adenylyl cyclase isoforms (i.e., AC3, AC5, and AC6) through direct interactions. However, there are more than twenty RGS proteins, and much remains to be learned regarding their actions and specificity for regulating adenylyl cyclase.[4]

GPCR regulation of group I adenylyl cyclases (AC1, AC3, and AC8)

Adenylyl cyclases in group I are all stimulated by Ca^{2+}, and studies with AC1 and AC8 have made it increasingly apparent that these two adenylyl cyclase isoforms are stimulated primarily by elevated cytoplasmic Ca^{2+} through capacitative Ca^{2+} entry (CCE). The mechanism of CCE was first proposed by Putney.[22] CCE is triggered by depletion of intracellular Ca^{2+} stores, primarily the endoplasmic reticulum (ER). Activation of subunits of the $G\alpha_q$ class stimulates membrane-bound PLC generating IP_3 and DAG from PIP_2. The main role of the second messenger IP_3 is to mobilize Ca^{2+} from intracellular stores via activation of the IP_3 receptor, which then allows for CCE. Thus, depletion of intracellular Ca^{2+} stores can be evoked experimentally by stimulating $G\alpha_q$-coupled receptors with agonists, such as carbachol-mediated stimulation of M_1-like muscarinic receptors. Alternatively, CCE can also be passively induced in a $G\alpha_q$-independent manner by treatment with the sarco/endoplasmic reticulum Ca^{2+} ATPase (SERCA) inhibitor, thapsigargin. The CCE machinery is composed of the Orai proteins that form the pore in the plasma membrane, and the STIM proteins that constitute the Ca^{2+} sensor and are located in the ER. Although heterologous expression of Orai1 and STIM1 functionally reconstitutes CCE,[23] the transient receptor potential (TRP) channels may also play a role in CCE.[10]

A number of additional divergent regulatory properties relevant to GPCR modulation are also present for group I adenylyl cyclases. For example, AC1 is markedly inhibited by $G\alpha_i$ and $G\alpha_i$-coupled receptors, whereas AC8 appears to have reduced sensitivity to $G\alpha_i$ and regulation by $G\alpha_{i/o}$-coupled receptors.[24-26] Additional diversity for AC1 and AC8 activation reflect their ability to respond to convergent activation by Ca^{2+}/calmodulin and $G\alpha_s$. Specifically, AC1 shows robust synergistic responses to activation by the Ca^{2+} ionophore, A23187, and $G\alpha_s$ activation, whereas the AC8 response is simply additive.[24,25] In a recent report directly comparing AC1 and AC8, it was also shown that AC1 is more sensitive to Ca^{2+} as well as to stimulation by carbachol-mediated Ca^{2+} release when compared to AC8.[27] Such divergent regulation translates to modulation by $G\alpha_q$-coupled receptors, where $G\alpha_q$-mediated activation of PLC precipitates the release of Ca^{2+} from intracellular stores and the activation of PKC as well as other Ca^{2+}-dependent pathways (e.g., calmodulin kinase activation). For example, in HEK293 cells stably expressing AC1, carbachol stimulation of endogenously expressed muscarinic M_1-like receptors has been shown to increase cAMP levels and synergistically enhance $G\alpha_s$ activation of AC1 in a Ca^{2+}-dependent manner.[28,29] A similar synergistic effect of $G\alpha_s$ activation and $G\alpha_q$-modulated Ca^{2+} have been observed on endogenous AC3 activity in cardiac fibroblasts.[30,31] These observations suggest that there are overlapping modes of AC1 and AC3 regulation in intact cellular models. However, the overall mechanisms for Ca^{2+}-stimulated adenylyl cyclase regulation involving calmodulin appear to differ across group I adenylyl cyclases. AC1 is inhibited by calmodulin kinase IV (CaMK IV) and AC3 is inhibited by calmodulin kinase II (CaMK II) providing unique negative feedback mechanisms.[4] Also, both AC1 and AC8 are directly activated by calmodulin,[27] whereas calmodulin activation of AC3 requires $G\alpha_s$ activation.[4] Further diversity of group I adenylyl cyclase regulation reflects their ability to be modulated by protein kinases. For example, AC8 appears insensitive to PKC modulation, and both AC1 and AC3 show modest activation in response to phorbol esters (i.e., phorbol 12-myristate 13-acetate; PMA).[32]

A common feature of group I adenylyl cyclases is their ability to be directly inhibited by $G\beta\gamma$.[4] The $G\beta\gamma$ inhibitory response blunts the activation following stimulation of $G\alpha_s$- and $G\alpha_q$-coupled receptors, whereas the $G\alpha_i$ response of AC1 is enhanced. In summary, group I adenylyl cyclases are sensitive to regulation by GPCRs linked to $G\alpha_s$, $G\alpha_{i/o}$, and $G\alpha_q$ as well as being inhibited by $G\beta\gamma$.

GPCR regulation of group II adenylyl cyclases (AC2, AC4, and AC7)

Group II adenylyl cyclases appear to be unique in that they are directly or indirectly sensitive to regulation by receptors linked to all families of G proteins ($G\alpha_s$, $G\alpha_{i/o}$, $G\alpha_q$, $G\alpha_{12/13}$, and $G\beta\gamma$). $G\alpha_s$ stimulates all group II adenylyl cyclases, and the $G\alpha_s$-stimulated activity is enhanced five- to tenfold by the $G\beta\gamma$ subunits.[4] The $G\beta\gamma$ enhancement has been demonstrated in vitro with exogenous purified $G\beta\gamma$ and in intact cells via the release of $G\beta\gamma$ subunits from $G\alpha_{i/o}$-coupled receptors.[33,34] The ability of $G\beta\gamma$ subunits to potentiate $G\alpha_s$ stimulation of AC2, AC4, and AC7 is the result of direct interaction of both $G\alpha_s$ and $G\beta\gamma$ subunits with these adenylyl cyclase isoforms, and may reflect the ability of $G\beta\gamma$ subunits to enhance $G\alpha_s$-adenylyl cyclase interactions.[35] Extensive work with AC2 has provided evidence that $G\beta\gamma$ binding may be shared across the enzyme involving both a well-characterized motif on the C1b domain as well as the C2 domain.[4]

AC2 and AC7 also show conditional $G\beta\gamma$ activation in the presence of PKC activation.[36-38] This regulatory property appears to be a result of the divergent regulation of the group II adenylyl cyclases to PKC. The PKC family of proteins consists of nine genes that encode for serine/threonine kinases that share a common kinase domain but differ in their regulatory regions.[39] The PKC isoforms can be subdivided into three groups based on their structure and activating characteristics. The classical/conventional (α, β, and γ) isoforms are Ca^{2+} dependent and are activated by diacylglycerol (DAG), phosphatidylserine (PS), or phorbol esters. Novel (δ, ε, η, ξ) isoforms are Ca^{2+}-independent, but are activated by DAG, PS, or phorbol esters. Atypical (ζ, λ) isoforms are Ca^{2+}-independent and do not require DAG for activation. Basal activity as well as forskolin-, $G\beta\gamma$-, or $G\alpha_s$-stimulated AC2 activity can be enhanced by the activation of PKC.[40, 41] Activation of PKC via $G\alpha_q$-coupled receptors (e.g., $5HT_{2A}$ and M_5) or following expression of a constitutive $G\alpha_q$ also enhances $G\alpha_s$-stimulated AC2 activity.[11] These receptor-mediated effects are blocked by the PKC inhibitor, bisindolylmaleimide I.[11] Previous studies suggest that AC2 is regulated by both conventional and novel PKCs as PKCα phosphorylates AC2 in vitro and PKCδ appears to stimulate AC2 in HEK293 cells.[40,41] AC7 is also activated by PKC; however, the effects of PKC appear to dissipate quickly.[42,43] Since those initial studies, it has been determined that AC7 is synergistically stimulated by $G\alpha_s$ and PKCδ.[4] Together, these observations suggest potential overlap for PKCδ stimulation of AC2 and AC7. The basal activity of AC2 is exquisitely sensitive to PKC activation, showing robust responses to phorbol esters and $G\alpha_q$ activation.[12,32] In contrast, both basal and forskolin-stimulated AC4 activity are resistant to PKC activation by phorbol esters or reconstituted PKCα.[32,40] It was also revealed that PKCα activation can inhibit both $G\alpha_s$ stimulation and conditional Gβγ activation of AC4.[40]

GPCR regulation of group III adenylyl cyclases (AC5 and AC6)

AC5 and AC6 are inhibited by submicromolar concentrations of Ca^{2+} (0.2–0.6 μM) and make up the Ca^{2+}-inhibited family of adenylyl cyclases.[4,10] This Ca^{2+} sensitivity contrasts all other adenylyl cyclases that are inhibited by high concentrations of Ca^{2+} (10–25 μM), which competes with magnesium at the active site.[4] It is proposed that the source of Ca^{2+} to regulate AC5 and AC6 is CCE that can be mediated by $G\alpha_q$-coupled receptor activation.[10,44] As discussed previously, two membrane proteins, STIM1 and Orai1, can mediate CCE, thus understanding their regulation by G proteins may have implications in GPCR-AC5/AC6 signaling.[45,46]

In many ways, AC5 and AC6 appear to be the prototypical adenylyl cyclases in that they are robustly activated by $G\alpha_s$ and inhibited by the inhibitory G proteins, $G\alpha_i$ and $G\alpha_z$.[4] Activated $G\alpha_s$ promotes the interaction between the C1 and C2 loops of adenylyl cyclase, whereas $G\alpha_i$ binds to the C1 domain and impairs the C1-C2 interaction.[4] AC5 and AC6 are the most sensitive isoforms to inhibition by inhibitory G proteins ($G\alpha_{i1,2,3/z}$).[4] In vitro studies suggest that AC5 and AC6 are insensitive to inhibition to $G\alpha_o$.[26,35] However, several lines of evidence suggest that $G\alpha_o$ can inhibit adenylyl cyclase activity in intact cells expressing group III adenylyl cyclases.[9] Studies with pertussis toxin-insensitive (PTXi) $G\alpha_o$ or $G\alpha_o$ knockdown studies suggest that the inhibitory GPCRs (i.e., μ opioid and D_2 dopamine) activate $G\alpha_o$ to inhibit AC6 activity in cultured cells. In addition, antibodies against $G\alpha_o$ decrease opioid receptor-mediated inhibition of adenylyl cyclase in striatal tissue where AC5 and AC6 are expressed at high levels.[9] Although the in vitro studies demonstrate that $G\alpha_o$ is not sufficient to inhibit AC5 or AC6, studies in cell lines and tissues suggest that receptor activation of $G\alpha_o$ signaling can inhibit cAMP accumulation in cells expressing those isoforms.[9]

The effect of Gβγ subunits on AC5 and AC6 activity appears to differ based on the experimental approach employed. Simple overexpression studies reveal that Gβγ subunits reduce cAMP accumulation in cells expressing AC5 or AC6.[47] In contrast, a recent study demonstrated that Gβγ subunits enhance $G\alpha_s$-stimulated AC5 and AC6 activity.[48] The mechanism for this stimulatory effect involved Gβγ binding to the N-terminus of AC5 or AC6, and it was suggested that the source of potentiating Gβγ is the $G\alpha_s$-Gβγ heterotrimer.[48] Previous truncation studies using recombinant AC6 demonstrated that the removal of the N-terminal 86 residues of AC6 enhances the ability of $G\alpha_i$ to inhibit AC6 activity.[49] This later observation may suggest that removal of portions of the Gβγ binding region on the N-terminus facilitates $G\alpha_i$ inhibition of AC6.

Protein kinases can also significantly regulate AC5 and AC6 activity by phosphorylation of several serine and threonine residues. PKA is directly regulated by cAMP, and both AC5 and AC6 are phosphorylated in vitro by PKA. This PKA-dependent phosphorylation decreases the ability of $G\alpha_s$ and forskolin to stimulate cAMP production in cells expressing AC5 or AC6.[9] Mutating a single serine in the C1 domain (S674) to alanine prevents phosphorylation and PKA-mediated inhibition of AC6.[50,51] The precise PKA phosphorylation sites in AC5 have not been identified, although AC5 contains fourteen putative PKA phosphorylation consensus sequences, including serine 788 that corresponds to serine 674 in AC6.[51,52]

Similar to AC2, phosphorylation of recombinant AC5 by PKCα or PKCζ enhances basal activity as well as forskolin- and $G\alpha_s$-stimulated cAMP accumulation.[53] Evidence for PKC (e.g., α, δ, or ζ) enhancement of AC5 activity has also been reported in gallbladder epithelium and MCF-7 cells expressing recombinant AC5.[9] However, one study indicates that the phorbol ester, PMA, does not stimulate AC5 activity.[32] The effects of PKC activation on AC6 activity are also inconsistent, revealing inhibition, stimulation, or no effect.[9] For example, in vitro studies with recombinant AC6 and a mixture of PKC isoforms isolated from rat brain (PKCα, βI, βII, γ, δ, ε, and μ) suggest a PKC-mediated decrease in forskolin-stimulated cAMP accumulation.[9] In contrast, phorbol esters (i.e., PMA) potentiate drug-stimulated cAMP accumulation in stably-transfected HEK-AC6 cells through a novel PKC isoform (i.e., δ, ε, η, or ξ) without altering basal cAMP levels.[50] PMA also potentiates drug-stimulated cAMP accumulation in other cell lines that express endogenous levels of AC6, including Chinese hamster ovary and Cath.a differentiated (CAD) cells.[9] These studies contrast the initial reports that that phorbol esters do not alter AC6 activity in transiently transfected HEK293 cells.[9,32]

The regulation of AC5 and AC6 likely involves coincidence detection and may reflect isoform specificity for the downstream signaling of $G\alpha_q$-coupled receptors. For example, in tissues that express AC5 and AC6, such as cardiac fibroblasts and gastric smooth muscle, activation of $G\alpha_q$-coupled receptors enhances adenylyl cyclase signaling.[9,30,31] Using stably transfected HEK-AC5 and HEK-AC6 cells, it was demonstrated that $G\alpha_q$-coupled receptor signaling enhances AC6 activation, but not AC5 activation.[54] Specifically, the M_1 muscarinic receptor agonist carbachol significantly enhances the activity of forskolin-stimulated AC6, but not AC5.[54] Subsequent experiments reveal a similar potentiation of AC6 by the $G\alpha_q$-coupled $5HT_{2A}$ receptor as well as by expression of constitutively active $G\alpha_q$ (Q209L). These $G\alpha_q$-mediated effects are not prevented by PKC inhibitors, and subsequent biochemical studies suggest the involvement of a novel Ca^{2+} calmodulin-dependent pathway.[54]

GPCR regulation of group IV adenylyl cyclases (AC9)

The least characterized and most divergent in sequence of the adenylyl cyclase isoforms is AC9.[12] Wild-type AC9 is insensitive to the small molecule forskolin; however, a single amino acid change (Y1021L) in the C2 loop can restore forskolin sensitivity.[55] AC9 is stimulated by $G\alpha_s$ and inhibited by Ca^{2+}/calcineurin.[56,57] Subsequent studies examining GPCR modulation of AC9 have identified several additional modes of AC9 regulation, including serving as a coincidence detector.[11,12] For example, $G\alpha_s$-stimulated AC9 activity is inhibited by activation of $G\alpha_{i/o}$-coupled receptors, and this inhibition is blocked by pertussis toxin.[11] $G\alpha_s$-stimulated AC9 activity is also reduced by more than 50% in the presence of phorbol esters, and this reduction appears to be the result of a novel PKC isoform as it was prevented by bisindolylmaleimide I but not Gö6976 (an inhibitor of conventional PKC isoforms). An additional series of experiments revealed that sequestering of $G\beta\gamma$ subunits enhances AC9 activity, suggesting that $G\beta\gamma$ subunits may negatively regulate AC9.[12] It was also determined that AC9 is regulated by $G\alpha_q$. For example, activation of $G\alpha_q$-linked receptors (e.g., M_1 or $5HT_{2A}$) or constitutively active $G\alpha_q$ (Q209L) enhances $G\alpha_s$-stimulated AC9 activity in a PKC-independent fashion. Biochemical studies suggested that this $G\alpha_q$ receptor-mediated response involves CamK II.[12] Although the work by Cumbay and Watts[11,12] was completed using an AC9 mutant where the last 101 amino acids were removed,[56,57] side-by-side experiments suggest that the full length AC9 behaves similar to the truncated AC9 showing robust inhibition by phorbol esters (Cumbay, M.G. and Watts, V.J., unpublished observations).

GPCR-MEDIATED HETEROLOGOUS SENSITIZATION OF ADENYLYL CYCLASE

Heterologous sensitization

Acute activation of $G\alpha_{i/o}$-coupled receptors inhibits adenylyl cyclase activity, whereas prolonged activation of these inhibitory receptors enhances subsequent drug-stimulated cAMP accumulation. This heterologous sensitization of adenylyl cyclase was first observed following persistent activation of the δ opioid receptor in the laboratory of Dr. Marshall Nirenberg who proposed that the increased adenylyl cyclase responsiveness is a mechanism of opiate tolerance and dependence.[58] This intriguing phenomenon has also been referred to as cAMP overshoot, supersensitivity, superactivation, supersensitization, and heterologous sensitization of adenylyl cyclase. Heterologous sensitization occurs following persistent activation of a number of $G\alpha_{i/o}$-coupled receptors (e.g., D_2 dopamine, $5HT_{1A}$ serotonin, μ and δ opioid, and CB_1 cannabinoid) in both neuronal and non-neuronal cellular models.[59] Previous studies support a hypothesis that persistent activation of a $G\alpha_{i/o}$-coupled receptor promotes the dissociation/rearrangement of $G\alpha$ and $G\beta\gamma$ subunits in a pertussis toxin-sensitive manner that induces sensitization of adenylyl cyclase through both $G\alpha_s$-dependent and $G\alpha_s$-independent mechanisms. The signaling events that follow the activation of the $G\alpha_{i/o}$ subunits and the release of the $G\beta\gamma$ subunits produce enhanced adenylyl cyclase responsiveness through a variety of mechanisms that may include phosphorylation events, Gβγ regulation of adenylyl cyclase, and Gα$_s$-adenylyl cyclase interactions.[59]

Gα$_{i/o}$ modulation and protein kinases in heterologous sensitization

Pertussis toxin treatment prevents heterologous sensitization of both endogenous and recombinant adenylyl cyclases in several cellular models[59] The promiscuous nature of pertussis toxin (i.e., preventing receptor coupling to Gα$_{i/o/t}$) prompted several studies examining the G protein specificity for GPCR-mediated heterologous sensitization. Investigators have employed genetically engineered pertussis toxin-insensitive G proteins (PTXi Gα$_x$) to determine the G protein specificity. Studies consistently show that Gα$_o$ is particularly efficient in rescuing D$_2$ dopamine and μ opioid receptor-induced sensitization.[60-62] In addition to Gα$_o$, other PTXi Gα$_i$ subunits may also support sensitization[59,60,63] consistent with a general role for multiple Gα$_{i/o}$ subunits in GPCR-mediated sensitization. The ability of multiple Gα$_{i/o}$ subunits to rescue sensitization may also suggest the involvement of the simultaneous activation of multiple Gα$_{i/o}$ proteins in sensitization.[59,60] One Gα$_{i/o}$ signaling pathway that may influence heterologous sensitization includes the modulation of protein kinases including PKC, PKA, and Raf-1 kinase.[59] Modulation of one or more of these protein kinases is likely to alter acute adenylyl cyclase activity and possibly sensitization. For example, several adenylyl cyclase isoforms (i.e., AC1, AC2, AC5, AC6, and AC7; Table 10.1) show enhanced functional activity in the presence of activators of PKC, suggesting a potential role of PKC-mediated phosphorylation in heterologous sensitization.[64,65] A role for activation of Raf-1 kinase in heterologous sensitization has also been suggested because inhibitors of Raf-1 kinase partially block sensitization of adenylyl cyclase.[59,66-68] These studies suggest a role for kinases in adenylyl cyclase sensitization; however, these effects appear to be isoform-dependent and may reflect the acute regulatory properties of the individual adenylyl cyclase isoforms.

Role of Gβγ in heterologous sensitization of adenylyl cyclase

Receptor-mediated activation of Gβγ signaling has also been implicated in adenylyl cyclase sensitization. That both the release/rearrangement of Gβγ subunits and heterologous sensitization occur in a pertussis toxin-sensitive manner suggests a potential relationship. Direct evidence supporting a role for Gβγ subunits in heterologous sensitization has been obtained using Gβγ sequestering agents, such as the C-terminus of G protein receptor kinase 2 (βARK-ct) or Gα$_t$. Expression of these Gβγ subunit scavengers attenuates the development of heterologous sensitization following the activation of μ opioid, CB$_1$ cannabinoid, and D$_2$ dopamine receptors in cultured cell systems.[59,69,70] The simplest explanation for these results is that prolonged activation of Gα$_{i/o}$ liberates Gβγ subunits that directly activate and sensitize adenylyl cyclase. This mechanism is unlikely, however, because adenylyl cyclase isoforms capable of undergoing heterologous sensitization show markedly different patterns of regulation by Gβγ subunits.[10,48,59,70] Despite the complexities for direct Gβγ regulation of adenylyl cyclase activity, there appears to be an absolute requirement for Gβγ subunits in the development of heterologous sensitization.

Gα$_s$ and adenylyl cyclase isoform differences in heterologous sensitization

A number of studies have provided evidence that agonist-induced sensitization is influenced by the complement of endogenous or recombinant adenylyl cyclase isoforms present within the cell.[59,64] A recent study also revealed that adenylyl cyclase-selective siRNAs reduce the magnitude of heterologous sensitization in neurons.[71] Although all adenylyl cyclase isoforms appear capable of being sensitized, several unique regulatory patterns have been reported.[59] For example, both of the Ca^{2+}-inhibited isoforms of adenylyl cyclase, AC5 and AC6, show a marked degree of heterologous sensitization following activation by Gα$_s$ or forskolin.[15,59] Persistent agonist treatment also causes sensitization of AC1 and AC8 to Ca^{2+} stimulation.[59] In contrast, the remaining Ca^{2+}-stimulated isoform, AC3, does not show robust sensitization to Ca^{2+} ionophores or Gα$_s$,[59] although sensitization to forskolin has been reported.[59,72] AC2, AC4, and AC7, which are conditionally activated by Gβγ subunits, show a unique pattern of heterologous sensitization. Specifically, it was observed that these isoforms of adenylyl cyclase either show no sensitization or have a reduced responsiveness to Gα$_s$-stimulated cAMP accumulation following agonist treatment.[59] In contrast, PKC-stimulated AC2 activity is robustly sensitized by persistent activation of D_2 dopamine receptors[25,73] or shows no change following μ opioid receptor activation.[74] AC9, the only forskolin-insensitive adenylyl cyclase, shows sensitization to Gα$_s$-coupled receptor stimulation following D_2 receptor activation.[25] Thus, the preponderance of evidence indicates that most isoforms are capable of undergoing sensitization; however, their distinctive patterns of regulation by GPCRs will likely play a role in their overall responsiveness following heterologous sensitization.[10,59]

In spite of their differential regulation, all isoforms of adenylyl cyclase are activated by Gα$_s$,[4] and several observations support the hypothesis that the expression of heterologous sensitization involves enhanced Gα$_s$ activity or enhanced Gα$_s$-adenylyl cyclase interactions.[59,75] For example, isoforms of adenylyl cyclase that are activated synergistically by Gα$_s$ together with isoform-selective activators (i.e., AC1, Ca^{2+}; AC2, phorbol esters; AC5, 100 nM forskolin) show robust short-term sensitization.[59] These observations are consistent with a role for Gα$_s$ in heterologous sensitization; however, the precise role for Gα$_s$ is unknown. Studies with a Gα$_s$-deficient cell model (Gα$_s$ knockout; GSKO cells) reveal that individual adenylyl cyclases have a differential requirement for Gα$_s$ in sensitization. Specifically, there are Gα$_s$-dependent (AC1 and AC5) and Gα$_s$-independent (AC1) mechanisms for heterologous sensitization.[15] The mechanisms for facilitating these Gα$_s$-adenylyl cyclase interactions are currently under investigation and may involve Gα$_s$ phosphorylation;[76] however, it does not appear to require increased palmitoylation or direct Gβγ subunit interactions (Ejendal, K.F.K. and Watts, V.J., unpublished observations).

POSTTRANSLATIONAL MODIFICATIONS AND GPCR SIGNALING

Adenylyl cyclases are also subject to a variety of posttranslational modifications that can alter their ability to be modulated by GPCRs. As described previously, a number of adenylyl cyclases are subject to regulation by phosphorylation as a result of protein kinase activation. In addition, all adenylyl cyclases have conserved putative N-glycosylation sites on the extracellular loops between TM 9–10 and/or TM 11–12. Several adenylyl cyclase isoforms have been shown to be glycosylated, and the effect of glycosylation on GPCR modulation of AC6 and AC9 have been examined in detail.[11,77–80] AC6 can be glycosylated on two asparagine residues located on extracellular loops 5 and 6 in the second membrane-spanning domain (M2), N805 and N890.[80] Disruption of AC6 glycosylation by mutagenesis of the asparagine residues or using tunicamycin as a chemical inhibitor does not alter $G\alpha_s$-stimulated AC6 activity. However, preventing glycosylation reduces inhibition by divalent cations, inhibitory G proteins, and by PKC.[80] A similar series of studies using mutagenesis as well as tunicamycin examined the role of glycosylation in GPCR modulation of AC9. These studies revealed that glycosylation is required for maximal AC9 stimulation by the $G\alpha_s$-coupled β_2 adrenoceptor (β_2AR) agonist, isoproterenol.[11] In contrast, glycosylation does not alter inhibition of AC9 by $G\alpha_{i/o}$ or in response to activators of PKC. Other adenylyl cyclases such as AC2, AC3, AC5, and AC8 are likely also glycosylated on extracellular asparagine residues; however, a detailed molecular analysis of GPCR modulation on these isoforms remains to be done.[77,78] The divergent effects of glycosylation on GPCR modulation of AC6 and AC9 suggest that unique patterns will be observed with other adenylyl cyclase isoforms.

Similar to glycosylation, the diffusible second messenger, nitric oxide (NO), appears to cause isoform-specific posttranslational modification of adenylyl cyclases.[81,82] Early studies used N18TG2 neuroblastoma cells that primarily express AC6 to reveal an NO-mediated inhibition of $G\alpha_s$-coupled receptor stimulation of adenylyl cyclase signaling.[81] It was revealed that NO inhibits adenylyl cyclase activity in N18TG2 cells via a covalent S-nitrosylation of AC6. Additional studies examining the ability of NO modulation of other adenylyl cyclase isoforms suggest cell-type differences. Both $G\alpha_s$- and forskolin-stimulated activity of recombinant AC5 and AC6 are reduced by NO donors in N18TG2 cell membranes.[82] The same study reported that NO donors fail to inhibit AC1 and AC2 activity. However, studies using intact COS-7 cells revealed that NO releasers selectively inhibit transfected AC1 and AC6, but not AC2 or AC5.[9] The mechanisms for NO modulation of adenylyl cyclase may be direct or indirect through attenuated expression of $G\alpha_i$.[83] There is still significant work remaining to examine the role that NO may play in modulating adenylyl cyclases in native tissues. Similarly, additional adenylyl cyclase glycosylation studies are also required to assess its role in regulating how GPCRs control cAMP signaling.

ADENYLYL CYCLASE OLIGOMERIZATION

Crystal structures of adenylyl cyclases

Due to the complexity of obtaining crystal structures of membrane proteins, only crystallographic data of the cytosolic domains are available, and two crystal structures of the soluble domains of adenylyl cyclase have been solved.[84,85] Zhang and colleagues expressed the C2 domain of AC2 from rat and observed a C2-C2 homodimer, with two dimers forming a tetramer.[85] Two forskolin molecules are present in the C2-C2 dimer interface, and this interface is also thought to comprise residues of the active site and for ATP binding. In contrast, the structure by Tesmer and colleagues shows a C1-C2 heterodimer of the C1 domain from canine AC5 with the C2 domain of rat AC2, in complex with one forskolin molecule and $G\alpha_s$ present in the dimer interface.[84] It is noteworthy to mention that $G\alpha_s$ stimulates all adenylyl cyclases by binding amino acid residues of both the C1 and C2 to enhance interactions between the catalytic halves. In contrast, $G\alpha_i$ binds only to the C1 catalytic domain, which prevents C1-C2 interaction and thereby catalysis. Although these two structures have differences, both show that the catalytic core of adenylyl cyclase is composed of a dimer between two catalytic domains. Collectively, the structural data suggest that homodimers (e.g., C2-C2 from AC2) as well as heterodimers (e.g., C1 from AC5 and C2 from AC2) of co-expressed adenylyl cyclases may form, which opens up the possibility that inter-AC dimerization may take place in vivo.[10]

Regulation of cyclic AMP signaling by oligomerization of adenylyl cyclases

In addition to the structural data, a limited number of experimental findings indicate that adenylyl cyclases, analogous to GPCRs, form intermolecular dimers and possibly multimers. By co-expressing an inactive, epitope-tagged AC1 molecule with an active, untagged AC1, Tang and colleagues showed that they could immunoprecipitate AC1 enzymatic activity suggesting that AC1 forms a homodimer.[86] In line with these findings, AC8 has also been shown to form homodimers. Co-transfection of full-length AC8 with an inactive AC8 truncation mutant results in a dramatic inhibition of activity, suggesting that AC8 may function as a dimer or oligomer.[87] The activities of full-length AC5 or AC6 are also inhibited by the inactive AC8 mutant, further suggesting that AC8 may form intermolecular heterodimers with these isoforms of adenylyl cyclase.[87]

The nine different isoforms of adenylyl cyclase have distinct regulatory characteristics (Table 10.1), but have overlapping expression patterns.[14] Hence, in nature, heterodimers of adenylyl cyclase may form and these oligomers may differ in function, regulation, and activity. In a recent study, the interaction between AC2 and AC5 was measured using bioluminescence resonance energy transfer (BRET). The interactions between AC2 and AC5 are specific, as measured by BRET, and are enhanced by the addition of $G\alpha_s$ or forskolin.[88] Interestingly, the putative heteromer has increased activity in response to forskolin and $G\alpha_s$ when compared to the homomers. It was also observed that endogenously expressed

AC2 and AC5 co-localize in mouse cardiac myocytes, thus AC2-AC5 heterodimers also have the potential to form in vivo.

Another study in CAD cells provided indirect evidence that adenylyl cyclase interactions can actually negatively influence adenylyl cyclase activity.[13] The primary adenylyl cyclase isoforms found in CAD cells are AC6 and AC9. Upon differentiation, the cells display a marked loss of expression of the AC9 isoform, which correlates with higher forskolin-stimulated cAMP accumulation.[13] Subsequent transient transfection studies revealed that the overexpression of AC9 reduce forskolin but not $G\alpha_s$-coupled receptor-stimulated cAMP accumulation. These observations with AC9 are consistent with early work with AC9 in HEK293 cells,[57] indirectly suggesting that AC6 and AC9 interact and that AC9 can negatively regulate the activity of AC6.

Collectively, heterodimers or heterooligomers of adenylyl cyclase may display unique functional properties when compared to the corresponding monomeric or homooligomeric adenylyl cyclases. In conjunction with the crystallographic data, it is intriguing to speculate about the endless possibilities of intra- and intermolecular dimers and oligomers of adenylyl cyclases that can be formed as well as the functional consequences such dimerization may have on coupling to G proteins as well as GPCRs.

ADENYLYL CYCLASE AND GPCR COMPLEXES

Compartmentalization of adenylyl cyclases and GPCRs in membrane rafts

To fine-tune and coordinate signaling pathways in the cell, signaling components are often co-localized into cholesterol and sphingolipid-rich membrane microdomains called membrane rafts or lipid rafts. A subset of membrane rafts are enriched in the scaffold protein caveolin, resulting in flask-shaped invaginations of the plasma membrane that are referred to as caveolae.[10] Fractionation of cellular membranes into nonraft and raft/caveolae have generated information about the localization of specific signaling components; however, the outcomes of fractionation studies may depend on the specific method used, as recently demonstrated for the β_2AR.[89] To further complicate interpretation of such fractionation data, there is evidence that receptors, and likely other signaling components such as adenylyl cyclases, may redistribute in the membrane depending on their activation state.[90]

Despite the limitations mentioned above, there are numerous lines of evidence showing that certain GPCRs and their ability to regulate adenylyl cyclase reflects their localization into raft and nonraft fractions of the membrane.[91] For example, closely related receptors such as the μ and δ opioid receptors have been shown to differ in membrane localization.[92] It was subsequently observed that μ receptor signaling to adenylyl cyclase is more sensitive to chemical disruption of rafts using methyl-β-cyclodextrin treatment when compared to the δ receptor, and that this signaling dependency on cholesterol is enhanced upon chronic agonist treatment.[92] These observations suggest that μ receptor signaling components, including the specific adenylyl cyclase, are localized in rafts. Thus, the spatial separation of the μ opioid receptor into rafts contributes to its signaling properties.

Similar to GPCRs, adenylyl cyclase localization in rafts appears to be isoform-specific.[10] Not surprisingly, the localization of adenylyl cyclase isoforms to raft or nonraft portions of the membrane is closely linked to their regulation. For instance, both AC1 and AC8 are stimulated by Ca^{2+} through CCE, and these isoforms co-localize with the CCE machinery to lipid rafts.[10,23] The third Ca^{2+}-stimulated adenylyl cyclase, AC3, is also localized to rafts. Moreover, the Ca^{2+} inhibited AC5 and AC6 are also localized to rafts, which may serve to functionally localize these enzymes with other important signaling molecules such as nitric oxide signaling proteins, phosphodiesterases, and the Na^+/H^+ exchanger that protects AC6 from changes in intracellular pH.[9] In summary, the spatial separation of receptors and adenylyl cyclases into rafts or nonraft fractions of the membrane allows for signaling through distinct pathways, involving distinct GPCRs and isoforms of adenylyl cyclases.

Coordination of adenylyl cyclase-containing signaling complexes

To appropriately integrate signals, adenylyl cyclases interact with other signaling proteins in multimeric signaling complexes. In addition to adenylyl cyclase, these multiprotein complexes may contain upstream regulators like GPCRs and G proteins, downstream effectors like PKA and ion channels, as well as scaffolding proteins.[10] A-kinase anchoring proteins (AKAPs) constitute a large and diverse family of scaffolding proteins that spatially coordinate cAMP signaling with other signaling events by simultaneously binding PKA regulatory subunits, GPCRs, and adenylyl cyclase.[10] AKAP interaction can regulate the activity of adenylyl cyclase. For instance, AKAP79 tethering of AC5 leads to phosphorylation of AC5 by PKA, which suppresses the isoproterenol-stimulated activity of AC5.[10] Direct protein-protein interactions between adenylyl cyclase isoforms and AKAP occur, and the functional consequences of these interactions appear to be adenylyl cyclase isoform specific. Whereas the activities of AC2 and AC3 are inhibited by binding to the AKAP9/Yotiao, AC1 and AC9 activities are unaffected.[93] In addition to AKAP tethering, many other proteins such as RGS2, SNAPIN, PAM, and PP2A have been shown to interact with adenylyl cyclase and alter their activity.[4,9,10]

Direct interactions between GPCR and adenylyl cyclase isoforms

Recently, the existence of direct GPCR-adenylyl cyclase interactions has been described.[5] Understanding the regulation of these signaling complexes is only now beginning through the application of fluorescent technologies. Specifically, the study of these complexes has been accelerated by the application of resonance energy transfer (RET) techniques (e.g., FRET, BRET) and protein complementation assays (PCAs) such as bimolecular fluorescence/luminescence complementation

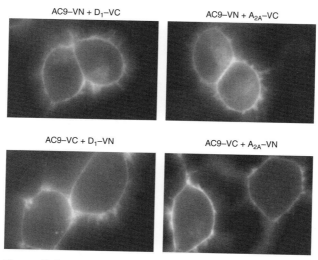

AC9–VN + D$_1$–VC   AC9–VN + A$_{2A}$–VC

AC9–VC + D$_1$–VN   AC9–VC + A$_{2A}$–VN

Figure 10-3: Visualization of adenylyl cyclase-GPCR interactions using Bimolecular Fluorescence Complementation, BiFC.[94] Briefly, the A$_{2A}$ adenosine, the D$_1$ dopamine receptor, and AC9 were C-terminally tagged with nonfluorescent BiFC fragments (i.e., the C terminus of Venus, VC or the N terminus of Venus, VN) as indicated. Neuronal CAD cells were transfected with the indicated cDNA for 24 hrs and imaged using fluorescence microscopy.[94] When the two proteins are co-expressed in close proximity, the fluorophore of the YFP variant Venus (VN+VC=YFP) is reconstituted and can be visualized by fluorescence microscopy.

(BiFC/BiLC).[94,95] These studies suggest that adenylyl cyclase-GPCR interactions can be influenced by receptor activation, Gα and Gβγ subunits, and RGS proteins.[5,96] It has been demonstrated that constitutive AC2-β$_2$AR interactions as measured by BRET and co-immunoprecipitation were prevented by the expression of a Gα$_s$ minigene as well as βARK-ct, a Gβγ sequestering agent.[96] These observations suggest that AC2-β$_2$AR interactions are modulated by Gα$_s$ and Gβγ. More recent studies have provided support for the hypothesis that AC5-β$_2$AR interactions are modulated by Gα$_s$ activation and lipid rafts.[89] Specifically, disruption of lipid rafts enhances basal and drug-stimulated AC5-β$_2$AR interactions. Very recent studies in our laboratory used BiFC to reveal AC9 interactions with two Gα$_s$-coupled receptors (i.e., D$_1$ dopamine and A$_{2A}$ adenosine) in CAD cells (Figure 10.3; Ejendal, K.F.K., Przybyla, J.A., and Watts, V.J., unpublished observations). The studies highlighted earlier set the stage for future studies examining the mechanisms involved in adenylyl cyclase-GPCR interactions.

Impact of GPCR heterooligomers on adenylyl cyclase signaling

Classically, a single GPCR has been considered to couple to specific G proteins, thereby mediating a distinct downstream effect on individual adenylyl cyclases. However, over the last decade, a number of reports have shown that GPCRs also function as dimers, or higher-order oligomers, and these homodimers and heterodimers may exhibit unique pharmacological properties affecting receptor signaling to adenylyl cyclase (see Chapters 3–6). For example, studies on the

CB$_1$ cannabinoid receptor and the D$_2$ dopamine receptor suggest that the CB$_1$-D$_2$ heterodimer may couple to Gα_s, thus stimulating adenylyl cyclase activity, although expression of either receptor alone inhibits adenylyl cyclase activity through G$\alpha_{i/o}$.[97,98] The D$_1$ and D$_2$ dopamine receptors activate G$\alpha_{s/olf}$ and G$\alpha_{i/o}$, respectively; however, the D$_1$-D$_2$ heterodimer complex displays unique pharmacology in that it couples to Gα_q.[99] Activation of the D$_1$-D$_2$ heteromer results in the stimulation of PLC that initiates a rapid Ca^{2+} signal, which in turn has the potential to activate group I adenylyl cyclases (e.g., AC1) and inhibit group III adenylyl cyclases (see group discussions). Further analysis of the D$_1$-D$_2$ heterodimer and comparing it to the D$_2$-D$_5$ heterodimer revealed differences in the Ca^{2+} signal evoked, where the D$_1$-D$_2$ signal is independent of extracellular Ca^{2+} but the D$_2$-D$_5$ signal requires influx of extracellular Ca^{2+}.[100] These observed differences in Ca^{2+} requirements could further translate into isoform-specific regulation of adenylyl cyclase activity upon activation of different dopamine receptor heteromers. A final example of receptor dimer modulation of adenylyl cyclase signaling also involves the dopamine receptor family. It has been reported that D$_1$ and D$_3$ receptor interactions enhance the ability of the D$_1$ dopamine receptor to stimulate adenylyl cyclase activity in HEK293 cells.[101, 102] Together, these observations suggest that receptor-receptor interactions may also play a role in modulating GPCR-adenylyl cyclase signaling including isoform specificity. More speculative is the possibility that the formation of specific GPCR complexes could ultimately promote the formation of specific adenylyl cyclase-adenylyl cyclase interactions, which would contribute to receptor-adenylyl cyclase isoform specific signaling events.

EXAMPLES OF GPCR-ADENYLYL CYCLASE SPECIFICITY

Effects of Gβ and Gγ modulation on GPCR-adenylyl cyclase signaling

The G$\beta\gamma$ subunits are thought to function as heterodimers; however, increasing evidence suggests that individual Gβ and Gγ subunits also have unique roles in modulating effectors.[5,20] Surprising examples of specificity and alterations in GPCR-adenylyl cyclase signaling have been demonstrated using methodologies exploring changes in individual G$\beta\gamma$ combinations or the expression of individual subunits. For example, an elegant pharmacological study examined the ability of unique G$\beta\gamma$ combinations to stimulate adenylyl cyclase activity in response to activation of A$_{2A}$ adenosine or β_1 adrenoceptor.[103] These studies revealed that cells expressing G$\beta_4\gamma_2$ show a tenfold enhanced potency in A$_{2A}$ adenosine receptor-stimulated adenylyl cyclase activity when compared to cells expressing G$\beta_1\gamma_2$. More subtle differences in G$\beta\gamma$ modulation of receptor-adenylyl cyclase coupling are seen with the β_1 adrenoceptor as well as studies examining Gα_s activation of AC1 and AC2.[103] Another report used lentivirus shRNA to silence individual and multiple Gβ subunits.[104] The results of these experiments suggest that silencing individual Gβ subunits (i.e., Gβ_1 or Gβ_2) has only a modest effect on of PGE$_2$- or isoproterenol-stimulated cAMP accumulation. In contrast, simultaneous elimination of $G\beta_1$ and $G\beta_2$ markedly reduce $G\alpha_s$-coupled receptor stimulation of adenylyl cyclase without altering the forskolin response.[104] A ribozyme approach was used to eliminate $G\gamma_7$ from HEK293 cells to show a selective loss in the ability of the D_1 but not the D_5 dopamine receptor to stimulate adenylyl cyclase activity.[105] Subsequent studies in $G\gamma_7$ knockout mice also revealed a loss of D_1 receptor-stimulated adenylyl cyclase activity in the striatum.[106] However, the $G\gamma_7$ knockout mice also had a selective reduction in the stimulatory G protein $G\alpha_{olf}$ that accompanied the reduction in adenylyl cyclase activity. These studies provide evidence for an important role of individual and specific combinations of $G\beta$ and $G\gamma$ subunits in regulating GPCR-adenylyl cyclase signaling.

Adenylyl cyclase knockout mice

Our knowledge of the relative contribution of adenylyl cyclase isoforms in various disease states is mainly derived from overexpression and knockout studies, and current findings on this topic have been summarized and discussed in two excellent recent reviews.[4,107] Several knockouts and transgenic mice of adenylyl cyclase isoforms from groups I, II, and III have been generated; however, only a few examples where GPCR-adenylyl cyclase signaling is altered are presented further in this chapter.

Mice deficient in either AC1 (AC1[-/-]) or AC8 (AC8[-/-]) or the double knockout (AC1[-/-]/AC8[-/-]) show no gross abnormalities, but further characterization of these animals revealed that they have alterations in memory and learning behavior. The AC1-deficient mice primarily exhibit loss of long-term potentiation (LTP) and spatial memory,[108] whereas AC8 knockout mice have alterations in memory as well as reduced anxiety behavior when subjected to stress.[109] The effects on learning is especially pronounced in mice deficient in both AC1 and AC8, as these animals completely lack both late-phase LTP and long-term memory.[110] Interestingly, administration of forskolin restored learning, suggesting that signaling components downstream of adenylyl cyclase are not altered in AC1[-/-]/AC8[-/-] mice. In addition to learning and memory, mice deficient in AC1 and AC8 also show alterations in response to the opioid receptor agonist, morphine, where development of morphine tolerance was attenuated in AC1[-/-]/AC8[-/-] mice when compared to wild-type mice.[111] In contrast, mice that overexpress AC7 are more sensitive to the analgesic effects of morphine and appear to more rapidly develop morphine tolerance than the wild-type controls.[112] These observations link alterations in opioid receptor-adenylyl cyclase signaling to tolerance following morphine administration.

A classic example of what knockout animals can teach us about GPCR-adenylyl cyclase signaling specificity is revealed by studies examining AC5 knockout mice (AC5[-/-]). AC5 is highly expressed in brain regions associated with motor control and is also expressed at high levels in the heart.[4] Elimination of AC5 markedly reduces the behavioral and biochemical effects of D_2 dopamine receptors that are co-expressed with AC5 in the striatum.[2] Additional studies implicate striatal AC5 as the primary isoform regulating opioid receptor function.[2] Specifically, there was a significant reduction in the behavioral and analgesic effects of morphine. The deletion of AC5 also reduces GPCR signaling associated with cardiovascular function.[2] Perhaps the most intriguing observations in the AC5$^{-/-}$ mice are their increased bone quality, reduced body weight, and increased lifespan.[113] The identification of the GPCR(s) involved in mediating these effects of AC5 is yet to be achieved, but the implications of finding modulators for these unknown GPCR(s) are extremely high.

GPCR-adenylyl cyclase cross-talk: D$_2$ dopamine receptor knockout mice

The unique intricacies of adenylyl cyclase-GPCR signaling can further be appreciated when one considers that a genetic deletion of one GPCR can markedly change the ability of another GPCR to modulate adenylyl cyclase. An example of this phenomenon is revealed in mice lacking D$_2$ dopamine receptors (D$_2^{-/-}$), which lost their ability to respond to caffeine, an A$_{2A}$ receptor antagonist.[114] A series of mechanistic studies demonstrated that the loss of the caffeine response reflects a functional uncoupling of the Gα_s-coupled A$_{2A}$ adenosine receptor from stimulation of adenylyl cyclase in the striatum. This effect is not readily explained by changes in other components of the signaling pathway (e.g., Gα_s, Gα_{olf}, and AC6). Thus, elimination of a G$\alpha_{i/o}$-coupled receptor ultimately prevents A$_{2A}$-adenylyl cyclase coupling. These observations suggest that administration of a receptor antagonist (i.e., D$_2$ antagonist; haloperidol) could markedly alter the ability of an off-target receptor (i.e., A$_{2A}$ adenosine) to modulate adenylyl cyclase signaling. The example described previously highlights further the complexities associated with GPCR-adenylyl cyclase signaling in a physiological animal model and provides additional impetus for an extensive evaluation of all signaling pathways in knockout or transgenic animals.

CONCLUDING REMARKS

Continued discoveries in the regulation of the membrane bound isoforms of adenylyl cyclase have provided the potential for modulation of the activity of these enzymes by several important GPCR-mediated pathways. Today, more than 50% of clinically used drugs target G protein-coupled receptors. Members in each group of adenylyl cyclase are subject to direct and indirect regulation by GPCRs linked to Gα_s, G$\alpha_{i/o}$, Gα_q, and G$\alpha_{12/13}$. Additional forms of modulation relevant to GPCRs signaling include heterologous sensitization, adenylyl cyclase oligomerization, and membrane compartmentalization. Such modes of regulation could involve posttranslational modifications and changes in subcellular localization that may afford each adenylyl cyclase with distinct and specific GPCR regulation. More complex forms of regulation involve GPCR heterodimers and alterations in the cellular components that regulate GPCR-adenylyl cyclase coupling. The number of GPCRs and adenylyl cyclase isoforms in combination with their unique forms of regulation provide for significant signaling diversity. Items on the forefront in this field include studies of how oligomerization influences GPCR-adenylyl cyclase specificity. Additional work with genetic deletion and transgenic animals is also of great interest as is the development of selective adenylyl cyclase modulators that could be used in conjunction with GPCR-selective ligands for in vivo studies. It is anticipated the new methodological approaches focused on novel animal models, luminescent and fluorescent studies, "omic" technologies, and high throughput and high content drug discovery efforts will continue to guide scientists in investigations of GPCR-adenylyl cyclase signaling.

ACKNOWLEDGMENTS

This work was supported by U.S. Public Health Service grant MH060397 (V.J.W.). We thank David M. Allen for assistance with the initial design of the figures. We are extremely grateful to previous laboratory members, Drs. Michael A. Beazely, Medhane G. Cumbay, Christopher A. Johnston, Pierre-Alexandre. Vidi, and Timothy A. Vortherms, who made significant contributions toward this chapter. In addition, we would like to thank Jason M. Conley for proofreading and making suggestions to improve the present manuscript. We also wish to acknowledge the outstanding scientists whose work is cited in this review for their contributions to our understanding of GPCR-adenylyl cyclase signaling. Finally, we apologize to those whose work we did not directly cite in our efforts to satisfy the editor's citation limit.

REFERENCES

1. Sutherland, E. W. and Rall, T. W. (1958) Fractionation and characterization of a cyclic adenine ribonucleotide formed by tissue particles, *J. Biol. Chem. 232*, 1077–1091.
2. Chester, J. A. and Watts, V. J. (2007) Adenylyl cyclase 5: a new clue in the search for the "fountain of youth"?, *Sci STKE. 2007*, e64.
3. Watts, V. J. (2007) Adenylyl cyclase isoforms as novel therapeutic targets: an exciting example of excitotoxicity neuroprotection, *Mol. Interv. 7*, 70–73.
4. Sadana, R. and Dessauer, C. W. (2009) Physiological roles for G protein-regulated adenylyl cyclase isoforms: insights from knockout and overexpression studies, *Neurosignals. 17*, 5–22.
5. Dupre, D. J., Robitaille, M., Rebois, R. V., and Hebert, T. E. (2009) The role of Gbetagamma subunits in the organization, assembly, and function of GPCR signaling complexes, *Annu. Rev. Pharmacol. Toxicol. 49*, 31–56.
6. Omori, K. and Kotera, J. (2007) Overview of PDEs and their regulation, *Circ. Res. 100*, 309–327.
7. Krupinski, J., Coussen, F., Bakalyar, H. A., Tang, W. J., Feinstein, P. G., Orth, K., Slaughter, C., Reed, R. R., and Gilman, A. G. (1989) Adenylyl cyclase amino acid sequence: possible channel- or transporter-like structure, *Science 244*, 1558–1564.
8. Buck, J., Sinclair, M. L., Schapal, L., Cann, M. J., and Levin, L. R. (1999) Cytosolic adenylyl cyclase defines a unique signaling molecule in mammals, *Proc. Natl. Acad. Sci. U. S. A 96*, 79–84.
9. Beazely, M. A. and Watts, V. J. (2006) Regulatory properties of adenylate cyclases type 5 and 6: A progress report, *Eur. J. Pharmacol. 535*, 1–12.
10. Willoughby, D. and Cooper, D. M. (2007) Organization and Ca2+ regulation of adenylyl cyclases in cAMP microdomains, *Phyisiol. Rev. 87*, 965–1010.

11. Cumbay, M. G. and Watts, V. J. (2004) Novel regulatory properties of human type 9 adenylate cyclase (AC9), *J. Pharmacol. Exp. Ther. 310*, 108–115.

12. Cumbay, M. G. and Watts, V. J. (2005) Galphaq potentiation of adenylate cyclase type 9 activity through a Ca2+/calmodulin-dependent pathway, *Biochem. Pharmacol. 69*, 1247–1256.

13. Johnston, C. A., Beazely, M. A., Bilodeau, M. L., Andrisani, O. M., and Watts, V. J. (2004) Differentiation-induced alterations in cyclic AMP signaling in the Cath.a Differentiated (CAD) neuronal cell line, *J. Neurochem. 88*, 1497–1508.

14. Visel, A., varez-Bolado, G., Thaller, C., and Eichele, G. (2006) Comprehensive analysis of the expression patterns of the adenylate cyclase gene family in the developing and adult mouse brain, *J. Comp Neurol. 496*, 684–697.

15. Vortherms, T. A., Nguyen, C. H., Bastepe, M., Juppner, H., and Watts, V. J. (2006) D(2) dopamine receptor-induced sensitization of adenylyl cyclase type 1 is Galpha(s) independent, *Neuropharmacology 50*, 576–584.

16. Chern, Y. (2000) Regulation of adenylyl cyclases in the central nervous system, *Cellular Signalling 12*, 195–204.

17. Defer, N., Best- Belpomme, M., and Hanoune, J. (2000) Tissue specificity and physiological relevance of various isoforms of adenylyl cyclase, *Am. J. Physiol. Renal Physiol. 279*, F400-F416.

18. Antoni, F. A., Wiegand, U. K., Black, J., and Simpson, J. (2006) Cellular localisation of adenylyl cyclase: a post-genome perspective, *Neurochem. Res. 31*, 287–295.

19. Ferguson, G. D. and Storm, D. R. (2004) Why calcium-stimulated adenylyl cyclases?, *Physiology (Bethesda.) 19*, 271–276.

20. McIntire, W. E. (2009) Structural determinants involved in the formation and activation of G protein betagamma dimers, *Neurosignals. 17*, 82–99.

21. Jiang, L. I., Collins, J., Davis, R., Fraser, I. D., and Sternweis, P. C. (2008) Regulation of cAMP responses by the G12/13 pathway converges on adenylyl cyclase VII, *J. Biol. Chem. 283*, 23429–23439.

22. Putney, J. W., Jr. (1986) A model for receptor-regulated calcium entry, *Cell Calcium 7*, 1–12.

23. Martin, A. C., Willoughby, D., Ciruela, A., Ayling, L. J., Pagano, M., Wachten, S., Tengholm, A., and Cooper, D. M. (2009) Capacitative Ca2+ entry via Orai1 and stromal interacting molecule 1 (STIM1) regulates adenylyl cyclase type 8, *Mol. Pharmacol. 75*, 830–842.

24. Nielsen, M. D., Chan, G. C. K., Poser, S. W., and Storm, D. R. (1996) Differential regulation of type I and type VIII Ca2+-stimulated adenylyl cyclases by Gi-coupled receptors in vivo, *J. Biol. Chem. 271*, 33308–33316.

25. Cumbay, M. G. and Watts, V. J. (2001) Heterologous sensitization of recombinant adenylate cyclases by activation of D2 dopamine receptors, *J. Pharmacol. Exp. Ther. 297*, 1201–1209.

26. Taussig, R., Iñiguez-Lluhi, J. A., and Gilman, A. (1993) Inhibition of adenylyl cyclase by Gi alpha, *Science 261*, 218–221.

27. Masada, N., Ciruela, A., Macdougall, D. A., and Cooper, D. M. (2009) Distinct mechanisms of regulation by Ca2+/calmodulin of type 1 and 8 adenylyl cyclases support their different physiological roles, *J. Biol. Chem. 284*, 4451–4463.

28. Choi, E. J., Wong, W. T., Hinds, T. R., and Storm, D. R. (1992) Calcium and muscarinic agonist stimulation of type I adenylylcyclase in whole cells, *J. Biol. Chem. 267*, 12440–12442.

29. Wayman, G. A., Impey, S., Wu, A., Kindsvogel, W., Prichard, L., and Storm, D. R. (1994) Synergistic activation of the type I adenylyl cyclase by Ca2+ and Gs-coupled receptors in vivo, *J. Biol. Chem. 269*, 25400–25405.

30. Ostrom, R. S., Naugle, J. E., Hase, M., Gregorian, C., Swaney, J. S., Insel, P. A., Brunton, L. L., and Meszaros, J. G. (2003) Angiotensin II enhances adenylyl cyclase signaling via Ca2+/calmodulin. Gq-Gs cross-talk regulates collagen production in cardiac fibroblasts, *J. Biol. Chem. 278*, 24461–24468.

31. Meszaros, J. G., Gonzalez, A. M., Endo- Mochizuki, Y., Villegas, S., Villarreal, F., and Brunton, L. L. (2000) Identification of G protein-coupled signaling pathways in cardiac fibroblasts: cross talk between G(q) and G(s), *Am. J. Physiol Cell Physiol 278*, 154–162.

32. Jacobowitz, O., Chen, J., Premont, R. T., and Iyengar, R. (1993) Stimulation of specific types of GS-stimulated adenylyl cyclases by phorbol ester treatment, *J. Biol. Chem. 268*, 3829–3832.

33. Tang, W. J. and Gilman, A. G. (1991) Type-specific regulation of adenylyl cyclase by G protein α subunits, *Science 254*, 1500–1502.

34. Federman, A. D., Conklin, B. R., Schrader, K. A., Reed, R. R., and Bourne, H. R. (1992) Hormonal stimulation of adenylyl cyclase through Gi-protein α subunits, *Nature 356*, 159–161.

35. Taussig, R., Tang, W. J., Hepler, J. R., and Gilman, A. G. (1994) Distinct patterns of bidirectional regulation of mammalian adenylyl cyclases, *J. Biol. Chem. 269*, 6093–6100.

36. Nelson, E. J., Hellevuo, K., Yoshimura, M., and Tabakoff, B. (2003) Ethanol-induced phosphorylation and potentiation of the activity of type 7 adenylyl cyclase. Involvement of protein kinase C delta, *J. Biol. Chem. 278*, 4552–4560.

37. Tsu, R. C. and Wong, Y. H. (1996) Gi-mediated stimulation of type II adenylyl cyclase is augmented by Gq-coupled receptor activation and phorbol ester treatment, *J. Neurosci. 16*, 1317–1323.

38. Watts, V. J. and Neve, K. A. (1997) Activation of type II adenylate cyclase by D2 and D4 but not D3 dopamine receptors, *Mol. Pharmacol. 52*, 181–186.

39. Mackay, H. J. and Twelves, C. J. (2007) Targeting the protein kinase C family: are we there yet?, *Nat. Rev. Cancer 7*, 554–562.

40. Zimmermann, G. and Taussig, R. (1996) Protein kinase C alters the responsiveness of adenylyl cyclases to G protein α and βγ subunits, *J. Biol. Chem. 271*, 27161–27166.

41. Nguyen, C. H. and Watts, V. J. (2006) Dexamethasone-induced Ras protein 1 negatively regulates protein kinase C delta: implications for adenylyl cyclase 2 signaling, *Mol. Pharmacol. 69*, 1763–1771.

42. Watson, P. A., Krupinski, J., Kempinski, A. M., and Frankenfield, C. D. (1994) Molecular cloning and characterization of the type VII isoform of mammalian adenylyl cyclase expressed widely in mouse tissues and in S49 mouse lymphoma cells, *J. Biol. Chem. 269*, 28893–28898.

43. Hellevuo, K., Yoshimura, M., Mons, N., Hoffman, P. L., Cooper, D. M., and Tabakoff, B. (1995) The characterization of a novel human adenylyl cyclase which is present in brain and other tissues, *J. Biol. Chem. 270*, 11581–11589.

44. Fagan, K. A., Smith, K. E., and Cooper, D. M. F. (2000) Regulation of the Ca2+-inhibitable adenylyl cyclase type VI by capacitative Ca2+ entry requires localization in cholesterol-rich domains, *J. Biol. Chem. 275*, 26530–26537.

45. Soboloff, J., Spassova, M. A., Dziadek, M. A., and Gill, D. L. (2006) Calcium signals mediated by STIM and Orai proteins – a new paradigm in inter-organelle communication, *Biochim. Biophys. Acta 1763*, 1161–1168.

46. Korzeniowski, M. K., Popovic, M. A., Szentpetery, Z., Varnai, P., Stojilkovic, S. S., and Balla, T. (2009) Dependence of stim1/orai1 mediated calcium entry on plasma membrane phosphoinositides, *J. Biol. Chem. 284*, 21027–21035.

47. Bayewitch, M. L., Avidor-Reiss, T., Levy, R., Pfeuffer, T., Nevo, I., Simonds, W. F., and Vogel, Z. (1998) Inhibition of adenylyl cyclase isoforms V and VI by various Gβγ subunits, *FASEB J. 12*, 1019–1025.

48. Gao, X., Sadana, R., Dessauer, C. W., and Patel, T. B. (2007) Conditional stimulation of type V and VI adenylyl cyclases by G protein betagamma subunits, *J. Biol. Chem. 282*, 294–302.

49. Kao, Y. Y., Lai, H. L., Hwang, M. J., and Chern, Y. (2004) An important functional role of the N terminus domain of type VI adenylyl cyclase in Galphai-mediated inhibition, *J. Biol. Chem. 279*, 34440–34448.

50. Beazely, M. A., Alan, J. K., and Watts, V. J. (2005) Protein kinase C and epidermal growth factor stimulation of Raf1 potentiates adenylyl cyclase type 6 activation in intact cells, *Mol. Pharmacol. 67*, 250–259.

51. Chen, Y., Harry, A., Li, J., Smit, M. J., Bai, X., Magnusson, R., Pieroni, J. P., Weng, G., and Iyengar, R. (1997) Adenylyl cyclase 6 is selectively regulated by protein kinase A phosphorylation in a region involved in Galphas stimulation, *Proc. Natl. Acad. Sci. U. S. A 94*, 14100–14104.

52. Iwami, G., Kawabe, J. I., Ebina, T., Cannon, P. J., Homey, C. J., and Ishikawa, Y. (1995) Regulation of adenylyl cyclase by protein kinase A, *J. Biol. Chem. 270*, 12481–12484.

53. Kawabe, J., Iwami, G., Ebina, T., Ohno, S., Katada, T., Ueda, Y., Homcy, C. J., and Ishikawa, Y. (1994) Differential activation of adenylyl cyclase by protein kinase C isoenzymes, *J. Biol. Chem. 269*, 16554–16558.

54. Beazely, M. A. and Watts, V. J. (2005) Galpha(q)-coupled receptor signaling enhances adenylate cyclase type 6 activation, *Biochem. Pharmacol. 70*, 113–120.

55. Yan, S. Z., Huang, Z. H., Andrews, R. K., and Tang, W. J. (1998) Conversion of forskolin-insensitive to forskolin-sensitive (mouse-type IX) adenylyl cyclase, *Molecular Pharmacology 53*, 182–187.

56. Paterson, J. M., Smith, S. M., Simpson, J., Grace, O. C., Sosunov, A., Bell, J. E., and Antoni, F. A. (2000) Characterization of human adenylyl cyclase IX reveals inhibition by Ca2+/calcineurin and differential mRNA polyadenylation, J . *Neurochem. 75*, 1358–1367.

57. Hacker, B. M., Tomlinson, J. E., Wayman, G. A., Sultana, R., Chan, G., Villacres, E., Disteche, C., and Storm, D. R. (1998) Cloning, chromosomal mapping, and regulatory properties of the human type 9 adenylyl cyclase (ADCY9), *Genomics 50*, 97–104.

58. Sharma, S. K., Klee, W. A., and Nirenberg, M. (1975) Dual regulation of adenylate cyclase accounts for narcotic dependence and tolerance, *Proc. Natl. Acad. Sci. U. S. A 72*, 3092–3096.

59. Watts, V. J. and Neve, K. A. (2005) Sensitization of adenylate cyclase by Galpha(i/o)-coupled receptors, *Pharmacol Ther 106*, 405–421.

60. Clark, M. J. and Traynor, J. R. (2006) Mediation of adenylyl cyclase sensitization by PTX-insensitive GalphaoA, Galphai1, Galphai2 or Galphai3, *J. Neurochem. 99*, 1494–1504.

61. Clark, M. J., Harrison, C., Zhong, H., Neubig, R. R., and Traynor, J. R. (2003) Endogenous RGS protein action modulates mu-opioid signaling through Galphao. Effects on adenylyl cyclase, extracellular signal-regulated kinases, and intracellular calcium pathways., *J. Biol. Chem. 278*, 9418–9425.

62. Watts, V. J., Wiens, B. L., Cumbay, M. G., Vu, M. N., Neve, R. L., and Neve, K. A. (1998) Selective activation of Gαo by D2L dopamine receptors in NS20Y neuroblastoma cells, *J. Neurosci. 18*, 8692–8699.

63. Zhang, L., Tetrault, J., Wang, W., Loh, H. H., and Law, P. Y. (2006) Short- and long-term regulation of adenylyl cyclase activity by delta-opioid receptor are mediated by Galphai2 in neuroblastoma N2A cells, *Mol. Pharmacol. 69*, 1810–1819.

64. Gintzler, A. R. and Chakrabarti, S. (2006) Post-opioid receptor adaptations to chronic morphine; altered functionality and associations of signaling molecules, *Life Sci 79*, 717–722.

65. Shy, M., Chakrabarti, S., and Gintzler, A. R. (2008) Plasticity of adenylyl cyclase-related signaling sequelae after long-term morphine treatment, *Mol. Pharmacol. 73*, 868–879.

66. Beazely, M. A. and Watts, V. J. (2005) Activation of a novel PKC isoform synergistically enhances D2L dopamine receptor-mediated sensitization of adenylate cyclase type 6, *Cell Signal 17*, 647–653.

67. Tumati, S., Yamamura, H. I., Vanderah, T. W., Roeske, W. R., and Varga, E. V. (2009) Sustained morphine treatment augments capsaicin-evoked CGRP release from primary sensory neurons in a PKA and Raf-1- dependent manner, *J. Pharmacol. Exp. Ther.* In press.

68. Yue, X., Varga, E. V., Stropova, D., Vanderah, T. W., Yamamura, H. I., and Roeske, W. R. (2006) Chronic morphine-mediated adenylyl cyclase superactivation is attenuated by the Raf-1 inhibitor, GW5074, *Eur. J. Pharmacol. 540*, 57–59.

69. Nguyen, C. H. and Watts, V. J. (2005) Dexras1 blocks receptor-mediated heterologous sensitization of adenylyl cyclase 1, *Biochem Biophys Res Commun 332*, 913–920.

70. Steiner, D., Avidor- Reiss, T., Schallmach, E., Saya, D., and Vogel, Z. (2005) Inhibition and superactivation of the calcium-stimulated isoforms of adenylyl cyclase: role of Gbetagamma dimers, *J Mol Neurosci 27*, 195–203.

71. Fan, P., Jiang, Z., Diamond, I., and Yao, L. (2009) Up-regulation of AGS3 during morphine withdrawal promotes cAMP superactivation via adenylyl cyclase 5 and 7 in rat nucleus accumbens/striatal neurons, *Mol. Pharmacol.* In press.

72. Rhee, M. H., Nevo, I., Avidor- Reiss, T., Levy, R., and Vogel, Z. (2000) Differential superactivation of adenylyl cyclase isozymes after chronic activation of the CB1 cannabinoid receptor, *Mol. Pharmacol. 57*, 746–752.

73. Watts, V. J. and Neve, K. A. (1996) Sensitization of endogenous and recombinant adenylate cyclase by activation of D2 dopamine receptors, *Mol. Pharmacol. 50*, 966–976.

74. Schallmach, E., Steiner, D., and Vogel, Z. (2006) Adenylyl cyclase type II activity is regulated by two different mechanisms: implications for acute and chronic opioid exposure, *Neuropharmacology 50*, 998–1005.

75. Vortherms, T. A., Nguyen, C. H., Berlot, C. H., and Watts, V. J. (2004) Using molecular tools to dissect the role of Gs in sensitization of AC1, *Mol. Pharmacol. 66*, 1617–1624.

76. Chakrabarti, S. and Gintzler, A. R. (2007) Phosphorylation of Galphas influences its association with the micro-opioid receptor and is modulated by long-term morphine exposure, *Mol. Pharmacol. 72*, 753–760.

77. Bol, G. F., Hulster, A., and Pfeuffer, T. (1997) Adenylyl cyclase type II is stimulated by PKC via C-terminal phosphorylation, *Biochim. Biophys. Acta 1358*, 307–313.

78. Cali, J. J., Parekh, R. S., and Krupinski, J. (1996) Splice variants of type VIII adenylyl cyclase. Differences in glycosylation and regulation by Ca2+/calmodulin, *J. Biol. Chem. 271*, 1089–1095.

79. Wei, J., Wayman, G., and Storm, D. R. (1996) Phosphorylation and inhibition of type III adenylyl cyclase by calmodulin-dependent protein kinase II in vivo, *J. Biol. Chem. 271*, 24231–24235.

80. Wu, G. C., Lai, H. L., Lin, Y. W., Chu, Y. T., and Chern, Y. (2001) N-Glycosylation and residues Asn805 and Asn890 are involved in the functional properties of type VI adenylyl cyclase, *J. Biol. Chem. 276*, 35450–35457.

81. McVey, M., Hill, J., Howlett, A., and Klein, C. (1999) Adenylyl cyclase, a coincidence detector for nitric oxide, *J. Biol. Chem. 274*, 18887–18892.

82. Hill, J., Howlett, A., and Klein, C. (2000) Nitric oxide selectively inhibits adenylyl cyclase isoforms 5 and 6, *Cell Signal. 12*, 233–237.

83. Arejian, M., Li, Y., and Anand- Srivastava, M. B. (2009) Nitric oxide attenuates the expression of natriuretic peptide receptor C and associated adenylyl cyclase signaling in aortic vascular smooth muscle cells: role of MAPK, *Am. J. Physiol Heart Circ. Physiol 296*, H1859-H1867.

84. Tesmer, J. J., Sunahara, R. K., Gilman, A. G., and Sprang, S. R. (1997) Crystal structure of the catalytic domains of adenylyl cyclase in a complex with Gsα GTPS, *Science 278*, 1907–1916.

85. Zhang, G. Y., Liu, Y., Ruoho, A. E., and Hurley, J. H. (1997) Structure of the adenylyl cyclase catalytic core, *Nature 386*, 247–253.

86. Tang, W. J., Stanzel, M., and Gilman, A. G. (1995) Truncation and alanine-scanning mutants of type I adenylyl cyclase, *Biochemistry 34*, 14563–14572.

87. Gu, C., Cali, J. J., and Cooper, D. M. F. (2002) Dimerization of mammalian adenylate cyclases, *Eur. J. Biochem. 269*, 413–421.

88. Baragli, A., Grieco, M. L., Trieu, P., Villeneuve, L. R., and Hebert, T. E. (2008) Heterodimers of adenylyl cyclases 2 and 5 show enhanced functional responses in the presence of Galpha s, *Cell Signal 20*, 480–492.

89. Pontier, S. M., Percherancier, Y., Galandrin, S., Breit, A., Gales, C., and Bouvier, M. (2008) Cholesterol-dependent separation of the beta2-adrenergic receptor from its partners determines signaling efficacy: insight into nanoscale organization of signal transduction, *J. Biol. Chem. 283*, 24659–24672.

90. Rybin, V. O., Xu, X., Lisanti, M. P., and Steinberg, S. F. (2000) Differential targeting of á-adrenergic receptor subtypes and adenylyl cyclase to cardiomyocyte caveolae, *J. Biol. Chem. 275*, 41447–41457.

91. Patel, H. H., Murray, F., and Insel, P. A. (2008) G-protein-coupled receptor-signaling components in membrane raft and caveolae microdomains, *Handb. Exp. Pharmacol. 167* –184.

92. Levitt, E. S., Clark, M. J., Jenkins, P. M., Martens, J. R., and Traynor, J. R. (2009) Differential effect of membrane cholesterol removal on mu and delta opioid receptors: A parallel comparison of acute and chronic signaling to adenylyl cyclase, *J. Biol. Chem. 284*, 22108–22122.

93. Piggott, L. A., Bauman, A. L., Scott, J. D., and Dessauer, C. W. (2008) The A-kinase anchoring protein Yotiao binds and regulates adenylyl cyclase in brain, *Proc Natl Acad Sci U S A 105*, 13835–13840.

94. Vidi, P. A. and Watts, V. J. (2009) Fluorescent and bioluminescent protein-fragment complementation assays in the study of G protein-coupled receptor oligomerization and signaling, *Mol. Pharmacol. 75*, 733–739.

95. Bouvier, M., Heveker, N., Jockers, R., Marullo, S., and Milligan, G. (2007) BRET analysis of GPCR oligomerization: newer does not mean better, *Nat Methods 4*, 3–4.

96. Dupre, D. J., Baragli, A., Rebois, R. V., Ethier, N., and Hebert, T. E. (2007) Signalling complexes associated with adenylyl cyclase II are assembled during their biosynthesis, *Cell Signal 19*, 481–489.

97. Jarrahian, A., Watts, V. J., and Barker, E. L. (2004) D2 dopamine receptors modulate Galpha-subunit coupling of the CB1 cannabinoid receptor, *J. Pharmacol. Exp. Ther. 308*, 880–886.

98. Kearn, C. S., Blake- Palmer, K., Daniel, E., Mackie, K., and Glass, M. (2005) Concurrent stimulation of cannabinoid CB1 and dopamine D2 receptors enhances heterodimer formation: a mechanism for receptor cross-talk?, *Mol. Pharmacol. 67*, 1697–1704.

99. Lee, S. P., So, C. H., Rashid, A. J., Varghese, G., Cheng, R., Lanca, A. J., O'Dowd B. F., and George, S. R. (2004) Dopamine D1 and D2 receptor Co-activation generates a novel phospholipase C-mediated calcium signal, *J. Biol. Chem. 279*, 35671–35678.

100. So, C. H., Verma, V., Alijaniaram, M., Cheng, R., Rashid, A. J., O'Dowd B. F., and George, S. R. (2009) Calcium signaling by dopamine D5 receptor and D5-D2 receptor hetero-oligomers occurs by a mechanism distinct from that for dopamine D1-D2 receptor hetero-oligomers, *Mol. Pharmacol. 75*, 843–854.

101. Fiorentini, C., Busi, C., Gorruso, E., Gotti, C., Spano, P., and Missale, C. (2008) Reciprocal regulation of dopamine D1 and D3 receptor function and trafficking by heterodimerization, *Mol. Pharmacol. 74*, 59–69.

102. Marcellino, D., Ferre, S., Casado, V., Cortes, A., Le, F. B., Mazzola, C., Drago, F., Saur, O., Stark, H., Soriano, A., Barnes, C., Goldberg, S. R., Lluis, C., Fuxe, K., and Franco, R. (2008) Identification of dopamine D1-D3 receptor heteromers. Indications for a role of synergistic D1-D3 receptor interactions in the striatum, *J. Biol. Chem. 283*, 26016–26025.

103. McIntire, W. E., MacCleery, G., and Garrison, J. C. (2001) The G protein β subunit is a determinant in the coupling of Gs to the β1-adrenergic and A2a adenosine receptors, *J. Biol. Chem. 276*, 15801–15809.

104. Hwang, J. I., Choi, S., Fraser, I. D., Chang, M. S., and Simon, M. I. (2005) Silencing the expression of multiple Gbeta-subunits eliminates signaling mediated by all four families of G proteins, *Proc. Natl. Acad. Sci. U. S. A 102*, 9493–9498.

105. Wang, Q., Jolly, J. P., Surmeier, J. D., Mullah, B. M., Lidow, M. S., Bergson, C. M., and Robishaw, J. D. (2001) Differential dependence of the D1 and D5 dopamine receptors on the G protein gamma 7 subunit for activation of adenylylcyclase, *J. Biol. Chem. 276*, 39386–39393.

106. Schwindinger, W. F., Betz, K. S., Giger, K. E., Sabol, A., Bronson, S. K., and Robishaw, J. D. (2003) Loss of G protein gamma 7 alters behavior and reduces striatal alphaolf level and cAMP production, *J. Biol. Chem. 278*, 6575–6579.

107. Pierre, S., Eschenhagen, T., Geisslinger, G., and Scholich, K. (2009) Capturing adenylyl cyclases as potential drug targets, *Nat. Rev. Drug Discov. 8*, 321–335.

108. Wu, Z. L., Thomas, S. A., Villacres, E. C., Xia, Z., Simmons, M. L., Chavkin, C., Palmiter, R. D., and Storm, D. R. (1995) Altered behavior and long-term potentiation in type I adenylyl cyclase mutant mice, *Proc. Natl. Acad. Sci. U. S. A 92*, 220–224.

109. Schaefer, M. L., Wong, S. T., Wozniak, D. F., Muglia, L. M., Liauw, J. A., Zhuo, M., Nardi, A., Hartman, R. E., Vogt, S. K., Luedke, C. E., Storm, D. R., and Muglia, L. J. (2000) Altered stress-induced anxiety in adenylyl cyclase type VIII-deficient mice, *J Neurosci 20*, 4809–4820.

110. Wong, S. T., Athos, J., Figueroa, X. A., Pineda, V. V., Scheafer, M. L., Chavkin, C. C., Muglia, L. J., and Storm, D. R. (1999) Calcium-stimulated adenylyl cyclase activity is critical for hippocampus-dependent long-term memory and late phase LTP, *Neuron 23*, 787–798.

111. Li, S., Lee, M. L., Bruchas, M. R., Chan, G. C., Storm, D. R., and Chavkin, C. (2006) Calmodulin-stimulated adenylyl cyclase gene deletion affects morphine responses, *Mol. Pharmacol. 70*, 1742–1749.

112. Yoshimura, M., Wu, P. H., Hoffman, P. L., and Tabakoff, B. (2000) Overexpression of type 7 adenylyl cyclase in the mouse brain enhances acute and chronic actions of morphine, *Mol. Pharmacol. 58*, 1011–1016.

113. Yan, L., Vatner, D. E., O' Connor J. P., Ivessa A., Ge H., Chen W., Hirotani S., Ishikawa Y., Sadoshima J., and Vatner S. F. (2007) Type 5 adenylyl cyclase disruption increases longevity and protects against stress, *Cell 130*, 247–258.

114. Zahniser, N. R., Simosky, J. K., Mayfield, R. D., Negri, C. A., Hanania, T., Larson, G. A., Kelly, M. A., Grandy, D. K., Rubinstein, M., Low, M. J., and Fredholm, B. B. (2000) Functional uncoupling of adenosine A2A receptors and reduced response to caffeine in mice lacking dopamine D2 receptors, *J. Neurosci.* 5949–5957.

11 G protein-independent and β arrestin-dependent GPCR signaling

Zhongzhen Nie and Yehia Daaka

INTRODUCTION

G protein-coupled receptors (GPCRs) constitute the largest family of cell surface receptors that regulate a wide array of cellular functions elicited by extracellular neurotransmitters, hormones, and light photons.[1] GPCRs have the signature structure of 7 transmembrane domains and are also termed 7 transmembrane receptors. The classical signaling GPCR unit includes the ligand-bound receptor, heterotrimeric αβγ G proteins, and plasma membrane-anchored effector.[1] Ligand agonist binding induces conformational changes in the GPCR thereby allowing it to function as a guanine nucleotide exchange factor to catalyze the exchange of GDP for GTP on the α subunit of the heterotrimeric G proteins. The Gα-GTP and Gβγ subunits independently but coordinately activate downstream effectors to generate specific cellular responses. For instance, $Gα_s$-GTP activates adenylyl cyclase to generate cyclic adenosine monophosphate (cAMP) that, in turn, activates protein kinase A (PKA) and the small GTPase Rap.

Duration of the GPCR signaling is regulated by two distinct groups of proteins namely the GPCR kinases (GRKs) and Arrestins.[2,3] There are four Arrestin proteins, with Arrestin1 and 4 expression being confined to the retinal rods and cones, and Arrestin2 and 3 (also known as β-Arrestin1 and β-Arrestin2, respectively) being ubiquitously expressed in mammalian tissues. The β-Arrestin1 and 2 proteins are 78% identical in their amino acid sequences and structurally differ mainly in the carboxyl termini. Functionally, the β-Arrestin1 and 2 can often substitute for each other, and mice with genetic deletion of either isoform are viable, although the double β-Arrestin1 and 2 knockout is embryonically lethal.[4] Emerging evidence demonstrates that in addition to their classical roles as terminators of GPCR/G protein signaling, the β-Arrestins may function as bona fide adaptors and signal transducers from a variety of activated receptor types.[5,6] In this minireview, we will highlight recent discoveries of β-Arrestin- regulated receptor signaling and their possible involvement in disease processes.

β-ARRESTIN INTERACTION WITH GPCRS

Receptor phosphorylation

Ligand agonist binding to GPCRs results in recruitment of GPCR kinases (GRKs) into close proximity to the agonist-bound receptor. The GRKs phosphorylate specific serine and threonine residues in the intracellular loops and carboxyl termini of the agonist-occupied receptors, thereby converting the receptors to high-affinity binders to β- Arrestins.[3] Seven GRKs have been functionally identified and molecularly cloned, with GRK2 and 3 being the most extensively studied. Both GRK2 and 3 bind free Gβγ subunits and thus are recruited to the plasma membrane following G protein activation by the GPCRs. The GRK2 and 3 also contain a pleckstrin homology domain that binds phosphatidylinositol 4,5-bis-phosphate, further contributing to their membrane localization upon receptor activation.[7] In addition to the regulation of their functions by subcellular distribution via interaction with partner lipids and proteins, the activation of GRKs is also regulated by protein phosphorylation. For example, phosphorylation of GRK2 by the Ca^{2+}-dependent protein kinase C (PKC) increases its activity, whereas phosphorylation of GRK2 or 3 by the mitogen-activated protein kinases (MAPKs) extracellular signal regulated kinase 1 and 2 (ERK) decreases their kinase activity and promotes their redistribution from the plasma membrane to the cytosol.

Receptor conformation

Ligand binding to GPCRs causes rearrangement of their transmembrane segments. In the case of the canonical GPCR β2 adrenoceptors (β2AR), the agonist binding induces rearrangement of helix 3 and 6 that, in turn, causes conformational changes in the intracellular domains with the consequent coupling to the G proteins and controlled activation of downstream effectors.[8] Evidence is accumulating that GPCRs do not necessarily activate their distinct effectors to the same extent – in other words, different ligands (both agonist and antagonist) may have collateral efficacies.[9] Explanation for this phenomenon lies in the various conformations that a ligand-occupied GPCR may adopt. A GPCR may have conformations that favor coupling to different subsets of G proteins, or to β-Arrestins.[10] As a result, the binding of a specific ligand can induce the G protein signaling, the β-Arrestin signaling, or the blockade of one pathway and activation of the other. An example is the βAR ligand carvedilol that functions as an inverse agonist for G protein-mediated adenylyl cyclase activation, but as an agonist for β-Arrestin-mediated ERK phosphorylation.[11]

CLASSICAL FUNCTIONS OF β-ARRESTIN PROTEINS

Receptor desensitization

Desensitization refers to the decreased signaling of GPCRs to repeated or sustained stimulation with ligand agonists, an adaptive property of cells to prevent potentially harmful responses due to excessive receptor activation.[12] Two critical components of GPCR desensitization are receptor phosphorylation and β-Arrestin coupling.[13] GRKs are recruited to the ligand-occupied receptors and phosphorylate them on serine and threonine residues that are frequently present in short stretches in the intracellular loops and the carboxyl terminal tails of the GPCRs. Phosphorylation of GPCRs by GRKs per se may not affect the GPCR signaling to downstream effectors, but instead increases the binding affinity to the β-Arrestins that confer sterical hindrance to prevent further G protein coupling and activation.[3] More recently, the β-Arrestins were also shown to facilitate the dampening of receptor/G protein signaling via increased breakdown of accumulated second messengers such as cAMP and diacylglycerol (Figure 11.1).

One prominent feature of GRKs is that they phosphorylate only ligand-bound GPCRs and thus induce homologous desensitization.[14] GPCRs can also be phosphorylated by second messenger-activated protein kinases such as PKA or PKC. Once activated, these kinases phosphorylate GPCRs regardless of their agonist occupancy status, resulting in a so-called heterologous desensitization. In addition to receptor desensitization, phosphorylation of GPCRs by second

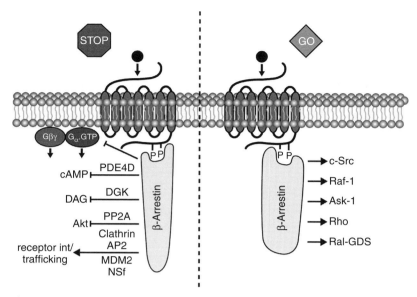

Figure 11-1: β-Arrestin-mediated signaling. Following the stimulation of a GPCR with ligand, the receptor is phosphorylated by a GRK thereby providing a high affinity binding site for β-Arrestins. Depending on the nature of ligand-receptor pair and post-translational modification, the β-Arrestins can adopt one of several possible conformations that, in turn, dictate the binding partner(s) and signals.

messenger-dependent PKA/PKC may lead to a switch in the coupling preference of the GPCR to a different heterotrimeric G protein, thereby resulting in altering the multitude of signaling networks initiated by the single agonist-receptor pair. For example, phosphorylation of βAR by PKA decreases the receptor coupling to G_s (thereby reducing cAMP production) and simultaneously enhances the receptor coupling to G_i that actively inhibits the G_s-mediated adenylyl cyclase signal.[15] Moreover, the regulated coupling of βAR to G_i initiates new signaling networks, such as the novel β-Arrestin-mediated ERK activation.[16]

Receptor endocytosis

GPCRs undergo constitutive and agonist-stimulated endocytosis and the molecular mechanisms governing the different endocytic pathways are being elucidated.[17] Endocytosed receptors can recycle back to the plasma membrane for reuse or can be targeted to lysosomes for degradation. The β-Arrestins play a pivotal role in the agonist-stimulated endocytosis of GPCRs via clathrin-coated or caveolar pits.[18] Association with the phosphorylated GPCRs through the phosphate sensor region results in conformational changes of β-Arrestin proteins, allowing them to bind clathrin and the adaptor protein complex AP-2, both of which are critical components of receptor-mediated endocytosis (Figure 11.1). The binding occurs between the amino terminal region of the clathrin heavy chain and the carboxyl terminal region of β-Arrestins, and between the β2 subunit of AP-2 and the extreme carboxyl terminal region of β-Arrestins.[19] The AP-2 can recognize different sorting signals on GPCRs such as di-leucine motifs or tyrosine-based signals, and clathrin is a critical component for membrane deformation and vesicle formation. Based on their affinity for β-Arrestins, GPCRs can be divided into two groups.[5] Class A receptors (e.g., β2AR) bind β-Arrestins transiently, have greater affinity for β-Arrestin2 than for β-Arrestin1, and dissociate from the β-Arrestins before the receptors are endocytosed. Class B receptors (e.g., angiotensin II type 1A or AT1A) bind β-Arrestins more tightly, have similar affinities for β-Arrestin1 and β-Arrestin2, and are endocytosed together with β-Arrestins. In general, class A receptors tend to recycle back to the plasma membrane whereas class B receptors tend to be targeted for lysosomal degradation leading to the idea that β-Arrestins regulate intracellular trafficking of GPCR-contained vesicles.

Additional mechanisms are involved in the regulation of the endocytic trafficking of GPCRs by β-Arrestins. For example, β-Arrestins bind ubiquitin ligases, such as MDM2, that add ubiquitin moieties to GPCRs to facilitate endocytosis, lysosomal targeting, and degradation of the affected receptor.[5] Other targets for β-Arrestins in the regulation of GPCR endocytosis include the ADP ribosylation factor Arf6 that regulates trafficking at the plasma membrane and on endosomes[20] and ARNO, a guanine nucleotide exchange factor that catalyzes Arf6 activation.[21] Indeed, internalization of the β2AR can be inhibited by the forced overexpression of GPCR kinase interactors GIT1 or GIT2, GTPase-activating proteins that inactivate Arf6,[22,23] further supporting a general β-Arrestin-regulated role in the endocytosis of GPCRs.

NEW FUNCTION OF β-ARRESTINS

Interaction with receptors other than GPCRs

In addition to their established character of interacting with agonist-occupied GPCRs to regulate the receptors' fate, emerging evidence establishes the ability of the β-Arrestins to form complexes with various partner proteins, including single membrane-anchored receptors.[3] The β-Arrestins can be recruited to insulin-like growth factor 1 (IGF-1) receptors and to couple the activated IGF-1 receptor to the phosphatidylinositol 3-kinase system in a manner independent of the tyrosine kinase activity of the receptor.[24] The β-Arrestins also regulate endocytosis of nonclassical GPCRs, including Frizzled and Smoothened, as well as the single membrane-spanning receptors for transforming growth factor-β and low-density lipoproteins.[3,25] Hence, by means of their ability to regulate the expression levels of various receptor subtypes on the plasma membrane, it is reasonable to propose that β-Arrestins dictate cellular response to the extracellular milieu.

Interaction with signaling network components

ERK MAP kinases

Like receptor tyrosine kinases (RTKs), GPCRs transduce mitogenic signals that control cellular proliferation and invasion, at least in part, via activation of the

ERK MAP kinases. The mechanisms of activation of ERK signaling pathways by GPCRs are complex and, depending on receptor subtype and cell context, may involve one of several cascades that are transduced by G proteins alone, β-Arrestins alone, or both. GPCR-mediated activation of the ERK cascade can be achieved with second messenger-dependent pathways, such as the PKA-mediated phosphorylation of Rap, PKC-dependent activation of Raf-1, or Ca^{2+}- and cell adhesion-dependent activation of focal adhesion localized kinase Pyk2. Often, however, the second messenger-dependent mechanisms are insufficient to account for the activation of ERK by GPCRs, and it is becoming evident that signal transduction by certain GPCRs converges with the pathway used by canonical RTKs. This pathway, known as RTK transactivation, may involve the G protein-dependent (e.g., $G\alpha_q$ and $G\beta\gamma$) and β-Arrestin-mediated phosphorylation of the target RTK, being mainly the epidermal growth factor receptor (EGFR). Exact determinants involved in the GPCR-mediated EGFR transactivation remain poorly defined, although it is accepted now that G protein-mediated (e.g., $G\alpha_q$ and $G\beta\gamma$) ectodomain shedding of EGFR ligands by members of the "A Disintegrin And Metalloproteinase" (ADAM) family of matrix metalloproteinases plays a critical role.[26]

The involvement of β-Arrestins in the regulation of ERK activity was first appreciated with the observation that a dominant negative form of β-Arrestin1 attenuated the βAR-stimulated ERK phosphorylation.[27] Shortly thereafter, the β-Arrestins were demonstrated as adaptors that mediate formation of a trimeric complex containing the activated βAR and nonreceptor tyrosine kinase c-Src.[28] Subsequent results demonstrated that the activated βAR-β-Arrestin-c-Src complex relays signals lead to the transactivation of EGFR and subsequent phosphorylation of ERK.[29] Hence, under this scenario, both G proteins and β-Arrestins contribute to EGFR transactivation and ERK phosphorylation.

Experimental evidence also shows that following GPCR activation, the β-Arrestins serve as scaffolds bringing into close proximity Raf-1 and MEK, leading to ERK activation.[30] It remains uncertain, however, whether this mode of β-Arrestin-dependent ERK activation requires the G protein signaling. Several lines of evidence have led to the conclusion that certain GPCRs, including those for μ-opioid, muscarinic acetylcholine, and angiotensin II, transduce the G protein-independent but β-Arrestin-mediated signals.[10] In the case of the μ-opioid receptor, endogenous ligand enkephalin peptides induce the G protein activation and receptor internalization, whereas the ligand morphine induces the G protein activation but not the β-Arrestin-mediated receptor internalization. For muscarinic acetylcholine, which promotes coupling of cognate receptor to both G_s and G_q proteins, the results establish that whereas some receptor agonists block the G_s pathway, others promote the G_q signals. This observation was extended to β-Arrestin where some compounds induced the G protein signaling but failed to stimulate the receptor phosphorylation, β-Arrestin recruitment, and receptor internalization. Likewise, several ligands were discovered to promote the recruitment of β-Arrestin and induce receptor internalization, but failed to show detectable G protein activation. Together, these observations have led to the proposition that the ligand-receptor complex can adopt multiple conformations and that each permits the coupling to a selective effector, specifically G protein and/or β- Arrestin.[10]

Direct demonstration for β-Arrestin-mediated and G protein-independent signaling by activated GPCRs to ERK came from studies in the Lefkowitz lab using the angiotensin II-AT1A receptor system as a model. In both fibroblasts and smooth muscle cells, angiotensin II signals through G_q-mediated activation of phospholipase C and induces receptor phosphorylation by GRKs and recruitment of β-Arrestins. At this juncture, β-Arrestin recruitment to the AT1A receptor predictably leads to receptor desensitization (with respect to G_q activation) and internalization. Stimulation with angiotensin II also leads to ERK phosphorylation, and three lines of evidence implicate a β-Arrestin-mediated but G protein-independent signal. First, a mutated AT1A receptor (DRY/AAY) incapable of coupling to G proteins transduces the angiotensin II-mediated ERK activation. Second, AT1A receptor ligand [Sar1, Ile4, Ile8]AngII is deficient in its ability to promote the detectable G protein activation, but can induce the ERK phosphorylation. Thirdly, the knockdown of β-Arrestin expression with RNAi abolishes ERK phosphorylation following stimulation of the mutated AT1A receptor (DRY/AAY) with angiotensin II, or the stimulation of wild-type AT1A receptor with [Sar1, Ile4, Ile8]AngII. Different from G protein-mediated ERK activation that usually functions in the nucleus, β-Arrestin-activated ERK remain in the cytosol, most probably as a result of the scaffolding effect of β-Arrestins that restrict activated ERK to endocytic structures.[31] Another marked difference between the G protein- and β-Arrestins-regulated modes of ERK activation is signal duration. Whereas G protein-mediated, like small GTPase Ras-mediated ERK activation has faster onset and shorter duration, β-Arrestin-mediated ERK activation has slower onset and persists longer.[5,6]

PDE

Activation of G_s by β2AR leads to the production of the second messenger cAMP that in turn activates PKA and the small GTPase Rap via the guanine nucleotide exchange factor "exchange protein directly activated by cAMP" (EPac). Termination of cAMP signal is mediated by its degradation through the phosphodiesterase (PDE) enzymes. The β-Arrestins form a complex with PDE4Ds in the cytosol and co-translocate with PDE4Ds to the plasma membrane upon activation of β2AR.[32] Hence β-Arrestins can curtail the β2AR-induced cAMP signaling via two mechanisms: by decreasing cAMP production in response to receptor desensitization, and by increasing cAMP degradation through the PDEs. In addition, PDE recruitment by β-Arrestins following β2AR activation may regulate the switch in G protein coupling from G_s to G_i[33] that results from β2AR phosphorylation by PKA.[15]

DGK

Similar to the recruitment of PDE following β2AR activation, β-Arrestins bind to and recruit diacylglycerol kinase (DGK) to the plasma membrane following activation of the G_q-coupled M1 muscarinic receptor. DGKs use diacylglycerol as substrate to generate phosphatidic acid, thereby turning off signaling through diacylglycerol, such as activation of PKC.[34] Phosphatidic acid itself is a signaling molecule involved in the regulation of several cellular functions, including vesicle trafficking and enzyme activation. Hence, the β-Arrestins may be viewed as functional switches that dampen signals through diacylglycerol and simultaneously permit signals through phosphatidic acid, again illustrating complexity of the β-Arrestins' roles in cellular signaling.

Akt

The β-Arrestins regulate Akt activation in response to stimulation of both GPCRs and RTKs. The α-thrombin receptor, for example, recruits β-Arrestin1 to elicit the rapid but transient activation of Akt that is not involved in G1-S phase progression,[35] and this effect of activated α-thrombin receptors seems to be dependent upon the release of βγ subunits from G_q proteins.[36] The D2 dopaminergic receptor recruits β-Arrestin2 that forms a complex with Akt and the serine/threonine phosphatase PP2A.[37] Dephosphorylation of Akt by PP2A results in the D2 receptor-mediated inactivation of Akt that contributes to the locomoter activity of dopaminergic neurotransmission. Also, following stimulation of the IGF-1 receptor, β-Arrestin1 is recruited to the plasma membrane where it regulates the activation of phosphatidylinositol 3- kinase and Akt in a manner independent of the tyrosine kinase activity of the IGF-1 receptor.[24]

REGULATION OF β-ARRESTIN FUNCTION

Receptor-biased agonism

In the classical view of receptor pharmacology, an agonist is assumed to possess the same efficacy for the different effectors through which the cognate receptor transduces a signal. This understanding has recently been challenged with the finding of receptor-biased agonism, that is, selective activation of a subset of activities of a particular receptor by its agonist.[10] Properties intrinsic to both the agonist and the receptor contribute to this biased agonism. For example, neither the mutant AT1A receptor (DRY/AAY) nor the agonist [Sar1,Ile4, Ile8]AngII could activate G protein and downstream effectors, but are competent to induce both β-Arrestin membrane recruitment and ERK activation.[38] Also, the βAR inverse agonist carvedilol can induce βAR phosphorylation, β-Arrestin membrane translocation, receptor internalization, and ERK activation that is dependent upon β-Arrestin2. Given the beneficial effects of carvedilol to patients with heart failure, the application of biased ligands may have therapeutic advantages.[9,10]

Phosphorylation

β-Arrestins are phosphoproteins present mainly in the cytosol and are recruited to the plasma membrane to associate with agonist-activated and GRK-phosphorylated GPCRs. β-Arrestin1 is phosphorylated on serine 412 by ERK and β-Arrestin2 is phosphorylated mainly on threonine 383 by casein kinase II.[39–41] Phosphorylation of β-Arrestins selectively regulates their functions: Whereas the binding of β-Arrestins to GPCRs and consequent GPCR desensitization is not affected, their binding to clathrin and subsequent receptor internalization requires dephosphorylation of the β-Arrestin proteins, albeit by as yet unidentified phosphatase(s).

Ubiquitinylylation

Ubiquitination refers to the post-translational modification of proteins by the addition of a 76 amino acid polypeptide ubiquitin via the sequential enzymatic reactions of E1 (ubiquitin-activating), E2 (ubiquitin-conjugating), and E3 (ubiquitin ligase) enzymes. Upon receptor activation by an agonist, β-Arrestins are polyubiquitinylated by RING finger E3 ligase MDM2.[42] Ubiquitinylation of β-Arrestins regulates GPCR internalization, and the duration of β-Arrestin ubiquitinylation correlates with the pattern of receptor trafficking. Activation of class A receptors such as β2AR, which dissociates from β-Arrestins before internalization, results in transient ubiquitinylation of the β-Arrestins. On the other hand, activation of class B receptors such as the V2 vasopressin receptor, which binds β-Arrestins tightly and internalizes together with β- Arrestins, induces prolonged ubiquitinylation of the β-Arrestins,[43] suggesting that the ubiquitinylation status of β-Arrestins determines stability of the receptor-β-Arrestin complex, as well as the trafficking pattern of the receptor. The enzyme ubiquitin-specific protease 33 was recently shown to catalyze the deubiquitinylation of β-Arrestins,[44] establishing feedback regulatory mechanisms to control the β-Arrestins functions. In addition to serving as substrates for ubiquitinylation, the β-Arrestins also serve as adaptors that bring the E3 ligase into close proximity to target GPCRs, leading to receptor ubiquitinylation that regulates the receptors' endocytic trafficking and lysosomal targeting.

S-Nitrosylation

S-nitrosylation of proteins is the modification of the thiol group of cysteine residues by covalent attachment of a nitrogen monoxide, which has emerged as an important post-translational mechanism whereby nitric oxide regulates cellular functions. Recently, β-Arrestin2, but not β-Arrestin1, was shown to be S-nitrosylated on cysteine residue 410 that provided an additional regulatory step in the life cycle of GPCR signaling.[45] The β-Arrestin2 was shown to form a complex with endothelial nitric oxide synthase (eNOS). S-nitrosylation of β-Arrestin2 was induced by activating β2AR, suggesting functional coupling between β2AR and eNOS. Interaction with eNOS was inhibited following S-nitrosylation of β-Arrestin2, whereas interactions with the endocytic components clathrin and AP-2 were augmented. Thus, agonist binding to β2AR results in activation of eNOS and consequent S-nitrosylation of β-Arrestin2 that, upon dissociation from eNOS, associates with the clathrin endocytic machinery, leading to increased internalization of β2AR, while β-Arrestin2 is denitrosylated and recycled.[45] Mechanisms involved in the regulated β-Arrestin2 denitrosylation remain to be discovered.

β-ARRESTINS AND DISEASE

Studies on clinical relevance of β-Arrestins have focused mainly on the heart and brain organs. For example, the μ-opioid receptor agonist that is negatively biased toward β-Arrestins may have the advantage of enhanced and prolonged analgesic effect and less undesirable side effects such as tolerance and constipation.[46,47] The β-Arrestin2 regulates D2 dopamine receptor-mediated neurotransmission and behavior through modulation of Akt activity.[37] The β-Arrestin1 was recently implicated in the cutaneous flushing but not the lipid lowering effects of nicotinic acid. Hence biased ligands for the nicotinic acid receptor GPR109A that selectively activate G protein signaling but not β-Arrestin signaling may have therapeutic advantages in the treatment of lipid disorders.[48] Also, β-Arrestin2 was shown to regulate some insulin signaling pathways, and deficiency in β-Arrestin2-mediated insulin signaling may contribute to the insulin resistance and type II diabetes.[49] Moreover, the β-Arrestin1 can induce experimental autoimmune uveitis[50] and causes the Oguchi disease, an autosomal recessive form of congenital night blindness that is usually seen in individuals of Japanese descent.[51]

Cell migration

The role of β-Arrestins in cell migration was first established based on the observation that lymphocytes from β-Arrestin2-deficient mice failed to migrate toward a chemotaxic gradient.[52] The β-Arrestin-regulated cell migration may underlie some of the pathophysiology of allergic asthma that cannot be induced in animals lacking the β-Arrestin2.[53] The β-Arrestins may mediate GPCR-induced cell migration that can be G protein-dependent or -independent, and both forms of β-Arrestins can be involved. Molecular mechanisms underlying this effect of β-Arrestins are being identified,[54] with emphasis on the β-Arrestin-mediated ERK activation.[55] In addition, β-Arrestins have been shown to activate Rho A[56] and to associate with filamin A,[57] both of which are involved in the remodeling of actin cytoskeleton critical for the cell movement.

Cell proliferation

GPCRs can induce the G protein- and Ras-dependent activation of ERK that is transient in nature, and the activated ERK proteins translocate to the nucleus where they regulate the transcriptional machinery and cellular proliferation.[58] GPCRs also induce the β-Arrestin-mediated ERK activation, although this pool of activated ERK remain in the cytosol and, therefore reduce the ERK transcriptional activity and resultant GPCR mitogenic effects.[59] For example, wild-type protease-activated receptors (PAR)2 activate ERK in the cytosol with no mito-genic effects, whereas a mutated PAR2 receptor deficient in β-Arrestin binding is able to promote the nuclear ERK distribution and cellular proliferation.[60] In atherosclerotic human arteries, β-Arrestin2 levels are increased in comparison to those of nonatherosclerotic arteries,[61] and it stimulates the proliferation of smooth muscle cells, leading to hyperplasia and atherosclerosis.[62] Interestingly, β-Arrestin1 inhibits the proliferation of smooth muscle cells, suggesting possible beneficial effects for the selective targeting of individual β-Arrestin isoforms in atherosclerosis.

Heart disease

In the heart, β-Arrestin2-mediated ERK activation may contribute to the benefi-cial effect of the β-Arrestin-biased ligand carvedilol for βAR[11] and the G protein-independent cardioprotective effects of angiotensin II.[63] The β-Arrestin-mediated cardioprotective effect may involve transactivation of the EGFR,[64] and recently the β-Arrestin1 was shown to mediate angiotensin II-induced aldosterone produc-tion by the adrenal gland that is independent of G protein signaling,[65] suggesting that blocking adrenal β-Arrestin1 activity may retard heart failure progression after myocardial infarction.

CONCLUSION AND PERSPECTIVE

In addition to their originally identified role in terminating GPCR signaling, β-Arrestins are now acknowledged to regulate signaling networks initiated by different receptors, including GPCRs, RTKs, and cytokine receptors. G protein- or β-Arrestin-biased ligands are promising therapeutic agents for a variety of human diseases. A recent proteomic study identified over 300 interacting proteins for β-Arrestins[66] and the functional consequence for many of these interactions awaits further investigation, and it will not be surprising if more interacting pro-teins are found. Future studies on the molecular mechanisms by which β-Arrestins regulate different cellular processes will definitely improve our understanding of some human diseases and their clinical outcomes.

ACKNOWLEDGMENTS

This work is supported in part by National Institute of Health Grants CA129155, CA131988 and AI079014 (to Yehia Daaka) and CA124578 (to Zhongzhen Nie).

REFERENCES

1.   Pierce KL, Premont RT, Lefkowitz RJ. Seven-transmembrane receptors. Nat. Rev. Mol. Cell Biol. (2002) 3: 639–50.

2. Premont RT, Gainetdinov RR. Physiological roles of G protein-coupled receptor kinases and Arrestins. Annu. Rev. Physiol. (2007) 69: 511–34.

3. Moore CAC, Milano SK, Benovic JL. Regulation of receptor trafficking by GRKs and Arrestins. Annu. Rev. Physiol. (2007) 69: 451–82.

4. Kohout TA, Lin FT, Perry SJ, Conner DA, Lefkowitz RJ. β-Arrestin 1 and 2 differentially regulate heptahelical receptor signaling and trafficking. Proc. Natl. Acad. Sci. U.S.A. (2001) 98: 1601–6.

5. Lefkowitz RJ, Shenoy SK. Transduction of receptor signals by β-Arrestins. Science (2005) 308: 512–7.

6. DeFea K. β-Arrestins and heterotrimeric G-proteins: collaborators and competitors in signal transduction. Br. J. Pharmacol. (2008) 153: S298-S309.

7. Penn RB, Pronin AN, Benovic JL. Regulation of G protein-coupled receptor kinases. Trends Cardiovasc. Med. (2000) 10: 81–9.

8. Ballesteros JA, Jensen AD, Liapakis G, et al. Activation of the β2-adrenergic receptor involves disruption of an ionic lock between the cytoplasmic ends of transmembrane segments 3 and 6. J. Biol. Chem. (2001) 276: 29171–7.

9. Kenakin T. New concepts in drug discovery: Collateral efficacy and permissive antagonism. Nat. Rev. Drug Discov. (2005) 4: 919–27.

10. Violin JD, Lefkowitz RJ. β-Arrestin-biased ligands at seven-transmembrane receptors. Trends Pharmacol. Sci. (2007) 28: 416–22.

11. Wisler JW, DeWire SM, Whalen EJ, et al. A unique mechanism of β-blocker action: Carvedilol stimulates β-Arrestin signaling. Proc. Natl. Acad. Sci. U.S.A. (2007) 104: 16657–62.

12. McDonald PH, Lefkowitz RJ. β-Arrestins: New roles in regulating heptahelical receptors' functions. Cell. Signal. (2001) 13: 683–9.

13. Reiter E, Lefkowitz RJ. GRKs and β-Arrestins: roles in receptor silencing, trafficking and signaling. Trends Endocrinol. Metab. (2006) 17: 159–65.

14. Pitcher JA, Freedman NJ, Lefkowitz RJ. G protein-coupled receptor kinases. Annu. Rev. Biochem. (1998) 67: 653–92.

15. Daaka Y, Luttrell LM, Lefkowitz RJ. Switching of the coupling of the β2-adrenergic receptor to different G proteins by protein kinase A. Nature (1997) 390: 88–91.

16. Shenoy SK, Drake MT, Nelson CD, et al. β-Arrestin-dependent, G protein-independent ERK1/2 activation by the β2 adrenergic receptor. J. Biol. Chem. (2006) 281: 1261–73.

17. Scarselli M, Donaldson JG. Constitutive internalization of G protein-coupled receptors and G proteins via clathrin-independent endocytosis. J. Biol. Chem. (2009) 284: 3577–85.

18. Marchese A, Paing MM, Temple BRS, Trejo J. G protein-coupled receptor sorting to endosomes and lysosomes. Annu. Rev. Pharmacol. Toxicol. (2008) 48: 601–29.

19. Gurevich VV, Gurevich EV. The structural basis of Arrestin-mediated regulation of G-protein- coupled receptors. Pharmacol. Ther. (2006) 110: 465–502.

20. Donaldson JG. Multiple roles for Arf6: Sorting, structuring, and signaling at the plasma membrane. J. Biol. Chem. (2003) 278: 41573–6.

21. Casanova JE. Regulation of Arf activation: the Sec7 family of guanine nucleotide exchange factors. Traffic (2007) 8: 1476–85.

22. Claing A, Perry SJ, Achiriloaie M, et al. Multiple endocytic pathways of G protein-coupled receptors delineated by GIT1 sensitivity. Proc. Natl. Acad. Sci. U.S.A. (2000) 97: 1119–24.

23. Randazzo PA, Hirsch DS. Arf GAPs: multifunctional proteins that regulate membrane traffic and actin remodelling. Cell. Signal. (2004) 16: 401–13.

24. Povsic TJ, Kohout TA, Lefkowitz RJ. β-Arrestin1 mediates insulin-like growth factor 1 (IGF-1) activation of phosphatidylinositol 3-kinase (PI3K) and anti-apoptosis. J. Biol. Chem. (2003) 278: 51334–9.

25. DeWire SM, Ahn S, Lefkowitz RJ, Shenoy SK. β-Arrestins and cell signaling. Annu. Rev. Physiol. (2007) 69: 483–510.

26. Prenzel N, Zwick E, Daub H, et al. EGF receptor transactivation by G-protein-coupled receptors requires metalloproteinase cleavage of proHB-EGF. Nature 1999; 402: 884–8.

27. Daaka Y, Luttrell LM, Ahn S, et al. Essential role for G protein-coupled receptor endocytosis in the activation of mitogen-activated protein kinase. J. Biol. Chem. (1998) 273: 685–8.

28. Luttrell LM, Ferguson SSG, Daaka Y, et al. β-Arrestin-dependent formation of β2 adrenergic receptor Src protein kinase complexes. Science (1999) 283: 655–61.

29. Maudsley S, Pierce KL, Zamah AM, et al. The β2-adrenergic receptor mediates extracellular signal-regulated kinase activation via assembly of a multi-receptor complex with the epidermal growth factor receptor. J. Biol. Chem. (2000) 275: 9572–80.

30. Luttrell LM, Roudabush FL, Choy EW, et al. Activation and targeting of extracellular signal-regulated kinases by β-Arrestin scaffolds. Proc. Natl. Acad. Sci. U.S.A. (2001) 98: 2449–54.

31. Shenoy SK, Lefkowitz RJ. Receptor-specific ubiquitination of β-Arrestin directs assembly and targeting of seven-transmembrane receptor signalosomes. J. Biol. Chem. (2005) 280: 15315–24.

32. Perry SJ, Baillie GS, Kohout TA, et al. Targeting of cyclic AMP degradation to β2-adrenergic receptors by β-Arrestins. Science (2002) 298: 834–6.

33. Baillie GS, Sood A, McPhee I, et al. β-Arrestin-mediated PDE4 cAMP phosphodiesterase recruitment regulates β-adrenoceptor switching from Gs to Gi. Proc. Natl. Acad. Sci. U.S.A. (2003) 100: 940–5.

34. Nelson CD, Perry SJ, Regier DS, Prescott SM, Topham MK, Lefkowitz RJ. Targeting of diacylglycerol degradation to M1 muscarinic receptors by β-Arrestins. Science (2007) 315: 663–6.

35. Goel R, Phillips-Mason PJ, Raben DM, Baldassare JJ. α-Thrombin induces rapid and sustained Akt phosphorylation by β-Arrestin1-dependent and -independent mechanisms, and only the sustained Akt phosphorylation is essential for G(1) phase progression. J. Biol. Chem. (2002) 277: 18640–8.

36. Goel R, Phillips-Mason PJ, Gardner A, Raben DM, Baldassare JJ. α-thrombin-mediated phosphatidylinositol 3-kinase activation through release of Gβγ dimers from Gαq and Gαi2. J. Biol. Chem. (2004) 279: 6701–10.

37. Beaulieu JM, Sotnikova TD, Marion S, Lefkowitz RJ, Gainetdinov RR, Caron MG. An Akt/β- arrestin 2/PP2A signaling complex mediates dopaminergic neurotransmission and behavior. Cell (2005) 122: 261–73.

38. Wei HJ, Ahn S, Shenoy SK, et al. Independent β-Arrestin 2 and G protein-mediated pathways for angiotensin II activation of extracellular signal-regulated kinases 1 and 2. Proc. Natl. Acad. Sci. U.S.A. (2003) 100: 10782–7.

39. Lin FT, Krueger KM, Kendall HE, et al. Clathrin-mediated endocytosis of the β-adrenergic receptor is regulated by phosphorylation/dephosphorylation of β-Arrestin1. J. Biol. Chem. (1997) 272: 31051–7.

40. Lin FT, Miller WE, Luttrell LM, Lefkowitz RJ. Feedback regulation of β-Arrestin1 function by extracellular signal-regulated kinases. J. Biol. Chem. (1999) 274: 15971–4.

41. Kim YM, Barak LS, Caron MG, Benovic JL. Regulation of Arrestin-3 phosphorylation by casein kinase II. J. Biol. Chem. (2002) 277: 16837–46.

42. Shenoy SK, McDonald PH, Kohout TA, Lefkowitz RJ. Regulation of receptor fate by ubiquitination of activated β2-adrenergic receptor and β-Arrestin. Science (2001) 294: 1307–13.

43. Shenoy SK, Lefkowitz RJ. Trafficking patterns of β-Arrestin and G protein-coupled receptors determined by the kinetics of β-arrestin deubiquitination. J. Biol. Chem. (2003) 278: 14498–506.

44. Shenoy SK, Modi AS, Shukla AK, et al. β-Arrestin-dependent signaling and trafficking of 7-transmembrane receptors is reciprocally regulated by the deubiquitinase USP33 and the E3 ligase Mdm2. Proc. Natl. Acad. Sci. U.S.A. (2009) 106: 6650–5.

45. Ozawa K, Whalen EJ, Nelson CD, et al. S-nitrosylation of β-Arrestin regulates β-adrenergic receptor trafficking. Mol. Cell (2008) 31: 395–405.

46. Bohn LM, Lefkowitz RJ, Gainetdinov RR, Peppel K, Caron MG, Lin FT. Enhanced morphine analgesia in mice lacking β-Arrestin 2. Science (1999) 286: 2495–8.

47. Bohn LM, Gainetdinov RR, Lin FT, Lefkowitz RJ, Caron MG. μ-Opioid receptor desensitization by β-Arrestin-2 determines morphine tolerance but not dependence. Nature (2000) 408: 720–3.

48. Walters RW, Shukla AK, Kovacs JJ, et al. β-Arrestin1 mediates nicotinic acid-induced flushing, but not its antilipolytic effect, in mice. J. Clin. Invest. (2009) 119: 1312–21.

49. Luan B, Zhao J, Wu HY, et al. Deficiency of a β-Arrestin-2 signal complex contributes to insulin resistance. Nature (2009) 457: 1146–U105.

50. Shinohara T, Singh VK, Tsuda M, Yamaki K, Abe T, Suzuki S. S-antigen from gene to autoimmune uveitis. Exp. Eye Res. (1990) 50: 751–7.

51. Baylor DA, Burns ME. Control of rhodopsin activity in vision. Eye (1998) 12: 521–5.

52. Fong AM, Premont RT, Richardson RM, Yu YRA, Lefkowitz RJ, Patel DD. Defective lymphocyte chemotaxis in β-Arrestin2-and GRK6-deficient mice. Proc. Natl. Acad. Sci. U.S.A. (2002) 99: 7478–83.

53. Walker JKL, Fong AM, Lawson BL, et al. β-Arrestin-2 regulates the development of allergic asthma. J. Clin. Invest. (2003) 112: 566–74.

54. DeFea KA. Stop that cell! β-Arrestin-dependent chemotaxis: A tale of localized actin assembly and receptor desensitization. Annu. Rev. Physiol. (2007) 69: 535–60.

55. Ge L, Shenoy SK, Lefkowitz RJ, DeFea K. Constitutive protease-activated receptor-2-mediated migration of MDA MB-231 breast cancer cells requires both β-Arrestin-1 and -2. J. Biol. Chem. (2004) 279: 55419–24.

56. Barnes WG, Reiter E, Violin JD, Ren XR, Milligan G, Lefkowitz RJ. β-Arrestin 1 and Gαq/11 coordinately activate RhoA and stress fiber formation following receptor stimulation. J. Biol. Chem. (2005) 280: 8041–50.

57. Scott MGH, Pierotti V, Storez H, et al. Cooperative regulation of extracellular signal-regulated kinase activation and cell shape change by filamin A and β-Arrestins. Mol. Cell. Biol. (2006) 26: 3432–45.

58. Morrison DK, Davis RJ. Regulation of MAP kinase signaling modules by scaffold proteins in mammals. Annu. Rev. Cell Dev. Biol. (2003) 19: 91–118.

59. Tohgo AK, Pierce KL, Choy EW, Lefkowitz RJ, Luttrell LM. β-Arrestin scaffolding of the ERK cascade enhances cytosolic ERK activity but inhibits ERK-mediated transcription following angiotensin AT1a receptor stimulation. J. Biol. Chem. (2002) 277: 9429–36.

60. DeFea KA, Zalevsky J, Thoma MS, Dery O, Mullins RD, Bunnett NW. β-Arrestin-dependent endocytosis of proteinase-activated receptor 2 is required for intracellular targeting of activated ERK1/2. J. Cell Biol. (2000) 148: 1267–81.

61. Archacki SR, Angheloiu G, Tian XL, et al. Identification of new genes differentially expressed in coronary artery disease by expression profiling. Physiol. Genomics (2003) 15: 65–74.

62. Kim J, Zhang L, Peppel K, et al. β-Arrestins regulate atherosclerosis and neointimal hyperplasia by controlling smooth muscle cell proliferation and migration. Cir. Res. (2008) 103: 70–9.

63. Zhai PY, Yamamoto M, Galeotti J, et al. Cardiac-specific overexpression of AT1 receptor mutant lacking Gαq/Gαi coupling causes hypertrophy and bradycardia in transgenic mice. J. Clin. Invest. (2005) 115: 3045–56.

64. Noma T, Lemaire A, Prasad SVN, et al. β-Arrestin-mediated β1-adrenergic receptor transactivation of the EGFR confers cardioprotection. J. Clin. Invest. (2007) 117: 2445–58.

65. Lymperopoulos A, Rengo G, Zincarelli C, Kim J, Soltys S, Koch WJ. An adrenal β-Arrestin 1- mediated signaling pathway underlies angiotensin II-induced aldosterone production in vitro and in vivo. Proc. Natl. Acad. Sci. U.S.A. (2009) 106: 5825–30.

66. Xiao K, McClatchy DB, Shukla AK, et al. Functional specialization of β-Arrestin interactions revealed by proteomic analysis. Proc. Natl. Acad. Sci. U.S.A. (2007) 104: 12011–6.

12 Assays to read GPCR modulation and signaling

Ralf Heilker and Michael Wolff

GPCR LIGANDS IN DRUG DISCOVERY

The GPCR protein family is the pharmaceutically most successful target class.[1] Approximately 30% of all marketed prescription drugs address GPCRs, and still GPCRs are taking a prominent place in the target portfolios of many pharmaceutical companies.

To reduce late-stage attrition rates, the pharmaceutical industry aims at a broader biological characterization of drug candidates in early preclinical research. Accordingly, the early pharmacological profiling has been expanded to support the medicinal chemists in lead optimization by putting the compounds into a disease-relevant rank order of potencies and efficacies. In case of GPCRs, this has technologically been enabled by the establishment of numerous medium or high throughput screening-compatible assay formats to measure ligand affinities,

G protein modulation, second messenger production, and downstream signaling events.

However, the connection of the target GPCR-dependent signaling pathway(s) and the disease-relevant processes is often poorly characterized. The broader testing of the GPCR ligands ultimately aims to deliver a bioactivity profile of these test compounds in several signaling assays. From this profile, the particular parameters of compound-modulated signaling may be selected that match best with the effects of the same compound for in vivo disease-relevant assays.

The concepts of constitutive activity, ligand-biased signaling and GPCR cross-talk through hetero-dimerization, discussed in this chapter, further complicate the interpretation of small-molecule GPCR ligand activities. Thus, the "inverse agonist" was established as a new paradigm for possible pharmacological intervention with the first demonstration of constitutive (i.e., ligand independent) activity of GPCRs.[2,3] An inverse agonist is defined as an antagonistic compound that reduces the resting state constitutive activity of a GPCR, whereas a neutral antagonist effect may only be detected in the presence of an agonist. Experimentally, the acceptance of "constitutive activity" had long remained contentious due to the suspicion of contaminating agonist molecules in the membrane receptor preparation or in the medium of the analyzed cells.[4] A further controversial issue arose about the question to which extent the paradigms of constitutive activity and inverse agonism are physiologically or therapeutically relevant.

Meanwhile, there are some clear examples illustrating the pharmacological significance of distinguishing between neutral antagonists and inverse agonists on, for example, histamine H3 receptor,[5] serotonin 5-HT2 receptors,[6] or β_1-adrenoceptors.[7] For that reason, there is demand for suitable assay techniques to analyze inverse agonism with regard to pharmacologically significant receptor signaling.

The usefulness of embracing the concept of ligand-biased signaling[8,9] in drug discovery is nicely exemplified by the work of Richman and colleagues.[10] They identified agonists of GPR109A that selectively elicit the therapeutic, anti-lipolytic pathway while avoiding the activation of the parallel flush-inducing pathway. In contrast to flushing agonists, exposure of cells expressing GPR109A to the non-flushing agonists failed to induce internalization of the receptor or to activate ERK 1/2 mitogen-activated protein kinase phosphorylation.

Likewise, the in vivo relevance of the long-debated concept of GPCR dimerization[11] is beautifully illustrated by the work of Javier Gonzalez-Maeso and colleagues,[12] showing that the metabotropic glutamate receptor 2 (mGluR2) interacts through specific transmembrane helix domains with the serotonin 5-HT2A receptor (2AR), a member of an unrelated GPCR family, to form a functional complex in brain cortex. The 2AR-mGluR2 complex triggers unique cellular responses when targeted by hallucinogenic drugs, and activation of mGluR2 abolishes hallucinogen-specific signaling and behavioral responses.

Another challenge in the context of GPCR assay development is that a significant number of these receptors have not yet been matched with biologically relevant ligands. To address these so-called "orphan receptors" in drug discovery, they must either first be "deorphanized" or dealt with in ignorance of the physiological ligands.[13,14] For deorphanization, the cellularly overexpressed orphan receptor is typically tested for a functional response to "orphan" biological mediators, natural peptides, lipids, and metabolic intermediates, but also complex biological mixtures such as tissue lysates.

If neither the physiological ligand nor a suitable reference agonist can be identified prior to a high throughput screening (HTS) campaign, the orphan receptor is typically analyzed by a generic functional readout that bears a high probability of responding to an agonistic ligand. In this context, promiscuous G proteins that couple the majority of GPCRs to Ca^{2+} signaling have been of major importance. More recently, the use of arrestin redistribution assays has been described as a format applicable to the majority of GPCRs, independent of their specific G protein coupling. However, the topic of ligand-biased signaling addressed previously in this chapter leaves some doubt whether such generic functional readouts will identify all test compounds that modulate the receptor of interest with regard to the disease-relevant signaling pathway(s).

"CLASSICAL" DRUG DISCOVERY TECHNIQUES TO ANALYZE GPCR-LIGAND INTERACTIONS AND GPCR SIGNALING

GPCR ligand binding

There are several biochemical assay formats to investigate direct ligand-GPCR interactions, for example scintillation proximity assays (SPA),[15] fluorescence polarization (FP) assays,[16] and fluorescence intensity distribution analysis (FIDA) assays.[17]

Currently, SPA is the most frequently employed HTS assay to identify inhibitors of ligand-GPCR binding. SPA employs microspheres that are doped with ß-emission sensitive scintillants. These so-called scintillation beads may be coated with bioreagents such as wheat germ agglutinin (WGA). WGA is a lectin that binds the carbohydrate groups of glycolipids and glycoproteins in standard cellular plasma membranes. Thus, membrane fragments from cells overexpressing a specific target GPCR may be immobilized on the surface of WGA-coated scintillation beads. If a radioligand binds to the target GPCR on the scintillation bead, the scintillant on the bead surface is induced to emit light. If radioligand binding is inhibited by a test compound, the radio-emission of the unbound ligand is absorbed by the surrounding aqueous solution and no light emission from the scintillation bead is induced. Unlike classical radioactive filtration assays or radio-immune assays, the unbound radioligand does not have to be removed prior to the scintillation measurement. An assay format like SPA that does not require a washing step is referred to as "homogeneous." In an HTS environment, this homogeneous format is time-saving and therefore attractive. Another time-saving aspect of state-of-the-art SPA is the application of low temperature charge-coupled devices (CCDs) that can measure scintillation from all wells of a microtiter plate (MTP) in parallel.[18–20] To improve the sensitivity of CCD-based detection, scintillation "imaging beads" were designed that emit light in the red region.[21] As many pharmaceutical test compounds from medicinal chemistry absorb light in the blue region of the spectrum, the red scintillation of these imaging beads is a further advantage to reduce compound-related assay artefacts.

As a nonradioactive alternative to the SPA format, some pharmaceutical companies have established an FP format for GPCR-directed drug discovery assays.[16,22] FP measurements are based on the fact that fluorophores absorb light along a particular direction with respect to their intramolecular axes. The rotation of the fluorophore during the excited-state lifetime drives the depolarization of the fluorescence emission. In a drug discovery assay application of this physical principle, the binding of a small fluorophore-labeled ligand to a large receptor may be investigated. If the small ligand binds to the large receptor, the rotational rate of the fluorophore is reduced, and the polarization of the fluorescence emission is high. If the binding of the ligand is inhibited by a test compound, the fluorophore maintains the high rotational rate of the free ligand-fluorophore conjugate: The fluorescence emission quickly becomes depolarized.

HTS groups worldwide have made attempts to reduce the sample volume in drug screening assays substantially to save bioreagents and to reduce the consumption of test compounds. All macroscopic scintillation counting or fluorescence methods such as the above SPA or FP face the problem of increasing background with decreasing assay volumes.[23] In contrast, confocal fluorescence drug discovery assays enable miniaturization without loss of data quality due to an only femtoliter-sized observation volume.[17,24,25] Further, confocal fluorescence studies provide a broad information spectrum on the identity, size, concentration, and environment of the fluorophore-labeled entity. This enables the development of assays with several simultaneous fluorescence readouts such as molecular brightness, diffusion time, anisotropy, lifetime, or cross-correlation. These multiplexed assay formats also support the exclusion of test compound-related assay artefacts.

Since all three techniques, SPA, FP, and FIDA, are homogeneous, they ease automated liquid handling requirements for HTS. However, a potential limitation of the SPA format in the context of an automated workflow is the need to dispense a relatively dense suspension of scintillation beads that have a propensity to settle under the influence of gravity. To some extent this challenge for SPA may be overcome by using appropriate bead stirring devices as reservoirs.[26]

All three techniques enable the use of long-wavelength-emitting scintillants or fluorophores that aim to avoid interference by the frequently yellow-to-red colored compounds from medicinal chemistry. Thus, colored test compounds may absorb fluorescence excitation and emission light for both FIDA and FP experiments, as well as scintillation light from SPA beads. Additionally, some test compounds display autofluorescence, which may again be avoided to some extent by the use of fluorophore biolabels with long wavelength emission. The application of SPA, FIDA, and FP to GPCR ligand binding assays has been reviewed in comparatively more detail earlier.[27]

Advantageously, all such biochemical binding assays enable equilibrium binding or kinetic displacement measurements so that both K_D-type and k_{on}/k_{off}-type values may be derived. A further benefit of the biochemical binding formats is that test compounds usually display fewer "off-target effects" than in the typically more complex functional formats.

The biochemical binding assays described earlier in the chapter measure the competition of the test compound with a labeled reference ligand. Disadvantageously, this leaves a risk of missing noncompetitive, allosteric ligands. Thus, all aspects of ligand-induced GPCR modulation such as full/partial agonism, inverse agonism versus neutral antagonism, negative or positive modulation must be analyzed in functional – often cellular – assay formats.[28]

G protein function and second messenger generation

When a GPCR becomes activated by an agonist, the associated heterotrimeric G protein catalyses the exchange of guanosine diphosphate (GDP) by guanosine triphosphate (GTP), enabling both Gα–GTP and βγ–dimers to interact with a variety of downstream effectors.[29] There are four major families of Gα proteins: G_s, $G_{i/o}$, G_q, and G_{12}.[30] One way to directly monitor the respective nucleotide exchange on the Gα-subunit is the use of [35S]GTPγS in an SPA-based format.[31] This assay principle theoretically provides a generic functional format to analyze GPCRs with different G protein couplings. Practically, however, the [35S]GTPγS exchange assay can be implemented much easier for GPCRs coupling to G proteins of the $G_{i/o}$ family. This is due to the high nucleotide exchange rate, the high GTPγS affinity, and the high expression rate of $G_{i/o}$ proteins, which partially suppresses the GTPγS exchange on G_s, G_q, and G_{12} proteins. The applicability of this format was expanded to the latter three G protein families by a variation of the [35S]GTPγS exchange assay where the activated G protein is antibody-captured to SPA beads.[32]

The preference of the GPCR for specific Gα isoforms[33] influences the downstream signaling: Proteins of the $Gα_{q/11}$ family stimulate phospholipase C (PLC), representatives of the $Gα_{i/o}$ and $Gα_s$ families mostly modulate adenylate cyclase (AC) activity. The more recently described $Gα_{12/13}$ family acts via stimulation of the small GTPase Rho, omitting a small molecule second messenger.

If the GPCR of interest signals via PLC, the most broadly applied cell-based technique to measure GPCR activation is the Ca^{2+} release assay, either measured in a fluorescent format using Ca^{2+}-sensitive fluorophores[34] or in a luminescent format using aequorin and a chemiluminescent substrate.[35] If the GPCR of interest signals via AC, the cytosolic cyclic adenosine monophosphate (cAMP) content may be determined using various detection technologies.[36]

GPCR downstream signaling

Bioluminescence-based reporter gene assays (RGAs) have successfully been employed to measure functional activity of GPCR ligands.[37] This type of RGAs is very sensitive due to the low signal background of the bioluminescent read-out and due to the signal amplification steps between GPCR activation and the cumulative reporter gene expression. A cAMP response element (CRE) in the promoter of the reporter gene enables the specific monitoring of $G_{s/i}$-dependent signaling: elevated cytoplasmic cAMP levels activate protein kinase A (PKA), which turns on the CRE binding protein and initiates CRE-dependent reporter expression. If a nuclear factor of activated T cells (NFAT) response element (RE) is contained in the promoter of the reporter gene, the respective RGA may monitor G_q-dependent signaling: In response to elevated cytoplasmic Ca^{2+} levels, NFAT is dephosphorylated by the Ca^{2+}-dependent phosphatase calcineurin and induces NFAT-RE-dependent reporter expression.[38] Since reporter gene assays typically require a few hours of incubation to accumulate the reporter gene product in the cells, potential side effects of a test compound with regard to, for example, toxicity or GPCR desensitization should be taken into account.

Another marker that has been applied in drug discovery to monitor the activation of G protein signaling since more recently is the phosphorylation status of ERK1/2.[39] Advantageously from a practical perspective, $G\alpha_q$, $G\alpha_s$, and $G\alpha_i$ signaling converge on the level of ERK phosphorylation. ERK phosphorylation may be quantified in the lysate of GPCR-stimulated cells by the so-called SureFire™ assay (TGR BioSciences Pty Ltd, Thebarton, Australia) or by High Content Screening (HCS) as described further in this chapter.

Arrestins are cytoplasmic proteins recruited to the plasma membrane by ligand-activated, GRK-phosphorylated GPCRs.[40] Arrestins then uncouple the GPCR from the cognate G protein and target the desensitized receptors to clathrin-coated pits for endocytosis. Apart from this indirect influence on G protein signaling by the physical interceding between GPCR and G protein, the arrestins have more recently been found to directly mediate G protein-independent signaling.[41,42] The latter mechanism regulates aspects of cell motility, chemotaxis, apoptosis, and likely other cellular functions through a rapidly expanding list of signaling pathways. Arrestin signaling may be monitored by HTS-applicable techniques like, for example, enzyme fragment complementation[43] or the Transfluor™ technique as described further in this chapter.

HIGH CONTENT SCREENING IN THE GPCR TARGET FAMILY APPROACH: ONE TECHNOLOGY PLATFORM TO ANALYZE ALL ASPECTS OF GPCR LIGAND FUNCTION

High content screening (HCS)

In recent years, a technique generally referred to as high content screening (HCS) has been introduced both to the early pharmaceutical drug discovery process[44–46] and to academic research.[47,48] HCS combines high-resolution fluorescence microscopy with automated image analysis. In the area of drug discovery, HCS provides several advantages over "classical" HTS as described earlier in the chapter. Cellular HTS conventionally monitors the mean response of the whole cell population of a microtiter plate (MTP)-well. In contrast, HCS can distinguish the individual response of many cells in an MTP-well, which may differ with respect to the differentiation, the stage of the cell cycle, the state of transfection, or due to natural variability.

As a result, heterogeneous pharmaceutical drug effects on mixed cell populations may be analyzed in a single MTP-well. "On-target" drug effects may be cross-correlated with other phenomena, such as cellular toxicity. Compound artefacts such as cell lysis or compound autofluorescence may be detected. HCS permits work with endogenous targets and/or primary cells[49] using specific antibodies or morphological image analysis.[50] In this way, HCS enables novel assay formats that do not depend on an overall change of fluorescence or luminescence intensity from the whole MTP well. In addition to these general benefits in the area of drug discovery, HCS is well suited to cover all aspects of GPCR ligand binding and function.[51]

HCS to measure GPCR ligand binding, internalization, and arrestin signaling

Ligand binding and ligand-induced internalization of a GPCR[52] may be monitored using a specific, fluorophore ("fluor")-labeled ligand.[53] Similar to the classic radioligand binding assay, the fluor-labeled ligand protocol investigates test compounds with regard to reference ligand displacement. However, beyond the possibilities of the radioligand displacement, the HCS protocol enables the identification of test compounds that allow for the binding of the fluor-labeled ligand but prevent its internalization. Image analysis allows us to distinguish inactive test compounds from displacing or from the nondisplacing antagonists. In case of inactive test compounds, an image analysis algorithm can be established to search for ligand fluorescence in perinuclear, granular spots (Figure 12.1A). For ligand-displacing antagonists, there should be no ligand fluorescence associated with the cells (Figure 12.1B). For nondisplacing antagonists of internalization, the ligand fluorescence is typically associated with the plasma membrane (Figure 12.1C). In the latter case, an image analysis algorithm can be established to search for cell peripheral ligand fluorescence.

Using a fluor-labeled antibody against the GPCR of interest,[54] test compounds can additionally be qualified as agonists or antagonists of internalization (Figure 12.1D-E). If combined in multiplexed or parallel experiments, the use of fluor-labeled ligand and antibody may serve to analyze test compound-induced/antagonized intracellular receptor internalization, recycling, or degradation. Depending on the specific disease biology and target, these data may help predict, for example, test compound-induced receptor desensitization and tachyphylaxis issues.

In parallel to receptor internalization, it has been demonstrated for numerous GPCRs[55] that the so-called arrestins undergo an intracellular redistribution process as described earlier in the chapter. In the Transfluor™ technology,[56] the redistribution of an arrestin-fluorescent protein (ArrFP) conjugate is monitored by HCS (Figure 12.1F-G). In nonstimulated cells, ArrFP is uniformly distributed

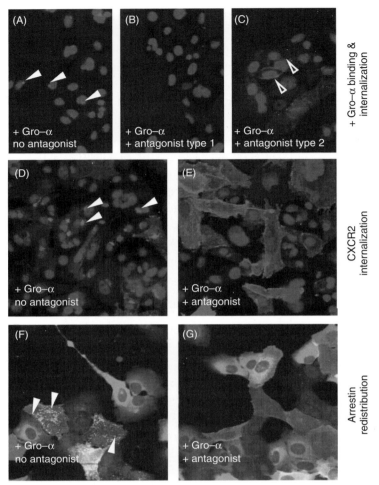

Figure 12-1: HCS to monitor GPCR ligand binding, internalization and arrestin redistribution. The nuclei of the formaldehyde-fixed cells were always stained with Hoechst 33342™ (shown in blue). (A/B/C) Fluorophore (Fluor)-labeled ligand binding and internalization. 10 nM fluor-labeled Gro-α (shown in red) was added to the supernatant of CXCR2/Gα₁₆ stably transfected Chinese Hamster Ovary (CHO) cells. (A) In the absence of an antagonist, the fluor-labeled Gro-α was accumulated in endosomal compartments; indicated by filled arrowheads. (B) A ligand-displacing antagonist prevented binding of fluor-labeled Gro–α to CXCR2. (C) An allosteric, not ligand-displacing antagonist allowed fluor-labeled Gro-α to bind to CXCR2 at the plasma membrane (indicated by open arrowheads) but inhibited Gro-α/CXCR2 internalization. (D/E) Fluor-labeled anti-CXCR2 antibody to detect CXCR2 internalization. 15 nM Gro-α was added to the supernatant, CXCR2 was immune-stained in the permeabilized cells with a fluorophore-labeled anti-CXCR2 antibody (shown in red). (D) In the absence of an antagonist, CXCR2 was accumulated in endosomal compartments; indicated by filled arrowheads. (E) An antagonist – independent if ligand-displacing or not – prevented Gro-α-stimulated CXCR2 internalization. Due to the permeabilization of the cells, the anti-CXCR2 antibody also detected steady-state intracellular CXCR2. (F/G) Arrestin redistribution. A human osteosarcoma cell line (U2-OS) stably transfected with arrestin-green fluorescent protein (ArrGFP; cell line is a kind gift of Molecular Devices/now part of MDS Analytical Technologies, Concord, ON, Canada) was transiently transfected with CXCR2. 15 nM Gro–α was added to the supernatant, ArrGFP is shown in green, CXCR2 was immune-stained (shown in red). (F) In the absence of an antagonist, the cells expressing CXCR2 showed granular spots in close proximity to the nuclei (green channel); indicated by filled arrowheads. (G) In the presence of an antagonist, the green fluorescence was more uniformly distributed across the cytosol. *See color plate 13.*

across the cytoplasm. Upon receptor stimulation, ArrFP is recruited to the activated GPCR, accompanies the GPCR to clathrin-coated pits and, for some GPCRs, even to the endocytic vesicle level.[57] In HCS, the corresponding change from uniform cytoplasmic fluorescence to granular fluorescence may be quantified. Depending on the FP in conjugation with the arrestin, this assay format can be adapted to be more or less sensitive to a constitutive activity of the employed GPCR.[58]

HCS to measure second messenger generation

To detect the rise of cytoplasmic Ca^2+ in response to the stimulation of a G_q-coupled receptor, NFAT may be employed in an HCS assay.[40] In response to elevated cytoplasmic calcium levels, NFAT is dephosphorylated by calcineurin and migrates from the cytoplasm to the nucleus.[38] If NFAT is expressed as a fusion protein with a fluorescent protein, this nuclear translocation may be monitored and quantified as a measure of GPCR activation (Figure 12.2A-B). Of course, all data generated in receptor-overexpressing systems and/or systems employing nonphysiological signaling chain components (e.g., a promiscuous $G\alpha_{16}$) should be handled with care and, if available, compared to other signaling assays.

Protein kinase A (PKA) may be employed to monitor intracellular cAMP levels. PKA is a heterotetrameric complex of two regulatory and two catalytic subunits. At low cAMP concentrations, the PKA holoenzyme is associated with various intracellular membranes through the so-called A kinase anchor proteins (AKAPs).[59] With rising cAMP levels, the PKA holoenzyme is dissociated and releases the catalytic subunit (PKAcat) into the cytoplasm (Figure 12.2C-D). In an HCS application of this cAMP-dependent process, the release of PKAcat is monitored by employing a fusion protein of PKAcat and GFP.[51]

HCS to measure various GPCR downstream signaling events

The monitoring of the ERK pathway is a broadly applicable assay format, because numerous activated GPCRs – independent of their G protein coupling – induce ERK signaling,[39] In an HCS application of this principle, a phosphoERK-specific antibody is employed (Figure 12.2E-F) in an immune stain procedure.[54] Advantageously, ERK phosphorylation is detectable by HCS already a few minutes after GPCR stimulation, so that the relatively long test compound incubation times as required, for example, in reporter gene assays may be avoided.

$G_{12/13}$–coupled GPCRs signal by activation of the small monomeric GTPase Rho. Using this signaling process in HCS, the cytosol-to-plasma membrane translocation of a GFP-tagged RhoGEF is employed to monitor and quantify GPCR/$G_{12/13}$ activation.[60] Alternative experimental techniques to quantify $G_{12/13}$-signaling are less pathway-specific (e.g., serum response element [SRE]-dependent reporter gene assays) or of lower throughput (e.g., specific affinity precipitation of GTP-bound Rho and subsequent immunoblot).

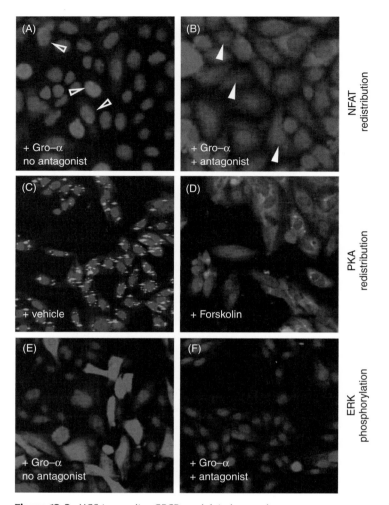

Figure 12-2: HCS to monitor GPCR-modulated second messenger responses and ERK signaling. The nuclei of the formaldehyde-fixed cells were always stained with Hoechst 33342™ (shown in blue). (A/B) NFAT nuclear translocation. CXCR2/Gα_{16} stably transfected CHO cells were transiently transfected with an NFAT-red fluorescent protein (RFP) fusion construct. 15 nM Gro-α was added, NFAT-RFP is shown in red. (A) In the absence of an antagonist, the cells expressing NFAT-RFP displayed NFAT to be in the nuclei; indicated by open arrowheads. (B) In the presence of an antagonist, the red NFAT fluorescence was localized to the cytosol; indicated by filled arrowheads. (C/D) PKA-GFP redistribution. PKAcat-EGFP transfected CHO cells were purchased from BioImage/ ThermoFisher Scientific. (C) In vehicle-stimulated cells, a PKA-GFP fusion protein (shown in green) was associated with various intracellular membranes. (D) Upon forskolin stimulation, the cytosolic cAMP level increased and PKA-GFP was redistributed homogeneously across the cytoplasm. (E/F) ERK phosphorylation. CXCR2/Gα_{16} stably transfected CHO cells were stimulated with 15 nM Gro-α, phospho-ERK was immune-stained (shown in red). (E) In the absence of an antagonist, there was a high level of ERK phosphorylation. (F) In the presence of an antagonist, ERK phosphorylation was reduced to background levels. *See color plate 14.*

HCS in comparison to label-free cell-based techniques

The GPCR-modulated reorganization of the cytoskeleton or morphological changes of the cells may be monitored in the HCS format. In this context, HCS competes with the so-called label-free techniques,[61] for example, resonant wave-guide grating and electrical biosensors, which have also been applied to measure cellular signaling in the recent past. In contrast to the label-free formats, however, specific cytoskeletal changes may be observed using an appropriate fluorophore-labeled bioreagent (e.g., phalloidin to stain the actin filaments in a cell). Another advantage of the HCS approach is, for example, that a GPCR target-specific signaling may be cross-correlated with the reorganization of the cytoskeleton or the morphological change on the single-cell level.

DISCUSSION

The GPCR assay formats enumerated earlier in this chapter serve to measure diverse aspects of GPCR ligand pharmacology like ligand-receptor binding, receptor activation/ desensitization, and receptor signaling. The individual assay formats differ with respect to information content, ease of assay development, cost, and throughput. Several assay formats deliver partially complementary information: A binding assay typically produces a data set that is focussed on the ligand-receptor interaction itself, the cellular assay provides a view on the cellular context of the ligand-induced receptor activation, and various functional assays provide different facets of the pharmacological effect spectrum. Many GPCR signaling pathways are identical in different cell types and redundantly important across diverse target and disease contexts, making the establishment of according assay formats worthwhile for several projects.

The particular power of HCS assays lies in the potential for monitoring diverse aspects of GPCR function (e.g., ligand binding, G protein signaling, and receptor internalization) based upon a single technology platform. HCS allows us to generate additional information by observing additional readouts in the same or parallel assays, such as rounding or loss of cells in case of a toxic compound effect, enhancement of plasma membrane GPCR levels by an inverse agonist, or the enhancement of pERK levels by a compound with a stress-inducing side effect.

However, HCS also bears certain limitations. A technical challenge in transferring an HCS signaling assay from one GPCR to another is that not all the bioreagents are generic. Thus, it may, for example, be quite cumbersome to screen for an HCS-suitable antibody directed against a GPCR of interest. Further, fluorophore-labeled ligands are not always commercially available. Another challenge is that particular signaling events may display differences in different model cell lines, based on, for example, diverse protein levels of the GPCR or the signaling chain proteins.

Due to the relatively fast signaling kinetics of some readouts and the relatively long measurement times for imaging, most medium-throughput HCS assays are carried out as end-point assays using formaldehyde fixation. In addition, end-point assays simplify the assay protocols since they decouple the liquid handling steps from the imaging process. As different signaling readouts have different kinetics, only particular signaling events can be multiplexed with one another (e.g., ERK phosphorylation and PKA redistribution at approximately 5 min after stimulation; or e.g., ligand/receptor internalization, arrestin and NFAT redistribution for many receptors at 15–60 min after stimulation).

High-content assay formats are feasible for high throughput, but are generally used as secondary assays due to the relatively high costs and due to the often more complex assay protocols. With regard to the assay reagents, the antibodies and the imaging-suitable MTPs are typically rather expensive. In particular, when an immune staining procedure is required, the assay protocols are labor-intensive and time-consuming. Currently, HCS is not much used for primary screening but rather employed to add new data after a primary uHTS campaign in a classic HTS format. The establishment of the HCS technology requires relatively high capital investment and depends on the generation of specialized technical skills, making a good case to launch HCS activities preferentially in a central servicing unit.

In summary, all discussed biochemical and cellular assays on GPCR ligands have their individual merits, with each approach providing advantages over the other under particular circumstances. The diversity of assay formats to characterize GPCR ligands delivers partially overlapping information, thereby cross-validating experimental answers and reducing the probability of technology-based assay artefacts. In our opinion, an appropriate assay panel for an extended pharmacological characterization of GPCR ligands should include data on binding, G protein signaling via various pathways and GPCR internalization.

ACKNOWLEDGMENTS

We are grateful to Ulrike Küfner-Mühl, Martin Valler, Barbara Kistler, Peter Gierschik, and Simone Kredel for helpful advice and discussions. We thank Susanne Schmidt and Michael Sulger for technical support using the HCS technology platform.

SUGGESTED READING

1. Jacoby, E., Bouhelal, R., Gerspacher, M., and Seuwen, K. (2006). The 7 TM G-protein-coupled receptor target family. ChemMedChem. *1*, 761–782.
2. Gierschik, P., Sidiropoulos, D., Steisslinger, M., and Jakobs, K.H. (1989). Na+ regulation of formyl peptide receptor-mediated signal transduction in HL 60 cells. Evidence that the cation prevents activation of the G-protein by unoccupied receptors. Eur. J. Pharmacol. *172*, 481–492.
3. Costa, T. and Herz, A. (1989). Antagonists with negative intrinsic activity at delta opioid receptors coupled to GTP-binding proteins. Proc. Natl. Acad. Sci. U. S. A *86*, 7321–7325.

4. Greasley, P.J. and Clapham, J.C. (2006). Inverse agonism or neutral antagonism at G-protein coupled receptors: a medicinal chemistry challenge worth pursuing? Eur. J. Pharmacol. *553*, 1–9.

5. Fox, G.B., Pan, J.B., Esbenshade, T.A., Bitner, R.S., Nikkel, A.L., Miller, T., Kang, C.H., Bennani, Y.L., Black, L.A., Faghih, R., Hancock, A.A., and Decker, M.W. (2002). Differential in vivo effects of H3 receptor ligands in a new mouse dipsogenia model. Pharmacol. Biochem. Behav. *72*, 741–750.

6. Harvey, J.A., Welsh, S.E., Hood, H., and Romano, A.G. (1999). Effect of 5-HT2 receptor antagonists on a cranial nerve reflex in the rabbit: evidence for inverse agonism. Psychopharmacology (Berl) *141*, 162–168.

7. Engelhardt, S., Grimmer, Y., Fan, G.H., and Lohse, M.J. (2001). Constitutive activity of the human beta(1)-adrenergic receptor in beta(1)-receptor transgenic mice. Mol. Pharmacol. *60*, 712–717.

8. Drake, M.T., Violin, J.D., Whalen, E.J., Wisler, J.W., Shenoy, S.K., and Lefkowitz, R.J. (2008). {beta}-Arrestin-biased Agonism at the {beta}2-Adrenergic Receptor. J. Biol. Chem. *283*, 5669–5676.

9. Sorg, G., Schubert, H.D., Buttner, F.H., Valler, M.J., and Heilker, R. (2002b). Comparison of photomultiplier tube- and charge coupled device-based scintillation counting. Life Science News *11*, 1–3.

10. Ramm, P. (1999). Imaging systems in assay screening. Drug Discov. Today *4*, 401–410.

11. Milligan, G. (2003). Principles: extending the utility of [35S]GTP gamma S binding assays. Trends Pharmacol. Sci. *24*, 87–90.

12. Gonzalez-Maeso, J., Ang, R.L., Yuen, T., Chan, P., Weisstaub, N.V., Lopez-Gimenez, J.F., Zhou, M., Okawa, Y., Callado, L.F., Milligan, G., Gingrich, J.A., Filizola, M., Meana, J.J., and Sealfon, S.C. (2008). Identification of a serotonin/glutamate receptor complex implicated in psychosis. Nature *452*, 93–97.

13. Osmond, R.I., Sheehan, A., Borowicz, R., Barnett, E., Harvey, G., Turner, C., Brown, A., Crouch, M.F., and Dyer, A.R. (2005). GPCR screening via ERK 1/2: a novel platform for screening G protein-coupled receptors. J. Biomol. Screen. *10*, 730–737.

14. Sullivan, E., Tucker, E.M., and Dale, I.L. (1999). Measurement of [Ca2+] using the Fluorometric Imaging Plate Reader (FLIPR). Methods Mol. Biol. *114*, 125–133.

15. Alouani, S. (2000). Scintillation proximity binding assay. Methods Mol. Biol. *138*, 135–141.

16. Banks, P. and Harvey, M. (2002). Considerations for using fluorescence polarization in the screening of g protein-coupled receptors. J. Biomol. Screen. *7*, 111–117.

17. Wolff, M., Kauschke, S.G., Schmidt, S., and Heilker, R. (2008). Activation and Translocation of Glucokinase in Rat Primary Hepatocytes Monitored by High Content Image Analysis. J. Biomol. Screen. *13*, 837–46.

18. Parmentier, M. and Detheux, M. (2006). Deorphanization of G-protein-coupled receptors. Ernst Schering Found. Symp. Proc. *2*, 163–186.

19. Sorg, G., Schubert, H.D., Buttner, F.H., and Heilker, R. (2002a). Automated high throughput screening for serine kinase inhibitors using a LEADseeker scintillation proximity assay in the 1536-well format. J. Biomol. Screen. *7*, 11–19.

20. Sorg, G. and Heilker, R. (2002). Automated high-throughput screening. European Pharmaceutical Review *3*, 25–35.

21. Jessop, R.A. (1998). Imaging proximity assays. Proc. SPIE *3259*, 228–233.

22. Harris, A., Cox, S., Burns, D., and Norey, C. (2003). Miniaturization of fluorescence polarization receptor-binding assays using CyDye-labeled ligands. J. Biomol. Screen. *8*, 410–420.

23. Haupts, U., Rudiger, M., and Pope, A.J. (2000). Macroscopic versus microscopic fluorescence techniques in (ultra)-high-throughput screening. Drug Discov. Today HTS Suppl *1*, 3–9.

24. Auer, M., Moore, K.J., Meyer-Almes, F.J., Guenther, R., Pope, A.J., and Stoeckli, K. (1998). Fluorescence correlation spectroscopy: lead discovery by miniaturized HTS. Drug Discov. Today *3*, 457–465.

25. Zemanova, L., Schenk, A., Valler, M.J., Nienhaus, G.U., and Heilker, R. (2003). Confocal optics microscopy for biochemical and cellular high-throughput screening. Drug Discov. Today *8*, 1085–1093.

26. Siehler, S. (2008). Cell-based assays in GPCR drug discovery. Biotechnol. J. *3*, 471–483.

27. Heilker, R., Zemanova, L., Valler, M.J., and Nienhaus, G.U. (2005). Confocal fluorescence microscopy for high-throughput screening of G-protein coupled receptors. Current Medicinal Chemistry *12*, 2551–2559.

28. Shenoy, S.K., Drake, M.T., Nelson, C.D., Houtz, D.A., Xiao, K., Madabushi, S., Reiter, E., Premont, R.T., Lichtarge, O., and Lefkowitz, R.J. (2006). beta-arrestin-dependent, G protein-independent ERK1/2 activation by the beta2 adrenergic receptor. J. Biol. Chem. *281*, 1261–1273.

29. Cabrera-Vera, T.M., Vanhauwe, J., Thomas, T.O., Medkova, M., Preininger, A., Mazzoni, M.R., and Hamm, H.E. (2003). Insights into G protein structure, function, and regulation. Endocr. Rev. *24*, 765–781.

30. Oakley, R.H., Laporte, S.A., Holt, J.A., Barak, L.S., and Caron, M.G. (2001). Molecular determinants underlying the formation of stable intracellular G protein-coupled receptor-beta-arrestin complexes after receptor endocytosis*. J. Biol. Chem. *276*, 19452–19460.

31. Meyer, B.H., Freuler, F., Guerini, D., and Siehler, S. (2008). Reversible translocation of p115-RhoGEF by G(12/13)-coupled receptors. J. Cell Biochem. *104*, 1660–1670.

32. Delapp, N.W. (2004). The antibody-capture [(35)S]GTPgammaS scintillation proximity assay: a powerful emerging technique for analysis of GPCR pharmacology. Trends Pharmacol. Sci. *25*, 400–401.

33. Kostenis, E., Waelbroeck, M., and Milligan, G. (2005). Techniques: Promiscuous Galpha proteins in basic research and drug discovery. Trends Pharmacol. Sci. *26*, 595–602.

34. Strange, P.G. (2008). Signaling mechanisms of GPCR ligands. Curr. Opin. Drug Discov. Devel. *11*, 196–202.

35. Dupriez, V.J., Maes, K., Le Poul, E., Burgeon, E., and Detheux, M. (2002). Aequorin-based functional assays for G-protein-coupled receptors, ion channels, and tyrosine kinase receptors. Receptors. Channels *8*, 319–330.

36. Gabriel, D., Vernier, M., Pfeifer, M.J., Dasen, B., Tenaillon, L., and Bouhelal, R. (2003). High throughput screening technologies for direct cyclic AMP measurement. Assay Drug Dev. Technol. *1*, 291–303.

37. Hill, S.J., Baker, J.G., and Rees, S. (2001). Reporter-gene systems for the study of G-protein-coupled receptors. Curr. Opin. Pharmacol. *1*, 526–532.

38. Berridge, M.J. (1993). Inositol trisphosphate and calcium signalling. Nature *361*, 315–325.

39. Offermanns, S. (2003). G-proteins as transducers in transmembrane signalling. Prog. Biophys. Mol. Biol. *83*, 101–130.

40. Wolff, M., Kredel, S., Haasen, D., Wiedenmann, J., Nienhaus, G.U., Kistler, B., Oswald, F., and Heilker, R. (2010). High content screening of CXCR2-dependent signalling pathways. Comb. Chem. High Throughput Screen. *13*, 3–15.

41. Lefkowitz, R.J. (1998). G protein-coupled receptors. III. New roles for receptor kinases and beta-arrestins in receptor signaling and desensitization. J. Biol. Chem. *273*, 18677–18680.

42. Richman, J.G., Kanemitsu-Parks, M., Gaidarov, I., Cameron, J.S., Griffin, P., Zheng, H., Guerra, N.C., Cham, L., iejewski-Lenoir, D., Behan, D.P., Boatman, D., Chen, R., Skinner, P., Ornelas, P., Waters, M.G., Wright, S.D., Semple, G., and Connolly, D.T. (2007). Nicotinic acid receptor agonists differentially activate downstream effectors. J. Biol. Chem. *282*, 18028–18036.

43. Eglen, R.M. (2005). Functional G protein-coupled receptor assays for primary and secondary screening. Comb. Chem. High Throughput Screen. *8*, 311–318.

44. Bullen, A. (2008). Microscopic imaging techniques for drug discovery. Nat. Rev. Drug Discov. *7*, 54–67.

45. Hoffman, A.F. and Garippa, R.J. (2007). A pharmaceutical company user's perspective on the potential of high content screening in drug discovery. Methods Mol. Biol. *356*, 19–31.

46. Giuliano, K.A., Johnston, P.A., Gough, A., and Taylor, D.L. (2006). Systems cell biology based on high-content screening. Methods Enzymol. *414*, 601–619.

47. Carpenter, A.E. (2007). Image-based chemical screening. Nat. Chem. Biol. *3*, 461–465.

48. Krausz, E. (2007). High-content siRNA screening. Mol. Biosyst. *3*, 232–240.

49. Wolff, M., Haasen, D., Merk, S., Kroner, M., Maier, U., Bordel, S., Wiedenmann, J., Nienhaus, G.U., Valler, M.J., and Heilker, R. (2006). Automated High Content Screening for phosphoinositide 3 kinase inhibition using an AKT1 redistribution assay. Comb. Chem. High Throughput Screen. *9*, 339–350.

50. Wigglesworth, M.J., Wolfe, L.A., and Wise, A. (2006). Orphan seven transmembrane receptor screening. Ernst Schering Found. Symp. Proc. *105*–143.

51. Heilker, R., Wolff, M., Tautermann, C.S., and Bieler, M. (2009). G-protein-coupled receptor-focused drug discovery using a target class platform approach. Drug Discov. Today *14*, 231–40.

52. Heilker, R. (2006). High Content Screening to monitor G-protein coupled receptor internalisation. Ernst Schering Foundation Symposium Proceedings *2*, 229–248.

53. Haasen, D., Schnapp, A., Valler, M.J., and Heilker, R. (2006). G protein-coupled receptor internalization assays in the high-content screening format. Methods Enzymol. *414*, 121–139.

54. Haasen, D., Merk, S., Seither, P., Martyres, D., Hobbie, S., and Heilker, R. (2008). Pharmacological profiling of chemokine receptor-directed compounds using high-content screening. J. Biomol. Screen. *13*, 40–53.

55. Ferguson, S.S. (2001). Evolving concepts in G protein-coupled receptor endocytosis: the role in receptor desensitization and signaling. Pharmacol. Rev. *53*, 1–24.

56. Milligan, G. (2004). G protein-coupled receptor dimerization: function and ligand pharmacology. Mol. Pharmacol. 2004. Jul.;*66*. (1):1–7. *66*, 1–7.

57. Oakley, R.H., Hudson, C.C., Cruickshank, R.D., Meyers, D.M., Payne, R.E., Rhem, S.M., and Loomis, C.R. (2002). The cellular distribution of fluorescently labeled arrestins provides a robust, sensitive, and universal assay for screening of G protein-coupled receptors. Assay Drug Dev. Technol. *1*, 21–30.

58. Kredel, S., Wolff, M., Wiedenmann, J., Moepps, B., Nienhaus, G.U., Gierschik, P., Kistler, B., and Heilker, R. (2009). CXCR2 inverse agonism detected by arrestin redistribution. J. Biomol. Screen. *14*, 1076–91.

59. Feliciello, A., Gottesman, M.E., and Avvedimento, E.V. (2001). The biological functions of A-kinase anchor proteins. J. Mol. Biol. *308*, 99–114.

60. Lefkowitz, R.J. and Shenoy, S.K. (2005). Transduction of receptor signals by beta-arrestins. Science *308*, 512–517.

61. Fang, Y., Frutos, A.G., and Verklereen, R. (2008). Label-free cell-based assays for GPCR screening. Comb. Chem. High Throughput Screen. *11*, 357–369.

62. Zemanova, L., Schenk, A., Valler, M.J., Nienhaus, G.U., and Heilker, R. (2005). High-throughput screening of interactions between G protein-coupled receptors and ligands using confocal optics microscopy. Methods Mol. Biol. *305*, 365–384.

13 Assessing allosteric ligand-receptor interactions

Ivan Toma Vranesic and Daniel Hoyer

INTRODUCTION

There are about 1,000 GPCRs of which a few hundreds have known endogenous ligands, with the others representing what is generically known as orphan receptors. GPCRs represent the target of about 40% of the known/registered medications/drugs.[1] Although "biologics" are gaining in importance, this high proportion may not fall so rapidly, as new concepts and knowledge emerge, such as the link of recently deorphanized receptors with disease (e.g., sphingolipids, NPS, Orexin, MCH, proton sensing receptors, to name a few) or the development of allosteric modulators that fill a vacuum left by failed attempts at orthosteric modulation of a number of GPCRs. Except for two recent additions (Cinacalcet, calcium sensing receptor agonist and Miravaroc, a CCR5 receptor antagonist), but this number will increase steadily as many allosteric modulators are in development, all registered GPCR-interacting drugs belong to the orthosteric type: These drugs act either as agonists, inverse agonists, or neutral antagonists at the recognition site of the endogenous hormone/peptide/neurotransmitter. Some of these compounds are partial agonists or partial inverse agonists, although these definitions become more fuzzy, because the degree of intrinsic activity (whether positive or negative) is largely system-dependent, as is the nature of any given compound (since stimulus trafficking allows to cross the borders between agonism and inverse agonism). Over the last two decades, a new concept has emerged for ligands interacting with GPCRs, that is, the notion of allosteric modulation.[2–6] This concept, which was initially limited to compounds interacting with oligomeric proteins,[7] such as hemoglobin or ligand-gated ion channels (e.g., GABA$_A$ receptors), has become applicable to an increasing number of drugs targeting GPCRs,[8–10] and not only those capable of forming dimers or more elaborate multimeric complexes (e.g., cholinergic muscarinic receptors, metabotropic glutamate receptors, GABA$_B$ receptors, calcium sensing receptors, and others). Such allosteric modulators may come in many different forms, as positive or negative enhancers, dual agonists or antagonists, and more neutral binders. This short paper will review some of these aspects as well as the concepts and methods used to recognize/characterize/assess allosteric modulators.

THE NATURE OF RECEPTOR ALLOSTERISM

The two state model as represented further in this chapter[11–15] is an extension of the model proposed by Del Castillo and Katz to explain agonist-induced activation (opening) of ligand-gated channels,[16] where the agonist A binds to the inactive state of the receptor/channel R, and the inactive complex AR then rapidly converts into the active complex AR*. The peculiarity of the two state model is that it postulates R* to exist in the absence of the agonist – in other words, it allows the expression of constitutive activity of the receptor in the absence of any agonist (endogenous or exogenous), as was first demonstrated by Costa and Herz with delta opiate receptors naturally expressed by neuroblastoma x glioma cells (NG 108–15).[17] These cells show basal or constitutive GTPase activity that could be modulated by agonists to produce more activity, but also by compounds such as ICI 174864 to reduce the basal or constitute activity seen in the absence of agonist. Therefore, ICI 174864 was described as an inverse agonist, since it inhibited constitutive activity of delta opiate receptors, as opposed to compounds known now as neutral/competitive antagonists, such as MR 2266. MR2266 was devoid of effect on its own, but was perfectly capable of inhibiting the effects of both positive and inverse delta opiate agonists, such as DADLE and ICI 174864, respectively. In the meantime, the concept of constitutive activity[18] and inverse agonism has been validated in many receptor systems, especially since the cloning era, which has allowed to recombinantly express receptors in cells with very high densities.[1,19]

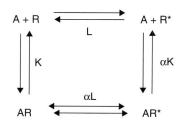

In the two-state model, an agonist will shift the equilibrium towards AR*, that is, the active state, whereas an antagonist will be neutral or silent; in contrast, the inverse agonist will shift the equilibrium toward AR, that is, the inactive state of the receptor. K is the expression of the binding affinity of the receptor for ligand A, whereas α expresses to some extent the ligand's ability to activate the receptor if it is an agonist (and the opposite for an inverse agonist) – in other words, it is some approximation of efficacy.

$\alpha > 1$: agonist
$\alpha = 1$: competitive antagonist (neutral / silent antagonist)
$\alpha < 1$: inverse agonist

The constant L is system-dependent (cell/tissue) and will define the level of constitutively active receptors R*. If L is very small, no constitutive activity will be observed, and the inverse agonist will have no effect in the absence of an agonist. Thus, in the presence of an agonist, an inverse agonist will act like a classical or competitive antagonist when no constitutive activity is observed. Some receptors are capable of showing rather elevated constitutive activity (e.g., opiate receptors, 5-HT$_4$, 5-HT$_7$, H$_3$ receptors, to name a few), whereas others show very low levels; this can change, however, depending on mutations that can render receptors extremely active as happens in some diseases,[20] or as a result of editing, as can be seen with the 5-HT$_{2C}$ receptor.[8,21] However, the two-state model *stricto sensu* does not explain allosteric modulation of GPCRs.

On the other hand, the "Ternary complex model"[22–24] (more further in the chapter), which was introduced to explain interactions of G proteins with their receptors (GPCRs) in the early 1980s, only accounts for allosteric effects on binding affinities, but not for effects on receptor activation.

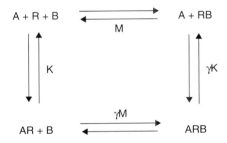

In this model, two ligands A and B bind to different sites on the receptor R to create the ternary complex ARB. As in the two-state model, the binding affinity of A for the nonliganded receptor R is expressed by the equilibrium constant K. However, the affinity of R for A can be modified (allosteric constant γ) by the previous binding of B. Since the thermodynamic stability of the ternary complex ARB is independent of the order of ligand binding (first A, then B, or vice versa), the binding affinity M of B is modified to the same extent γ by the precedent binding of A. This phenomenon is known as reciprocal modulation. The allosteric constant γ thus always is a characteristic of the pair of ligands present.

To allow for both the effects of ligand binding on receptor activation and those of allosteric effectors on ligand affinity and efficacy, an "Allosteric two state model" has been proposed,[25] which can be seen as an adaptation of the cubic ternary complex proposed by Weiss and collaborators,[26–29] albeit with some differences.[25] The Allosteric two state model is essentially a combination of the "two state" and "ternary complex" models[22,23] and could be expanded to the three state model [12,13,30] to accommodate biased signaling that has also been observed with some allosteric ligands. For the purpose of our considerations, we will regard A as the *orthosteric* ligand and B as the *allosteric* ligand or modulator, although these assignments are arbitrary and interchangeable. In the extended model, the front and left-hand faces of the cube shown below, represent the ternary complex and the two state models, respectively. In the latter, the modulator B needs to be introduced for formalistic reasons, but remains unbound to the receptor in all states. To allow for all possible ligand binding and isomerization steps to take place in any order, the borders of the different rectangles need to be completed to result in a cube. In this cube, all inactive receptor states (R) are on the front side, whereas the active states (R*) are on the rear side.

K: association constant of the agonist A
L: receptor isomerization constant
M: association constant of the modulator B
α: intrinsic efficacy of A
β: intrinsic efficacy of B
γ: binding cooperativity between A and B
δ: activation cooperativity between A and B

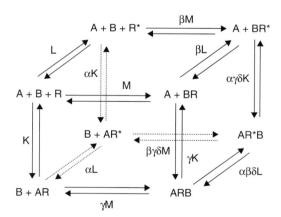

This model has several interesting features: first, it accounts for the ability of a modulator B, not only to modulate the affinity (like in the ternary complex model, "γ"), but also the efficacy ("δ") of an agonist A (shown in bold):

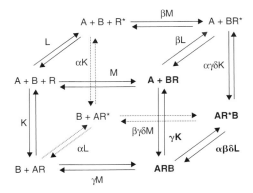

Second, it becomes clear that a modulator may also be able to influence the stabilization of the active receptor form R* and thereby have an intrinsic efficacy of its own (term β ≠ 1, shown in bold below), both as an *allosteric agonist* or an *inverse allosteric agonist*:

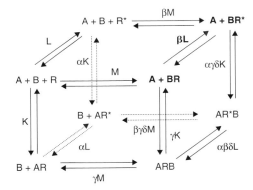

Further, the model explains the phenomenon of co-agonism: the two ligands A and B, which in their own right have no intrinsic efficacy (α = β = 1, shown in italics), can act in a positively cooperative manner to produce activation (δ > 1, shown in bold). Therefore, the binding of both A and B is necessary for receptor activation. An example of such a mechanism is the combined activation of the NMDA receptor by glutamate and glycine.

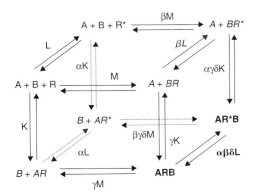

A full mathematical formalization of this model is provided by Hall.[25] We have used the equations reported therein to simulate the data that will be illustrated in the graphs throughout the text.

It should be emphasized, however, that this model relies on the assumption that the receptor is the limiting factor in determining the magnitude of the functional response, and not the effector system. In other words, it is only strictly valid in a system without receptor reserve. Thus, just as an orthosteric ligand can be pharmacologically defined by two variables, its binding affinity K and its intrinsic efficacy α (which will define agonism for $\alpha > 1$, inverse agonism for $\alpha < 1$, or neutral antagonism for $\alpha = 1$), an allosteric modulator is described by its affinity M and the magnitude of its allosteric effect $\gamma\delta$, in addition to its possible intrinsic efficacy ß. Also, although the model differentiates between two receptor states, the number of such states is not limited; strictly speaking, it is a multistate model, because allosteric modulators can modify the affinity of the orthosteric ligands for active or inactive states, similarly to the cubic ternary complex described by Weiss and colleagues.[26]

PROPERTIES OF ALLOSTERIC MODULATORS

The work of Birdsall, Lazareno and collaborators has introduced the concept of allosteric modulators in the GPCR field, initially restricted to cholinergic muscarinic receptors.[2–6] It is fair to say that initial skepticism was unjustified and there are now many examples of allosteric ligands for and allosteric sites on GPCRs, which encompass all types of GPCR families (A, B, C) and orthosteric ligands (ions, amino acids, monamines, peptides, etc …). Currently, much research is dedicated to the discovery of allosteric modulators in the fields of mGlu, GABA$_B$, chemokine, adenosine, CGRP, or glucagon receptors, to name a few. The Allosteric two state model has been applied to a number of allosteric receptor modulators as illustrated further in the chapter.

Effects of β, the intrinsic efficacy of the allosteric modulator

Allosteric modulators can affect, positively or negatively, the apparent affinity of an orthosteric ligand depending on their level of intrinsic activity, but may also have effects on their own, in the absence of an orthosteric ligand: with $\beta > 1$, the orthosteric ligand affinity will increase, with $\beta < 1$ it will decrease, whereas at $\beta = 1$ no effect on the apparent affinity of the orthosteric ligand will be observed (see Figure 13.1, left). Indeed, when $\beta > 1$, the active state R* is favored, whereas at $\beta < 1$, the inactive R is the favored receptor species, and at $\beta = 1$, the allosteric modulator has no effect on the receptor states in its own right.

In saturation experiments performed with the orthosteric ligand A in the presence of the allosteric modulator B, $\beta = 1$ will have no effect on apparent affinity, whereas at $\beta < 1$, the apparent affinity will be decreased, that is, saturation curves are shifted to the right, whereas at $\beta > 1$, the curves will be shifted to the left.

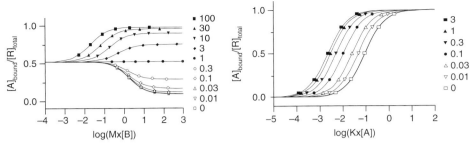

Figure 13-1: Left: Effects of the intrinsic efficacy β of the modulator B on the binding properties of an orthosteric ligand A expressed in terms of receptor occupancy. Shown are competition binding curves of the modulator with β varying from 100 to 0. Symbols on the curves represent $IC_{20, 50, 75, 80, 85, 90}$. Right: Effects of the modulator B (β = 100, B varying from 0.01 to 3 x M) on the saturation curves of an orthosteric ligand A represented in a logarithmic scale (note that B_{max} in not affected by the modulator). Symbols on the curves represent $EC_{10, 20, 50, 80, 90}$.

Variations in β have no effect on B_{max}; they only affect the apparent affinity (see Figure 13.1, right).

Since β is the expression of the intrinsic activity of the allosteric ligand B, it becomes evident that by increasing its concentration, the allosteric compound starts to produce its own effects, either positive or negative, on top of the orthosteric agonist A, whose concentration response curves (CRCs) will be shifted to the left in the case of an allosteric agonist, with no effect on maximal effect (E_{max}), but an increase in baseline activity (Figure 13.2, left), or to the right in the case of an inverse agonist modulator accompanied by a gradual decrease in E_{max} of the orthosteric agonist (Figure 13.2, right).

Effect of γ, the binding cooperativity of the allosteric modulator

Positive binding cooperativity of the allosteric ligand B (γ > 1) results in an increase in the level of binding of the orthosteric radioligand A, whereas negative cooperativity (γ < 1) results in a decrease in binding (see Figure 13.3). When γ =1, there is no interaction. The γ factor used here is analogous to what Ehlert described as the cooperativity factor α.[31]

Functionally speaking, the effects of γ are clearly different from those of β (compare Figure 13.4 and Figure 13.2). Positive or negative cooperativity affects the concentration response curves by a shift to the left (γ >1) or to the right (γ < 1), but this shift is limited and will saturate with high concentrations of the allosteric modulator whether positive (Figure 13.4, left), or negative modulator (Figure 13.4, right). As can be seen further, there is no difference in the effect on the CRCs of A whether the concentration of the modulator B is three-hundred-fold its affinity or is infinitely high (Figure 13.4).

In addition, the modulator B, whether positive or negative, will have no effects on its own, that is, neither at baseline activity nor on maximal effects of the orthosteric ligand A will be affected.

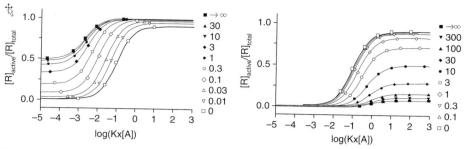

Figure 13-2: Effects of allosteric agonist (left) or inverse agonist (right) modulator on concentration effect curves of an orthosteric agonist. The concentrations of B vary between 0 and infinity. Symbols on the curves represent $EC_{10, 20, 30, 50, 70, 80, 90}$ (left) or $EC_{20, 30, 50, 80, 90}$ (right).

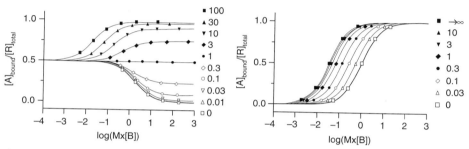

Figure 13-3: Effects of γ on the binding properties of a neutral antagonist. Left: When $\gamma > 1$, the binding is increased, and inversely so when $\gamma < 1$ (γ varies between 100 and 0). Symbols on the curves represent $IC_{20, 50, 70, 80, 85, 90}$. Right: a positive allosteric modulator B ($\gamma = 30$) when co-incubated at increasing concentrations (from 0 to infinity), will increase the affinity of the orthosteric radioligand, without affecting B_{max} in a saturation experiment. Symbols on the curves at 0.5 represent the apparent affinity K of the orthosteric ligand A.

Effect of δ, the activation cooperativity between A and B ligands

The positive and negative activation cooperativity δ of the allosteric ligand B will affect the binding curves of the orthosteric agonist A, either positively or negatively, and the effects will be more pronounced as the intrinsic activity of the orthosteric ligand is higher – that is, effects will be marked on high efficacy agonists or inverse agonists and much less visible on the low efficacy compounds (Figure 13.5 left and right, respectively). On the other hand, the binding of the neutral antagonist A is almost not affected by the activation cooperativity δ of the allosteric ligand B.

The allosteric activator will shift the CRCs of A to the left and upward to a certain degree, whereas the allosteric inhibitor will shift the CRCs of A to the right as well as downward (Figure 14.6).

EXAMPLES OF ALLOSTERIC MODULATORS

From the previously described examples, it becomes clear that allosteric modulators may affect apparent affinity or efficacy of orthosteric ligands in several ways.

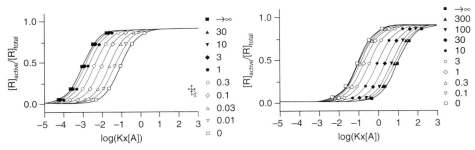

Figure 13-4: Effects of γ on the activation curves of an orthosteric agonist. Left: the positive allosteric modulator B ($\gamma = 100$) is present at increasing concentrations from 0 to infinity. Right: negative allosteric modulator B ($\gamma = 0.01$) is present at increasing concentrations from 0 to infinity. Note that in both cases the shift of the activation curves reaches a ceiling effect. Symbols on the curves represent $EC_{10, 20, 50, 80, 90}$.

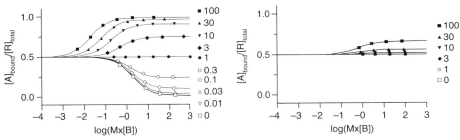

Figure 13-5: Effects of δ of the modulator B on binding curves of an orthosteric agonist (left) and neutral antagonist (right). The activation cooperativity δ varies from 0 to 100. Note that with the agonist radioligand, δ can both increase and decrease binding, whereas with the neutral ligand, B may stimulate, but not inhibit binding. Symbols on the curves represent $IC_{20, 50, 80, 90}$.

It should be realized though, that allosteric modulators can produce a combination of effects (depending on varying degrees of binding cooperativity, activation cooperativity, and intrinsic efficacy characteristic of a single modulator); some allosteric modulators may even bind to some extent to the orthosteric site (e.g., McN-A-343 at muscarinic, PD81723 at adenosine, L-692.429 at ghrelin receptors[32]), and these compounds were named as ago-allosteric modulators.[32] In any case, the many combination possibilities binding cooperativity, activation cooperativity, intrinsic efficacy, and ago-allosteric binding make each ligand almost unique and at times very complex to study.

Table 13.1 is a list of allosteric modulators covering all three families of GPCRs.[33,34] Many of these ligands have been reported / characterized in recombinant systems (over)expressing the GPCR of interest. There are examples of compounds that have been initially described as orthosteric agonists (McN-A-343, PD81723, L-692.429) or antagonists, which when investigated more thoroughly have revealed allosteric properties sometimes in addition of orthosteric binding. For instance, the chemokine receptor CXCR2 antagonist SB265610 was reported to bind competitively with the endogenous agonist. However, in a recent elegant and extensive study, Bradley and colleagues demonstrate that SB265610 behaves as an allosteric inverse agonist at the CXCR2 receptor, binding at a region distinct

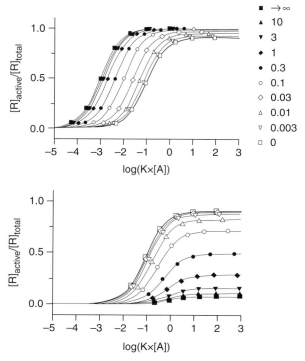

Figure 13-6: Effects of δ on activation curves of an orthosteric agonist (left δ = 100, right δ = 0.01). The allosteric modulator is present at increasing concentrations, from 0 to infinity. Note the differences between the two situations. Symbols on the curves represent $EC_{10, 20, 50, 80, 90}$ (top) or $EC_{20, 50, 75, 80, 85, 90}$ (bottom).

from the orthosteric agonist binding site to prevent receptor activation.[35] Such examples will probably become commonplace if and when compounds are investigated more carefully. In other cases, however, allosteric modulators have been recognized as such very early on and designed consequently. This is certainly the case for allosteric modulators of, for example, mGlu or $GABA_B$ receptors, for which mechanism of action and even allosteric sites have been identified rather early.

Thus, in the case of $GABA_B$ receptors, the "flytrap module" of the $GABA_{B1}$ subunit is known to bind the orthosteric agonist, whereas when it comes to the $GABA_{B2}$ subunit responsible for the transduction mechanism,[36,37] there is now convincing evidence that the $GABA_{B2}$ subunit is targeted by allosteric ligands such as CGP7930 and GS39783,[38–40] making the point that allosteric and orthosteric sites can be very different and located in totally different parts of the receptor. This is also the case for other GPCRs of the C family, such as mGluR1 or mGluR5,[41] where allosteric and orthosteric sites are clearly located in different parts of the receptor, the orthosteric site being in the large extracellular domain (similar to $GABA_B$), whereas the allosteric sites is clearly located with the transmembrane spanning regions.[42]

Somewhat disturbing is the fact that, although there is now convincing evidence that the allosteric and orthosteric binding sites can be quite different in

Table 13-1. Examples of Allosteric Modulators for GPCRs

Receptor Type and Family	Allosteric Modulators
GPCR Family A	
5HT $_{1B/1D}$	5-HT moduline (Leu-Ala-Ser-Ala)
5HT$_{2A}$, 5HT$_7$	Oleamide
5HT$_{2C}$	Oleamide; PNU-69176E
Adenosine A$_1$	PD 81723; LUF 5484
Adenosine A$_2$	Amiliorides
Adenosine A$_3$	VUFF 5455; VUFF 8504; DU124183,
Adrenoceptor α_1	Amiliorides; ρ-TIA
Adrenoceptor α_{2A}, α_{2B}	Amiliorides
Adrenoceptor $_{2D}$	Agmatine
Adrenoceptor β_2	Zinc
Cannabinoid CB$_1$	Org27569; Org27759; PSNCBAM-1
Chemokine CXCR1	Reparixin; SCH527123
Chemokine CXCR2	Reparixin; SCH527123; SB656933, SB265610
Chemokine CXCR3	IP-10; I-TAC
Chemokine CCR1	RSVM; ASLW; prichosanthin; plerixafor
Chemokine CCR3	BX-471; CP-481−715; UCB35625
Chemokine CCR5	Trichosanthin; AK602; AK530; TAKK220; TAK779; SCH 351125; ancriviroc; vicriviroc; aplaviroc; UK 427,857 (maraviroc)
Cholecystokinin CCK1	benzodiazepines
Dopamine D$_1$	Zinc
Dopamine D$_2$	Amilorides; zinc; L-prolyl-L-leuclylglycin-amide
Endothelin ET$_A$	Aspirin; sodium salicylate
Gonadotropin-releasing hormone receptor	Furan-1
Ghrelin receptor	L-692,429; MK677; GHRP- 6
Luteinizing hormone	Org 41841; [^{3}H]Org 43553
Melanocortin MC1	Zinc
mAChR M1	Brucine; W84; BQCA; TBPB; AC-42; MT3; MT7; 77-LH-28−1; NDMC; tacrine; McN-A-343; staurosporine,
mAChR M2	Gallamine, PG135, strychnine, THRX160209
mAChR M3	WIN 62,577, N chloromethyl-Brucine, WIN62,577
mAChR M4	Alcuronium; LY2033298; VU10010; VU0152099; VU0152100 ; thiochrome
mAChR M5	VU0238429
Neurokinin NK$_1$	Heparin, zinc
Opioid μ, δ	Cannabidiol
Purine P2$_{Y1}$	2,2O-pyridylsatogen tosylate

(*continued*)

Table 13-1. (*continued*)

Receptor Type and Family	Allosteric Modulators
GPCR Family B	
CRF1	Antalarmin; NBI 35965; DMP696; NBI 27914
CGRP	BIBN4096BS
Glucagon	Bay27−9955; L-168049
GLP1 receptor	T-0632; NovoNordisk Compounds 1−6
GPCR Family C	
Calcium sensing receptor	Fendeline; cinacalcet; NPS 467; NPS 568; L-amino acids; NPS 2143; calhex 231
$GABA_B$	CGP7930; CGP13501; GS39783
$mGluR_1$	(-)-CPCCOEt; Ro 67−7476; Ro 01−6128; BAY36−7620; [^{3}H]R214127; NPS 2390; cis-64a; JNJ 16259685, LY456236, EM-TBPC
$mGluR_2$	LY487379; BINA; LY181837; Ro 67−6221
$mGluR_4$	SIB-1893; MPEP; (-)-PHCCC; VU0155041; VU0080421
$mGluR_5$	SIB-1757; SIB-1893; MPEP; MTEP; DFB; DCB; DMeOB; CPPHA; CDPPB: VU-29; Fenobam; ADX-47273; ABP688; YM298198
$mGluR_7$	AMN082; MMPIP

5-HT, 5-hydroxytryptamine; CGRP, calcitonin gene related peptide; CRF1, corticotrophin releasing factor 1; GABA, γ-aminobutyric acid; GH, growth hormone; GLP1, glucagon-like peptide 1; mAChR, muscarinic acetylcholine receptor; mGluR, metabotropic glutamate receptor. (Adapted from May et al., 2007; Conn et al., 2009)

sequence and location, as evidenced by systematic mutation analysis of GPCRs targeted by allosteric ligands,[40,43,44] and more recently based on structural analysis using advanced solution nuclear magnetic resonance (NMR), there are nevertheless examples of overlaps between allosteric and orthosteric sites. In family A GPCRs, the second extracellular loop is involved in the binding of both orthosteric and allosteric G protein-coupled receptor ligands. Thus, it appears that a number of ligands are able to interact both at the orthosteric and allosteric sites[44–47] as reviewed recently by Schwartz and Holst.[32] This makes the quantitative analysis of the receptor-ligand interactions of such compounds complex, involving the generation of large sets of binding and functional data with a variety of orthosteric agonists or neutral ligands. There are even cases, such as muscarinic receptors, where several allosteric binding modes have been documented.[4,5,48]

In spite of this complexity, the data obtained with allosteric modulators can be analyzed fairly precisely according to the model proposed above, provided sufficient experimental data combining both binding and function are collected.[33,34] For instance, CPCCOEt is an antagonist at mGlu1 receptors that does not affect glutamate binding[49] but blocks considerably glutamate-induced IP_1 production. The data are best explained by a compound that is an allosteric antagonist, devoid of intrinsic activity ($\beta =1$), whose binding cooperativity γ is positive, whereas the activating cooperativity δ is well below 1 (see Figure 13.7). Along the same

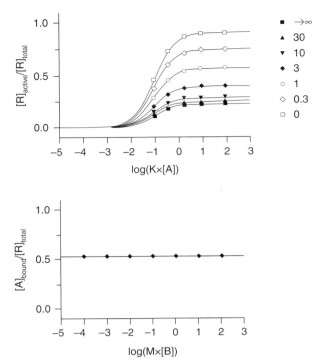

Figure 13-7: Effects of CPCCOEt on glutamate-induced IP$_1$ production via mGluR1 receptors and binding. Top: activation curves of glutamate (A) on IP$_1$ in the presence of varying concentrations of CPCCOEt (0 to infinity). Symbols on the curves represent EC$_{50, 75, 80, 85, 90}$. When fitted in binding terms, $\gamma = 8.5$, $\delta = 0.03$, $\beta = 1$, $\alpha = 10000$. Bottom: CPCCOEt (B) does not affect glutamate binding over a large concentration range (redrawn from Litschig et al, 1999, and Hall, 2000).

lines, AMN082 is a recently discovered mGluR7 positive modulator[50,51] devoid of effects on binding of orthosteric ligands such as [^{3}H]LY341495. Additionally, Urwyler and colleagues have reported detailed characterizations of GABA$_B$ positive modulators such as CGP7930 or GS39783.[52–54]

Some compounds behave rather surprisingly: For instance, Org27569 is peculiar in that it acts as a CB1 receptor antagonist, but increases the apparent affinity of orthosteric ligands.[55] Org27569 has mixed properties, combining positive modulation in terms of affinity (positive binding cooperativity) with negative modulation of the efficacy of the orthosteric agonist CP55940 (negative activation cooperativity). Depending on the system, such features may not be immediately obvious for any ligand, unless a combination of detailed binding and functional experiments are performed and the data analysed according to the Allosteric two state model.

Do allosteric ligands work in vivo? They do in many cases, although this cannot be taken for granted, since in vivo, the levels of the endogenous modulator is rarely known. In a microdialysis study, measuring inhibition of cAMP formation, Gjoni and colleagues have shown that GS39783 enhances GABA$_B$ receptor–mediated responses in vivo.[56] There is also ample evidence for the in vivo efficacy of allosteric modulators from preclinical behavioral models.[34] A case has been made for GABA$_B$ positive modulators[52–54] to be active in anxiety and depression and possibly addiction[36,57,58] based on in vivo data.[59–63] Similarly, the mGluR7 selective allosteric agonist AMN082 has shown activity in a number of animal models predictive for neuropsychiatric disorders.[50,51,64–66] Chan and colleagues[67] make a case for muscarinic allosteric ligands to treat schizophrenia. On the other hand, there have been positive clinical trials with mGluR2/3 agonists in schizophrenia as well,[68,69] but since it is notoriously difficult to get selective mGlu2 or mGlu3 orthosteric agonists, allosteric modulators may well provide a better approach. mGluR5 receptors have been a major focus for allosteric inhibitors.[70–76] For instance, Fenobam which was developed as an anxiolytic but whose mechanism of action was unknown at the time, is a rather selective allosteric mGluR5 modulator, and is now in development for other diseases such as Fragile X syndrome.[77–79] ABP688 an allosteric antagonist, is a very good PET ligand to study mGluR5 receptor distribution and occupancy.[74] In the same receptor family, there is a range of compounds based primarily (but not only) on the MPEP series of allosteric antagonists, in development for a number of diseases, such as anxiety, Parkinson's disease L-DOPA induced dyskinesias (PD LID), Fragile X, which is an autistic type behavior, or gastro esophageal reflux disorders (GERD).[77] Finally, at least two allosteric modulators of GPCRs have now entered the market. One of these, cinacalcet (Sensipar/Mimpara[80]), is a positive allosteric modulator of the calcium sensing receptor that increases sensitivity to circulating calcium. The CaSR is involved in the regulation of calcium homeostasis and renal calcium resorption, as well as in the maintenance of intracellular inositol triphosphate levels. More recently, maraviroc (Celsentri/Selzentry; Pfizer), a negative allosteric modulator of the chemokine receptor CCR5, was launched for the treatment of HIV infections. Maraviroc binds to CCR5, stabilizing a receptor conformation that has lower affinity for the HIV virus, thereby blocking CCR5-dependent entry of HIV-1 into cells,[81] and there are a number of ongoing clinical trials involving allosteric modulators for GPCRs.

POTENTIAL (THERAPEUTIC) ADVANTAGES OF ALLOSTERIC MODULATORS

Currently, based on what has been learned from enzymes, oligomeric receptors, and other complex peptides, it is suggested that binding to an allosteric site on a GPCR can induce a range of different conformational changes which may: (1) affect the affinity of orthosteric ligands either positively or negatively without affecting their relative efficacy, (2) change the signaling efficiency of such ligands, or (3) modulate the signaling of the GPCR even in the absence of an orthosteric ligand, that is, by acting directly as agonists or inverse agonists, although such action takes place at a site different from the orthosteric binding site. Thus, some allosteric modulators may be agonists or inverse agonists in their own right, in addition to modulating positively or negatively the effects of orthosteric ligands. Practically, any combination is possible for allosteric modulators. The range of intrinsic efficacy of allosteric modulators is similar to that of orthosteric ligands and can range from very low partial to full agonism or inverse agonism.

There is hope, and in specific cases a well-founded basis, that allosteric ligands/ sites do represent attractive drug targets: This is particularly true for receptors where the orthosteric binding site is highly conserved within a family (e.g., mGluRs/GABA$_B$ receptors, some dopamine or muscarinic receptor subtypes), and/or is spatially very restricted (e.g., calcium or proton sensing receptors, receptor for excitatory amino acids, Glutamate, GABA ...). It is rather clear by now that the chemical space to develop orthosteric ligands that mimic excitatory or inhibitory amino acids or small amine-like ligands is very restricted, and even when potent and selective orthosteric ligands have been found, it was only to realize that their PK properties, bioavailability, and brain penetration were less than ideal. Thus, it should not be surprising that most viable orthosteric mGluR$_{2/3}$ agonists are pro-drugs, because the parent compound has almost no chance to reach its targets when applied orally.[68] Similar limitations may apply to, for example, GABA$_B$ receptor ligands that have very narrow binding pockets. For the same reasons, GPCRs like calcium or proton sensing receptors represent almost impossible targets for synthetic orthosteric agonists or antagonists. In these cases, allosteric ligands represent the better approach.

It has been suggested that allosteric modulators have a number of structural and functional advantages. Since many allosteric ligands function in a state-dependent manner (i.e., are inactive in the absence of endogenous agonist), it has been argued that they will have limited side effects. Thus, a glutamate receptor allosteric activator is expected be devoid of the toxic potential of glutamate. Similarly, we have seen from the discussion in this chapter that positive or negative cooperativity can reach a ceiling effect even at very high modulator concentrations. However, the perceived advantages of an allosteric ligand only apply if the compound is indeed entirely state-dependent, that is, displays no intrinsic efficacy ($\beta = 1$). Also by the same reasoning, one would predict benzodiazepines that modulate GABA$_A$ receptors to be devoid of side effects: This is far from being the case. On the other hand, it has been shown that the positive GABA$_B$ receptor allosteric modulator GS39783, in contrast to classical agonists, does not induce desensitization.[82]

Another potential advantage of allosteric modulators relates to the higher selectivity that may be reached with such molecules as compared with orthosteric ligands: This assumption relates to the presumed diversity of allosteric sites when compared to the usually conserved pockets of orthosteric sites within a given receptor subfamily (e.g., Dopamine D$_2$, D$_3$, D$_4$, or Dopamine D$_1$ and D$_5$ have highly similar pharmacological signatures due to a relative identity in their respective orthosteric binding pockets). This is even more evident for receptors recognizing small amino acids, such as Glutamate or GABA receptors, or those responding to ions such as calcium or proton sensing receptors, where the chemical space is much reduced. Clearly, the orthosteric site in mGluRs is much conserved across one subfamily, and truly receptor subtype selective orthosteric ligands are rare.

Another note of caution is required here; as mentioned previously in this chapter, there is increasing evidence that some ligands are able to interact both at the allosteric as well at the orthosteric site, and this certainly does not speak in favor of selectivity. Further, as can be seen from the table listing selected allosteric ligands, some of these interact with more than one receptor: For instance, oleamide modulates the activity of $5\text{-}HT_{2A}$, $5\text{-}HT_{2C}$, and $5\text{-}HT_7$ receptors (yet the structural homology between $5\text{-}HT_2$ and $5\text{-}HT_7$ receptors is very low). Along the same lines, 5-HT moduline interacts equally well with $5\text{-}HT_{1B}$ and $5\text{-}HT_{1D}$ receptors that show high sequence homology and pharmacological profiles. There are further examples for allosteric ligands interacting at more than one receptor: amiliorides at α_1 / α_2 and adenosine A_1 receptors; repaxirin and SCH527123 at CXCR1 and CXCR2; and the list will undoubtedly extend as more allosteric regulators will be known.

SUMMARY AND CONCLUSIONS

This chapter has shown that allosteric ligands come in a variety of classes (endogenous ligands, or new synthetic species) and affect GPCRs of all three families in different ways (positive/negative modulators devoid of intrinsic activity in the absence of a ligand acting at the orthosteric sites), and/or allosteric agonists or antagonists that may produce effects on their own. The different activities can occur concomitantly in the same molecule or may be totally separate. Historically, it is the muscarinic receptor modulators that were first to be recognized as allosteric modulators, although it has taken time to get the concept generally accepted.[2–6] Thus, although GPCRs were for long considered not to be the target for allosteric modulators, a feature of enzyme/receptors forming oligomers or heteromers (e.g., ligand gated channels), an increasing number of GPCRs is now recognized to have one or more allosteric binding sites; allosteric binding sites are often topographically distinct from the orthosteric site recognized by the endogenous agonist and/or classical competitive antagonists/inverse agonists, although overlapping sites have also been reported. Such distinctions are not always documented definitively because neither the orthosteric nor the allosteric sites are well defined for the majority of GPCRs. On the other hand, the orthosteric binding site for class 2 GPCRs is well defined – for example, metabotropic glutamate or $GABA_B$ receptors (in spite of the absence of crystal or tridimensional structure of the whole receptor), and in these cases, distinct orthosteric sites have been well documented. Of course, such knowledge still relies largely on the availability of appropriate tools (ligands/radioligands/point mutagenesis …).

The difficulty lies in the definition of the allosteric effects and/or target at the GPCR proper, since potentially, any downstream site of action (at G proteins, transduction pathways, interacting proteins, even protein-protein interactions between the different partners of the receptor signal transduction pathway) may be confounded with true allosterism at the receptor level. Structural studies are only at the beginning of identifying the molecular determinants of allosteric sites on GPCRs, because even the determination of the actual orthosteric sites remains in its infancy for most receptors due to lack of conformational information (crystal or solution NMR-based structure of GPCRs has only been reported for β adrenoceptors and adenosine receptors). Thus, domains defining the orthosteric pocket on one type of GPCR may or may not contribute to an allosteric pocket for another GPCR. There are examples where allosteric sites appear well conserved across species, but given the known species variations at the level of orthosteric sites/ligands, a note of caution is required.

The progress made both in the chemistry, molecular biology, and structural biology of GPCRs over the last twenty years has been tremendous; as a consequence, the field of allosteric modulators for GPCR, which was still considered as an oddity a decade ago, has matured to the extent that preclinical and clinical data fully support an approach that may well provide better and safer medications for a number of diseases where orthosteric ligands have intrinsic issues. There are, however, technical issues in the assessment of the various parameters fully characterizing an allosteric ligand, and the data analysis is not trivial. Further, the final proof of the value of such compounds comes from their effects in vivo, and more specifically in human pathologies. In other words, the perceived advantages of allosteric modulators need to be verified in the clinical context. Finally, as with all chemicals, classical side effects are to be expected with allosteric modulators; however, this is a whole new field, which will expand the repertoire of possible interventions at GPCRs.

ACKNOWLEDGMENTS

We thank Dr. Stephan Urwyler for stimulating discussions and very helpful comments on an early version of this manuscript.

SUGGESTED READING

1. Bartfai, T., Benovic, J. L., Bockaert, J., Bond, R. A., Bouvier, M., Christopoulos, A., Civelli, O., Devi, L. A., George, S. R., Inui, A., Kobilka, B., Leurs, R., Neubig, R., Pin, J. P., Quirion, R., Roques, B. P., Sakmar, T. P., Seifert, R., Stenkamp, R. E . & Strange, P. G. (2004) The state of GPCR research in 2004. *Nature Reviews Drug Discovery*, 3, 574–626.
2. Birdsall, N. J. M., Cohen, F., Lazareno, S. & Matsui, H. (1995) Allosteric regulation of g-protein-linked receptors. *Biochemical Society Transactions*, 23, 108–111.
3. Lazareno, S. & Birdsall, N. J. M. (1995) Detection, quantitation, and verification of allosteric interactions of agents with labeled and unlabeled ligands at G-protein-coupled receptors – interactions of strychnine and acetylcholine at muscarinic receptors. *Molecular Pharmacology*, 48, 362–378.
4. Lazareno, S., Popham, A. & Birdsall, N. J. M. (2000) Allosteric interactions of staurosporine and other indolocarbazoles with N-[methyl-h-3] scopolamine and acetylcholine at muscarinic receptor subtypes: Identification of a second allosteric site. *Molecular Pharmacology*, 58, 194–206.

5. Lazareno, S., Popham, A. & Birdsall, N. J. M. (2002) Analogs of WIN 62,577 define a second allosteric site on muscarinic receptors. *Molecular Pharmacology*, 62, 1492–1505.

6. Lazareno, S., Dolezal, V., Popham, A. & Birdsall, N. J. M. (2004) Thiochrome enhances acetylcholine affinity at muscarinic m-4 receptors: Receptor subtype selectivity via cooperativity rather than affinity. *Molecular Pharmacology*, 65, 257–266.

7. Monod, J., Wyman, J. & Changeux, J. P. (1965) On nature of allosteric transitions – a plausible model. *J. Mol. Biol.*, 12, 88-&.

8. Christopoulos, A. (2006) Non-classical modes of signaling by 5ht(2c) receptors. *Journal of Pharmacological Sciences*, 101, 55–55.

9. Gilchrist, A. (2007) Modulating G-protein-coupled receptors: From traditional pharmacology to allosterics. *Trends in Pharmacological Sciences*, 28, 431–437.

10. Christopoulos, A. (2009) The impact of allosteric modulation of G-protein-coupled receptors on drug discovery. *Journal of Biomolecular Screening*, 14, 901–901.

11. Leff, P. (1995) The 2-state model of receptor activation. *Trends in Pharmacological Sciences*, 16, 89–97.

12. Scaramellini, C. & Leff, P. (1998) A three-state receptor model: Predictions of multiple agonist pharmacology for the same receptor type. *Advances in Serotonin Receptor Research – Molecular Biology, Signal Transduction, and Therapeutics*, 861, 97–103.

13. Strange, P. G. (1998) Three-state and two-state models. *Trends in Pharmacological Sciences*, 19, 85–86.

14. Black, J. W. & Leff, P. (1983) Operational models of pharmacological agonism. *Proc. R. Soc. Lond. Ser. B-Biol. Sci.*, 220, 141–162.

15. Black, J. W., Leff, P., Shankley, N. P. & Wood, J. (1985) An operational model of pharmacological agonism – the effect of e/[a] curve shape on agonist dissociation-constant estimation. *British Journal of Pharmacology*, 84, 561–571.

16. Del Castillo, J. & Katz, B. (1957) Interaction at end-plate receptors between different choline derivatives. *Proc. R. Soc. Lond. Ser. B-Biol. Sci.*, 146, 369–380.

17. Costa, T. & Herz, A. (1989) Antagonists with negative intrinsic activity at delta-opioid receptors coupled to GTP-binding proteins. *Proceedings of the National Academy of Sciences of the United States of America*, 86, 7321–7325.

18. Kjelsberg, M. A., Cotecchia, S., Ostrowski, J., Caron, M. G. & Lefkowitz, R. J. (1992) Constitutive activation of the alpha-1b-adrenergic receptor by all amino-acid substitutions at a single site – evidence for a region which constrains receptor activation. *Journal of Biological Chemistry*, 267, 1430–1433.

19. Neubig, R. R., Spedding, M., Kenakin, T. & Christopoulos, A. (2003) International union of pharmacology committee on receptor nomenclature and drug classification. Xxxviii. Update on terms and symbols in quantitative pharmacology. *Pharmacological Reviews*, 55, 597–606.

20. Smit, M. J., Vischer, H. F., Bakker, R. A., Jongejan, A., Timmerman, H., Pardo, L. & Leurs, R. (2007) Pharmacogenomic and structural analysis of constitutive G protein-coupled receptor activity. *Annual Review of Pharmacology and Toxicology*, 47, 53–87.

21. Herrick-Davis, K., Grinde, E. & Niswander, C. M. (1999) Serotonin 5-HT2c receptor RNA editing alters receptor basal activity: Implications for serotonergic signal transduction. *Journal of Neurochemistry*, 73, 1711–1717.

22. DeLean, A., Stadel, J. M. & Lefkowitz, R. J. (1980a) A ternary complex model explains the agonist-specific binding-properties of the adenylate cyclase-coupled beta-adrenergic-receptor. *Journal of Biological Chemistry*, 255, 7108–7117.

23. DeLean, A., Stadel, J. M. & Lefkowitz, R. J. (1980b) Ternary complex model explains the specific binding-properties of beta-adrenergic agonists. *Federation Proceedings*, 39, 517–517.

24. Kent, R. S., Delean, A. & Lefkowitz, R. J. (1980) Quantitative-analysis of beta-adrenergic-receptor interactions – resolution of high and low affinity states of the receptor by computer modeling of ligand-binding data. *Molecular Pharmacology*, 17, 14–23.

25. Hall, D. A. (2000) Modeling the functional effects of allosteric modulators at pharmacological receptors: An extension of the two-state model of receptor activation. *Molecular Pharmacology*, 58, 1412–1423.

26. Weiss, J. M., Morgan, P. H., Lutz, M. W. & Kenakin, T. P. (1996a) The cubic ternary complex receptor-occupancy model. 1. Model description. *Journal of Theoretical Biology*, 178, 151–167.

27. Weiss, J. M., Morgan, P. H., Lutz, M. W. & Kenakin, T. P. (1996b) The cubic ternary complex receptor-occupancy model. 2. Understanding apparent affinity. *Journal of Theoretical Biology*, 178, 169–182.

28. Weiss, J. M., Morgan, P. H., Lutz, M. W. & Kenakin, T. P. (1996c) The cubic ternary complex receptor-occupancy model. 3. Resurrecting efficacy. *Journal of Theoretical Biology*, 181, 381–397.

29. Christopoulos, A. & Kenakin, T. (2002) G protein-coupled receptor allosterism and complexing. *Pharmacological Reviews*, 54, 323–374.

30. Leff, P., Scaramellini, C., Law, C. & Mckechnie, K. (1997) A three-state receptor model of agonist action. *Trends in Pharmacological Sciences*, 18, 355–362.

31. Ehlert, F. J. (1988) Estimation of the affinities of allosteric ligands using radioligand binding and pharmacological null methods. *Molecular Pharmacology*, 33, 187–194.

32. Schwartz, T. W. & Holst, B. (2007) Allosteric enhancers, allosteric agonists and ago-allosteric modulators: Where do they bind and how do they act? *Trends in Pharmacological Sciences*, 28, 366–373.

33. May, L. T., Leach, K., Sexton, P. M. & Christopoulos, A. (2007) Allosteric modulation of G protein-coupled receptors. *Annual Review of Pharmacology and Toxicology*, 47, 1–51.

34. Conn, P. J., Christopoulos, A. & Lindsley, C. W. (2009) Allosteric modulators of GPCRs: A novel approach for the treatment of CNS disorders. *Nature Reviews Drug Discovery*, 8, 41–54.

35. Bradley, M. E., Bond, M. E., Manini, J., Brown, Z. & Charlton, S. J. (2009) SB265610 is an allosteric, inverse agonist at the human cxcr2 receptor. *British Journal of Pharmacology*, 158, 328–338.

36. Bettler, B., Kaupmann, K., Mosbacher, J. & Gassmann, M. (2004) Molecular structure and physiological functions of GABA(b) receptors. *Physiological Reviews*, 84, 835–867.

37. Deriu, D., Gassmann, M., Firbank, S., Ristig, D., Lampert, C., Mosbacher, J., Froestl, W., Kaupmann, K., Bettler, B. & Grutter, M. G. (2005) Determination of the minimal functional ligand-binding domain of the GABA(b(1b)) receptor. *Biochemical Journal*, 386, 423–431.

38. Pagano, A., Ruegg, D., Litschig, S., Stoehr, N., Stierlin, C., Heinrich, M., Floersheim, P., Prezeau, L., Carroll, F., Pin, J. P., Cambria, A., Vranesic, I., Flor, P. J., Gasparini, F. & Kuhn, R. (2000) The non-competitive antagonists 2-methyl-6-(phenylethynyl)pyridine and 7-hydroxyiminocyclopropan [b]chromen-1 alpha-carboxylic acid ethyl ester interact with overlapping binding pockets in the transmembrane region of group 1 metabotropic glutamate receptors. *Journal of Biological Chemistry*, 275, 33750–33758.

39. Binet, V., Goudet, C., Brajon, C., Le Corre, L., Acher, F., Pin, J. P. & Prezeau, L. (2004) Molecular mechanisms of GABA(b) receptor activation: New insights from the mechanism of action of CGP7930, a positive allosteric modulator. *Biochemical Society Transactions*, 32, 871–872.

40. Dupuis, D. S., Relkovic, D., Lhuillier, L., Mosbacher, J. & Kaupmann, K. (2006) Point mutations in the transmembrane region of GABA(b2) facilitate activation by the positive modulator n,n '-dicyclopentyl-2-methylsulfanyl-5-nitro-pyrimidine-4,6-diamine (GS39783) in the absence of the GABA(b1) subunit. *Molecular Pharmacology*, 70, 2027–2036.

41. Leach, K., Sexton, P. M. & Christopoulos, A. (2007) Allosteric GPCR modulators: Taking advantage of permissive receptor pharmacology. *Trends in Pharmacological Sciences*, 28, 382–389.

42. Parmentier, M. L., Prezeau, L., Bockaert, J. & Pin, J. P. (2002) A model for the functioning of family 3 gpcrs. *Trends in Pharmacological Sciences*, 23, 268–274.

43. Ott, D., Floersheim, P., Inderbitzin, W., Stoehr, N., Francotte, E., Lecis, G., Richert, P., Rihs, G., Flor, P. J., Kuhn, R. & Gasparini, F. (2000) Chiral resolution, pharmacological

characterization, and receptor docking of the noncompetitive mglu1 receptor antagonist (+/-)-2-hydroxyimino-1a,2-dihydro-1h-7-oxacyclopropa[b]naphthalene-7a-carboxylic acid ethyl ester. *Journal of Medicinal Chemistry*, 43, 4428–4436.

44. Avlani, V. A., Gregory, K. J., Morton, C. J., Parker, M. W., Sexton, P. M. & Christopoulos, A. (2007) Critical role for the second extracellular loop in the binding of both orthosteric and allosteric G protein-coupled receptor ligands. *Journal of Biological Chemistry*, 282, 25677–25686.

45. Valant, C., Gregory, K. J., Hall, N. E., Scammells, P. J., Lew, M. J., Sexton, P. M. & Christopoulos, A. (2008) A novel mechanism of G protein-coupled receptor functional selectivity muscarinic partial agonist MCN-a-343 as a bitopic orthosteric/allosteric ligand. *Journal of Biological Chemistry*, 283, 29312–29321.

46. Valant, C., Sexton, P. M. & Christopoulos, A. (2009) Orthosteric/allosteric bitopic ligands going hybrid at GPCRs. *Molecular Interventions*, 9, 125–135.

47. Antony, J., Kellershohn, K., Mohr-Andra, M., Kebig, A., Prilla, S., Muth, M., Heller, E., Disingrini, T., Dallanoce, C., Bertoni, S., Schrobang, J., Trankle, C., Kostenis, E., Christopoulos, A., Holtje, H. D., Barocelli, E., De Amici, M., Holzgrabe, U. & Mohr, K. (2009) Dualsteric GPCR targeting: A novel route to binding and signaling pathway selectivity. *Faseb Journal*, 23, 442–450.

48. Lanzafame, A. A., Sexton, P. M. & Christopoulos, A. (2006) Interaction studies of multiple binding sites on m-4 muscarinic acetylcholine receptors. *Molecular Pharmacology*, 70, 736–746.

49. Litschig, S., Gasparini, F., Rueegg, D., Stoehr, N., Flor, P. J., Vranesic, I., Prezeau, L., Pin, J. P., Thomsen, C. & Kuhn, R. (1999) CPCCOET, a noncompetitive metabotropic glutamate receptor 1 antagonist, inhibits receptor signaling without affecting glutamate binding. *Molecular Pharmacology*, 55, 453–461.

50. Mitsukawa, K., Mombereau, C., Lotscher, E., Uzunov, D. P., Van Der Putten, H., Flor, P. J. & Cryan, J. F. (2006) Metabotropic glutamate receptor subtype 7 ablation causes dysregulation of the HPA axis and increases hippocampal BDNF protein levels: Implications for stress-related psychiatric disorders. *Neuropsychopharmacology*, 31, 1112–1122.

51. Mitsukawa, K., Yamamoto, R., Ofner, S., Nozulak, J., Pescott, O., Lukic, S., Stoehr, N., Mombereau, C., Kuhn, R., Mcallister, K. H., Van Der Putten, H., Cryan, J. F. & Flor, P. J. (2005) A selective metabotropic glutamate receptor 7 agonist: Activation of receptor signaling via an allosteric site modulates stress parameters in vivo. *Proceedings of the National Academy of Sciences of the United States of America*, 102, 18712–18717.

52. Urwyler, S., Mosbacher, J., Lingenhoehl, K., Heid, J., Hofstetter, K., Froestl, W., Bettler, B. & Kaupmann, K. (2001) Positive allosteric modulation of native and recombinant,gamma-aminobutyric acid(b) receptors by 2,6-di-tert-butyl-4-(3-hydroxy-2,2-dimethyl-propyl)-phenol (CGP7930) and its aldehyde analog cgp13501. *Molecular Pharmacology*, 60, 963–971.

53. Urwyler, S., Pozza, M. F., Lingenhoehl, K., Mosbacher, J., Lampert, C., Froestl, W., Koller, M. & Kaupmann, K. (2003) N,n '-dicyclopentyl-2-methylsulfanyl-5-nitropyrimidine-4,6-diamine (GS39783) and structurally related compounds: Novel allosteric enhancers of gamma-aminobutyric acid(b) receptor function. *Journal of Pharmacology and Experimental Therapeutics*, 307, 322–330.

54. Guery, S., Floersheim, P., Kaupmann, K. & Froestl, W. (2007) Syntheses and optimization of new GS39783 analogues as positive allosteric modulators of GABAb receptors. *Bioorg. Med. Chem. Lett.*, 17, 6206–6211.

55. Price, M. R., Baillie, G. L., Thomas, A., Stevenson, L. A., Easson, M., Goodwin, R., Mclean, A., Mcintosh, L., Goodwin, G., Walker, G., Westwood, P., Marrs, J., Thomson, F., Cowley, P., Christopoulos, A., Pertwee, R. G. & Ross, R. A. (2005) Allosteric modulation of the cannabinoid cb1 receptor. *Molecular Pharmacology*, 68, 1484–1495.

56. Gjoni, T., Desrayaud, S., Imobersteg, S. & Urwyler, S. (2006) The positive allosteric modulator GS39783 enhances GABA(b) receptor-mediated inhibition of cyclic amp formation in rat striatum in vivo. *Journal of Neurochemistry*, 96, 1416–1422.

57. Cryan, J. F., Kelly, P. H., Chaperon, F., Gentsch, C., Mombereau, C., Lingenhoehl, K., Froestl, W., Bettler, B., Kaupmann, K. & Spooren, W. (2004) Behavioral characterization of the novel GABA(b) receptor-positive modulator GS39783 (n,n '-dicyclopentyl-2methylsulfanyl-5-nitro-pyrimidine-4,6-diamine): Anxiolytic-like activity without side effects associated with baclofen or benzodiazepines. *Journal of Pharmacology and Experimental Therapeutics*, 310, 952–963.

58. Cryan, J. F. & Kaupmann, K. (2005) Don't worry 'b' happy!: A role for GABA(b) receptors in anxiety and depression. *Trends in Pharmacological Sciences*, 26, 36–43.

59. Jacobson, L. H., Bettler, B., Kaupmann, K. & Cryan, J. F. (2006a) GABA(b(1)) receptor subunit isoforms exert a differential influence on baseline, but not GABA(b) receptor agonist-induced changes in mice. *Journal of Pharmacology and Experimental Therapeutics*, 319, 1317–1326.

60. Jacobson, L. H., Kelly, P. H., Bettler, B., Kaupmann, K. & Cryan, J. F. (2006b) GABA(b(1)) receptor isoforms differentially mediate the acquisition and extinction of aversive taste memories. *Journal of Neuroscience*, 26, 8800–8803.

61. Mombereau, C., Lhuillier, L., Kaupmann, K. & Cryan, J. F. (2007) GABA(b) receptor-positive modulation-induced blockade of the rewarding properties of nicotine is associated with a reduction in nucleus accumbens delta FOSb accumulation. *Journal of Pharmacology and Experimental Therapeutics*, 321, 172–177.

62. Paterson, N. E., Vlachou, S., Guery, S., Kaupmann, K., Froestl, W. & Markou, A. (2008) Positive modulation of GABA(b) receptors decreased nicotine self-administration and counteracted nicotine-induced enhancement of brain reward function in rats. *Journal of Pharmacology and Experimental Therapeutics*, 326, 306–314.

63. Maccioni, P., Carai, M. A. M., Kaupmann, K., Guery, S., Froestl, W., Leite-Morris, K. A., Gessa, G. L. & Colombo, G. (2009) Reduction of alcohol's reinforcing and motivational properties by the positive allosteric modulator of the GABA(b) receptor, BHF177, in alcohol-preferring rats. *Alcoholism-Clinical and Experimental Research*, 33, 1749–1756.

64. Palucha, A., Klak, K., Branski, P., Van Der Putten, H., Flor, P. J. & Pilc, A. (2007) Activation of the mglu7 receptor elicits antidepressant-like effects in mice. *Psychopharmacology*, 194, 555–562.

65. Fendt, M., Schmid, S., Thakker, D. R., Jacobson, L. H., Yamamoto, R., Mitsukawa, K., Maier, R., Natt, F., Husken, D., Kelly, P. H., Mcallister, K. H., Hoyer, D., Van Der Putten, H., Cryan, J. F. & Flor, P. J. (2008) mGlur7 facilitates extinction of aversive memories and controls amygdala plasticity. *Molecular Psychiatry*, 13, 970–979.

66. Stachowicz, K., Branski, P., Klak, K., Van Der Putten, H., Cryan, J. F., Flor, P. J. & Pilc, A. (2008) Selective activation of metabotropic G-protein-coupled glutamate 7 receptor elicits anxiolytic-like effects in mice by modulating gabaergic neurotransmission. *Behavioural Pharmacology*, 19, 597–603.

67. Chan, W. Y., Mckinzie, D. L., Bose, S., Mitchell, S. N., Witkin, J. M., Thompson, R. C., Christopoulos, A., Lazareno, S., Birdsall, N. J. M., Bymaster, F. P. & Felder, C. C. (2008) Allosteric modulation of the muscarinic m-4 receptor as an approach to treating schizophrenia. *Proceedings of the National Academy of Sciences of the United States of America*, 105, 10978–10983.

68. Patil, S. T., Zhang, L., Martenyi, F., Lowe, S. L., Jackson, K. A., Andreev, B. V., Avedisova, A. S., Bardenstein, L. M., Gurovich, I. Y., Morozova, M. A., Mosolov, S. N., Neznanov, N. G., Reznik, A. M., Smulevich, A. B., Tochilov, V. A., Johnson, B. G., Monn, J. A. & Schoepp, D. D. (2007) Activation of mglu2/3 receptors as a new approach to treat schizophrenia: A randomized phase 2 clinical trial. *Nature Medicine*, 13, 1102–1107.

69. Conn, P. J., Tamminga, C., Schoepp, D. D. & Lindsley, C. (2008) Schizophrenia: Moving beyond monoamine antagonists. *Molecular Interventions*, 8, 99–107.

70. Varney, M. A., Cosford, N. D. P., Jachec, C., Rao, S. P., Sacaan, A., Lin, F. F., Bleicher, L., Santori, E. M., Flor, P. J., Allgeier, H., Gasparini, F., Kuhn, R., Hess, S. D., Velicelebi, G. & Johnson, E. C. (1999a) SIB-1757 and SIB-1893: Selective, noncompetitive

antagonists of metabotropic glutamate receptor type 5. *Journal of Pharmacology and Experimental Therapeutics*, 290, 170–181.

71. Varney, M. A., Cosford, N., Jachec, C., Rao, S., Sacaan, A., Santori, E., Allgeier, H., Gasparini, F., Flor, P. J., Kuhn, R., Hess, S. D., Velicelebi, G. & Johnson, E. (1999b) Characterization of SIB-1757 and SIB-1893: Highly selective non-competitive antagonists of metabotropic glutamate receptor subtype 5 metabotropic (mglur5). *Neuropharmacology*, 38, 147.

72. Gasparini, F., Lingenhohl, K., Stoehr, N., Flor, P. J., Heinrich, M., Vranesic, I., Biollaz, M., Allgeier, H., Heckendorn, R., Urwyler, S., Varney, M. A., Johnson, E. C., Hess, S. D., Rao, S. P., Sacaan, A. I., Santori, E. M., Velicelebi, G. & Kuhn, R. (1999) 2-methyl-6-(phenylethynyl)-pyridine (MPEP), a potent, selective and systemically active mglu5 receptor antagonist. *Neuropharmacology*, 38, 1493–1503.

73. Gasparini, F., Andres, H., Flor, P. J., Heinrich, M., Inderbitzin, W., Lingenhohl, K., Muller, H., Munk, V. C., Omilusik, K., Stierlin, C., Stoehr, N., Vranesic, I. & Kuhn, R. (2002) [H-3]-m-MPEP, a potent, subtype-selective radioligand for the metabotropic glutamate receptor subtype 5. *Bioorg. Med. Chem. Lett.*, 12, 407–409.

74. Hintermann, S., Vranesic, I., Allgeier, H., Brulisauer, A., Hoyer, D., Lemaire, M., Moenius, T., Urwyler, S., Whitebread, S., Gasparini, F. & Auberson, Y. P. (2007) ABP688, a novel selective and high affinity ligand for the labeling of mglu5 receptors: Identification, in vitro pharmacology, pharmacokinetic and biodistribution studies. *Bioorganic & Medicinal Chemistry*, 15, 903–914.

75. Kuhn, R., Pagano, A., Stoehr, N., Vranesic, I., Flor, P. J., Lingenhohl, K., Spooren, W., Gentsch, C., Vassout, A., Pilc, A. & Gasparini, F. (2002) In vitro and in vivo characterization of MPEP, an allosteric modulator of the metabotropic glutamate receptor subtype 5: Review article. *Amino Acids*, 23, 207–211.

76. O'Brien, J. A., Lemaire, W., Chen, T.-B., Chang, R. S. L., Jacobson, M. A., Ha, S. N., Lindsley, C. W., Schaffhauser, H. J., Sur, C., Pettibone, D. J., Conn, P. J. & Williams, D. L. (2003) A family of highly selective allosteric modulators of the metabotropic glutamate receptor subtype 5. *Molecular Pharmacology*, 64, 731–740.

77. Jaeschke, G., Wettstein, J. G., Nordquist, R. E. & Spooren, W. (2008) Mglu5 receptor antagonists and their therapeutic potential. *Expert Opinion on Therapeutic Patents*, 18, 123–142.

78. Montana, M. C., Cavallone, L. F., Stubbert, K. K., Stefanescu, A. D., Kharasch, E. D. & Gereau, R. W. (2009) The metabotropic glutamate receptor subtype 5 antagonist fenobam is analgesic and has improved in vivo selectivity compared with the prototypical antagonist 2-methyl-6-(phenylethynyl)-pyridine. *Journal of Pharmacology and Experimental Therapeutics*, 330, 834–843.

79. Berry-Kravis, E., Hessl, D., Coffey, S., Hervey, C., Schneider, A., Yuhas, J., Hutchison, J., Snape, M., Tranfaglia, M., Nguyen, D. V. & Hagerman, R. (2009) A pilot open label, single dose trial of fenobam in adults with fragile x syndrome. *Journal of Medical Genetics*, 46, 266–271.

80. Harrington, P. E. & Fotsch, C. (2007) Calcium sensing receptor activators: Calcimimetics. *Curr. Med. Chem.*, 14, 3027–3034.

81. Dorr, P., Westby, M., Dobbs, S., Griffin, P., Irvine, B., Macartney, M., Mori, J., Rickett, G., Smith-Burchnell, C., Napier, C., Webster, R., Armour, D., Price, D., Stammen, B., Wood, A. & Perros, M. (2005) Maraviroc (UK-427,857), a potent, orally bioavailable, and selective small-molecule inhibitor of chemokine receptor ccr5 with broad-spectrum anti-human immunodeficiency virus type 1 activity. *Antimicrobial Agents and Chemotherapy*, 49, 4721–4732.

82. Gjoni, T. & Urwyler, S. (2008) Receptor activation involving positive allosteric modulation, unlike full agonism, does not result in GABA(b) receptor desensitization. *Neuropharmacology*, 55, 1293–1299.

83. Baker, J. G., Hall, I. P. & Hill, S. J. (2003) Agonist actions of "Beta-blockers" Provide evidence for two agonist activation sites or conformations of the human beta(1)-adrenoceptor. *Molecular Pharmacology*, 63, 1312–1321.

84. Berg, K. A., Maayani, S., Goldfarb, J., Scaramellini, C., Leff, P. & Clarke, W. P. (1998) Effector pathway-dependent relative efficacy at serotonin type 2a and 2c receptors: Evidence for agonist-directed trafficking of receptor stimulus. *Molecular Pharmacology*, 54, 94–104.

85. Gjoni, T. & Urwyler, S. (2009) Changes in the properties of allosteric and orthosteric GABA(b) receptor ligands after a continuous, desensitizing agonist pretreatment. *European Journal of Pharmacology*, 603, 37–41.

86. Heitzler, D., Crépieux, P., Poupon, A., Clément, F., Fages, F. & Reiter, E. (2009) Towards a systems biology approach of g protein-coupled receptor signalling: Challenges and expectations. *Comptes Rendus Biologies*, 332, 947–957.

87. Langmead, C. J. & Christopoulos, A. (2006) Allosteric agonists of 7TM receptors: Expanding the pharmacological toolbox. *Trends in Pharmacological Sciences*, 27, 475–481.

88. Maj, M., Bruno, V., Dragic, Z., Yamamoto, R., Battaglia, G., Inderbitzin, W., Stoehr, N., Stein, T., Gasparini, F., Vranesic, I., Kuhn, R., Nicoletti, F. & Flor, P. J. (2003) (-)-PHCCC, a positive allosteric modulator of mglur4: Characterization, mechanism of action, and neuroprotection. *Neuropharmacology*, 45, 895–906.

89. Miles, C. & Fiona, M. (2009) The impact of GPCR structures on pharmacology and structure-based drug design. *British Journal of Pharmacology*, 9999.

90. Rang, H. P. (2006) The receptor concept: Pharmacology's big idea. *British Journal of Pharmacology*, 147, S9-S16.

91. Stephenson, R. P. (1956) Modification of receptor theory. *British Journal of Pharmacology and Chemotherapy*, 11, 379–393.

92. Urwyler, S., Gjoni, T., Koljatic, J. & Dupuis, D. S. (2005) Mechanisms of allosteric modulation at GABA(b) receptors by CGP7930 and GS39783: Effects on affinities and efficacies of orthosteric ligands with distinct intrinsic properties. *Neuropharmacology*, 48, 343–353.

93. Bokoch, M. P., Zou, Y. Z., Rasmussen, S. G. F., Liu, C. W., Nygaard, R., Rosenbaum, D. M., Fung, J. J., Choi, H. J., Thian, F. S., Kobilka, T. S., Puglisi, J. D., Weis, W. I., Pardo, L., Prosser, R. S., Mueller, L. & Kobilka, B. K. (2010) Ligand-specific regulation of the extracellular surface of a g-protein-coupled receptor. *Nature*, 463, 108–114.

14 7TM receptor functional selectivity

Terry Kenakin

INTRODUCTION: 7TM RECEPTORS

In general, the concept of the receptor has brought order to physiology by relating chemically-driven but diverse physiological processes to single points of origin. For example, epinephrine produces a myriad of physiological effects in the body, yet these all begin with the interaction of epinephrine with a defined collection of proteins called α- and β-adrenoceptors. Identifying these receptors as sorting points for physiological signals has allowed the orderly exploration of the chemical modification of these processes for therapeutic advantage. Classical receptor theory has provided the backbone of pharmacological classifications of receptors and ligands driving this exploration. Specifically, by defining the receptor as the recognition point of chemical information, in vitro assays quantitatively describing effects at the receptor allow the quantification of drug effect in terms dependent only on the receptor and the ligand. For example, a mainstay of receptor classification studies is the agonist potency ratio, that is, within this theoretical framework rank orders of potency are immutable scales uniquely identifying agonists and their respective receptors. A requirement for this scale to be cell-type independent is the fact that the receptor must be the minimal unit for agonist identification (Figure 14.1). In such a system, every agonist would qualitatively produce all of the responses mediated by that receptor type. A new phenomenon in receptor pharmacology has been identified recently whereby this automatic association of all responses linked to a given receptor is not observed. In these systems, certain agonists produce only portions of the

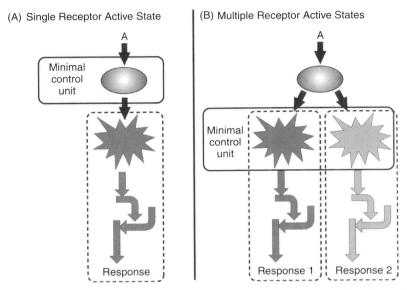

(A) Single Receptor Active State

(B) Multiple Receptor Active States

Minimal control unit

Response

Minimal control unit

Response 1 | Response 2

Figure 14-1: Two opposing views of receptor activation. (A) Linear agonist efficacy dictates that the receptor itself is the minimal control unit for signaling. Within this scheme, all agonists can activate all signaling processes controlled by the receptor. (B) In this scheme, the ligand-receptor complex (receptor active state) is the minimal control unit, thus different agonists, if they produce different active states, can mediate different sets of cellular signals.

repertoire of signaling activities mediated by a particular receptor; this phenomeon has been given the name 'functional selectivity'. If every agonist produces an identical receptor active state, and differences in intrinsic efficacy would be accommodated only by a scale of intensity, then functional selectivity would be impossible. However, published data clearly showing that various agonists have reversed orders of potency for a single receptor linked to different signaling pathways (*vide infra*) is incompatible with the notion that all agonists produce identical active states. Instead, these data compel the notion that different agonists can produce different receptor active states.[1] This makes the agonist-receptor complex, referred to as the 'receptor active state', the minimal pharmacological recognition unit (Figure 14.1). It is worth considering functional selectivity in terms of classical receptor theory as a preface to a discussion of its applicability to new therapies.

FUNCTIONAL SELECTIVITY: HISTORICAL CONTEXT

A precept of conventional receptor theory is that the relative activity of agonists is governed by their relative affinity and efficacy; this negates the effects of tissues and allows pharmacological testing in experimental systems with prediction of activity in therapeutic ones. However, exceptions to this general rule have been noted over the years,[2–5] Specifically, certain agonists were found to be especially active in certain tissues. For a number of years, it was not clear to what extent such apparent aberrations in activity were artifacts of experimental systems or whether this was a real pharmacological phenomenon. For instance, some reports of 'functional selectivity' were cases of weak agonists of multiple cellular processes producing activation of only the most sensitive of those processes in some poorly receptor-coupled or low receptor density tissues. This 'strength of signal' effect[1] produces cell-dependent agonist profiles that are the result of different efficiencies of coupling of the receptor to various cellular signaling pathways. Such behavior is consistent with single receptor active state systems and does not constitute what now is separately regarded as true conformationally based functional selectivity. However, truly incompatible data for uniform receptor active states systems have been published to show actual reversal of potency ratios of agonists for different cell pathways activated by the same receptor; this cannot be the result of receptor coupling efficiency but rather necessitates the existence of agonist-specific receptor conformations.[1]

As noted previously, a major prediction of a single receptor active-state model is that agonist potency ratios will be system-independent. In light of this fact, it is worth considering the meaning of agonist potency ratios in terms of the Black/Leff operational model[6] for receptor activation. Thus, cellular response resulting from receptor ([R]) activation by agonist [A] is given by the following scheme:

$$A+R \underset{\longleftarrow}{\overset{K_A}{\longrightarrow}} AR^*+E \underset{\longleftarrow}{\overset{K_E}{\longrightarrow}} AR^*E \longrightarrow RESPONSE \qquad [4-1]$$

where K_A is the equilibrium dissociation constant of the agonist-receptor complex and K_E is the virtual equilibrium dissociation constant of the activated-receptor and cell stimulus-response machinery complex. The equation for the production of response by an agonist with this model is given by Black and Leff[6] who defined a term τ that encompasses the intrinsic efficacy of the agonist and the sensitivity of the system with the substitution $\tau=[R]/K_E$:

$$Response = \frac{[A]/K_A \tau E_{max}}{[A]/K_A (1+\tau)1} \qquad [4-2]$$

The expression for the potency ratio of two agonists I and II from equation 4–2 is given by:

$$Pontency\,Ratio = K_{AI}(1+\tau_{II})/K_{AII}(1+\tau_I) \qquad [4-3]$$

Substituting for τ leads to:

$$Pontency\,Ratio = (K_{AI}K_{EI}(K_{EII} + [R]))/(K_{AII}K_{EII}(K_{EI} + [R])) \qquad [4-4]$$

It can be seen from equation 4 that the potency ratio for all processes mediated by binding of agonists I and II to the same receptor (ratio K_{AI}/K_{AII} must be constant) will be constant as long as the efficacy and tissue sensitivity terms

(K_{EI} and K_{EII}) do not change. Within a framework where efficacy is 'linear', meaning activation of a receptor necessarily leads to all activities mediated by that receptor, this will be true and it follows from equation 4 that the ratio of efficacies should be constant for two agonists acting on a common receptor.

Historically, the definition of 'efficacy' utilized in pharmacological studies was tissue activation and/or biochemical second messenger production. With the advent of more detailed biochemical dissection of the consequences of receptor agonism came the realization that efficacy is not always linear. The instances where this was determined was for pleiotropic receptors, that is, receptors that interact with more than one G protein or other cytosolic signaling proteins. With detailed study of these pleiotropic signaling pathways came revelations totally inconsistent with the previously described classification scheme for agonists. An early example of such inconsistencies was reported for the PACAP receptor, a pleiotropic receptor that activates pathways to elevate cAMP and insositol triphosphate (IP_3). Two PACAP peptide fragments ($PACAP_{1-27}$ and $PACAP_{1-38}$) produce elevated cyclic AMP and IP_3 in cells but the relative potency of these two agonists for these pathways are reversed,[7] that is, the relative efficacy of $PACAP_{1-27}$ for cAMP elevation is higher than that for $PACAP_{1-38}$ but lower for elevation of IP_3. These data are completely incompatible with the definition of potency ratios given by equation 4 unless the K_E values (efficacy) for activation of the IP_3 and cAMP pathways vary for the two agonists. The most simple way this can occur is to propose that each agonist produces a different receptor active state (R* in scheme 1), which then has a unique K_E value for the two signaling pathways (IP_3 and cAMP). Thus, the receptor active state, and not just the receptor, becomes the minimal recognition unit for agonist action (see Figure 14.1).

The first formal model proposed to describe such deviations from linear efficacy was based on a system of one receptor interacting with two Gproteins.[8]

$$\frac{[ARG_1]}{[R_{total}]} = \frac{\alpha\gamma L([A]/K_A)([G_1]/K_1)}{[A]/K_A\,(\alpha L(1+\gamma[G_1]/K_1+\beta[G_2]/K_2)+1)+L(1+[G_1]/K_1+[G_2]/K_2+1} \qquad [4\text{--}5]$$

where a receptor R in equilibrium with an active state R* (with equilibrium dissociation constant L) can interact with two G proteins G_1 and G_2 with respective equilibrium dissociation constants K_1 and K_2. An agonist A binds to R with equilibrium dissociation constant K_A and promotes the active state through selective affinity for R* of αK_A. Selective coupling to either G_1 or G_2 is defined by the constants β and γ respectively. This model allows for selective coupling to either G_1 or G_2 with different agonists through selective differences in β and γ. This would occur through differences in the R* species formed by different agonists – in other words, different agonists stabilized different receptor active states. Specifically, one agonist can confer selectivity to G protein G_1 through a higher value for γ over β, whereas another agonist may produce a receptor active state preferring G_2 through $\beta>\gamma$. Similarly, through the effect of varying $[G_1]/K_1$ and $[G_2]/K_2$ levels, it can be seen that cellular relative stoichiometries of cellular signaling components also could change the relative potency of agonists.

This model accommodates the confounding experimental data with agonist potency ratios and accounts for different agonist potency ratios for different signaling pathways mediated by the same receptor through different unique values of β and γ for different receptors.[1] Within this context, $PACAP_{1-27}$ produces an active state with higher efficacy for cAMP stimulus components (relative to $PACAP_{1-38}$) and $PACAP_{1-38}$ produces an active state with higher efficacy for IP_3 stimulus components. This phenomenon was named 'stimulus trafficking'[1] and shortly after this model was proposed, a large body of data that diverged from the simple prediction of a single receptor active state was re-evaluated in terms of this selective active state model.[9, 10] It should be noted that in independent studies not involving these formal receptor models, agonist and antagonist selectivity based on tissue factors also were noted in the literature as well.[2-5]

THERMODYNAMIC MECHANISM(S)

A mechanism requiring that agonists can stabilize a unique receptor conformation presupposes that the receptor can adopt more than one active (in the sense that it mediates a cellular function) conformation. In fact, molecular dynamics suggest that a range of receptor conformations can be adopted by 7TM receptors. A useful idea in this regard is to consider proteins, such as 7TM receptors, as being intrinsically disordered and traversing an 'energy landscape' to adopt many isoenergetic conformations in response to changes in thermal energy.[11-14] There are data to suggest that cell signaling proteins contain regions of 'intrinsic disorder' in disproportionately higher levels than other proteins,[14,15] suggesting that allosteric proteins like 7TM receptors are especially prone to producing collections of conformations (called protein ensembles). The idea that some of these can be stabilized through selective binding (described by Burgen[19] as 'conformational selection') defines the mechanism by which ligands can influence such 7TM ensembles. Specifically, through a bias in ligand binding, ligands enrich the conformations for which they have high affinity at the expense of others (Le Chatelier's principle), thereby shifting the entire ensemble to a unique composition. Advancing technology has enabled direct observation of receptor conformational changes in response to ligand binding,[17-22] and these experiments have confirmed the notion that different ligands can stabilize different receptor conformations. In addition, 7TM receptors have been shown to be much more than their common namesake, GPCR (G protein-coupled receptors). Specifically, 7TM receptors are now known to interact with many more cytosolic components than G proteins,[23-27] and many of these interactants (e.g., GRKs and β-arrestin) are involved in 7TM receptor behaviors that can be ligand driven. The fact that a 7TM receptor does not exercise its full repertoire of behaviors spontaneously, but rather only does so at the bidding of ligands such as agonists and antagonists, indicates that conformations of the receptor can be linked to different facets of 7TM receptor cell behavior.

The cytosolic proteins that can interact with receptors do so at different regions of the receptor protein. For example, mutation studies on the human α_2-adrenoceptor identify the second cytosolic loop as essential for activation of G_s and the third cytosolic loop for activation of G_i/G_o.[28] In contrast, interaction with β-arrestin for the thyrotropin-releasing hormone receptor utilizes the C-terminal tail.[29] It is relevant to note that changes in receptor conformation are not concerted (i.e., regions of the protein change with thermal energy and not all regions change in an identical fashion).[11,12,30–33] The data to indicate that 7TM receptors can form numerous ligand-stabilized conformations and the fact that different cytosolic interactants with 7TM receptors do so in different regions of the receptor lead to the conclusion that not all receptor conformations will expose these regions in an identical fashion with each new conformation that the receptor adopts. Evidence consistent with this view is found in ligand trafficking of stimulus in calcitonin receptors. Specifically, it has been observed that different ligands are heterogeneously sensitive to cellular background levels of G_i and G_s protein, that is, differentially stabilized conformations produce preferential G_i or G_s signaling.[34] Therefore, this supports the notion that different ligands, by stabilizing different receptor conformations, can target specific 7TM receptor behaviors; this is the essence of functional selectivity. Early data identified multiple active states not directly but rather through differences in receptor function, indicating that different receptor active states mediate different functions, that is, activation of different cytosolic signaling pathways and functions such as internalization.[10] The promise of this pharmacological phenomenon is that chemical structures can be devised to stabilize selected receptor active states to produce selective activation of cytosolic processes for therapeutic benefit.

Functional selectivity is based on the biochemical behavior known to as protein allosterism. From the description of allosterism for enzymes forty years ago[35,36] has come descriptions of receptors as multistate functional proteins[37–39] and the description of allosteric ligands for various receptors (e.g., muscarinic receptors[40–42]). These data precede the formal description of functional selectivity apparently separating this latter behavior of receptors as being special. However, it should be noted that functional selectivity is simply an extension of the natural allosteric nature of 7TM receptors. Whereas conventional allosteric behaviors describe the effects of allosteric modulators on the affinities and efficacies of probe ligands (e.g., endogenous agonists), functional selectivity describes this same behavior toward cytosolic interacts with the receptor. Receptor allosterism must be viewed in terms of relationships, namely a modulator, conduit (in this case the receptor), and guest (the molecule upon which the allosteric effect is imposed) (see Figure 14.2). It should be noted that modulators and guests are essentially interchangeable in their roles because allosteric effect is reciprocal. For example, if a modulator produces a tenfold increase in the affinity of a given guest, it is also true that the presence of the guest on the conduit imposes a tenfold increase in the affinity of the modulator. Receptor allosterism also should be viewed as vectorial. Thus, the classical model of guest allosterism is defined as a modulator changing the behavior of a given conduit toward another ligand

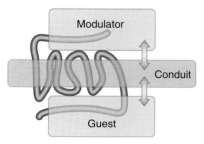

Figure 14-2: Vectorial flow of allosteric energy. An allosteric modulator binds to the receptor (protein conduit) to modify conformation and change the interaction of the receptor with a guest molecule. The flow of energy is bidirectional in that the effects of the modulator on the guest is mirrored by the same effect of the guest on the modulator.

(ligand binding–domain directed allostery). If the allosterism is expressed along the plane of the cell membrane, it can result in receptor dimerization and/or association with other membrane-bound proteins (laterally directed allostery). Finally, an allosteric effect directed toward the cytosol is functional selectivity. In these cases, various modulators (functionally selective agonists and antagonists) interact with protein conduits (receptors) to impose selective effects on various guests – in this case, cytosolic signaling proteins such as G proteins or β-arrestin.

A key feature of allosteric mechanisms is probe dependence whereby a given modulator may have very different effects on various guests occupying the receptor. For example, the modulator eburnamonine acts through the conduit of the muscarinic M2 receptor to change the responsiveness of the receptor to various muscarinic agonists. Thus, the affinity for acetylcholine is increased by a factor of fifteen, whereas the affinity for pilocarpine is reduced by a factor of 25.[43] When such allosteric probe dependence is directed towards the cytosol, it can result in agonist functional selectivity. Thus, in these terms, the selective agonists are the modulators, the receptor the conduit, and various cytosolic signaling proteins the guests.

There are two obvious points of control over these signaling systems; one, namely the chemical structure of the ligand, is fortuitous to the drug discovery process because it enables medicinal chemists to selectively stabilize various receptor conformations to reduce or at least change the signaling capability of the receptors. However, the other point of control, namely the cell, presents a difficulty to the drug-discovery process. Functionally selective effects are notoriously cell-type dependent, and this poses a hurdle to discovery efforts that utilize model systems (in one cell type) to predict drug effects in other (possibly unspecified) systems utilizing different cell types. Cell type dependence for functionally selective effects is an expected hazard if it is viewed in the modulator/conduit/guest model. Specifically, the various cytosolic signaling proteins (i.e., G proteins, β-arrestin, etc.) would constitute varying identities and quantities (variant stoichiometries between receptors and signaling proteins) of guest molecules, thus providing a variation in the thermodynamic allosteric effects imposed by different modulators on the conduit receptor.

It is worth examining functional selectivity in terms of points of pharmacological control and the factors relevant to harnessing this phenomenon for possible therapeutic utility. This can be done within the context of three requirements for attainment of pharmacological control of functional selectivity:

- Use of appropriate assays to detect and quantify functional selectivity
- Use of system-independent quantitative scales to measure and compare biased ligand effects
- De-emphasis and nullification of cell-type variability in functional selectivity.

It is worth discussing these issues within the context of new drug discovery.

PHARMACOLOGICAL ASSAYS

One approach to harnessing functional selectivity for therapeutics is to use appropriate pharmacological assays optimized to detect it and subsequently to yield scales of relative selectivity that can be used to guide medicinal chemists in the process of optimizing selective pathway activation. Pharmacological assays are a major tool pharmacologists use to define efficacies for ligands, and it has been demonstrated that some interactions may not be observed without the use of specific assays designed to see them – in other words, if you don't ask the question you will not learn the answer. For example, propranolol, a β-blocker first described over forty-five years ago,[44] is possibly one of the most, if not the most, studied drugs in the world (Ovidweb reference disclosures numbering 52,693 with 3,812 clinical trials [all Medline and Current Contents sources]), yet forty-one years after discovery, it was found that propranolol is a β-arrestin dependent extracellular signal-related kinase (ERK) agonist.[45] This discovery was facilitated through the ability to see the effect through the eyes of a new assay.

A basic tenet of this approach is that functionally multiplexed response assays, namely those allowing the maximal interaction of receptors with multiple cytosolic components, are the best for characterizing functional selectivity. The optimal system for detecting differences in receptor states in different regions is to be able to monitor multiple probes for these regions. Thus, it would be assumed that a live cell that interrogates the receptor with multiple cytosolic probes would be better suited to detect functional selectivity than a binding assay utilizing a single radioligand probe. Technology is providing the means to affect multiple probing of 7TM receptors in the form of whole cell functional assays. Thus, a number of systems designed to monitor receptor function through interactions with multiple cellular proteins based on imaging,[46–48] enzyme complementation,[49, 50] protease activation of transcriptional enzyme fragments,[51, 52] electrical impedance of cell monolayers,[53–55] and optical observation of dynamic mass redistribution in cells[56–61] are now available to measure whole-cell response to agonists. Whereas these systems are optimal for detecting functionally selective effects, they also unveil the problems in quantifying such effects with scales that are independent of cell type.

A QUANTITATIVE SCALE FOR FUNCTIONAL SELECTIVITY

If it is assumed that the receptor is the minimal recognition unit for molecules as they transmit information from the extracellular to the intracellular space, then the assay format used to monitor receptor activation should be immaterial. In practice, however, functional selectivity has been seen to be cell-type dependent. For example, there are a number of opioid receptor agonists that have been shown to be functionally selective with respect to G protein activation and receptor internalization. In addition, the ability of these opioid agonists to cause internalization of δ-opioid receptors is notoriously cell-type dependent,[62–66] presumably reflecting variation in receptor/β-arrestin stoichiometries in different cell types. This observed cell dependence of a functionally selective agonist effect in turn hinders quantitative description of agonism for structure-activity relationships. On the surface, cell-type variability for an agonist effect negates the concept of 'efficacy' as a useful system-independent parameters for drug classification. Whereas efficacy originally was described in terms of whole cell effect,[67] there is a component of efficacy that describes thermodynamic differences between the unliganded and ligand-bound receptor; this was termed by Furchgott[68] as 'intrinsic efficacy'. This parameter defines the isomerization of the receptor induced by agonist binding to become an active state capable of mediating cellular processes. So whereas intrinsic efficacy is a thermodynamic constant, cell-based efficacy can be variable. Cell dependence occurs as a result of the various stoichiomentries of coupling components in the cell which in turn vary according to requirements of the cell for normal function. The question, therefore, is not whether efficacy exists but rather if it can be utilized effectively to quantify agonist response in a system independent manner.

One possible approach is to measure a surrogate of 'isomerization' efficacy (true agonist-receptor efficacy) by measuring an immediate consequence of the interaction between the isomerized receptor and cellular signaling components. For example, the measurement of a second messenger such as cAMP may be suitably dependent on the receptor active state because it interacts with G_s protein so as to be a receptor-dependent, and not cell dependent scale of activation. Thus, a surrogate of isomerization efficacy would be 'pathway efficacy' (see Figure 14.3). These various pathway stimuli would then combine to yield cellular effect, and it would be at this point that the impact of cell type would be imposed (Figure 14.3). Variances in the quantity of the principle receptor coupling component would only alter the maximal scale of the pathway signal but not the relationships between different agonists.

The operational model[6] can be used to quantify pathway efficacy in the form of τ values for a given receptor-linked response system. The efficacy (τ) and affinity (K_A) would be unique identifiers of the agonist-dependent receptor active state and guest response coupler (i.e., G_s protein, β-arrestin, etc.). When a receptor is

Table 14-1. Biased Agonists for AA Release and IP$_3$ Production via 5-HT$_{2C}$ Receptors

Agonist	Arachidonate Release		IP$_3$ Production	
	τ/K_A (M^{-1})	Rel. 5-HT	τ/K_A (M^{-1})	Re. 5-HT
5-HT	1.5×10^7	1	1×10^7	1
DOI	7.92×10^6	0.528	1.38×10^6	0.138
TFMPP	3.67×10^6	0.244	1.75×10^6	0.175
AA Release vs IP$_3$ Production Pathway Bias (Relative to 5-HT)				
	DOI=	0.528/0.138	3.82	
	TFMPP=	0.244/0.175	1.39	

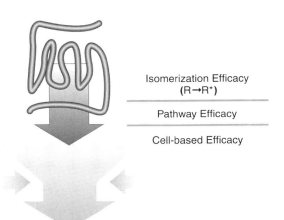

Isomerization Efficacy
(R→R*)

Pathway Efficacy

Cell-based Efficacy

Figure 14-3: Biased antagonism. A natural ligand A activates multiple signaling pathways in the cell until an allosteric antagonist B binds to the receptor and modifies this pattern. Specifically, the natural agonist will produce biased agonism in the presence of the antagonist.

promiscuous with respect to the number of response pathways associated with it, then there will be some relative emphasis of activation of these pathways by the natural endogenous agonist. The 'bias' of activation of other agonists then can be compared to that of the natural agonist. From these types of measurements, the bias for a given agonist for a given pathway (in the form of τ/K_A ratios) can be determined in a cell-independent way to guide medicinal chemists in structure activity studies. Table 14.1 shows the bias calculated with the operational model for three 5-HT$_C$ receptor agonists; the natural agonist serotonin and two functionally selective agonists DOI and TFMPP.[70] In this functional system, 5-HT$_C$ receptors couple separately to pathways mediating arachidonate acid (AA) release and IP$_3$ production. Calculation of τ/K_A values for each pathway for DOI and TFMPP, relative to values for serotonin, can be used to furnish cell-type independent estimates of the bias of the synthetic agonists for each pathway. It can be seen from the calculations shown in Table 14.1, TFMPP is 1.39-fold and DOI 3.82-fold more biased toward AA release (than IP$_3$ production) than is serotonin. These bias ratios then can be used by medicinal chemists to gauge the functional selectivity of these agonists for 5-HT$_C$ receptors for these two pathways in a cell type-independent manner.

In general, there is a great deal of evidence to support the notion that synthetic agonists show bias in receptor activation compared to natural endogenous agonists. It is still an open question whether this powerful mechanism is utilized by natural systems for fine control. It could be postulated that for receptors with redundant endogenous agonists such as chemokines, functionally selective signaling would be a way to differentiate signals, and some evidence for this has been reported. For instance, CCL19 and CCL21 are two natural agonists for the CCR7 chemokine receptor. While they activate G proteins through a common receptor, they differ in the type of pathway stimulation they elicit.

Specifically, only CCL19 (not CCL21) produces receptor agonist-dependent phosphorylation and recruitment of β-arrestin to terminate the G protein stimulus.[71] Similarly, there are indications that functional selectivity may furnish a means to achieve unique therapeutic phenotypes for the treatment of CNS disorders through dopamine[72] and serotonin receptors.[73] Selective coupling also has been associated with thyrotropin physiology. Specifically, the thyrotropin receptor couples to G_s and G_q proteins but each mediates separate effects, the G_s protein coupling may be associated with thyroid growth and differentiation, whereas the G_q coupling may be more associated with thyroid hormone synthesis.[74] The orexin receptor also has been shown to signal selectively for the production of adrenal steroids and their release (G_s protein) and catecholamine release (G_q protein).[75,76]

ANTAGONIST BIAS

The ongoing discussion considers direct trafficking of receptor stimulus to different signaling pathways and, as such, involves agonists. However, allosteric antagonists also can cause natural agonists to adopt biased profiles – see Figure 14.4. For example, the antagonist Nα-tosyltryptophan imposes biased agonism onto the natural agonist prostaglandin D2 on CRTH2 receptors. Specifically, while the natural agonist normally activates G_i protein and β-arrestin, binding of the antagonist changes this profile to one of sole activation of G_i (with no concomitant β-arrestin interaction[77]). Similarly, the natural agonist neurokinin A activates G_s and G_q through neurokinin (NK2) receptors, but the allosteric modulator LP1805 changes this pattern to one of enhanced G_q and blockade of G_s activation.[78] In general, it can be seen from the previous discussion that natural signaling processes can be biased through the intervention of allosteric (biased) antagonists. This adds yet another dimension to the therapeutic use of antagonists.

CONCLUSIONS

The past fifteen years of pharmacological research have revealed that 7TM receptors are much more than rheostats for cell stimulus triggered by hormones, autacoids, and neurotransmitters. Through judicious use of biased agonists and

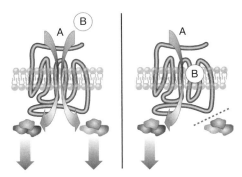

Figure 14-4: Different views of receptor/agonist efficacy. When a ligand binds to the receptor, the protein becomes a different species (is isomerized); this process is termed isomerization efficacy and is formally identical to Furchgott's 'intrinsic efficacy'.[68] The activated receptor directly interacts with cytosolic components such as G proteins and β-arrestin. The immediate result of this interaction is termed pathway efficacy and may be a surrogate for isomerization efficacy. The cell combines all existing pathway efficacies for its own needs and yields an agonist-mediated cellular behavior. This is termed 'cell-based' efficacy and is formally identical to Stephenson's[67] concept of isolated tissue efficacy.

antagonists, complex patterns of signaling can be obtained; the new frontier in this area of pharmacological research now is not to show that these phenomena can occur but rather which of these biased signal patterns offer unique therapeutic outcomes in the treatment of disease. In this regards, the clinical effects of functionally selective ligands, as evaluated with translational medicine, will be the key to fully evaluating this pharmacological receptor behavior.

REFERENCES

1. Kenakin, T.P. (1995) Agonist-receptor efficacy II: agonist trafficking of receptor signals. *Trends Pharmacol. Sci.* 16: 232–238
2. Roth, B.L., Chuang, D.-M. (1987) Multiple mechanisms of serotonergic signal transduction. *Life Sci.* 41: 1051–1064
3. Lawler, C.P., Prioleua, C., Lewis, M.M., Mak, C., Jiang, D., Schetz, J.A., Gonzalez, A.M., Sibley, D.R., and Mailman, R.B. (1999) Interactions of the novel antipsychotic aripiprazole (OPC-14597) with dopamine and serotonin receptor subtypes *Neuropsychopharmacol.* 20: 612–627
4. Ward, J.S., Merrit, L., Calligaro, D.O., Bymaster, F.P., Shannon, H.E., Sawyer, B.D., Mitch, C.H., Deeter, J.B., Peters, S.C., Sheardown, M.J., Olesen, P.H., Swedberg, M.D.B., and Sauerberg, P. (1995) Functionally selective M1 muscarinic agonists. 3. Side chains and azacycles contributing to functional muscarinic selectivity among pyrazacycles. *J. Med. Chem.* 38: 3469–3481
5. Heldman, E., Barg, J., Fisher, A., Levy, R., Pittel, Z., Zimlichman, R., Kushnir, M. and Vogel, Z. (1996) Pharmacological basis for functional selectivity of partial muscarinic receptor agonists. *Eur. J. Pharmacol.* 297: 283–291
6. Black J.W., Leff P. (1983) Operational models of pharmacological agonist. *Proc. R. Soc. Lond.* [Biol.] 220:141–162
7. Spengler, D., Waeber, C., Pantoloni, C., Holsboer, F., Bockaert, J., Seeburg, P.H., and Journot, L. (1993) Differential signal transduction by five splice variants of the PACAP receptor. *Nature* 365: 170–175

8. Kenakin, T.P., Morgan, P.H. (1989) The theoretical effects of single and multiple transducer receptor coupling proteins on estimates of the relative potency of agonists. *Mol. Pharmacol.* 35:214–222

9. Kenakin, T.P. (2002) Efficacy at G protein coupled receptors. *An. Rev. Pharmacol. Toxicol.* 42: 349–379

10. Kenakin, T.P. (2003) Ligand-selective receptor conformations revisited: the promise and the problem. *Trends Pharmacol. Sci.* 24: 346–354

11. Fraunfelder, H., Parak, F., and Young, R.D. (1988) Conformational substrates in proteins. *Annu. Rev. Biophys. Biophys. Chem.* 17: 451–479

12. Fraunfelder, H., Sligar, S.G., and Wolynes, P.G. (1991) The energy landscapes and motions of proteins. *Science* 254: 1598–1603

13. Hilser, V.J., Garcia-Moreno, B., Oas, T.G., Kapp, G., and Whitten, S.T. (2006) A statistical thermodynamic model of the protein ensemble. *Chem. Rev.* 106: 1545–1558

14. Hilser, V.J., Thompson, E.B. (2007) Intrinsic disorder as a mechanism to optimize allosteric coupling in proteins. *Porc. Nat. Acad. Sci. USA* 104: 8311–8315

15. Liu, J., Perumal, N.B., Oldfield, J., Su, E.W., Uversky, V.N., Dunker, A.K. (2006) Intrinsic disorder in transcription factors. *Biochem.* 45: 6873–6888

16. Burgen, A.S.V. (1966) Conformational changes and drug action. *Fed. Proc.* 40: 2723–2728

17. Gether, U., Sansan, L., and Kobilka, B.K. (1995) Fluorescent labeling of purified β2-adrenergic receptor: evidence for ligand specific conformational changes. *J. Biol. Chem.* 270: 28268–28275.

18. Ghanouni, P., Gryczynski, Z., Steenhuis, J.J., Lee, T.W., Farrens, D.L., Lakowicz, J.R., Kobilka, B.K. (2001) Functionally different agonists produce distinct conformations in G-protein coupling domains of the b2-adrenergic receptor. *J. Biol. Chem.* 276: 24433–24436

19. Hruby, V.J., Tollin, G. (2007) Plasmon-waveguide resonance (PWR) spectroscopy for directly viewing rates of GPCR/G-protein interactions and quantifying affinities. *Curr. Opin Pharmacol.* 7: 507–514

20. Okada, T., Palczewski, K. (2001) Crystal structure of rhodopsin: implications for vision and beyond. *Curr. Opin. Struc. Biol.* 11: 420–426

21. Palanche, T., Ilien, B., Zoffmann, S., Reck, M.-P., Bucher, B., Edelstein, S.J., and Galzi, J.-L. (2001) The neurokinin A receptor activates calcium and cAMP responses through distinct conformational states. *J. Biol. Chem.* 276: 34853–34861

22. Swaminath, G., Xiang, Y., Lee, T.W., Steenhuis, J., Parnot, C., Kobilka, B.K. (2004) Sequential binding of agonists to the β2 adrenoceptor: kinetic evidence for intermediate conformational states. *J. Biol. Chem.* 279: 686–691

23. Lefkowitz, R.J., Shenoy, S.K. (2005) Transduction of receptor signals by β-arrestins. *Science* 308: 512–517

24. Luttrell, L.M. (2005) Composition and function of G protein-coupled receptor signalsomes controlling mitogen-activated protein kinase activity. *J. Mol. Neurosci.* 26: 253–263

25. Tilakaratne, N., Sexton, P.M. (2005) G-protein-coupled receptor-protein interactions: basis for new concepts on receptor structure and function. *Clin. Expt. Pharmacol. Physiol.* 32: 979–987

26. Wang, Q., Limbird, L.E. (2007) Regulation of alpha(2)AR trafficking and signaling by interacting proteins. *Biochem. Pharmacol.* 73: 1135–1145

27. Brady, A.E., Limbird, L.E. (2002) G protein-coupled receptor interacting proteins: emerging roles in localization and signal transduction [review]. *Cell Signal.* 14: 297–309

28. Ikezu, T., Okamoto, T., Ogata, E., and Nishimoto, I. (1992) Amino acids constitute a Gi-activator sequence of the α2-adrenergic receptor and have a Phe substitute in the G-protein-activator sequence motif. *FEBS Lett.* 311: 29–32

29. Jones, B.W., Hinkle, P.M. (2008) Arrestin binds to different phosphorylated regions of the thyrotropin-releasing hormone receptor with distinct functional consequences. *Mol. Pharmacol.* 74: 195–202

30. Hilser, J. and Freire, E. (1997) Predicting the equilibrium protein folding pathway: structure-based analysis of staphylococcal nuclease. *Protein Struct. Funct. Bioinform.* 27: 171–183

31. Hilser, V.J., Dowdy, D., Oas, T.G., and Freire, E. (1998) The structural distribution of cooperative interactions in proteins: analysis of the native state ensemble. *Proc. Natl. Acad. Sci. USA* 95: 9903–9908

32. Woodward, C., Simon, I., and Tuchsen, E. (1982) Hydrogen exchange and the dynamic structure of proteins. *Mol. Cell. Biochem.* 48:135–141

33. Woodward, C. (1993) Is the slow-exchange core the protein folding core? *Trends Biochem. Sci.* 18: 359–360

34. Watson, C., Chen, G., Irving, P.E., Way, J., Chen, W.-J. and Kenakin, T.P. (2000) The use of stimulus-biased assay systems to detect agonist-specific receptor active states: implications for the trafficking of receptor stimulus by agonists. *Mol. Pharmacol.* 58: 1230–1238

35. Koshland, D.E. (1960) The active site of enzyme action. *Adv. Enzymol.* 22: 45–97

36. Monod, J., Wyamn, J., and Changeux, J.P. (1965) On the nature of allosteric transitions. *J. Biol. Chem.* 12: 88–118

37. Thron, C.D. (1973) On the analysis of pharmacological experiments in terms of an allosteric receptor model. *Mol. Pharmacol.* 9: 1–9

38. Karlin, A. (1967) On the application of 'a plausible model' of allosteric proteins to the receptor for acetylcholine. *J. Theoret. Biol,* 16: 306–320

39. Colquhoun, D. (1973) The relationship between classical and cooperative models for drug action. In: A Symposium on Drug Receptors, ed. by H.P. Rang, pp 149–182, Baltimore: University Park Press.

40. Birdsall, N.J., Hulme, E.C., and Stockton, J.M. (1983) Muscarinic receptor subclasses: allosteric interactions. *Cold Spring Harbor Symposia on Quantitative Biology.* 48 Pt 1: 53–56

41. Stockton, J.M., Birdsall, N.J., Burgen, A.S., and Hulme, E.C. (1983) Modification of the binding properties of muscarinic receptors by gallamine. *Molecul Pharmacol.* 23: 551–557

42. Jakubik, J., el-Fakahany, E.E. (1997) Positive cooperativity of acetylcholine and other agonists with allosteric ligands on muscarinic acetylcholine receptors. *Mol. Pharmacol.* 52: 172–179

43. Jakubik, J., Bacakova, L., Lisa, V., el-Fakahany, E.E., and Tucek, S. (1996) Activation of muscarinic acetylcholine receptors via their allosteric binding sites. *Proc. Nat Acad Sci USA* 93: 8705–8709

44. Black, J.W., Duncan, W.A., and Shanks, R.G. (1965) Comparison of some properties of pronethalol and propranolol. *Br. J. Pharmacol. Chemo.* 25:577–591

45. Azzi, M., Charest, P.G., Angers, S., Rousseau, G., Kohout, T., Bouvier, M., and Piñeyro, G. (2003) β-arrestin-mediated activation of MAPK by inverse agonists reveals distinct active conformations for G-protein-coupled receptors. *Proc. Natl. Acad. Sci. USA* 100: 11406–11411

46. Milligan, G. (2003) High-content assays for ligand regulation of G-protein coupled receptors. *Drug Disc. Today* 8: 579–585

47. Lefkowitz, R.J., Whalen, E.J. (2004) β-arrestins: traffic cops of cell signaling. *Curr. Opin. Cell Biol.* 16: 162–168

48. Fredriksson, R., Schioth, H.B. (2005) The repertoire of G-protein receptors in fully sequenced genomes. *Mol. Pharmacol.* 67: 1414–1425

49. Olson, K.R., Eglen, R.M. (2007) Beta galactosidase complementation: a cell-based luminescent assay platform for drug discovery. *Assay Drug Dev Technol.* 5: 137–144

50. Zhao, X., Jones, A., Olson, K.R., Peng, K., Wehrman, T., Park, A., mallari, R., Nebalasca, D., Young, S.W., and Xiao, S.-H. (2008) A homogeneous enzyme fragment complementation-based b-arrestin translocation assay for high-throughput screening of G-protein-coupled receptors. *J. Biomol. Screen* 13: 737–747

51. Barnea, G., Strapps, W., Herrada, G., Berman, Y., Ong, J., Kloss, B., Axesl, R., and Lee, K.J. (2008) The genetic design of signaling cascades to record receptor activation. *Proc. Natl. Acad. Sci USA* 105: 64–69

52. Verkaar, F., van Rosmalen, J.W.G., Blomenrohr, M., van Koppen, C.J., Blankesteijn, W.M., Smits, J.F.M., and Zaman, G.J.R. (2008) G-protein independent cell-based assays for drug discovery on seven-transmembrane receptors. *Biotechnol. Ann. Rev.* 14: 253–274

53. Verdonk, E., Johnson, K., McGuinness, R., Leung, G., Chen, Y.-W., Tang, H.R., Michelotti, J.M., and Liu, V.F. (2006) Cellular dielectric spectroscopy: a label-free comprehensive platform for functional evaluation of endogenous receptors. *Assay Drug Dev. Technol.* 4: 609–619

54. McGuinness, R. (2007) Impedance-based cellular assay technologies: recent advances. *Curr. Opin. Pharmacol.* 7: 535–540

55. Peter, M.F., Knappenberger, K.S., Wilkins, D., Sygowski, L.A., Lazor, L.A., Liu, J., and Scott, C.W. (2007) Evaluation of cellular dielectric spectroscopy, a whole-cell label-free technology for drug discovery on Gi-coupled GPCRs. *J. Biomol. Screen* 12: 312–319

56. Fang, Y., Ferrie, A.M., Fontaine, N.H., and Yuen, P.K. (2005) Characteristics of dynamic mass redistribution of EGF receptor signaling in living cells measured with label free optical biosensors. *Anal. Chem.* 77: 5720–5725

57. Fang, Y., Li, G., and Peng, J. (2005) Optical biosensor provides insights for bradykinin B2 receptor signaling in A431 cells. *FEBS Lett* 579: 6365–6374

58. Fang, Y. (2006) Label-free cell-based assays with optical biosensors in drug discovery. *Assay Drug Dev. Technol.* 4: 583–595

59. Cunningham, B.T., Li, P., Schultz, S., Lin, B., Baird, C., Gerstenmaier, J., et al. (2004) Label free assays on the BIND system. *J Biomolec. Screen.* 9: 481–490

60. Yu, N., Atienza, J.M., Bernard, J., Blanc, S., Zhu, J., Wang, X., et al. (2006) Real-time monitoring of morphological changes in living cells by electronic cell sensor assays: an approach to study G-protein coupled receptors. *Anal. Chem.* 78: 35–43

61. Fang, Y., Ferrie, A.M., Fontaine, N., Mauro, J., and Balikrishnan, J. (2006) Resonant waveguide grating biosensors for living cell sensing. *Biophys. J.* 91: 1925–1940

62. Whistler, J.L., Van Zastrow, M. (1998) Morphine-activated opioid receptors elude desensitization by β-arrestin. *Proc. Nat. Acad. Sci. USA* 95: 9914–9919

63. Zhang Jie Zhang, Ferguson, S.S.G., Brak, L.S., Bodduluri, S.R., Laporte, S.A., Law, P-Y., and Caron, M.G. (1998) Role for G protein-coupled receptor kinase in agonist-specific regulation of μ-opioid receptor responsiveness. *Proc. Natl. Acad. Sci. USA* 95: 7157–7162

64. Bailey, C. P., Couch, D., Johnson, E., Griffiths, K., Kelly, E., and Henderson G. (2003) μ-Opioid receptor desensitization in mature rat neurons: lack of interaction between DAMGO and morphine. *J. Neurosci.* 23: 10515–10520

65. Bohn, L.M., Dykstra, L.A., Lefkowitz, R.J., Caron, M.G., and Barak, L.S. (2004) Relative opioid efficacy is determined by the complements of the G protein coupled receptor desensitization machinery. *Mol. Pharmacol.* 66: 106–112

66. Koch, T., Widera, A., Bartzsch, K., Schulz, S., Brandenburg, L.-O., Wundrack, N., Beyer, A., Grecksch, G., and Höllt V. (2005) Receptor endocytosis counteracts the development of opioid tolerance. *Mol. Pharmacol.* 67: 280–287

67. Stephenson, R.P. (1956) A modification of receptor theory. *Br. J. Pharmacol.* 11: 379–393

68. Furchgott, R.F. (1966) The use of b-haloalkylamines in the differentiation of receptors and in the determination of dissociation constants of receptor-agonist complexes. In Advances in Drug Research, vol 3 ed. by N.J. Harper and A.B. Simmonds, pp 21–55, Academic Press, London, New York

69. Berg, K.A. et al. (1998) Effector pathway-dependent relative efficacy at serotonin type 2A and 2C receptors: evidence for agonist-directed trafficking of receptor stimulus. *Mol. Pharmacol.* 54, 94–104

70. Kenakin, T.P. (2009) 7TM receptor allostery: putting numbers to shapeshifting proteins. *Trends Pharmacol. Sci.* (in press)

71. Kohout, T.A., Nicholas, S.L., Perry, S.J., Reinhart, G., Junger, S., and Struthers, R.S. (2004) Differential desensitization, receptor phosphorylation, β-arrestin recruitment, and ERK1/2 activation by the two endogenous ligands for the CC chemokine receptor 7. *J Biol Chem.* 279: 23214–23222

72. Mailman, R.B. (2007) GPCR functional selectivity has therapeutic impact. *Trends Pharmacol Sci.* 28: 390–397

73. Schmid, C.L., Raehal, K.M., and Bohn L.M. (2008) Agonist-directed signaling of the serotonin 2A receptor depends on β-arrestin-2 interactions in vivo. *Proc Natl Acad Sci USA* 105: 1079–1084

74. Vassart, G., Dumont, D. (1992)The thyrotropin receptor and the regulation of thyrocyte function and growth. *Endocr Rev.* 13: 596–611

75. Mazzocchi, G., Malendowicz, L.K., Aragona, F., and Nussdorfer, G.G. (2001) Human pheochromocytomas express orexin receptor type 2 gene and display an in vitro secretory response to orexins A and B. *J Clin Endocrinol Metab.* 86: 4818–4821

76. Mazzocchi, G., Malendowicz, L.K., Gottardo, A.F., and Nussdorfer, G.G. (2001) Orexin A stimulates cortisol secretion from human adrenocortical cells through activation of the adenylate cyclase-dependent signaling cascade. *J Clin Endocrinol Metab.* 86: 778–782

77. Mathiesen, J.M., Ulven, T., Martini, L., Gerlach, L.O., Heineman, A., and Kostenis, E. (2005) Identification of indole derivatives exclusively interfering with a G protein-independent signaling pathway of the prostaglandin D2 receptor CRTH2. *Mol Pharmacol.* 68: 393–402

78. Maillet, E.L., Pellegrini, N., Valant, C., Bucher, B., Hibert, M., Bourguignon, J-J., and Galzi, J-L. (2007) A novel, conformation-specific allosteric inhibitor of the tachykinin NK2 receptor (NK2R) with functionally selective properties. *FASEB J* 21: 2124–2134.

PART V: PHYSIOLOGICAL FUNCTIONS AND DRUG TARGETING OF GPCRS

15 β-Adrenoceptors in cardiovascular and respiratory diseases

Michele Ciccarelli, J. Kurt Chuprun, and Walter J. Koch

FUTURE PERSPECTIVES OVERVIEW

β-Adrenoceptors (βARs) of which there are three subtypes (β_1, β_2 and β_3) belong to the superfamily of seven transmembrane-spanning receptors or G protein-coupled receptors (GPCRs) and are the binding sites triggering the physiological actions of the catecholamine neurotransmitters, epinephrine and norepinephrine. Classically, the actions of catecholamines via βAR stimulation occur following activation of heterotrimeric G proteins and subsequent intracellular signal transduction pathways. βARs play a profound role in cardiovascular and pulmonary function because they are highly expressed on cardiac, vascular smooth muscle, and airway smooth muscle cells. This chapter will highlight the latest translational research that has recently provided important insight into how normal and abnormal βAR signaling can impact the cardiovascular and respiratory systems, including how mutations of these receptors, present in the human population, impact health and disease.

INTRODUCTION

β-adrenoceptors receptors (βARs) have been extensively studied in the last forty years. Genetic and molecular studies have confirmed three subtypes (β_1-, β_2- and β_3-) based on their structure and function as well as tissue localization. Discoveries of the mechanisms governing βAR signaling has allowed development of therapeutic strategies to combat cardiovascular and respiratory diseases including βAR antagonists for hypertension and heart failure and selective β_2AR agonists for asthma and airway disease. Further, targeting specific regulation of these receptors may lead to future novel therapeutic strategies for heart and pulmonary diseases. More recently, human genetics and pharmacogenomic studies have identified single nucleotide polymorphisms (SNPs) in the genes encoding βARs that appear to have significant impact on cardiopulmonary disease outcomes and define initial paths for "personalized medicine."

PHYSIOLOGY AND PHARMACOLOGY OF β-ADRENOCEPTORS RECEPTORS (βARs)

βAR Structure and Function

βAR Structure and Subtypes

βARs belong to the superfamily of seven transmembrane receptors, also known as GPCRs.[1-4] Models for βARs indicate seven relatively α-helical hydrophobic domains that anchor the receptors in the plasma membrane with the amino-terminus extracellular and the carboxyl-terminus intracellular. Interaction with heterotrimeric G proteins occurs within the intracellular loops between transmembrane α-helices while ligand binding occurs within pockets of these α-helices.[2–4] The three known βAR subtypes, $β_1$, $β_2$, and $β_3$, differ primarily in their location: $β_1$ARs predominate in the heart, cerebral cortex, and kidney.[5,6] The major βAR subtype in the lung, cerebellum, uterus, skeletal muscle, and blood vessels is the $β_2$AR.[5,7–9] The $β_3$AR is highly expressed in brown and white adipose tissue, but has also been detected in brain, stomach, and gallbladder.[10–12]

The classification of $β_1$- and $β_2$ARs was initially based upon their relative affinities for epinephrine and norepinephrine.[13] $β_1$ARs have similar affinity for both catecholamines, whereas the $β_2$AR binds epinephrine with about thirtyfold greater affinity than it does norepinephrine. Therefore, the first differentiation between $β_1$- and $β_2$ARs was essentially functional, based on the use of a prototypical tissue (i.e., a tissue that predominantly has a single AR-subtype as evidenced by its response). Subsequent to this initial pharmacological classification, various βAR agonists and antagonists have been developed that distinguish between the two βAR subtypes, and these compounds have allowed further quantification and characterization.[14]

Multiple radioligand binding and signaling studies conducted in rat adipocyte suggested the presence of an additional βAR that was ultimately classified as the $β_3$AR.[15–18] The $β_3$AR is the predominant βAR subtype in adipose tissue but it is also present in some regions of the gut, and perhaps in skeletal and cardiac muscle.[17] This receptor is pharmacologically different from $β_1$ and $β_2$ARs in that its structure in the fourth intracellular loop has fewer sites for inactivation by phosphorylation.[16–18] Another peculiar characteristic of this receptor is that in some cases, several antagonists of $β_1$ and $β_2$ARs are partial or full agonists for the $β_3$AR.[19,20]

Importantly, molecular cloning studies confirmed the existence of three distinct βAR subtypes, having different sized RNA transcripts, protein molecular weights, and unique amino acid sequences. The three βAR subtypes have approximately 50% homology in amino acid sequence within a given species, whereas an individual βAR subtype has 75% or more homology across species. A summary of βAR properties are listed in Table 15.1.

Table 15-1. βARs Subtypes

	$β_1$AR	$β_2$AR	$β_3$AR
Prototypical tissue	Heart	Trachea	Adipocyte
Selective agonist	–	Fenoterol	CGP 12,177
Selective antagonist	CGP 20,712A	ICI 118,551	SR 59,230A
Glycosylated molecular weight	65,000	65,000	65,000
No. of amino acids (human)	477	413	408
mRNA, approx. kb	2.8	2.0	2.2
Introns	No	No	Yes
Phosphorylation sites	Yes	Yes	Few or none

βAR activation and signaling

Classical βAR signaling is mediated by activation of heterotrimeric G proteins.[2–4] Recently, G protein-independent signaling has been described that is mediated by phosphorylated receptors (see below) and signaling through β-arrestin scaffolds.[21] Details of this latter pathway is beyond the scope of this chapter. βAR-G protein activation occurs through the interaction of agonist-occupied receptor and the α-subunit of the G protein (Gα), causing activation through GDP-GTP exchange and dissociation of Gα and Gβγ subunits.[22–25] Both Gα and Gβγ then serve as signaling mediators to directly interact with a variety of effector proteins, including enzymes and ion channels.[26,27] Historically, the specificity and selectivity in GPCR signaling was thought to be achieved by coupling of a given GPCR to a single class of G proteins.[2] As a prototypical GPCR, βARs were found to interact with G_s, which in turn activated adenylyl cyclase (AC), catalyzing 3′,5′-adenosine monophosphate (cAMP) formation and activation of cAMP-dependent protein kinase (PKA). In the heart, activation of the βAR-G_s-AC-cAMP-PKA signaling axis results in the increased rate (chronotropy) and force (inotropy) of contraction of the cardiomyocyte, as well as increased rate of relaxation (lusitropy).[28,29] Primarily, these effects of this signal transduction pathway are mediated via PKA phosphorylation of several downstream molecules involved in calcium (Ca^{2+})-dependent excitation-contraction coupling, including the L-type voltage-dependent Ca^{2+} channel, the sarcoplasmic reticulum Ca^{2+}-ATPase inhibitory protein, phospholamban, and the myofilament protein, troponin I.[28–31]

The paradigm that βARs exclusively couple to the -G_s-AC-cAMP pathway is not true because βARs, like other GPCRs, appear to be able to couple to more than one G protein. This is especially true for the βARs in the heart because although the $β_1$AR appears to primarily couple to -G_s, the $β_2$AR can couple to -G_i in addition to -G_s.[28,29] Of interest, the distinct G protein coupling of $β_1$- and $β_2$ARs fulfill distinct, sometimes even opposite physiological and pathological roles. This has been shown specifically in the heart because $β_2$AR-G_i coupling appears to influence myocyte survival and can actually protect against $β_1$AR-mediated cellular apoptosis.[32–35] Further, the coupling of $β_2$ARs to -G_s appears to mediate the functional compartmentalization of these receptors,[36] a process potentially mediated via association with specific phosphatases[36] or PKA anchoring proteins.[37] βARs, like most other GPCRs, are tightly regulated by mechanisms that rapidly quench the G protein signal. This occurs primarily via the phosphorylation of activated receptors leading to the process of desensitization.[38] This process can be ascertained acutely with the reversible uncoupling of activated receptor with G protein[39] and also chronically with the specific loss of receptors on the plasma membrane, which is better classified as down-regulation.[40] Although desensitization can occur for βARs through PKA,[38] the primary mechanisms of desensitization is due to phosphorylation by a specific family of kinases known as the GPCR kinases (GRKs).[1,41,42] GRK phosphorylation of GPCRs such as βARs specifically promotes the binding of β-arrestin molecules to receptors, which sterically blocks further G protein activation.[41–43] β-arrestin binding to GRK-phosphorylated βARs dictates their subsequent fate because it initializes receptor internalization that can lead to receptor degradation, receptor resensitization via dephosphorylation, or stimulate novel intracellular kinase signaling that can be independent of G protein activation.[44,45] Thus, the importance of GRK action on GPCRs not only includes the ending of the G protein signal but also in the induction of β-arrestin signaling, which, although outside the scope of this chapter, may be a critical area of cardiovascular biology. Seven GRK family members have been identified (GRK1–7). In the heart, GRK2 (also known as the βAR kinase or βARK1) is the most abundantly expressed GRK, with GRK5 also present at significant levels.[28, 29] As detailed later in this chapter, the expression and activity of GRKs in the heart is extremely important in cardiovascular pathologies, and GRK2 particularly is emerging as a novel target to treat heart failure.[28]

Role of βARs in regulation of cardiovascular and respiratory function

A critical function of the cardiac adrenergic system is to regulate myocardial function on a beat-beat or short-term basis. βAR signaling and signal transduction systems are what allow the heart to adapt quickly to workloads that may vary dramatically, allowing the heart to increase its output by several-fold within a matter of seconds. This system, which has probably evolved to deal with trauma and blood loss, confers a considerable short-term survival advantage. However, these powerful compensatory mechanisms are also utilized to deal with hemodynamic overload, and they are also activated when myocardial contractile function is compromised for any reason. Indeed, increased adrenergic stimulation in the form of sympathetic nervous system hyperactivity is a component of several cardiac disorders. Cardiac function modulated by adrenergic neural discharge and circulating catecholamines is generally referred to as the extrinsic regulatory pathway. Increased adrenergic stimulation markedly increases heart rate, contractile strength, and relaxation rate through the βAR-dependent mechanisms discussed above. Since β_1ARs predominate in the normal heart (~75–80%), much of the adrenergic control of heart function appears to be through these receptors, although β_2ARs can also play a role.[46,47] Interestingly, β_3ARs in the human heart have been shown to be coupled to the nitric oxide (NO) system,[48] which can have inotropic and lusitropic actions mediated by cGMP.[48,49] β_2ARs (and β_3ARs) may play a more important role in compromised myocardium where there is a specific loss of β_1AR levels.[28,29,46,47]

Outside the myocardium, β_2ARs are important in cardiovascular homeostasis because they regulate vascular tone. In fact, the sympathetic nervous system has a significant influence on vascular tone, acting through α_1ARs and β_2ARs both localized on vascular smooth muscle cells.[50] Activation of α_1ARs produces vasoconstriction whereas stimulation of β_2ARs induces vasodilation. Given the predominant effect of norepinephrine on α_1ARs, the sympathetic system activation produces mainly an increase of the vascular tone. However, circulating epinephrine from the adrenal gland has a higher affinity for β_2ARs, and this results in a vasorelaxation effect. β_2ARs have also recently been shown to be of potential importance in the vascular endothelium. Endothelial cells, which line the lumen of vessels, integrate and elaborate several stimuli producing a response in vascular smooth muscle that can alter vessel tone. This occurs primarily through the release of endothelin (causing vasoconstriction) or NO (causing vasorelaxation). Recently, β_2ARs have been also localized on the endothelial cells regulating the activity of endothelial NO synthase (eNOS) and release of NO.[51-53]

β_2ARs are also critical in the regulation of respiratory function via airway smooth muscle and alveolar epithelium.[54] The adrenergic system regulates airway smooth muscle tone through β_2ARs because cAMP-dependent mechanisms are generally believed to be the key trigger for eliciting βAR-mediated relaxation of smooth muscle.[55] Downstream effectors activated via a cAMP-dependent mechanism include plasma membrane K^+ channels, such as the large-conductance, Ca^{2+}-activated K^+ (MaxiK) channel.[55] βAR-mediated relaxant mechanisms also include cAMP-independent signaling pathways. This view is supported by numerous pharmacological and electrophysiological lines of evidence. In airway smooth muscle, direct activation of the MaxiK channel by $G\alpha_s$ is a mechanism by which stimulation of β_2AR elicits muscle relaxation independently of the elevation of cAMP.[56] The increased conductance to K^+ produces cellular hyperpolarization in smooth muscle cells and bronchial relaxation. In addition to effecting airway smooth muscle relaxation by decreasing Ca^{2+} intracellular level, via hyperpolarization and/or other mechanisms, there is evidence that adrenergic agents may diminish Ca^{2+} sensitivity of the contractile elements.[57] β_2ARs are also located in the alveolar epithelium where they regulate alveolar Na^+ active transport through proteins such as the amiloride-sensitive epithelial Na^+ channel, cystic fibrosis transmembrane conductance regulator, and Na^+,K^+-ATPase.[54] Receptor-mediated increases in alveolar active Na^+ transport are likely due to direct and indirect upregulation of the epithelial Na^+ channels via recruitment from intracellular pools to the apical cell membrane.[54] These effects of β_2AR-agonists and cAMP appear to be mediated via PKA, which phosphorylates cytoskeleton proteins and promotes exocytosis to the cell membrane[58] and direct phosphorylation of epithelial Na^+ channel β and γ subunits. In addition to exocytosis from intracellular pools, signaling via β_2ARs increases the expression of epithelial Na^+ channel α-subunit mRNA and protein.[59] These findings indicate a role for β_2AR in maintaining the alveolar fluid balance and adaptation to pulmonary edema.

THERAPEUTIC USES OF βAR MODULATORS DRUGS IN CARDIOVASCULAR AND RESPIRATORY DISEASES

Therapeutic uses of βAR agonists

Cardiovascular diseases

βARs in human hearts are activated by the endogenous catecholamine agonists norepinephrine and epinephrine. These and other synthetic drugs provide an option for increasing cardiac output in acute conditions, including shock states, and in chronic heart failure when other means of support have failed.[60] However, the use of βAR agonists has been stopped for chronic cardiac diseases such as congestive heart failure due to decreased survival of patients and to the widely accepted concept that excessive catecholamine stimulation leads to a worsening of βAR function.[60] Nevertheless, βAR agonists are currently used therapeutically in acute cardiac decompensation states.

Cardiogenic shock is a condition that is characterized by a decreased pumping ability of the heart causing a shock-like state (i.e., global hypoperfusion). It most commonly occurs in association with, and as a direct result of, acute myocardial infarction. Similar to other such states, cardiogenic shock is considered to be a clinical diagnosis characterized by decreased urine output, altered mentation, and hypotension. The most recent prospective study of cardiogenic shock defines cardiogenic shock as sustained hypotension (systolic blood pressure less than 90 mm Hg lasting more than 30 min) with evidence of tissue hypoperfusion with adequate left ventricular (LV) filling pressure.[61] Intravenous infusion of inotropic drugs is required to support cardiac contractility and adequate perfusion to the vital organs.[62] Dobutamine (5–20 mcg/kg/min IV continuous infusion) as a β_1AR agonist has strong inotropic effect and less chronotropic effects, although higher doses may increase heart rate, exacerbating myocardial ischemia. Dobutamine is also indicated as short-term treatment of the impaired cardiac function after surgery.[63,64]

Hypotensive patients with severe sepsis frequently have low systemic vascular resistance. Hypotension caused by vasodilation in patients with septic shock can be refractory to infusion of inotropic drugs such as dobutamine, and thus potentially lethal. Several clinical studies have indicated that norepinephrine has merit in these patients given both its inotropic and vasoconstrictive effects on cardiovascular system.[65] In addition, norepinephrine may have renal benefits in these patients probably related to differential vasoconstriction of the vasculature to less critical peripheral tissue.[65]

β_2AR agonists are generally not used as inotropic agents in heart disease, although data has recently emerged demonstrating that the use of clenbuterol in end-stage heart failure sufferers who have been placed on a mechanical assist device has benefit.[66,67] Interestingly, clenbuterol addition to standard of care with a mechanical pump appears to lead to novel regression of LV dysfunction and improved recovery that actually may negate cardiac transplantation.[66] The mechanism of how this β_2AR agonist promotes heart failure regression is not understood but certainly worthy of future investigation.

Respiratory diseases

β_2AR agonists are currently the drugs of choice for treatment of respiratory diseases such asthma and chronic obstructive pulmonary disease (COPD).[68] Asthma is a chronic disease of the respiratory system in which the airway occasionally constricts, becomes inflamed, and is lined with excessive amounts of mucus, often in response to one or more injurious triggers. These acute episodes may be evoked by such things as exposure to an environmental stimulant (or allergen), cold air, exercise or exertion, or emotional stress. This airway narrowing causes symptoms such as wheezing, shortness of breath, chest tightness, and coughing, all of which respond to β_2AR agonists. β_2AR agonists are effective as bronchodilators through the mechanisms described earlier in the chapter and also improve alveolar function.

COPD, also known as chronic obstructive airway disease (COAD), is actually a group of diseases characterized by limitation of airflow in the airway that is not fully reversible, such as the case with asthma. It is most often due to tobacco smoking but can be due to other airborne irritants such as coal dust or solvents. Salbutamol is the β_2AR agonist that took the place of epinephrine and isoproterenol for the treatment of respiratory diseases characterized by airway constriction due to selectivity for β_2ARs and higher resistance to metabolism.[68] Salbutamol is usually given by the inhaled route for direct effect on bronchial smooth muscle, and its effect takes place rapidly within 5 to 20 minutes. Salmeterol is a long-acting β_2AR agonist that is also currently prescribed for the treatment of asthma and COPD. Salmeterol can last up to 12 hours compared to 4–6 hours with the use of salbutamol.[68,69] Other β_2AR agonists used in asthma and COPD include terbutaline and fenoterol. Of note, some studies have described an increased risk for adverse cardiovascular events in patients with obstructive airway disease and use of β_2AR agonists. The initiation of treatment seems to increase heart rate and reduce K^+ concentration.[70] Through these mechanisms, and other effects of β_2AR stimulation, β_2AR-agonists may precipitate ischemia, congestive heart failure, arrhythmias, and sudden death.

Finally, anaphylactic shock, the most serious of allergic reactions, is a life-threatening medical emergency because of rapid constriction of the airway, often within minutes of onset. Anaphylactic shock requires immediate advanced medical care; first-aid measures include rescue of breathing (part of Cardiopulmonary Resuscitation/ CPR) and administration of epinephrine as a rapid-acting βAR agonist.[71]

Therapeutic uses of βAR antagonists

Hypertension

βAR antagonists, commonly referred to as β-blockers, are widely used for several cardiovascular disorders and are generally avoided in respiratory diseases. One of the first uses of β-blockers was in hypertension, because they are effective at lowering blood pressure. The use of β-blockers to lower blood pressure is interesting from a scientific perspective, because the exact mechanism is unknown. Several mechanisms have been proposed to explain the antihypertensive effects of β-blockers. The "cardiac output theory" proposes that β-blockers primarily reduce heart rate and myocardial contractility and as a result lead to reduced cardiac output lowering blood pressure.[72] However, βAR antagonists possessing intrinsic sympathomimetic activity and not cardioselective still effectively lower blood pressure.[72] Therefore, a decrease in cardiac output may not be the only mechanism involved in the reduction of blood pressure. The "renin theory" proposes that the reduction in blood pressure is secondary to a reduction in plasma renin activity by β-blockers. Hypertensive patients with increased plasma renin activity tend to be more responsive to βAR antagonists (possibly by reducing renin levels) than hypertensive patients with initial low plasma renin levels.[73] Plasma renin levels may predict responsiveness to β-blockers;[73] to date, however, no definitive correlation between the alteration in plasma renin levels and the degree of blood pressure reduction by β-blockers has been definitely demonstrated.

There is also a "central nervous system theory" describing a putative mechanism of action of β-blockers to reduce blood pressure because they can enter the CNS. Indeed, in animal experiments, the βAR antagonist propranolol, when injected into cerebral ventricles, reduced blood pressure.[74, 75] However, the amount of propranolol that must be given intraventricularly to lower blood pressure is substantially greater than the amount of propranolol given systemically to lower blood pressure in human subjects. In addition, β-blockers such as atenolol, which penetrate the brain poorly due to poor lipophilic properties, effectively lower blood pressure in humans, and thus an exclusive CNS-mediated mechanism is doubtful. The "baroreceptor hypothesis" attributes the antihypertensive effect of β-blockers to a resetting of baroreceptor sensitivity. However, recent studies have found that a decrease in blood pressure with a low dose of propranolol did not depend on baroreceptors.[76]

Overall, the scientific literature does not unanimously support any of these, and mechanism for the antihypertensive action of these drugs and the exact mechanism is unclear. Nonetheless, they do decrease blood pressure effectively but apparently not in all patient populations because β-blockers are more effective in lowering blood pressure in whites than in the black population.[77] Of further interest, recent publications have found that β-blockers are less effective than other antihypertensive drugs in preventing cardiovascular outcomes in hypertensive patients, potentially due to lower heart rates. Bangalore and colleagues recently reported that a lower heart rate was associated with a greater risk for all-cause mortality, cardiovascular mortality, myocardial infarction (MI), stroke, or heart failure.[78] They concluded that β-blocker–associated reduction in heart rate increased the risk of cardiovascular events and death in hypertensive patients.[79,80]

βAR antagonists in acute and chronic myocardial ischemia

Myocardial ischemia is characterized by an imbalance between the demand for oxygen to the myocardium and the supply via the coronary arteries.[81] Clinical conditions associated with myocardial ischemia are unstable angina, MI, and stable angina. β-blockers are widely accepted as anti-ischemic drugs. They reduce the demand of O_2 by myocardium via a reduction of heart rate and also can reduce cardiac wall-stress.[82,83] Moreover, they increase coronary perfusion by increasing the diastolic phase of the cardiac cycle. Unstable angina and MI are classified as acute coronary syndromes that are considered conditions of reduced oxygen supply or low-flow. Acute coronary syndromes are the most common cause of hospital admission in patients with coronary artery disease. The pathophysiology of acute angina or MI generally involves the rupture of a coronary atherosclerotic plaque and subsequent thrombus formation.[84–86] Importantly, several studies have shown the benefits of β-blockers in managements of MI syndromes in particular to prevent recurrent episodes of ischemia.[59] Moreover, the use of β-blockers reduces infarct size, re-infarction and mortality.[87–90]

The effects of β-blockers in MI can be divided by their indication for early administration (early after or even during an MI episode) and also their later administration for secondary prevention. During MI, infusion of a β-blocker such as metoprolol or atenolol reduces cardiac index and rate, as well as blood pressure with a following decrease of O_2 myocardial demand per minute and beat. Further, β-blockers seem to also have some beneficial metabolic effects on myocardium by reducing lypolysis and free fatty acid consumption.[91,92] Generally, $β_1AR$-selective antagonists are used, but nonselective β-blockers have been effective. Esmolol is particularly useful in these acute conditions because it has a short half-life and rapid onset. Esmolol can also be used generally in patients with contraindications such as asthma because its βAR blockade is gone within 30 min.[93,94] Importantly, β-blockers are contraindicated in patients with low heart rate (<60 beats/min), systolic blood pressure <100 mmHg, and with moderate to severe LV dysfunction.

Oral administration of β-blockers is currently employed for secondary prevention to reduce mortality and morbidity in patients with diagnosis of MI. A low ejection fraction is not considered a contraindication for β-blockers in this scenario because they also reduce mortality in patients with chronic congestive heart failure (see further in this chapter). In chronic, stable angina, β-blockers have a dual aim to both reduce morbidity and mortality and also eliminate angina with minimal adverse effects, allowing patients to return to normal activities. To reach these goals, β-blockers need to be carefully titrated using heart rate and symptomatic relief as parameters dictating changes in β-blocker dose.[95]

Applications of β-blockers in heart failure and cardiomyopathy

Heart failure is a cardiac condition that occurs when a problem with the structure or function of the heart impairs its ability to supply sufficient blood flow to meet the body's needs. Neuro-adrenergic activation is part of a compensatory mechanism that allows maintaining an adequate cardiac output through stimulation of cardiac βARs, but it is now generally accepted that chronically elevated stimulation of the cardiac β-adrenergic system is toxic to the heart and may contribute to the pathogenesis of congestive heart failure of various causes.[96,97] In fact, transgenic mouse models overexpressing $β_1ARs$ can cause cardiomyopathy and heart failure.[98] In human heart failure, as well as in several animal models, elevated circulating catecholamines lead, via various compensatory mechanisms, to decreased levels and functional activity of cardiac $β_1ARs$ and thus to marked desensitization of the heart to inotropic stimulation.[99] These biochemical and physiological changes appear to be mediated by elevated levels of the enzyme GRK2 in the heart[100] (see further in this chapter). Interestingly, β-blockers used to be contraindicated in heart failure due to the already low-responsiveness of cardiac βARs; however, their use has breathed new life into the armamentarium available for physicians to provide decreased morbidity and mortality in patients suffering from chronic heart failure.[101] Overall, the successful use of β-blockers in treating chronic heart failure is generally explained by their ability to block the noxious effects of chronic endogenous sympathetic stimulation of the failing heart, but they also cause a molecular normalization and improvement of βAR signaling when the receptors are not blocked.[28]

Metoprolol and bisoprolol are second-generation β-blockers because they are relatively selective for β_1ARs and do not block pre- and post-synaptic β_2ARs.[102] The clinical trial MERIT-HF showed the benefits of metoprolol in terms of reduced mortality in particular in patients with HF due to ischemic cardiomyopathy and in younger versus older patients.[101] Third-generation β-blockers such as carvedilol, which also have αAR antagonist activity and thus vasodilatative properties, also are effective at reducing mortality in heart failure patients.[103] The latter property of these drugs may produce a reduction of after-load allowing a better tolerance of the reduced cardiac output due to the initial effect of β-blocker activity. Carvedilol has low selectivity for β_1AR, but has vasodilatory and antioxidant properties.[103] At higher doses, carvedilol produces complete adrenergic antagonism, which seems to be associated to a better cardiac performance in heart failure.[104] The possibility of a better outcome in presence of complete adrenergic antagonism versus selective β_1AR antagonism has been clarified by the COMET trial comparing carvedilol with metoprolol.[105] Carvedilol reduced cardiovascular mortality, sudden death, death caused by circulatory failure, death caused by stroke, as well as fatal and nonfatal MIs with respect to metoprolol, whereas no difference in all-cause hospitalizations or in worsening heart failure were found between treatment groups.[106]

β-blockers are also commonly used in dilatative and hypertrophic cardiomyopathy. β-blockers are a cornerstone for treatment of these diseases since they allow to reduce episodes of angina, syncope, and dyspnea that are associated with these structural conditions of the heart.[107]

Use of β-blockers for arrhythmias

β-blockers are classified as class II anti-arrhythmic drugs. Their use in prevention or treatment of several kinds of tachyarrhythmias is mainly due to two different mechanisms: anti-adrenergic effect on cardiac βARs and membrane stabilization. β-blockers counteract the pro-arrhythmic effect of the activated adrenergic system, especially in pathological conditions such as heart failure and/or ischemia. Adrenergic stimulation promotes arrhythmogenesis through a variety of mechanisms. It enhances the pacemaker current, which increases automaticity.[108] It also increases the inward Ca^{2+} current thereby potentially enhancing delays after depolarizations and triggered activity.[108] In patients with ischemic cardiomyopathy, dobutamine (β_1AR agonist) has been shown to decrease the action-potential duration to a significantly greater extent in potentially ischemic zones compared to normal zones, suggesting circulating catecholamines can further exacerbate electrical heterogeneity.[109] Membrane stabilization is a local anesthetic effect mainly due to reduction of inward Na$^+$ current. However, this latter appears to be involved at higher concentration of β-blocker, therefore the anti-arrhythmic effect at the doses used in clinical practice is overall given by the antagonism of βARs.

Several compounds have been approved in United States for treatment of arrhythmias and prevention of sudden death (acebutolol, metoprolol, esmolol, atenolol, propanolol, timolol), but all are considered efficient at the right dosage, without particular advantages among these. They are indicated in particular to arrythmias associated to adrenergic activation such as with physical or emotional stress, or through the use of cocaine and conditions with hypersympathetic nervous system activity. β-blockers can be used to normalize ventricular frequency in the presence of flutter and atrial fibrillation and in treatment of supraventricular tachycardia induced by extra-conduction pathway such in Wolff-Parkinson-White syndrome.[110,111]

βAR REGULATORY MOLECULES IN CARDIOVASCULAR AND RESPIRATORY DISEASES

GRKs and cardiac diseases

As mentioned above, GRKs induce the desensitization of GPCRs, and in the heart, the actions of GRK2 and GRK5 appear extremely important in the regulation of βAR function. Importantly, GRK2 is pathological in the failing heart because its deletion in mice after MI dramatically improves survival.[112] Several studies by us using a peptide GRK2 inhibitor (named the βARKct) has also shown positive effects in small and larger animal models of heart failure and clinical gene therapy trials are being planned.[113] Up-regulation of cardiac GRK2 expression and activity in end-stage human HF was first found in 1993.[114] Increased cardiac GRK activity appears at least partially responsible for the chronic desensitization, down-regulation, and dysfunctional signaling of myocardial βARs, which is a cardinal characteristic of the failing heart.[115] As detailed earlier in this chapter, the most important minute-by-minute regulator of normal cardiac function is βAR tone, governed by systemically circulating or locally released sympathetic catecholamines. In heart failure, these acute compensatory events are pathologically activated in response to chronically diminished cardiac output (Figure 15.1). Unremitting sympathetic stimulation results in GRK2- and GRK5-mediated phosphorylation of myocardial βARs, which diminishes catecholamine responsiveness by prompting receptor desensitization and down-regulation. Importantly, over the last fifteen years, our laboratory has shown that desensitization of βARs via GRK2 is critical for normal and compromised cardiac function.[115]

In addition to increased GRK2, GRK5 up-regulation has also been found in human HF.[116] This is important because we have recently showed that GRK5 has the unique ability to accumulate into the nucleus of myocytes where it acts as a

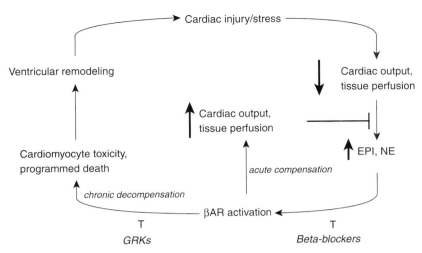

Figure 15-1: Sympathetic activation in heart failure. Positive feedback loop for catecholamine (*EPI* epinephrine, *NE* norepinephrine) release in heart failure. Enhanced cardiac function can interrupt cycle by diminishing sympathetic activity. Beta-blockers and GRKs interrupt cycle by impairing β-adrenergic receptor (βAR) signaling. Reproduced with permission from Dorn, GW, 2009, J. Mol. Med.
   With kind permission from Springer Science+Business Media: Dorn.[1]

pro-hypertrophic kinase regulating gene transcription.[117] Therefore, like GRK2, GRK5 may be a target for treating cardiomyopathies. A more detailed review of GRKs and their role in cardiovascular disorders are beyond the scope of this chapter but several recent reviews have covered this topic;[1,29,118] however, it is interesting to note that chronic βAR blockade can result in the decreased expression of GRK2,[119] and perhaps this can also contribute to the successful use of β-blockers.[28,44]

GRKs in respiratory diseases

Since several GPCRs exert control over airway smooth muscle, GRK activity may be important in the dysfunctional signaling apparent in certain respiratory diseases such as asthma. This includes ARs as well as other GPCRs such as cholinergic receptors. This is especially important for β_2ARs because they are targets for GRKs, and the loss of βAR-mediated relaxation in asthma may have a GRK component. Severe asthmatics demonstrate reduced responsiveness to bronchodilators that act through β_2ARs, and some studies suggest that this impaired β_2AR-mediated relaxation is due to desensitization and/or increased cholinergic muscarinic antagonism of β_2AR.[120,121] Interestingly, studies with GRK5 knockout mice have shown changes in tracheal smooth muscle physiology.[122] This study showed that GRK5 is likely an important contributor to the desensitization of M2 muscarinic receptors on airway smooth muscle, thereby regulating the ability of these receptors to oppose airway smooth muscle relaxation.[122] Therefore, GRKs appear as possible target for development of new therapeutics approaches in treatment of respiratory disease.

HUMAN βAR POLYMORPHISMS AND THEIR ROLE IN CARDIOVASCULAR AND RESPIRATORY HEALTH AND DISEASE

Human β_1AR polymorphisms

One of the goals of human genetics and pharmacogenomic research is to provide clinicians with potential tools to individualize therapy based on a person's genetic make-up. Elucidation of clinically relevant gene mutations and polymorphisms has shown that allelic variance of specific genes can significantly influence pathophysiology as well treatment of a particular disease. Importantly, several single nucleotide polymorphisms (SNPs) for β_1AR and β_2AR genes in humans have significant clinical consequences in the realm of cardiovascular and respiratory diseases, and these will be discussed further in this chapter.

Two major SNPs in the human β_1AR coding region have significant ramifications for clinical medicine (Figure 15.2). The first at position 49 in the amino-terminus of the receptor is a serine is substituted by a glycine [123–126] with the minor allele (Gly) occurring with an allele frequency of 12–16% in Caucasians and Asian people and 13–28% in African Americans.[127–130] The second SNP is found at position 389 in the proximal part of the carboxyl-terminus (a region that is within a G_s coupling domain) where an arginine is substituted by a glycine.[123–125] The minor allele Gly is found at an allele frequency of 24–34% in Caucasians and Asians and 39–46% in African Americans.[127–130] These two SNPs are in linkage disequilibrium so that the haplotype Gly49Gly/Gly389Gly occurs very rarely, if at all.[127] Interestingly, the Gly389 allele was present in the first clone of the β_1AR and was therefore – although being the minor allele – for a long time considered to be the 'wild-type (WT)' β_1AR.[131]

In view of the critical role of the β_1AR in cardiac function and disease, the presence of β_1AR polymorphisms has raised questions concerning whether they may be associated with differing susceptibility to cardiac diseases, including heart failure, and/or whether they may contribute to the prognosis of cardiac disease. Intrinsically related to these questions is whether β_1AR agonists or β-blockers interact differently with the variant receptors.

Several attempts have been made to determine the functional impact of these polymorphisms. In vitro, the Gly49 β_1AR, the Arg389 β_1AR, and the Gly49-Arg389 diplotype are down-regulated to a significantly greater extent upon long term agonist exposure compared with the Ser49 and Gly389 variants.[132, 133] Whereas agonist- and antagonist-binding affinities and adenylyl cyclase coupling were not different between the Ser49 and Gly49 variants of the β_1AR,[133] the Arg389 β_1AR variant has been reported to display increased agonist-promoted binding in the absence of GTP; this is consistent with enhanced coupling to G_s.[124] In addition, the Arg389 β_1AR displays higher basal and isoproterenol-induced AC activity and cAMP responses, indicating hyperfunctionality of this receptor variant.[124,134,135] Sandilands and colleagues[136] have confirmed these data by studying in HEK 293 cells the functional importance of four possible β_1AR haplotypes. They found that isoproterenol increased cAMP production with a haplotype rank

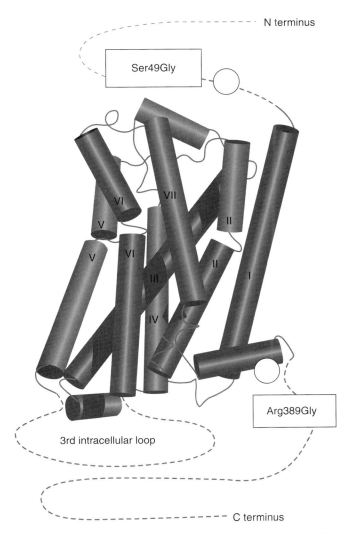

Figure 15-2: Localization of the two most frequent polymorphisms within the human β₁-adrenoceptor. The two polymorphisms are represented by circles. Reproduced with permission from Ahles, A, 2009, Trends in Pharmacological Science.

Reprinted from Trends in Pharmacological Sciences, Ahles and Englehardt, Polymorphisms determine B-adrenoceptors conformation (2009) with permission from Elsevier.

order: Gly49Arg389 > Ser49Arg389 > Gly49Gly389 ‡ Ser49Gly389. Moreover, they also observed that isoproterenol-induced receptor desensitization was largest for the Gly49-Arg389 haplotype and smallest for the Ser49-Gly389 haplotype. These studies in recombinant systems indicate that the Arg389Gly β₁AR polymorphism determines the functional responsiveness of the system with Arg389.

Studies have also been done in vivo to study the functional characteristics of the Arg389Gly β₁AR polymorphism on cardiac β₁AR responsiveness. Several groups have studied dynamic exercise-evoked increases in heart rate in young and middle-aged subjects homozygous for Arg389 or Gly389 β₁AR, however, in all these studies exercise-induced increases in heart rate were nearly identical in

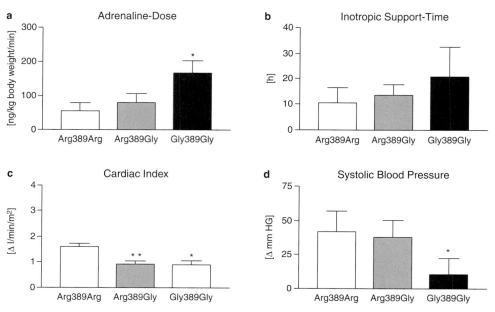

Figure 15-3: Inotropic support in patients with the Arg389Gly-;β₁AR polymorphism. **a** Adrenaline dose (ng/kg/min), **b** inotropic support time (h), **c** increases in cardiac index (Δ l/min/m2), and **d** increases in systolic blood pressure (Δ mmHg) in post-coronary artery bypass surgery (CABG) patients with the Arg389Gly-β1AR polymorphism.

With kind permission from Springer Science+Business Media: Leineweber et al.[145]

Arg389 and Gly389 β₁AR subjects; no genotype-dependent difference could be observed.[137–142] In contrast, infusion of the β₁AR agonist dobutamine into subjects homozygous for Arg389 β₁AR caused significantly larger increases in heart rate and contractility than in subjects carrying one[143] or two alleles of Gly389 β₁AR.[144] Moreover, patients undergoing coronary artery bypass grafting, who were homozygous for Arg389 β₁AR, required post-surgery for a shorter time and a lower dose of epinephrine to support inotropic function of the heart compared to patients with one or two Gly389 alleles[145] (Figure 15.3). Thus, it appears that in vivo, Arg389Gly β₁AR does not respond in a genotype-dependent manner to a more physiologic stimulus (exercise-induced increase in catecholamines) but does so to a more pharmacologic stimulus (dobutamine).

Human β₂AR polymorphisms

The human β₂AR gene is located on chromosome 5q31–32,[146] and in the coding region, twelve single-base substitutions have been identified. Five of these are nonsynonymous SNPs and three appear to be most important in terms of cardiovascular and respiratory health; at position 16, a Gly for Arg substitution (Arg16Gly); at position 27, a Gln for Glu (Gln27Glu), and at position 164, an Ile for a Thr (Thr164Ile). These three SNPs occur at significant allele frequencies and have significant functional effects in vitro and in vivo.[147] Of the other two SNPs, Val34Met appears to not alter receptor function,[148] and the functional consequence of the Ser220Cys polymorphism has not been studied. There are also eight additional SNPs within the 1.5 kb 5′ untranslated region upstream from the start codon.[149] There is a mutation within a short open reading frame, called the beta upstream peptide or the 5′ leader cistron (Cys19Arg), and this appears functionally important because it can affect β_2AR expresssion at a translational level.[150]

There is strong linkage disequilibrium between codon 16 and codon 27.[151] Thus, subjects homozygous for Glu27Glu are nearly always homozygous for Gly16Gly; haplotype Arg16Glu27 occurs naturally but is extremely rare (present in less than 1% of the population).[128–130,148] Moreover, the Thr164Ile β_2AR SNP is closely associated with Gly at position 16 and Gln at position 27. Finally, there is a tight linkage disequilibrium between Glu 27 allele of the Gln27Glu polymorphism and the Arg-19 allele of the Cys 19Arg polymorphism, so that subjects homozygous for Arg-19Arg are nearly almost also homozygous for Glu27Glu.[150]

The three functionally important SNPs have been studied in detail to learn how they may alter β_2AR signaling and function. For example, the amino acid exchange from Thr to Ile at position 164 creates a receptor variant with extensive signaling defects. This includes fourfold lower binding affinities of the endogenous catecholamines and reduced basal and agonist-induced AC activation.[132, 134] Importantly, in recombinant cell systems as well as in humans, subjects heterozygous for the Thr164Ile β_2AR allele have blunted heart rate and contractility responses to terbutaline compared with Thr164Thr β_2AR subjects.[152–154] In contrast to this loss of coupling, the two amino-terminal SNPs (Arg16Gly and Gln27Glu) do not result in altered ligand binding and agonist-stimulated AC activity.[155] Several studies have confirmed this result as β_2AR-mediated increases in heart rate and contractility were not different in subjects with the Arg16Gly or Gln27Glu β_2AR variants.[156–160] Thus, it appears to be clear that the Arg16Gly and Gln27Glu β_2AR polymorphisms have no considerable influence on the functional activity of β_2AR, at least in the heart.

In contrast to the cardiac responses, divergent results have been obtained for the impact of the Arg16Gly and Gln27Glu β_2AR SNPs on vascular responses. Systemic infusion of the β_2AR agonists salbutamol, terbutaline, or epinephrine resulted in larger vasodilation in subjects homozygous for Arg16 β_2AR than in subjects carrying one or two Gly16 alleles.[156,161] On the other hand, local infusion of βAR agonists into brachial artery or hand vein evoked larger vasodilator responses in subjects homozygous for Gly16 [162,163] or Glu27 β_2AR.[162,164] Thus, it remains to be clearly understood which haplotype is specifically associated with an enhanced vasodilator responsiveness.

It appears that – at least in vascular and bronchial smooth muscle – the Arg16Gln27 β_2AR haplotype is rather susceptible to agonist-induced desensitization, more than Gly haplotype. However, these findings might be explained by a model of regulation proposed by Liggett,[165] where endogenous catecholamines dynamically desensitize β_2ARs in their basal state, and this occurs to a greater extent for the Gly16 than for the Arg16 β_2AR variant. Accordingly, exogenous agonist-induced desensitization should then be greater for the Arg16 than for Gly16 β_2AR variant (that already is endogenously desensitized).

Role of βAR polymorphisms on cardiovascular and pulmonary diseases and therapy

Risk of developing heart failure and therapeutic responses in cardiac diseases with βAR SNPs

As detailed above, cardiac βARs are chronically activated in heart failure due to increased sympathetic activity. Since the Arg389Gly βAR variant is essentially a "gain of function" mutation (the Arg389 variant is three to four times more responsive[166] to agonist stimulation than the Gly389 variant),[134] chronic cate-cholamine exposure in patients with this form of the β_1AR may have enhanced heart failure progression. Therefore, attempts have been made experimentally to find out whether or not there is an association between the Arg389Gly SNP and heart failure. No association of this SNP or the Ser49 or Gly49 alleles to heart failure patients have been found,[125,167–171] suggesting that the Ser49Gly and Arg389Gly β_1AR polymorphisms do not play a causal role in the development of heart failure. Moreover, in a large study with 1,554 subjects, White and colleagues[172] did not find significant differences in the frequency of the Arg389 or Gly389 allele between patients with coronary artery disease and controls, and Kanki and colleagues[173] did not find any difference in allele frequencies for the Ser49Gly or Arg389Gly β_1AR in patients with acquired long QT syndrome and control subjects. However, patients with dilated cardiomyopathy and homozygous for Arg389 β_1AR had significantly more LV tachycardia than patients with one or two Gly389 alleles.[174]

Another approach to study the role of the Arg389Gly and Ser49Gly β_1AR SNPs in heart failure is to investigate possible associations with the outcome of heart failure (e.g., five-year mortality, death, or time to transplantation).[126,167,175–177] The Arg389Gly β_1AR SNP did not affect outcome.[167,175–177] The Ser49Gly β_1AR SNP has led to variable results as some studies have showed that outcomes were better in patients carrying one or two alleles of Gly49 β_1AR than in patients homozygous for Ser49 β_1AR.[126,167] However, two additional studies did not find an association between the Ser49Gly β_1AR SNP and heart failure outcomes.[175,177]

Importantly, Small and colleagues[166] assessed a possible association between the combination of the Arg389Gly β_1AR with a deletion "loss of function" mutant of the α_{2c}AR and they found that in African Americans, the combination of the deletion mutant of the α_{2c}AR together with the Arg389 allele of β_1AR significantly increased the risk of heart failure.[166] This probably occurs both through the hyperfunctional Arg389 mutation coupled with enhanced catecholamine secretion due to the loss of the α_2AR-mediated negative feedback.[166] Confirming the above studies, the Arg389 polymorphism alone had no predictive value for heart failure. It is important to note that in contrast to the findings of Small and colleagues,[166] three subsequent studies ??? Canham and colleagues[178] studying 1,121 African Americans (patients of the Dallas Heart study), Nonen and colleagues[170] studying 91 Japanese patients, and Metra and colleagues[171] studying 260 Italian Caucasians – did not find an enhanced risk of heart failure for patients with the combination of the α_{2c}AR deletion mutant and the Arg389 β_1AR variant. This continues to be the focus of heart failure investigations.

Finally for β_1AR mutations and heart failure, studies have observed that peak VO_2 is different in heart failure patients with the Arg389Gly β_1AR variant. Wagoner and colleagues exercised heart failure patients and found that peak VO_2 was significantly higher in patients homozygous for the Arg389 variant than in patients homozygous for the Gly389 β_1AR variant, and patients heterozygous for Arg389Gly β_1AR had an intermediate peak VO_2.[179] Similar results were obtained by Sandilands et al.[136]

Similar to β_1AR SNPs, β_2AR polymorphisms do not seem to be disease-causing genes in hypertension or in heart failure, but these specific mutations appear to affect responses to drugs. For example, the Arg16Gln27 β_2AR haplotype appears to be rather susceptible to agonist-induced desensitization,[155] and patients carrying this haplotype in asthma have been shown to suffer from rapid loss of therapeutic efficacy of chronic β_2AR agonist therapy.[180–184] Interestingly, in a population of hypertensive patients, it has been reported that a positive relationship exists between the Glu27 β_2AR allele and myocardial hypertrophy independent by the haplotype (Arg or Gly16).[185,186] A possible explanation is based on the finding that the Glu27 variant of β_2AR is characterized by a gain of function due to a reduced agonist-induced desensitization and therefore it could enhance the hypertrophic effects of catecholamines. A report shows that the Ile164 β_2AR variant is much more prevalent in a population of coronary artery disease patients compared to control patients (12% versus 3%).[187] Further, Ile164 carriers exhibit a more severe pathology than those with Thr164 genotype. Similarly, a group of patients with peripheral artery disease also exhibited a high prevalence of the Ile164 genotype (7%) and a more severe clinical phenotype than those with Thr164.[123] Overall, these data demonstrate that the impact of a given polymorphism can change according to the selection of subjects and the pre-existing disease state. Importantly for the β_2AR, the above studies have found a strong relationship between β_2AR SNPs and therapeutic outcomes in subjects with coronary artery disease and hypertension.

Pharmacogenetic interaction in heart failure patients between β-blockers and βAR SNPs

As detailed earlier in this chapter, β-blocker treatment can have positive therapeutic effects in heart failure and is part of the "standard of care."[188] As part of understanding the known SNPs of βARs and their role in cardiovascular signaling, studies have attempted to determine if heart failure patients with the β_1AR mutations, Ser49Gly or Arg389Gly, may have altered responses to β-blockers. A key study providing valuable insight into this field examined a mixed population of ischemic and nonischemic heart failure patients who were treated with carvedilol. Interestingly, after six months, patients with the Arg389Arg β_1AR had a greater improvement in LV ejection fraction (LVEF, as a measure of cardiac function) (8.7%) compared with patients with Gly389Gly β_1ARs (0.93%).[189] In addition, Terra and colleagues treated sixty-one heart failure patients with long-acting metoprolol for three months and found that patients homozygous for Arg389 β_1AR had a significant improvement in LVEF, whereas patients carrying one or two alleles of Gly389 β_1AR did not show any significant LVEF improvement.[190] According to these studies, the Arg389Gly β_1AR polymorphism seems to have a strong impact on response to treatment with β_1AR agonists and antagonists. Thus, it might be possible, by assessing the β_1AR genotype, to predict responsiveness to β_1AR ligands; patients homozygous for the Arg389 β_1AR polymorphism should be good responders, whereas patients homozygous for the Gly389 β_1AR variant should be poor or nonresponders.

Survival in heart failure patients after βAR blockade has also been examined. In a retrospective study on 600 patients from the MERIT-HF study, White and colleagues examined whether the Arg389Gly β_1AR polymorphism might affect the effects of long-acting metoprolol on outcomes and found that this SNP did not significantly affect the number of patients suffering from adverse effects (all cause mortality or hospitalization).[191] On the other hand, Liggett and colleagues found in 1,040 patients from the BEST trial (a placebo-controlled study of bucindolol in heart failure) that bucindolol-treated patients homozygous for Arg389 β_1AR had a significant reduction in five-year mortality and hospitalization compared with placebo-treated Arg389 β_1AR patients.[192] Moreover, bucindolol-treated patients carrying one or two Gly389 alleles did not exhibit this clinical benefit from this β-blocker.

The SNP at position 49 of the β_1AR also has shown significant differences in treatment as a study with 184 patients on β_1AR blockers showed that the five-year survival was significantly better when patients carried one or two alleles of Gly49 β_1AR compared to those homozygous for Ser49 β_1AR [126]. In a follow-up study, Magnusson and colleagues suggested that the influence of codon 49 in the β_1AR on outcome after heart failure treatment with β-blockers is more pronounced than codon 389.[167]

Effects of β_2AR SNPs on asthma therapy

Due to the importance of β_2ARs in regulating airway smooth muscle and the use of agonists as common bronchodilators used to treat airflow limitations associated with obstructive airway diseases, including asthma and COPD, it is obviously important to assess whether the described β_2AR SNPs play any role in therapeutic responses. Certainly, besides any specific β_2AR mutation, variation in individual responsiveness to β_2AR agonists may be determined by other genetic or environmental factors such as smoking status, age, and so on. Importantly, studies have shown that genetic variants of the β_2AR in humans play a significant role in responses to exogenous β_2AR agonists.[193,194] Indeed, clinical studies have shown the genetic contribution of the Arg16Gly β_2AR polymorphism to short-term bronchodilatative response (BDR) to β_2AR agonists in patients with asthma.[193,194] One aspect that may be important is that β_2AR mutations may affect the level of receptor desensitization and down-regulation after agonist exposure.[195] Other studies indicate that β_2AR SNPs may alter bronchoprotection.[196]

The majority of clinical studies have concentrated on the Arg16Gly polymorphism because this mutation appears to alter β_2AR agonist responses. In a Japanese study in patients with COPD, the Arg16 β_2AR allele was associated with lower responses to β_2AR agonist inhalation.[197] This effect of the Arg16Gly polymorphism was independent of airflow limitation, age, and smoking status.[197] Further, a prospective trial, called BARGE (Beta-Adrenergic Response by Genotype) included patients who use the β_2AR agonist albuterol regularly.[181] Asthmatic patients with the Arg16Arg β_2AR genotype did not respond to treatment as well as the Gly variant, and the authors concluded that these patients should discontinue the use of a β_2AR agonist and switch to alternative treatments for their asthma. Since albuterol and other short-term β_2AR agonists are not used in mild asthma, studies need to be done to examine the effect of β_2AR genotype on patients who take inhaled corticosteroids and long-acting β_2AR agonists. Retrospective data again suggest that individuals homozygous for Arg16 may do worse on these long-acting bronchodilators than the rest of the population, and these data were found to not involve any effects from inhaled steroids.[184,196] All of the above effects need to be studied in individual populations as ethnic origin appears extremely important if not for only the fact that the rates of the βAR SNPs vary among populations. For example, the prevalence of the β_2AR SNPs at codons 16 and 27 is different in the African American population, which might at least in part explain some of the differences seen in asthma patients treated with long-term β_2AR agonists.[198]

FUTURE PERSPECTIVES

Our understanding about the biology and genetics of βARs is rapidly evolving and provides the possibility to develop and evaluate new therapeutic approaches and strategies for cardiovascular and respiratory diseases. These include development of molecules that inhibit GRKs targeting the desensitization of βARs, as well as other key properties of these GPCR regulatory molecules. The advancement of genetic research and technology has put βAR mutations on the leading edge of personalized medicine because it is clear that genetic variation within the genes for these receptors can alter outcomes and perhaps guide future treatment strategies. Clearly, these exhaustively studied receptors still have much to tell us in the realm of translational medicine and could lead the way to providing specific treatment options in heart failure or pulmonary disease based on a patient's individual βAR gene make-up.

REFERENCES

1. Dorn GW, 2nd. GRK mythology: G-protein receptor kinases in cardiovascular disease. *Journal of molecular medicine (Berlin, Germany)*. May 2009;87(5):455–463.
2. Dohlman HG, Caron MG, Lefkowitz RJ. A family of receptors coupled to guanine nucleotide regulatory proteins. *Biochemistry*. May 19 1987;26(10):2657–2664.

3. Lefkowitz RJ. Seven transmembrane receptors: something old, something new. *Acta physiologica (Oxford, England)*. May 2007;190(1):9–19.
4. Caron MG, Lefkowitz RJ. Catecholamine receptors: structure, function, and regulation. *Recent progress in hormone research*. 1993;48:277–290.
5. McPherson GA, Malta E, Molenaar P, Raper C. The affinity and efficacy of the selective beta 1-adrenoceptor stimulant RO363 at beta 1- and beta 2-adrenoceptor sites. *British journal of pharmacology*. Aug 1984;82(4):897–904.
6. Krstew E, McPherson GA, Malta E, Molenaar P, Raper C. Is Ro 03–7894 an irreversible antagonist at beta-adrenoceptor sites? *British journal of pharmacology*. Jun 1984;82(2):501–508.
7. Carswell H, Nahorski SR. Beta-adrenoceptor heterogeneity in guinea-pig airways: comparison of functional and receptor labelling studies. *British journal of pharmacology*. Aug 1983;79(4):965–971.
8. Minneman KP, Dibner MD, Wolfe BB, Molinoff PB. beta1- and beta2-Adrenergic receptors in rat cerebral cortex are independently regulated. *Science*. May 25 1979;204(4395):866–868.
9. Jensen J, Brors O, Dahl HA. Different beta-adrenergic receptor density in different rat skeletal muscle fibre types. *Pharmacol Toxicol*. Jun 1995;76(6):380–385.
10. Evans BA, Papaioannou M, Bonazzi VR, Summers RJ. Expression of beta 3-adrenoceptor mRNA in rat tissues. *British journal of pharmacology*. Jan 1996;117(1): 210–216.
11. Guillaume JL, Petitjean F, Haasemann M, Bianchi C, Eshdat Y, Strosberg AD. Antibodies for the immunochemistry of the human beta 3-adrenergic receptor. *Eur J Biochem*. Sep 1 1994;224(2):761–770.
12. Summers RJ, Papaioannou M, Harris S, Evans BA. Expression of beta 3-adrenoceptor mRNA in rat brain. *British journal of pharmacology*. Nov 1995;116(6):2547–2548.
13. Lands AM, Arnold A, McAuliff JP, Luduena FP, Brown TG, Jr. Differentiation of receptor systems activated by sympathomimetic amines. *Nature*. May 6 1967;214(5088):597–598.
14. Stiles GL, Lefkowitz RJ. Cardiac adrenergic receptors. *Annu Rev Med*. 1984;35:149–164.
15. Arch JR, Kaumann AJ. Beta 3 and atypical beta-adrenoceptors. *Med Res Rev*. Nov 1993;13(6):663–729.
16. Strosberg D. [Biotechnology of beta-adrenergic receptors]. *Pathol Biol (Paris)*. Oct 1992;40(8):767–772.
17. Emorine L, Blin N, Strosberg AD. The human beta 3-adrenoceptor: the search for a physiological function. *Trends Pharmacol Sci*. Jan 1994;15(1):3–7.
18. Giacobino JP. Beta 3-adrenoceptor: an update. *Eur J Endocrinol*. Apr 1995;132(4):377–385.
19. Nisoli E, Tonello C, Landi M, Carruba MO. Functional studies of the first selective beta 3-adrenergic receptor antagonist SR 59230A in rat brown adipocytes. *Mol Pharmacol*. Jan 1996;49(1):7–14.
20. Lafontan M, Berlan M. Fat cell adrenergic receptors and the control of white and brown fat cell function. *J Lipid Res*. Jul 1993;34(7):1057–1091.
21. Lefkowitz RJ, Rajagopal K, Whalen EJ. New roles for beta-arrestins in cell signaling: not just for seven-transmembrane receptors. *Molecular cell*. Dec 8 2006;24(5):643–652.
22. Hamm HE. The many faces of G protein signaling. *The Journal of biological chemistry*. Jan 9 1998;273(2):669–672.
23. Bourne HR. G proteins. The arginine finger strikes again. *Nature*. Oct 16 1997;389(6652):673–674.
24. Wess J. G-protein-coupled receptors: molecular mechanisms involved in receptor activation and selectivity of G-protein recognition. *Faseb J*. Apr 1997;11(5):346–354.
25. Lambright DG, Sondek J, Bohm A, Skiba NP, Hamm HE, Sigler PB. The 2.0 A crystal structure of a heterotrimeric G protein. *Nature*. Jan 25 1996;379(6563):311–319.

26. Ford CE, Skiba NP, Bae H, Daaka Y, Reuveny E, Shekter LR, Rosal R, Weng G, Yang CS, Iyengar R, Miller RJ, Jan LY, Lefkowitz RJ, Hamm HE. Molecular basis for interactions of G protein betagamma subunits with effectors. *Science*. May 22 1998;280(5367):1271–1274.

27. Clapham DE, Neer EJ. G protein beta gamma subunits. *Annual review of pharmacology and toxicology*. 1997;37:167–203.

28. Rockman HA, Koch WJ, Lefkowitz RJ. Seven-transmembrane-spanning receptors and heart function. *Nature*. Jan 10 2002;415(6868):206–212.

29. Keys JR, Koch WJ. The adrenergic pathway and heart failure. *Recent progress in hormone research*. 2004;59:13–30.

30. Kaumann A, Bartel S, Molenaar P, Sanders L, Burrell K, Vetter D, Hempel P, Karczewski P, Krause EG. Activation of beta2-adrenergic receptors hastens relaxation and mediates phosphorylation of phospholamban, troponin I, and C-protein in ventricular myocardium from patients with terminal heart failure. *Circulation*. Jan 5–12 1999;99(1):65–72.

31. Molenaar P, Bartel S, Cochrane A, Vetter D, Jalali H, Pohlner P, Burrell K, Karczewski P, Krause EG, Kaumann A. Both beta(2)- and beta(1)-adrenergic receptors mediate hastened relaxation and phosphorylation of phospholamban and troponin I in ventricular myocardium of Fallot infants, consistent with selective coupling of beta(2)-adrenergic receptors to G(s)-protein. *Circulation*. Oct 10 2000;102(15):1814–1821.

32. Chesley A, Lundberg MS, Asai T, Xiao RP, Ohtani S, Lakatta EG, Crow MT. The beta(2)-adrenergic receptor delivers an antiapoptotic signal to cardiac myocytes through G(i)-dependent coupling to phosphatidylinositol 3'-kinase. *Circulation research*. Dec 8 2000;87(12):1172–1179.

33. Communal C, Singh K, Sawyer DB, Colucci WS. Opposing effects of beta(1)- and beta(2)-adrenergic receptors on cardiac myocyte apoptosis: role of a pertussis toxin-sensitive G protein. *Circulation*. Nov 30 1999;100(22):2210–2212.

34. Zhu WZ, Zheng M, Koch WJ, Lefkowitz RJ, Kobilka BK, Xiao RP. Dual modulation of cell survival and cell death by beta(2)-adrenergic signaling in adult mouse cardiac myocytes. *Proceedings of the National Academy of Sciences of the United States of America*. Feb 13 2001;98(4):1607–1612.

35. Xiao RP, Avdonin P, Zhou YY, Cheng H, Akhter SA, Eschenhagen T, Lefkowitz RJ, Koch WJ, Lakatta EG. Coupling of beta2-adrenoceptor to Gi proteins and its physiological relevance in murine cardiac myocytes. *Circulation research*. Jan 8–22 1999;84(1):43–52.

36. Kuschel M, Zhou YY, Cheng H, Zhang SJ, Chen Y, Lakatta EG, Xiao RP. G(i) protein-mediated functional compartmentalization of cardiac beta(2)-adrenergic signaling. *The Journal of biological chemistry*. Jul 30 1999;274(31):22048–22052.

37. Schillace RV, Scott JD. Association of the type 1 protein phosphatase PP1 with the A-kinase anchoring protein AKAP220. *Curr Biol*. Mar 25 1999;9(6):321–324.

38. Hausdorff WP, Caron MG, Lefkowitz RJ. Turning off the signal: desensitization of beta-adrenergic receptor function. *Faseb J*. Aug 1990;4(11):2881–2889.

39. Mukherjee C, Caron MG, Lefkowitz RJ. Catecholamine-induced subsensitivity of adenylate cyclase associated with loss of beta-adrenergic receptor binding sites. *Proceedings of the National Academy of Sciences of the United States of America*. May 1975;72(5):1945–1949.

40. Su YF, Harden TK, Perkins JP. Isoproterenol-induced desensitization of adenylate cyclase in human astrocytoma cells. Relation of loss of hormonal responsiveness and decrement in beta-adrenergic receptors. *The Journal of biological chemistry*. Jan 10 1979;254(1):38–41.

41. Pitcher JA, Freedman NJ, Lefkowitz RJ. G protein-coupled receptor kinases. *Annu Rev Biochem*. 1998;67:653–692.

42. Inglese J, Freedman NJ, Koch WJ, Lefkowitz RJ. Structure and mechanism of the G protein-coupled receptor kinases. *The Journal of biological chemistry*. Nov 15 1993;268(32):23735–23738.

43. Penela P, Ribas C, Mayor F, Jr. Mechanisms of regulation of the expression and function of G protein-coupled receptor kinases. *Cellular signalling.* Nov 2003;15(11):973–981.

44. Premont RT, Gainetdinov RR. Physiological roles of G protein-coupled receptor kinases and arrestins. *Annual review of physiology.* 2007;69:511–534.

45. Lefkowitz RJ, Shenoy SK. Transduction of receptor signals by beta-arrestins. *Science.* Apr 22 2005;308(5721):512–517.

46. Bristow MR, Ginsburg R, Umans V, Fowler M, Minobe W, Rasmussen R, Zera P, Menlove R, Shah P, Jamieson S, et al. Beta 1- and beta 2-adrenergic-receptor sub-populations in nonfailing and failing human ventricular myocardium: coupling of both receptor subtypes to muscle contraction and selective beta 1-receptor down-regulation in heart failure. *Circulation research.* Sep 1986;59(3):297–309.

47. Brodde OE. Beta-adrenoceptors in cardiac disease. *Pharmacology & therapeutics.* Dec 1993;60(3):405–430.

48. Angelone T, Filice E, Quintieri AM, Imbrogno S, Recchia A, Pulera E, Mannarino C, Pellegrino D, Cerra MC. Beta3-adrenoceptors modulate left ventricular relaxation in the rat heart via the NO-cGMP-PKG pathway. *Acta physiologica (Oxford, England).* Jul 2008;193(3):229–239.

49. Schultz HD. Nitric oxide regulation of autonomic function in heart failure. *Curr Heart Fail Rep.* Jun 2009;6(2):71–80.

50. Piascik MT, Perez DM. Alpha1-adrenergic receptors: new insights and directions. *The Journal of pharmacology and experimental therapeutics.* Aug 2001;298(2):403–410.

51. Iaccarino G, Ciccarelli M, Sorriento D, Cipolletta E, Cerullo V, Iovino GL, Paudice A, Elia A, Santulli G, Campanile A, Arcucci O, Pastore L, Salvatore F, Condorelli G, Trimarco B. AKT participates in endothelial dysfunction in hypertension. *Circulation.* Jun 1 2004;109(21):2587–2593.

52. Iaccarino G, Cipolletta E, Fiorillo A, Annecchiarico M, Ciccarelli M, Cimini V, Koch WJ, Trimarco B. Beta(2)-adrenergic receptor gene delivery to the endothelium corrects impaired adrenergic vasorelaxation in hypertension. *Circulation.* Jul 16 2002;106(3):349–355.

53. Ciccarelli M, Cipolletta E, Santulli G, Campanile A, Pumiglia K, Cervero P, Pastore L, Astone D, Trimarco B, Iaccarino G. Endothelial beta2 adrenergic signaling to AKT: role of Gi and SRC. *Cellular signalling.* Sep 2007;19(9):1949–1955.

54. Mutlu GM, Koch WJ, Factor P. Alveolar epithelial beta 2-adrenergic receptors: their role in regulation of alveolar active sodium transport. *American journal of respiratory and critical care medicine.* Dec 15 2004;170(12):1270–1275.

55. Tanaka Y, Yamashita Y, Yamaki F, Horinouchi T, Shigenobu K, Koike K. MaxiK channel mediates beta2-adrenoceptor-activated relaxation to isoprenaline through cAMP-dependent and -independent mechanisms in guinea-pig tracheal smooth muscle. *J Smooth Muscle Res.* Dec 2003;39(6):205–219.

56. Kume H, Hall IP, Washabau RJ, Takagi K, Kotlikoff MI. Beta-adrenergic agonists regulate KCa channels in airway smooth muscle by cAMP-dependent and -independent mechanisms. *The Journal of clinical investigation.* Jan 1994;93(1):371–379.

57. Nelson MT, Quayle JM. Physiological roles and properties of potassium channels in arterial smooth muscle. *The American journal of physiology.* Apr 1995;268(4 Pt 1):C799–822.

58. Berdiev BK, Prat AG, Cantiello HF, Ausiello DA, Fuller CM, Jovov B, Benos DJ, Ismailov, II . Regulation of epithelial sodium channels by short actin filaments. *The Journal of biological chemistry.* Jul 26 1996;271(30):17704–17710.

59. Minakata Y, Suzuki S, Grygorczyk C, Dagenais A, Berthiaume Y. Impact of beta-adrenergic agonist on Na+ channel and Na+-K+-ATPase expression in alveolar type II cells. *The American journal of physiology.* Aug 1998;275(2 Pt 1):L414–422.

60. Bristow MR, Shakar SF, Linseman JV, Lowes BD. Inotropes and beta-blockers: is there a need for new guidelines? *Journal of cardiac failure.* Jun 2001;7(2 Suppl 1): 8–12.

61. Hochman JS, Tamis JE, Thompson TD, Weaver WD, White HD, Van de Werf F, Aylward P, Topol EJ, Califf RM. Sex, clinical presentation, and outcome in patients with acute coronary syndromes. Global Use of Strategies to Open Occluded Coronary Arteries in Acute Coronary Syndromes IIb Investigators. *The New England journal of medicine.* Jul 22 1999;341(4):226–232.

62. McGhie AI, Golstein RA. Pathogenesis and management of acute heart failure and cardiogenic shock: role of inotropic therapy. *Chest.* Nov 1992;102(5 Suppl 2):626S–632S.

63. Farid I, Litaker D, Tetzlaff JE. Implementing ACC/AHA guidelines for the preoperative management of patients with coronary artery disease scheduled for noncardiac surgery: effect on perioperative outcome. *Journal of clinical anesthesia.* Mar 2002;14(2):126–128.

64. Jakob SM, Tenhunen JJ, Heino A, Pradl R, Alhava E, Takala J. Splanchnic vasoregulation during mesenteric ischemia and reperfusion in pigs. *Shock (Augusta, Ga.* Aug 2002;18(2):142–147.

65. Martin C, Viviand X, Arnaud S, Vialet R, Rougnon T. Effects of norepinephrine plus dobutamine or norepinephrine alone on left ventricular performance of septic shock patients. *Critical care medicine.* Sep 1999;27(9):1708–1713.

66. Birks EJ, Tansley PD, Hardy J, George RS, Bowles CT, Burke M, Banner NR, Khaghani A, Yacoub MH. Left ventricular assist device and drug therapy for the reversal of heart failure. *The New England journal of medicine.* Nov 2 2006;355(18):1873–1884.

67. Hall JL, Birks EJ, Grindle S, Cullen ME, Barton PJ, Rider JE, Lee S, Harwalker S, Mariash A, Adhikari N, Charles NJ, Felkin LE, Polster S, George RS, Miller LW, Yacoub MH. Molecular signature of recovery following combination left ventricular assist device (LVAD) support and pharmacologic therapy. *European heart journal.* Mar 2007;28(5):613–627.

68. Campbell LM. From adrenaline to formoterol: advances in beta-agonist therapy in the treatment of asthma. *International journal of clinical practice.* Dec 2002;56(10):783–790.

69. Rabe KF. State of the art in beta2-agonist therapy: a safety review of long-acting agents. *International journal of clinical practice.* Oct 2003;57(8):689–697.

70. Crane J, Burgess C, Beasley R. Cardiovascular and hypokalaemic effects of inhaled salbutamol, fenoterol, and isoprenaline. *Thorax.* Feb 1989;44(2):136–140.

71. Simons FE, Clark S, Camargo CA, Jr. Anaphylaxis in the community: Learning from the survivors. *The Journal of allergy and clinical immunology.* Jun 18 2009.

72. Frishman W. Clinical pharmacology of the new beta-adrenergic blocking drugs. Part 13. The beta-adrenoceptor blocking drugs: a perspective. *American heart journal.* May 1980;99(5):665–670.

73. Laragh JH, Sealey JE, Buhler FR, Vaughan ED, Brunner HR, Gavras H, Baer L. The renin axis and vasoconstriction volume analysis for understanding and treating renovascular and renal hypertension. *The American journal of medicine.* Jan 1975;58(1):4–13.

74. Kleinrok Z, Ksiazek A. The effect of beta-adrenergic receptor blocking agents on hypertensive action of noradrenaline injected into the lateral ventricle of rat brain. *Polish journal of pharmacology and pharmacy.* Jul-Aug 1977;29(4):405–409.

75. Tuck ML. The sympathetic nervous system in essential hypertension. *American heart journal.* Oct 1986;112(4):877–886.

76. Lohmeier TE, Hildebrandt DA, Dwyer TM, Iliescu R, Irwin ED, Cates AW, Rossing MA. Prolonged activation of the baroreflex decreases arterial pressure even during chronic adrenergic blockade. *Hypertension.* May 2009;53(5):833–838.

77. Ferdinand KC, Armani AM. The management of hypertension in African Americans. *Critical pathways in cardiology.* Jun 2007;6(2):67–71.

78. Bangalore S, Sawhney S, Messerli FH. Relation of beta-blocker-induced heart rate lowering and cardioprotection in hypertension. *Journal of the American College of Cardiology.* Oct 28 2008;52(18):1482–1489.

79. Bangalore S, Wild D, Parkar S, Kukin M, Messerli FH. Beta-blockers for primary prevention of heart failure in patients with hypertension insights from a meta-analysis. *Journal of the American College of Cardiology*. Sep 23 2008;52(13):1062–1072.

80. Messerli FH, Bangalore S. Resting heart rate and cardiovascular disease: the beta-blocker-hypertension paradox. *Journal of the American College of Cardiology*. Jan 22 2008;51(3):330–331; author reply 331–332.

81. Ardehali A, Ports TA. Myocardial oxygen supply and demand. *Chest*. Sep 1990;98(3):699–705.

82. Parker JD, Testa MA, Jimenez AH, Tofler GH, Muller JE, Parker JO, Stone PH. Morning increase in ambulatory ischemia in patients with stable coronary artery disease. Importance of physical activity and increased cardiac demand. *Circulation*. Feb 1994;89(2):604–614.

83. Hoffman BB, Lefkowitz RJ. Adrenergic receptors in the heart. *Annual review of physiology*. 1982;44:475–484.

84. Fuster V. Mechanisms of arterial thrombosis: foundation for therapy. *American heart journal*. Jun 1998;135(6 Pt 3 Su):S361–366.

85. Fuster V. 50th anniversary historical article. Acute coronary syndromes: the degree and morphology of coronary stenoses. *Journal of the American College of Cardiology*. Dec 1999;34(7):1854–1856.

86. Davies MJ. Stability and instability: two faces of coronary atherosclerosis. The Paul Dudley White Lecture 1995. *Circulation*. Oct 15 1996;94(8):2013–2020.

87. Timolol-induced reduction in mortality and reinfarction in patients surviving acute myocardial infarction. *The New England journal of medicine*. Apr 2 1981;304(14):801–807.

88. Hjalmarson A, Elmfeldt D, Herlitz J, Holmberg S, Malek I, Nyberg G, Ryden L, Swedberg K, Vedin A, Waagstein F, Waldenstrom A, Waldenstrom J, Wedel H, Wilhelmsen L, Wilhelmsson C. Effect on mortality of metoprolol in acute myocardial infarction. A double-blind randomised trial. *Lancet*. Oct 17 1981;2(8251): 823–827.

89. A randomized trial of propranolol in patients with acute myocardial infarction. I. Mortality results. *Jama*. Mar 26 1982;247(12):1707–1714.

90. Randomised trial of intravenous atenolol among 16 027 cases of suspected acute myocardial infarction: ISIS-1. First International Study of Infarct Survival Collaborative Group. *Lancet*. Jul 12 1986;2(8498):57–66.

91. Metoprolol in acute myocardial infarction. Enzymatic estimation of infarct size. The MIAMI Trial Research Group. *The American journal of cardiology*. Nov 22 1985;56(14):27G–29G.

92. Metoprolol in acute myocardial infarction. Arrhythmias. The MIAMI Trial Research Group. *The American journal of cardiology*. Nov 22 1985;56(14):35G–38G.

93. Kirshenbaum JM, Kloner RA, Antman EM, Braunwald E. Use of an ultra short-acting beta-blocker in patients with acute myocardial ischemia. *Circulation*. Oct 1985;72(4):873–880.

94. Yusuf S, Peto R, Lewis J, Collins R, Sleight P. Beta blockade during and after myocardial infarction: an overview of the randomized trials. *Progress in cardiovascular diseases*. Mar-Apr 1985;27(5):335–371.

95. Borzak S, Fenton T, Glasser SP, Shook TL, MacCallum G, Young PM, Stone PH. Discordance between effects of anti-ischemic therapy on ambulatory ischemia, exercise performance and anginal symptoms in patients with stable angina pectoris. The Angina and Silent Ischemia Study Group (ASIS). *Journal of the American College of Cardiology*. Jun 1993;21(7):1605–1611.

96. Packer M. The development of positive inotropic agents for chronic heart failure: how have we gone astray? *Journal of the American College of Cardiology*. Oct 1993;22(4 Suppl A):119A–126A.

97. Fowler MB, Laser JA, Hopkins GL, Minobe W, Bristow MR. Assessment of the beta-adrenergic receptor pathway in the intact failing human heart: progressive receptor down-regulation and subsensitivity to agonist response. *Circulation*. Dec 1986;74(6):1290–1302.

98. Engelhardt S, Hein L, Wiesmann F, Lohse MJ. Progressive hypertrophy and heart failure in beta1-adrenergic receptor transgenic mice. *Proceedings of the National Academy of Sciences of the United States of America*. Jun 8 1999;96(12):7059–7064.

99. Vatner SF, Vatner DE, Homcy CJ. beta-adrenergic receptor signaling: an acute compensatory adjustment-inappropriate for the chronic stress of heart failure? Insights from Gsalpha overexpression and other genetically engineered animal models. *Circulation research*. Mar 17 2000;86(5):502–506.

100. Koch WJ, Milano CA, Lefkowitz RJ. Transgenic manipulation of myocardial G protein-coupled receptors and receptor kinases. *Circulation research*. Apr 1996;78(4):511–516.

101. Packer M, Bristow MR, Cohn JN, Colucci WS, Fowler MB, Gilbert EM, Shusterman NH. The effect of carvedilol on morbidity and mortality in patients with chronic heart failure. U.S. Carvedilol Heart Failure Study Group. *The New England journal of medicine*. May 23 1996;334(21):1349–1355.

102. Bauman JL, Talbert RL. Pharmacodynamics of beta-blockers in heart failure: lessons from the carvedilol or metoprolol European trial. *Journal of cardiovascular pharmacology and therapeutics*. Jun 2004;9(2):117–128.

103. Noguchi N, Nishino K, Niki E. Antioxidant action of the antihypertensive drug, carvedilol, against lipid peroxidation. *Biochemical pharmacology*. May 1 2000;59(9):1069–1076.

104. Bristow MR, Gilbert EM, Abraham WT, Adams KF, Fowler MB, Hershberger RE, Kubo SH, Narahara KA, Ingersoll H, Krueger S, Young S, Shusterman N. Carvedilol produces dose-related improvements in left ventricular function and survival in subjects with chronic heart failure. MOCHA Investigators. *Circulation*. Dec 1 1996;94(11):2807–2816.

105. Poole-Wilson PA, Swedberg K, Cleland JG, Di Lenarda A, Hanrath P, Komajda M, Lubsen J, Lutiger B, Metra M, Remme WJ, Torp-Pedersen C, Scherhag A, Skene A. Comparison of carvedilol and metoprolol on clinical outcomes in patients with chronic heart failure in the Carvedilol Or Metoprolol European Trial (COMET): randomised controlled trial. *Lancet*. Jul 5 2003;362(9377):7–13.

106. Cleland JG, Charlesworth A, Lubsen J, Swedberg K, Remme WJ, Erhardt L, Di Lenarda A, Komajda M, Metra M, Torp-Pedersen C, Poole-Wilson PA. A comparison of the effects of carvedilol and metoprolol on well-being, morbidity, and mortality (the "patient journey") in patients with heart failure: a report from the Carvedilol Or Metoprolol European Trial (COMET). *Journal of the American College of Cardiology*. Apr 18 2006;47(8):1603–1611.

107. Spirito P, Seidman CE, McKenna WJ, Maron BJ. The management of hypertrophic cardiomyopathy. *The New England journal of medicine*. Mar 13 1997;336(11): 775–785.

108. Reiter MJ, Reiffel JA. Importance of beta blockade in the therapy of serious ventricular arrhythmias. *The American journal of cardiology*. Aug 20 1998;82(4A): 9I–19I.

109. John RM, Taggart PI, Sutton PM, Ell PJ, Swanton H. Direct effect of dobutamine on action potential duration in ischemic compared with normal areas in the human ventricle. *Journal of the American College of Cardiology*. Oct 1992;20(4):896–903.

110. Luedtke SA, Kuhn RJ, McCaffrey FM. Pharmacologic management of supraventricular tachycardias in children. Part 2: Atrial flutter, atrial fibrillation, and junctional and atrial ectopic tachycardia. *The Annals of pharmacotherapy*. Nov 1997;31(11):1347–1359.

111. Luedtke SA, Kuhn RJ, McCaffrey FM. Pharmacologic management of supraventricular tachycardias in children. Part 1: Wolff-Parkinson-White and atrioventricular nodal reentry. *The Annals of pharmacotherapy.* Oct 1997;31(10):1227–1243.

112. Raake PW, Vinge LE, Gao E, Boucher M, Rengo G, Chen X, DeGeorge BR, Jr., Matkovich S, Houser SR, Most P, Eckhart AD, Dorn GW, 2nd, Koch WJ. G protein-coupled receptor kinase 2 ablation in cardiac myocytes before or after myocardial infarction prevents heart failure. *Circulation research.* Aug 15 2008;103(4): 413–422.

113. Rengo G, Lymperopoulos A, Zincarelli C, Donniacuo M, Soltys S, Rabinowitz JE, Koch WJ. Myocardial adeno-associated virus serotype 6-betaARKct gene therapy improves cardiac function and normalizes the neurohormonal axis in chronic heart failure. *Circulation.* Jan 6 2009;119(1):89–98.

114. Ungerer M, Bohm M, Elce JS, Erdmann E, Lohse MJ. Altered expression of beta-adrenergic receptor kinase and beta 1-adrenergic receptors in the failing human heart. *Circulation.* Feb 1993;87(2):454–463.

115. Dzimiri N, Basco C, Moorji A, Afrane B, Al-Halees Z. Characterization of lymphocyte beta 2-adrenoceptor signalling in patients with left ventricular volume overload disease. *Clinical and experimental pharmacology & physiology.* Mar 2002;29(3):181–188.

116. Dzimiri N, Muiya P, Andres E, Al-Halees Z. Differential functional expression of human myocardial G protein receptor kinases in left ventricular cardiac diseases. *European journal of pharmacology.* Apr 12 2004;489(3):167–177.

117. Martini JS, Raake P, Vinge LE, DeGeorge BR, Jr., Chuprun JK, Harris DM, Gao E, Eckhart AD, Pitcher JA, Koch WJ. Uncovering G protein-coupled receptor kinase-5 as a histone deacetylase kinase in the nucleus of cardiomyocytes. *Proceedings of the National Academy of Sciences of the United States of America.* Aug 26 2008;105(34):12457–12462.

118. Hata JA, Williams ML, Koch WJ. Genetic manipulation of myocardial beta-adrenergic receptor activation and desensitization. *Journal of molecular and cellular cardiology.* Jul 2004;37(1):11–21.

119. Iaccarino G, Tomhave ED, Lefkowitz RJ, Koch WJ. Reciprocal in vivo regulation of myocardial G protein-coupled receptor kinase expression by beta-adrenergic receptor stimulation and blockade. *Circulation.* Oct 27 1998;98(17):1783–1789.

120. Hakonarson H, Herrick DJ, Grunstein MM. Mechanism of impaired beta-adrenoceptor responsiveness in atopic sensitized airway smooth muscle. *The American journal of physiology.* Nov 1995;269(5 Pt 1):L645–652.

121. Wills-Karp M, Gilmour MI. Increased cholinergic antagonism underlies impaired beta-adrenergic response in ovalbumin-sensitized guinea pigs. *J Appl Physiol.* Jun 1993;74(6):2729–2735.

122. Walker JK, Gainetdinov RR, Feldman DS, McFawn PK, Caron MG, Lefkowitz RJ, Premont RT, Fisher JT. G protein-coupled receptor kinase 5 regulates airway responses induced by muscarinic receptor activation. *American journal of physiology.* Feb 2004;286(2):L312–319.

123. Maqbool A, Hall AS, Ball SG, Balmforth AJ. Common polymorphisms of beta1-adrenoceptor: identification and rapid screening assay. *Lancet.* Mar 13 1999;353(9156):897.

124. Mason DA, Moore JD, Green SA, Liggett SB. A gain-of-function polymorphism in a G-protein coupling domain of the human beta1-adrenergic receptor. *The Journal of biological chemistry.* Apr 30 1999;274(18):12670–12674.

125. Tesson F, Charron P, Peuchmaurd M, Nicaud V, Cambien F, Tiret L, Poirier O, Desnos M, Jullieres Y, Amouyel P, Roizes G, Dorent R, Schwartz K, Komajda M. Characterization of a unique genetic variant in the beta1-adrenoceptor gene and evaluation of its role in idiopathic dilated cardiomyopathy. CARDIGENE Group. *Journal of molecular and cellular cardiology.* May 1999;31(5):1025–1032.

126. Borjesson M, Magnusson Y, Hjalmarson A, Andersson B. A novel polymorphism in the gene coding for the beta(1)-adrenergic receptor associated with survival in patients with heart failure. *European heart journal*. Nov 2000;21(22): 1853–1858.

127. Johnson JA, Terra SG. Beta-adrenergic receptor polymorphisms: cardiovascular disease associations and pharmacogenetics. *Pharmaceutical research*. Dec 2002;19(12):1779–1787.

128. Small KM, McGraw DW, Liggett SB. Pharmacology and physiology of human adrenergic receptor polymorphisms. *Annual review of pharmacology and toxicology*. 2003;43:381–411.

129. Kirstein SL, Insel PA. Autonomic nervous system pharmacogenomics: a progress report. *Pharmacological reviews*. Mar 2004;56(1):31–52.

130. Leineweber K, Buscher R, Bruck H, Brodde OE. Beta-adrenoceptor polymorphisms. *Naunyn-Schmiedeberg's archives of pharmacology*. Jan 2004;369(1):1–22.

131. Frielle T, Collins S, Daniel KW, Caron MG, Lefkowitz RJ, Kobilka BK. Cloning of the cDNA for the human beta 1-adrenergic receptor. *Proceedings of the National Academy of Sciences of the United States of America*. Nov 1987;84(22):7920–7924.

132. Levin MC, Marullo S, Muntaner O, Andersson B, Magnusson Y. The myocardium-protective Gly-49 variant of the beta 1-adrenergic receptor exhibits constitutive activity and increased desensitization and down-regulation. *The Journal of biological chemistry*. Aug 23 2002;277(34):30429–30435.

133. Rathz DA, Brown KM, Kramer LA, Liggett SB. Amino acid 49 polymorphisms of the human beta1-adrenergic receptor affect agonist-promoted trafficking. *Journal of cardiovascular pharmacology*. Feb 2002;39(2):155–160.

134. Rathz DA, Gregory KN, Fang Y, Brown KM, Liggett SB. Hierarchy of polymorphic variation and desensitization permutations relative to beta 1- and beta 2-adrenergic receptor signaling. *The Journal of biological chemistry*. Mar 21 2003;278(12): 10784–10789.

135. Joseph SS, Lynham JA, Grace AA, Colledge WH, Kaumann AJ. Markedly reduced effects of (-)-isoprenaline but not of (-)-CGP12177 and unchanged affinity of beta-blockers at Gly389-beta1-adrenoceptors compared to Arg389-beta1-adrenoceptors. *British journal of pharmacology*. May 2004;142(1):51–56.

136. Sandilands A, Yeo G, Brown MJ, O'Shaughnessy KM. Functional responses of human beta1 adrenoceptors with defined haplotypes for the common 389R>G and 49S>G polymorphisms. *Pharmacogenetics*. Jun 2004;14(6):343–349.

137. Xie HG, Dishy V, Sofowora G, Kim RB, Landau R, Smiley RM, Zhou HH, Wood AJ, Harris P, Stein CM. Arg389Gly beta 1-adrenoceptor polymorphism varies in frequency among different ethnic groups but does not alter response in vivo. *Pharmacogenetics*. Apr 2001;11(3):191–197.

138. Buscher R, Belger H, Eilmes KJ, Tellkamp R, Radke J, Dhein S, Hoyer PF, Michel MC, Insel PA, Brodde OE. In-vivo studies do not support a major functional role for the Gly389Arg beta 1-adrenoceptor polymorphism in humans. *Pharmacogenetics*. Apr 2001;11(3):199–205.

139. Sofowora GG, Dishy V, Muszkat M, Xie HG, Kim RB, Harris PA, Prasad HC, Byrne DW, Nair UB, Wood AJ, Stein CM. A common beta1-adrenergic receptor polymorphism (Arg389Gly) affects blood pressure response to beta-blockade. *Clinical pharmacology and therapeutics*. Apr 2003;73(4):366–371.

140. Liu J, Liu ZQ, Tan ZR, Chen XP, Wang LS, Zhou G, Zhou HH. Gly389Arg polymorphism of beta1-adrenergic receptor is associated with the cardiovascular response to metoprolol. *Clinical pharmacology and therapeutics*. Oct 2003;74(4):372–379.

141. Defoor J, Martens K, Zielinska D, Matthijs G, Van Nerum H, Schepers D, Fagard R, Vanhees L. The CAREGENE study: polymorphisms of the beta1-adrenoceptor gene and aerobic power in coronary artery disease. *European heart journal*. Apr 2006;27(7):808–816.

142. Nieminen T, Lehtimaki T, Laiho J, Rontu R, Niemela K, Koobi T, Lehtinen R, Viik J, Turjanmaa V, Kahonen M. Effects of polymorphisms in beta1-adrenoceptor and alpha-subunit of G protein on heart rate and blood pressure during exercise test. The Finnish Cardiovascular Study. *J Appl Physiol*. Feb 2006;100(2):507–511.

143. La Rosee K, Huntgeburth M, Rosenkranz S, Bohm M, Schnabel P. The Arg389Gly beta1-adrenoceptor gene polymorphism determines contractile response to catecholamines. *Pharmacogenetics*. Nov 2004;14(11):711–716.

144. Bruck H, Leineweber K, Temme T, Weber M, Heusch G, Philipp T, Brodde OE. The Arg389Gly beta1-adrenoceptor polymorphism and catecholamine effects on plasma-renin activity. *Journal of the American College of Cardiology*. Dec 6 2005;46(11):2111–2115.

145. Leineweber K, Bogedain P, Wolf C, Wagner S, Weber M, Jakob HG, Heusch G, Philipp T, Brodde OE. In patients chronically treated with metoprolol, the demand of inotropic catecholamine support after coronary artery bypass grafting is determined by the Arg389Gly-beta 1-adrenoceptor polymorphism. *Naunyn-Schmiedeberg's archives of pharmacology*. Jul 2007;375(5):303–309.

146. Kobilka BK, Dixon RA, Frielle T, Dohlman HG, Bolanowski MA, Sigal IS, Yang-Feng TL, Francke U, Caron MG, Lefkowitz RJ. cDNA for the human beta 2-adrenergic receptor: a protein with multiple membrane-spanning domains and encoded by a gene whose chromosomal location is shared with that of the receptor for platelet-derived growth factor. *Proceedings of the National Academy of Sciences of the United States of America*. Jan 1987;84(1):46–50.

147. Taylor MR, Bristow MR. The emerging pharmacogenomics of the beta-adrenergic receptors. *Congestive heart failure (Greenwich, Conn*. Nov-Dec 2004;10(6):281–288.

148. Brodde OE, Leineweber K. Beta2-adrenoceptor gene polymorphisms. *Pharmacogenetics and genomics*. May 2005;15(5):267–275.

149. Scott MG, Swan C, Wheatley AP, Hall IP. Identification of novel polymorphisms within the promoter region of the human beta2 adrenergic receptor gene. *British journal of pharmacology*. Feb 1999;126(4):841–844.

150. McGraw DW, Forbes SL, Kramer LA, Liggett SB. Polymorphisms of the 5′ leader cistron of the human beta2-adrenergic receptor regulate receptor expression. *The Journal of clinical investigation*. Dec 1 1998;102(11):1927–1932.

151. Dewar JC, Wheatley AP, Venn A, Morrison JF, Britton J, Hall IP. Beta2-adrenoceptor polymorphisms are in linkage disequilibrium, but are not associated with asthma in an adult population. *Clin Exp Allergy*. Apr 1998;28(4):442–448.

152. Brodde OE, Buscher R, Tellkamp R, Radke J, Dhein S, Insel PA. Blunted cardiac responses to receptor activation in subjects with Thr164Ile beta(2)-adrenoceptors. *Circulation*. Feb 27 2001;103(8):1048–1050.

153. Bruck H, Leineweber K, Ulrich A, Radke J, Heusch G, Philipp T, Brodde OE. Thr164Ile polymorphism of the human beta2-adrenoceptor exhibits blunted desensitization of cardiac functional responses in vivo. *Am J Physiol Heart Circ Physiol*. Nov 2003;285(5):H2034–2038.

154. Barbato E, Penicka M, Delrue L, Van Durme F, De Bruyne B, Goethals M, Wijns W, Vanderheyden M, Bartunek J. Thr164Ile polymorphism of beta2-adrenergic receptor negatively modulates cardiac contractility: implications for prognosis in patients with idiopathic dilated cardiomyopathy. *Heart (British Cardiac Society)*. Jul 2007;93(7):856–861.

155. Green SA, Turki J, Innis M, Liggett SB. Amino-terminal polymorphisms of the human beta 2-adrenergic receptor impart distinct agonist-promoted regulatory properties. *Biochemistry*. Aug 16 1994;33(32):9414–9419.

156. Hoit BD, Suresh DP, Craft L, Walsh RA, Liggett SB. beta2-adrenergic receptor polymorphisms at amino acid 16 differentially influence agonist-stimulated blood pressure and peripheral blood flow in normal individuals. *American heart journal*. Mar 2000;139(3):537–542.

157. Bruck H, Leineweber K, Buscher R, Ulrich A, Radke J, Insel PA, Brodde OE. The Gln27Glu beta2-adrenoceptor polymorphism slows the onset of desensitization of cardiac functional responses in vivo. *Pharmacogenetics.* Feb 2003;13(2):59–66.

158. Trombetta IC, Batalha LT, Rondon MU, Laterza MC, Frazzatto E, Alves MJ, Santos AC, Brum PC, Barretto AC, Halpern A, Villares SM, Negrao CE. Gly16 + Glu27 beta2-adrenoceptor polymorphisms cause increased forearm blood flow responses to mental stress and handgrip in humans. *J Appl Physiol.* Mar 2005;98(3):787–794.

159. Eisenach JH, Barnes SA, Pike TL, Sokolnicki LA, Masuki S, Dietz NM, Rehfeldt KH, Turner ST, Joyner MJ. Arg16/Gly beta2-adrenergic receptor polymorphism alters the cardiac output response to isometric exercise. *J Appl Physiol.* Nov 2005;99(5):1776–1781.

160. Eisenach JH, McGuire AM, Schwingler RM, Turner ST, Joyner MJ. The Arg16/Gly beta2-adrenergic receptor polymorphism is associated with altered cardiovascular responses to isometric exercise. *Physiological genomics.* Feb 13 2004;16(3):323–328.

161. Gratze G, Fortin J, Labugger R, Binder A, Kotanko P, Timmermann B, Luft FC, Hoehe MR, Skrabal F. beta-2 Adrenergic receptor variants affect resting blood pressure and agonist-induced vasodilation in young adult Caucasians. *Hypertension.* Jun 1999;33(6):1425–1430.

162. Cockcroft JR, Gazis AG, Cross DJ, Wheatley A, Dewar J, Hall IP, Noon JP. Beta(2)-adrenoceptor polymorphism determines vascular reactivity in humans. *Hypertension.* Sep 2000;36(3):371–375.

163. Garovic VD, Joyner MJ, Dietz NM, Boerwinkle E, Turner ST. Beta(2)-adrenergic receptor polymorphism and nitric oxide-dependent forearm blood flow responses to isoproterenol in humans. *The Journal of physiology.* Jan 15 2003;546(Pt 2):583–589.

164. Dishy V, Landau R, Sofowora GG, Xie HG, Smiley RM, Kim RB, Byrne DW, Wood AJ, Stein CM. Beta2-adrenoceptor Thr164Ile polymorphism is associated with markedly decreased vasodilator and increased vasoconstrictor sensitivity in vivo. *Pharmacogenetics.* Aug 2004;14(8):517–522.

165. Liggett SB. The pharmacogenetics of beta2-adrenergic receptors: relevance to asthma. *The Journal of allergy and clinical immunology.* Feb 2000;105(2 Pt 2):S487–492.

166. Small KM, Wagoner LE, Levin AM, Kardia SL, Liggett SB. Synergistic polymorphisms of beta1- and alpha2C-adrenergic receptors and the risk of congestive heart failure. *The New England journal of medicine.* Oct 10 2002;347(15):1135–1142.

167. Magnusson Y, Levin MC, Eggertsen R, Nystrom E, Mobini R, Schaufelberger M, Andersson B. Ser49Gly of beta1-adrenergic receptor is associated with effective beta-blocker dose in dilated cardiomyopathy. *Clinical pharmacology and therapeutics.* Sep 2005;78(3):221–231.

168. Iwai C, Akita H, Kanazawa K, Shiga N, Terashima M, Matsuda Y, Takai E, Miyamoto Y, Shimizu M, Kajiya T, Hayashi T, Yokoyama M. Arg389Gly polymorphism of the human beta1-adrenergic receptor in patients with nonfatal acute myocardial infarction. *American heart journal.* Jul 2003;146(1):106–109.

169. Covolo L, Gelatti U, Metra M, Nodari S, Picciche A, Pezzali N, Zani C, Alberti A, Donato F, Nardi G, Dei Cas L. Role of beta1- and beta2-adrenoceptor polymorphisms in heart failure: a case-control study. *European heart journal.* Sep 2004;25(17):1534–1541.

170. Nonen S, Okamoto H, Akino M, Matsui Y, Fujio Y, Yoshiyama M, Takemoto Y, Yoshikawa J, Azuma J, Kitabatake A. No positive association between adrenergic receptor variants of alpha2cDel322–325, beta1Ser49, beta1Arg389 and the risk for heart failure in the Japanese population. *British journal of clinical pharmacology.* Oct 2005;60(4):414–417.

171. Metra M, Zani C, Covolo L, Nodari S, Pezzali N, Gelatti U, Donato F, Nardi G, Dei Cas L. Role of beta1- and alpha2c-adrenergic receptor polymorphisms and their combination in heart failure: a case-control study. *Eur J Heart Fail.* Mar 2006;8(2):131–135.

172. White HL, Maqbool A, McMahon AD, Yates L, Ball SG, Hall AS, Balmforth AJ. An evaluation of the beta-1 adrenergic receptor Arg389Gly polymorphism in individuals at risk of coronary events. A WOSCOPS substudy. *European heart journal*. Jul 2002;23(14):1087–1092.

173. Kanki H, Yang P, Xie HG, Kim RB, George AL, Jr., Roden DM. Polymorphisms in beta-adrenergic receptor genes in the acquired long QT syndrome. *Journal of cardiovascular electrophysiology*. Mar 2002;13(3):252–256.

174. Iwai C, Akita H, Shiga N, Takai E, Miyamoto Y, Shimizu M, Kawai H, Takarada A, Kajiya T, Yokoyama M. Suppressive effect of the Gly389 allele of the beta1-adrenergic receptor gene on the occurrence of ventricular tachycardia in dilated cardiomyopathy. *Circ J*. Aug 2002;66(8):723–728.

175. de Groote P, Lamblin N, Helbecque N, Mouquet F, Mc Fadden E, Hermant X, Amouyel P, Dallongeville J, Bauters C. The impact of beta-adrenoreceptor gene polymorphisms on survival in patients with congestive heart failure. *Eur J Heart Fail*. Oct 2005;7(6):966–973.

176. Forleo C, Resta N, Sorrentino S, Guida P, Manghisi A, De Luca V, Romito R, Iacoviello M, De Tommasi E, Troisi F, Rizzon B, Guanti G, Rizzon P, Pitzalis MV. Association of beta-adrenergic receptor polymorphisms and progression to heart failure in patients with idiopathic dilated cardiomyopathy. *The American journal of medicine*. Oct 1 2004;117(7):451–458.

177. Shin J, Lobmeyer MT, Gong Y, Zineh I, Langaee TY, Yarandi H, Schofield RS, Aranda JM, Jr., Hill JA, Pauly DF, Johnson JA. Relation of beta(2)-adrenoceptor haplotype to risk of death and heart transplantation in patients with heart failure. *The American journal of cardiology*. Jan 15 2007;99(2):250–255.

178. Canham RM, Das SR, Leonard D, Abdullah SM, Mehta SK, Chung AK, Li JL, Victor RG, Auchus RJ, Drazner MH. Alpha2cDel322–325 and beta1Arg389 adrenergic polymorphisms are not associated with reduced left ventricular ejection fraction or increased left ventricular volume. *Journal of the American College of Cardiology*. Jan 16 2007;49(2):274–276.

179. Wagoner LE, Craft LL, Zengel P, McGuire N, Rathz DA, Dorn GW, 2nd, Liggett SB. Polymorphisms of the beta1-adrenergic receptor predict exercise capacity in heart failure. *American heart journal*. Nov 2002;144(5):840–846.

180. Taylor DR, Drazen JM, Herbison GP, Yandava CN, Hancox RJ, Town GI. Asthma exacerbations during long term beta agonist use: influence of beta(2) adrenoceptor polymorphism. *Thorax*. Sep 2000;55(9):762–767.

181. Israel E, Chinchilli VM, Ford JG, Boushey HA, Cherniack R, Craig TJ, Deykin A, Fagan JK, Fahy JV, Fish J, Kraft M, Kunselman SJ, Lazarus SC, Lemanske RF, Jr., Liggett SB, Martin RJ, Mitra N, Peters SP, Silverman E, Sorkness CA, Szefler SJ, Wechsler ME, Weiss ST, Drazen JM. Use of regularly scheduled albuterol treatment in asthma: genotype-stratified, randomised, placebo-controlled cross-over trial. *Lancet*. Oct 23–29 2004;364(9444):1505–1512.

182. Israel E, Drazen JM, Liggett SB, Boushey HA, Cherniack RM, Chinchilli VM, Cooper DM, Fahy JV, Fish JE, Ford JG, Kraft M, Kunselman S, Lazarus SC, Lemanske RF, Martin RJ, McLean DE, Peters SP, Silverman EK, Sorkness CA, Szefler SJ, Weiss ST, Yandava CN. The effect of polymorphisms of the beta(2)-adrenergic receptor on the response to regular use of albuterol in asthma. *American journal of respiratory and critical care medicine*. Jul 2000;162(1):75–80.

183. Palmer CN, Lipworth BJ, Lee S, Ismail T, Macgregor DF, Mukhopadhyay S. Arginine-16 beta2 adrenoceptor genotype predisposes to exacerbations in young asthmatics taking regular salmeterol. *Thorax*. Nov 2006;61(11):940–944.

184. Wechsler ME, Lehman E, Lazarus SC, Lemanske RF, Jr., Boushey HA, Deykin A, Fahy JV, Sorkness CA, Chinchilli VM, Craig TJ, DiMango E, Kraft M, Leone F, Martin RJ, Peters SP, Szefler SJ, Liu W, Israel E. beta-Adrenergic receptor polymorphisms and response to salmeterol. *American journal of respiratory and critical care medicine*. Mar 1 2006;173(5):519–526.

185. Iaccarino G, Lanni F, Cipolletta E, Trimarco V, Izzo R, Iovino GL, De Luca N, Trimarco B. The Glu27 allele of the beta2 adrenergic receptor increases the risk of cardiac hypertrophy in hypertension. *Journal of hypertension*. Nov 2004;22(11):2117–2122.

186. Iaccarino G, Izzo R, Trimarco V, Cipolletta E, Lanni F, Sorriento D, Iovino GL, Rozza F, De Luca N, Priante O, Di Renzo G, Trimarco B. Beta2-adrenergic receptor polymorphisms and treatment-induced regression of left ventricular hypertrophy in hypertension. *Clinical pharmacology and therapeutics*. Dec 2006;80(6):633–645.

187. Piscione F, Iaccarino G, Galasso G, Cipolletta E, Rao MA, Brevetti G, Piccolo R, Trimarco B, Chiariello M. Effects of Ile164 polymorphism of beta2-adrenergic receptor gene on coronary artery disease. *Journal of the American College of Cardiology*. Oct 21 2008;52(17):1381–1388.

188. Hunt SA, Baker DW, Chin MH, Cinquegrani MP, Feldman AM, Francis GS, Ganiats TG, Goldstein S, Gregoratos G, Jessup ML, Noble RJ, Packer M, Silver MA, Stevenson LW, Gibbons RJ, Antman EM, Alpert JS, Faxon DP, Fuster V, Jacobs AK, Hiratzka LF, Russell RO, Smith SC, Jr. ACC/AHA guidelines for the evaluation and management of chronic heart failure in the adult: executive summary. A report of the American College of Cardiology/American Heart Association Task Force on Practice Guidelines (Committee to revise the 1995 Guidelines for the Evaluation and Management of Heart Failure). *Journal of the American College of Cardiology*. Dec 2001;38(7):2101–2113.

189. Chen L, Meyers D, Javorsky G, Burstow D, Lolekha P, Lucas M, Semmler AB, Savarimuthu SM, Fong KM, Yang IA, Atherton J, Galbraith AJ, Parsonage WA, Molenaar P. Arg389Gly-beta1-adrenergic receptors determine improvement in left ventricular systolic function in nonischemic cardiomyopathy patients with heart failure after chronic treatment with carvedilol. *Pharmacogenetics and genomics*. Nov 2007;17(11):941–949.

190. Terra SG, Hamilton KK, Pauly DF, Lee CR, Patterson JH, Adams KF, Schofield RS, Belgado BS, Hill JA, Aranda JM, Yarandi HN, Johnson JA. Beta1-adrenergic receptor polymorphisms and left ventricular remodeling changes in response to beta-blocker therapy. *Pharmacogenetics and genomics*. Apr 2005;15(4):227–234.

191. White HL, de Boer RA, Maqbool A, Greenwood D, van Veldhuisen DJ, Cuthbert R, Ball SG, Hall AS, Balmforth AJ. An evaluation of the beta-1 adrenergic receptor Arg389Gly polymorphism in individuals with heart failure: a MERIT-HF sub-study. *Eur J Heart Fail*. Aug 2003;5(4):463–468.

192. Liggett SB, Mialet-Perez J, Thaneemit-Chen S, Weber SA, Greene SM, Hodne D, Nelson B, Morrison J, Domanski MJ, Wagoner LE, Abraham WT, Anderson JL, Carlquist JF, Krause-Steinrauf HJ, Lazzeroni LC, Port JD, Lavori PW, Bristow MR. A polymorphism within a conserved beta(1)-adrenergic receptor motif alters cardiac function and beta-blocker response in human heart failure. *Proceedings of the National Academy of Sciences of the United States of America*. Jul 25 2006;103(30):11288–11293.

193. Choudhry S, Ung N, Avila PC, Ziv E, Nazario S, Casal J, Torres A, Gorman JD, Salari K, Rodriguez-Santana JR, Toscano M, Sylvia JS, Alioto M, Castro RA, Salazar M, Gomez I, Fagan JK, Salas J, Clark S, Lilly C, Matallana H, Selman M, Chapela R, Sheppard D, Weiss ST, Ford JG, Boushey HA, Drazen JM, Rodriguez-Cintron W, Silverman EK, Burchard EG. Pharmacogenetic differences in response to albuterol between Puerto Ricans and Mexicans with asthma. *American journal of respiratory and critical care medicine*. Mar 15 2005;171(6):563–570.

194. Martinez FD, Graves PE, Baldini M, Solomon S, Erickson R. Association between genetic polymorphisms of the beta2-adrenoceptor and response to albuterol in children with and without a history of wheezing. *The Journal of clinical investigation*. Dec 15 1997;100(12):3184–3188.

195. Green SA, Cole G, Jacinto M, Innis M, Liggett SB. A polymorphism of the human beta 2-adrenergic receptor within the fourth transmembrane domain alters ligand binding and functional properties of the receptor. *The Journal of biological chemistry*. Nov 5 1993;268(31):23116–23121.

196. Lee DK, Currie GP, Hall IP, Lima JJ, Lipworth BJ. The arginine-16 beta2-adreno-ceptor polymorphism predisposes to bronchoprotective subsensitivity in patients treated with formoterol and salmeterol. *British journal of clinical pharmacology.* Jan 2004;57(1):68–75.

197. Hizawa N, Makita H, Nasuhara Y, Betsuyaku T, Itoh Y, Nagai K, Hasegawa M, Nishimura M. Beta2-adrenergic receptor genetic polymorphisms and short-term bronchodila-tor responses in patients with COPD. *Chest.* Nov 2007;132(5):1485–1492.

198. Bleecker ER, Postma DS, Lawrance RM, Meyers DA, Ambrose HJ, Goldman M. Effect of ADRB2 polymorphisms on response to longacting beta2-agonist therapy: a pharmacogenetic analysis of two randomised studies. *Lancet.* Dec 22 2007;370(9605):2118–2125.

16 Role of metabotropic glutamate receptors in CNS disorders

Richard M. O'Connor and John F. Cryan

INTRODUCTION

G protein-coupled receptors (GPCRs) have been one of the most studied, exploited, and lucrative therapeutic targets in modern neuroscience drug discovery efforts. GPCRs constitute the largest structural class of hormone and neurotransmitter receptors. The amino acid glutamate functions as the major excitatory neurotransmitter in the mammalian central nervous system (CNS). Its abundance and ubiquitous distribution throughout the brain highlights its importance in maintaining homeostatic functioning of the CNS and in the manifestation of brain disorders.[1] The action of glutamate on cellular processes is mediated through a diverse family of glutamate receptors.[2] These are broadly subdivided into the ionotropic and metabotropic glutamate receptors. Ionotropic receptors are multisubunit ion channels that mediate the rapid effects of glutmatergic neurotransmission through their opening and closing, altering the intracelullar ionic concentration.[3] The metabotropic glutamate receptors (mGluRs) are typical members of the GPCR superfamily, which mediate their cellular effects by altering the concentrations of second messenger proteins.[4] The GPCR superfamily is divided into three families based on sequence similarity, with mGluRs classified as family 3 GPCRs, which also includes $GABA_B$ and certain pheromone receptors.[5] The effects of mGluR activation are slower and longer lived than those of the ionotropic glutamate receptors, and allow the fine-tuning of the effects of glutamate at synapses in which ionotropic mediate a more rapid short-lived response.[2] Although mGluRs mediate their effects through the altering of second messenger systems, their ultimate cellular influence can be diverse and wide-ranging, including regulation of ion channels, mediation of excitatory and inhibitory responses, and influencing the actions of other receptors such as ionotropic glutamate receptors.[3] In this review, we will focus on the pharmacology and physiology of mGluRs and how these receptors can be potential therapeutic targets for a host of neuropsychiatric diseases.

GLUTAMATE IN THE BRAIN

To appreciate the actions of mGluR in the brain, it is important to delineate the basic neurobiology of glutamatergic neurotransmission. Glutamate is stored in vesicles in presynaptic nerve terminals and is released in a Ca^{2+} dependent manner.[6] Synaptic glutamate levels are regulated by its synthesis, release, and uptake back into neurons and glial cells. Glutamate cannot cross the blood brain barrier and thus to be available to act as a neurotransmitter, it must be synthesised from glucose, α-ketoglutarate, an intermediate of the citric acid cycle (TCA), and other

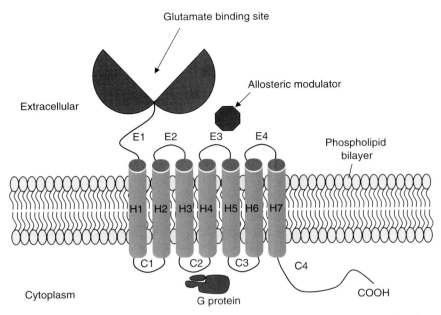

Figure 16-1: Schematic representation of the general structure of metabotropic glutamate receptors.

mGluR Pharmacology

precursors.[7] Glutamate is then transported into synaptic vesicles by the vesicular glutamate transporter, VGlut. The arrival of an action potential at the synaptic membrane leads to fusion of the synaptic vesicles and release of glutamate into the synaptic cleft. Glutamate is largely not metabolized extracellularly and is cleared from the synaptic cleft by sodium-dependent transport proteins termed excitatory amino acid transporters (EAATs); five different EAATs (named EAAT 1–5) have been characterized to date.[8] These transporters are expressed in both neurons and glia; however, it is believed that the uptake ability of glia (neuronal support cells) and especially astrocytes far exceeds that of neurons.[1] When glutamate is taken up by astrocytes, it is converted, in an ATP-dependent manner, to glutamine by glutamine synthetase (GS).[9] Glutamine is then transported into the extracellular fluid where it is taken back into neurons and converted back into glutamate.[10] It is essential that glutamate levels are regulated at an appropriate level in order to maintain proper communication between neurons and prevent overexcitation and excitotoxicity due to excess glutamate.

The topology of the mGluRs is typical of GPCRs, consisting of seven hydrophobic transmembrane domains joined by three extracellular loops and three intracellular loops, a large extracellular N-terminal domain, and a cytoplasmic C-terminal domain[3] (Figure 16.1). There are eight known subtypes of mGluRs, numbered 1–8, which are further broken down into three separate groups, I-III. Further diversity is generated by the alternative splicing of the different receptor subtypes. The mGluRs are divided into the three different groups based on amino acid similarity, pharmacology, and the preferred signal transduction mechanisms

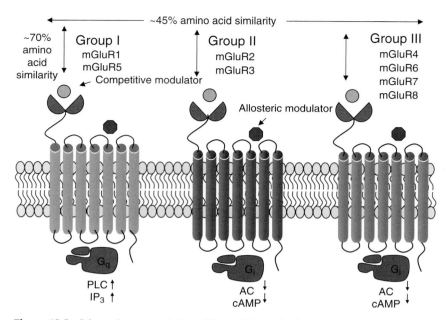

Figure 16-2: Schematic representation of the metabotropic glutamate receptor classification and signaling pathway. mGluRs are classified based on sequence similarity (showing a 70% amino acid similarity within groups and a 45% similarity between groups), preferential in vitro transduction mechanisms and pharmacological responses.

they couple to in cell culture systems[4] (Figure 16.2). Group I consists of mGluR1 and 5, group II contains mGluR2 and 3, and group III is made up of mGluR4, 6, 7 and 8.[11] mGluRs of the same group show about 70% amino acid similarity, whereas between groups this falls to about 45%.[2] mGluRs are very highly conserved throughout mammals and other species;[12] this high level of conservation indicates their importance in correct physiological functioning and indicates that an impairment in their functioning may have serious adverse consequences. Several pharmacological agonists and antagonists have been developed that have not only, as previously mentioned, allowed the separating of the eight mGluR subtypes into three distinct groups, but have also aided in the unraveling of their function in the CNS.[2,4] Regarding the second messenger systems that the different mGluR groups are coupled to, group I mGluRs are positively coupled to phospholipase C (PLC). The increased activity of PLC results in the cleavage of the phospholipid, PIP2, into diacylglycerol (DAG) and inositol 1,4,5-triphosphate (IP_3). DAG goes on to activate protein kinase C (PKC), whereas increased levels of IP_3 result in activation of IP_3 receptors increasing intracellular Ca^{2+} levels. Group II and group III mGluRs are negatively coupled to adenylyl cyclase in cell culture systems, which decreases cAMP; these decreased cAMP levels regulate K^+ and Ca^{2+} ion channels and deactivates protein kinase A (PKA).[4] The distribution of the different mGluR groups and subtypes is extremely heterogenous with overlapping and distinct regions of expression for each subtype, with the mGluRs facilitating glutamatergic neurotransmission on virtually all excitatory synapses.[13]

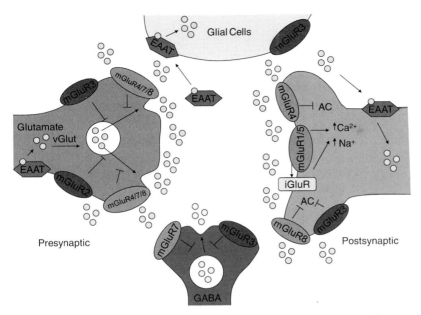

Figure 16-3: General synaptic localization of the different metabotropic glutamate receptor subtypes.

GROUP I MGLURS

Group I mGluRs (mGluR1 and mGluR5) are mainly located postsynaptically on glutamatergic synapses (see Figure 16.3) and have been implicated in a wide-ranging number of physiological events including synaptic plasticity,[4] neuronal development, neurodegeneration, and the induction of reactive astrocytes.[14] Group I mGluRs are coupled to PLC in recombinant reporter systems; however, in vivo, the activity of these receptors may be far more complex. It has been postulated that they are coupled to PLC through a G_q protein and adenylyl cylclase through a G_i protein, producing divergent effects of increasing Ca2+ levels and lowering cAMP levels, whereas the receptor may also be physically associated with $Ca^{2+,}$ and K^+ ion channels directly influencing intracellular ion concentrations.[15] They have also been shown to regulate AMPA and NMDA receptors further modulating neuronal excitability.[14] Moreover, mGluR1 and mGluR5 may be differentially associated with specific ion channels and coupled to different G proteins allowing these two receptors to produce diverse physiological functions.[14] Activation of group I mGluRs can result in divergent effects in different regions in the brain; for example, in the global pallidus nucleus, mGluR1 directly affects the depolarization of neurons, whereas mGluR5 acts to modulate mGluR1 function.[14] Distinct downstream effects of the group I mGluR activation have also been shown to be differentially regulated supporting the proposal that their coupling to different mechanisms produces different downstream effects.[14] mGluR1 and mGluR5 are differentially regulated by regulator of G protein signalling (RGS) proteins; these proteins enhance GTPase activity therefore inhibiting G protein signaling. This highlights another mechanism by which mGluR1 and mGluR5 can be differentially regulated in different cells.[14] The group I mGluRs also can influence gene expression through the downstream activation of different transcription factors. mGluR1 but not mGluR5 activation resulted in increased DNA binding of different members of the NF-κB family of transcription factors.[16] On the other hand, selective activation of mGluR5 leads to rapid and transient phosphorylation of JNK (and thus to activation of the AF-1 transcription factor through the phosphorylation of c-jun) independent of the typical mechanism of mGluR group I signaling of release of Ca^{2+} through IP_3 and activation of PKC, but rather through a transactivation of the epidermal growth factor (EGF) receptor, which in turn triggered downstream phosphorylation of JNK.[17] The above data indicates the diverse and wide-ranging cellular functions of the group I mGluRs, which allow them to have many diverse and complex physiological functions. Due to group I mGluRs being involved in such widespread physiological processes in the brain, dysfunction of these receptors or their signaling pathways have the potential to result in a wide range of CNS disorders. The development of mGluR pharmacological agonists and antagonists not only allowed the classification of these receptors, but also created invaluable tools allowing for the elucidation of the roles the different receptors play in normal physiological functioning.[4,18] In the following sections, we will summarize some of the main roles of the group I mGluRs in psychiatric and neurological disorders.

GROUP I MGLURS IN ANXIETY DISORDERS

Anxiety and stress disorders are the most common form of mental illness experienced by people today.[4] Anxiety disorders affect 40 million American adults and can have devastating effects on an individual's life (http://www.nimh.nih.gov/health/publications/anxiety-disorders.shtml). In Europe, recent evidence collected shows that anxiety disorders show a 12% prevalence in the adult population.[19] The most commonly diagnosed anxiety disorders are generalized anxiety disorder (GAD), panic disorder (with or without agoraphobia), specific phobia, social phobia, obsessive-compulsive disorder (OCD), and post-traumatic stress disorder (PTSD). These disorders are distinguished from one another mainly by the specific nature of the anxiety and by the provoking stimulus, if one exists. They also have distinguishing features based on the temporality of symptom manifestation and in the underlying and/or coincident neurobiological and neuroendocrine responses and markers.[20] Benzodiazepines – anxiolytics that function by increasing inhibitory GABAergic neurotransmission in the CNS – remain one of the most effective anxiolytics prescribed today despite being hindered with many unwanted side effects including withdrawal syndrome, tolerance, sedation, and abuse liability. Thus in principle, a similar anxiolytic mechanism could result by reducing excitatory glutamatergic neurotransmission within the same circuits. Given that group I mGluR activation increases glutamate release,[21] group I mGIuR antagonists reduce glutamate transmission and therefore have been postulated as potential anxiolytic agents.

A wealth of experimental evidence has been collected, which implicates the group I mGluRs in anxiety disorders (Table 16.1). Administration of the Group I selective antagonists (S)-4-carboxy-3-hydroxyphenyl-glycine (S-4C3-PHG), CPCCOEt and (S)-4CPG direct to the hippocampus produced an anxiolytic effect as measured by the Vogel test.[22,23] MPEP, an antagonist specific for mGluR5, displayed anxiolytic properties in several behavioral tests.[24,25] It was also reported that antagonism of mGluR5 by MPEP was more effective at relieving anxiety than mGluR1 antagonism by LY456236.[25] These data show that although both group I mGluR subtypes play a role in anxiety, mGluR5 may play a greater role than mGluR1. Given that antagonism of group I mGluRs produces anxiolytic effects, it is logical to hypothesize that activation of the group I mGluRs should produce an anxiogenic effect. Indeed, activation of mGluR1 and 5 in the amygdala by the agonist trans-ACPD of rats resulted in an increase in the startle response; however, trans-ACPD also is an agonist of group II mGluRs, and the results found may be due to action at these receptors.[26] Moreover, a group I-specific agonist tADA produced anxiogenic effects when injected directly into the dorsolateral periaqueductal gray (dlPAG); the dlPAG is an area of the brain believed to be involved in defensive behavior responses.[27] mGluR5 antagonists also function to block both the electrophysiological and behavioral correlates of amygdala-dependent fear learning.[28,29] Interestingly, mGluR5 expression has been shown to be up-regulated due to fear conditioning.[30]

The glutamatergic system is tightly linked with other neurotransmitter systems including the serotonergic, noradrenaline, and GABAergic systems,[31] all of which are critical to the manifestation of anxiety states. mGluR5 antagonists reduced stress-induced increases in extracellular noradrenaline similar to the benzodiazepine, diazepam.[32] Glutamate receptors are present on hypothalamic nuclei and play a role in the activation of the hypothalamic-pituitary adrenal[33] axis and subsequent release of corticosterone, one of the main hormones involved in the stress response.[34] Group I mGluR agonists and antagonists resulted in a paradoxical increase in corticosterone levels;[35] this may contribute to the effects of the mGluRs on anxiety.

Perhaps the most striking evidence for a role of mGluR5 in anxiety emerges from studies with the non-benzodiazepine fenobam, which for many years remained a clinically validated anxiolytic of unknown mechanism of action.[36] It turns out fenobam is a potent inhibitor of mGluR5.[37] Fenobam's development was halted due to psychotomimetic side effects. It is currently unclear if such effects are specific to fenobam or to mGluR5 antagonists as a whole. Taken together, the host of preclinical and preliminary clinical findings indicate that group I mGluRs play a key role in anxiety, with mGluR5 possibly having a greater role.

GROUP I MGLURS IN COGNITIVE DISORDERS

The group I mGluRs have been implicated in learning and memory, therefore they may potentially be a useful therapeutic target for cognitive disorders. As

Table 16-1. Pharmacological Evidence Implicating Group I mGluRs in Anxiety

Receptor	Modulation	Selectivity	Route	Name	Effect	Test	Reference
mGluR1 & 5	Antagonist	Non-selective	Intrahippocampal	S-4C3-HPG	Anxiolytic	Vogel test	[22]
mGluR1 & 5	Agonist	Non-selective	Intraamygdalar	trans-ACPD	Anxiogenic	Startle response	[26]
mGluR1 & 5	Antagonist	Group I	Intrahippocampal	CPCCOEt	Anxiolytic	Conflict drinking test	[23]
mGluR1& 5	Antagonist	Group I	Intrahippocampal	(S)-4CPG	Anxiolytic	Conflict drinking test	[23]
mGluR1& 5	Agonist	Group I	Intradorsolateral periaqueductal grays	tADA	Anxiogenic	EPM and Vogel Test	[27]
mGluR5	Agonist	mGluR5	Oral gavage	MPEP	Anxiolytic	Fear aquisition	[29]
mGluR5	Antagonist	mGluR5	Intrahippocampal	MPEP	Anxiolytic	Conflict drinking test and elevated plus maze	[24]
mGluR5	Antagonist	mGluR5	Oral gavage	MPEP	Anxiolytic	Elevated plus maze, social exploration, marble burying	[232]
mGluR5	Antagonist	mGluR5	Intraamygdala	MPEP	Fear aquisition	Fear aquisition	[28]
mGluR1	Antagonist	mGluR3	Intraperitoneal	LY456236	Anxiolytic	Vogel test and conditioned lick supression	[25]
mGluR5	Antagonist	mGluR5	Intraperitoneal	MPEP	Anxiolytic	Formalin induced nociception and spinal nerve ligation	[25]
mGluR5	Antagonist	mGluR5	Intraperitoneal	MTEP	Anxiolytic	Formalin induced nociception and spinal nerve ligation	[25]
mGluR5	Antagonist	mGluR5	Per os (Oral)	MPEP	Anxiolytic	Stress induced hyperthermia	[233]

mentioned, the group I mGluRs can lead to an increase in extracellular Ca^{2+}, inhibiting K^+ channels, allowing them to regulate neuronal excitability; they also may be coupled to the iontotropic glutamate receptors, further broadening their scope to regulate excitability. This ability to regulate excitability may be important in the mGluRs role in memory formation. Studies show that whereas mGluRs may not be required for the formation of memories, they do have a role in the consolidation and recall of memories.[13]

Much pharmacological evidence has been collected that demonstrates the important role group I mGluRs play in learning and memory (Table 16.2). In vivo studies have shown that infusion of MCPG, a mixed group I and group II antagonist, into the amygdala inhibited the electrophysiological correlate of memory, long-term potentiation (LTP) in rats.[38–40] MCPG also produces deficits in spatial learning[41] and impairs memory retention.[42] Specific pharmacological agents, such as the group I specific antagonist 4-CPG, and the mGluR5 specific antagonist MPEP, have also resulted in disruptions to memory retention.[42–44] In line with these results, the mGluR group I agonist trans-azetidine-2, 4-dicarbox-ylic acid (tADA) facilitated LTP; however, no improvement in memory reten-tion was found.[45,46] Interestingly post-training infusion with tADA did result in improved memory retention. Many additional studies have reported that altera-tion of the group I mGluRs results in altered memory retention as opposed to dis-rupted memory formation.[41,42,44] These results point to group I mGluRs playing an important role in the retention of memories rather than in the initial forma-tion of such cognitive processes. This is an important consideration when devel-oping pharmacological interventions for cognitive-related anxiety disorders such as post-traumatic stress disorder and phobias where fear-memories are already encoded. In agreement with the above pharmacological studies, mGlur1- and mGluR5-deficient mice display a poorer performance and impaired LTP com-pared to controls in a memory based water maze.[47,48]

Overall, more research should be done in this area to help elucidate the molec-ular mechanisms underlying cognitive processes, but the data so far suggest that Group I mGluR positive modulators could prove to be valuable tools for improv-ing cognition.

GROUP I MGLURS IN SCHIZOPHRENIA

Schizophrenia is a severe and disabling brain disorder; symptoms include altered perception of the world, including hallucinations and the hearing of voices, and cognitive effects (www.nimh.nih.gov). Aberrations in the dopamine neurotrans-mitter system previously dominated the literature concerning the biochemical theories underlying the pathophysiology of schizophrenia. However, over the past few decades, an increasing attention has been focused on a broader collection of neurotransmitter systems implicated in the disease, including the glutamater-gic system.[49,50] The more recent hypothesis describing the pathology of schizo-phrenia focuses on complex interactions between different neurotransmitter

Table 16-2. Pharmacological Evidence Implicating Group I mGluRs in Cognitive Disorders

Receptor	Modulation	Selectivity	Route	Name	Effect	Test	Reference
mGluR1 & 5	Agonist	Group I	Intracerebroventricularly	tADA	Defecit inspatial learning	Y-maze	[46]
mGluR1 & 5	Agonist	Group I	Intracerebroventricularly	tADA	Enhanced LTP		[45]
mGluR1& 5	Antagonist	Non-selective	Intracerebroventricularly	MCPG	Inhibited LTP and produced a deficit in spatial learning	Y-maze	[137]
mGluR1 & 5	Antagonist	Non-selective	Intracerebroventricularly	MCPG	Inhibited LTP		[38]
mGluR1& 5	Antagonist	Non-selective	Intracerebroventricularly	MCPG	Inhibited LTP		[40]
mGluR1 & 5	Antagonist	Group I	Intracerebroventricularly	4-CPG	Deficit in memory retention	Y-maze	[42]
mGluR1& 5	Antagonist	Group I	Intracerebroventricu larly	4-CPG	Deficit in memory formation and synaptic plasticity	Y-maze	[43]
mGluR1 & 5	Antagonist	Non-selective	Intracerebroventricularly	MCPG	Deficit in memory retention	Water maze	[41]
mGluR5	Antagonist	mGluR5	Intracerebroventricularly	MPEP	Deficit in memory retention	Y-maze	[44]

systems. The glutamatergic system can act to increase or decrease the release of other neurotransmitters, including dopamine,[51] linking glutamate to the original dopamine hypothesis. Glutamate's suggested involvement in the pathology of schizophrenia is strengthened by the fact that the administration of NMDA uncompetitive antagonists produces schizophrenia-like symptoms.[52,53] Furthermore, many genes involved in glutamatergic neurotransmission are candidate genes for schizophrenia.[54,55]

The analysis of the role of the mGluRs in schizophrenia, as in all disorders, required the development of appropriate animal models. Animal models are based on a combination of behavioral paradigms and either genetic or pharmacological tools that induce psychotomimetic effects.[56] The behavioral paradigms include those based on pre-pulse inhibition (PPI) and stereotypical behaviors such as circling motion and head movements. PPI occurs when a weak pre-stimulus (usually acoustic) occurs before a stronger startling stimulus. The presentation of the weaker stimulus before the stronger stimulus results in a decrease of the subject's startle response to the stronger stimulus. This phenomenon is commonly known as sensorimotor gating and is an organism's way of blocking out unnecessary information, which is deficient in many patients with schizophrenia.[57] The most commonly used method to produce a schizophrenia model is the administration of uncompetitive NMDA receptor antagonists that have been shown to produce schizophrenia-like symptoms in healthy volunteers.[56] Pharmacological evidence shows the possible involvement of mGluRs in schizophrenia (Table 16.3). The NMDA receptor antagonist phencyclidine, when administered, produces schizophrenia-like symptoms including pre-pulse inhibition deficits; and it has been found that administration of the mGluR5 antagonist MPEP exacerbates this disruption of PPI.[58,59] Administration of amphetamine also results in disrupted PPI; the mGluR5 agonist CHPG was also found to potentiate the amphetamine-induced disruption of PPI. The mGluR1 antagonist, (3aS,6aS)-6a-naphtalen-2-ylmethyl-5-methyliden-hexahydro-cyclopenta[c]furan-1-on (BAY 36–7620), eliminated some of the stereotypical behaviors produced by NMDA antagonists.[60] The mGluR5 selective agonist 3-cyano-N-(1,3-diphenyl-1H-pyrazol-5-yl)benzamide (CDPPB) partially reversed the PPI inhibition produced by amphetamine.[61] These studies show that antagonism of the group I mGluRs, particularly mGluR5, contributes to schizophrenia-like symptoms, whereas activation of these mGluRs attenuates schizophrenia-like symptoms and may represent the potential direction for the development of novel therapeutics.

Genetic evidence also points to group I mGluRs playing a role in schizophrenia. mGluR 5 has been mapped to chromosome 11q14 corresponding to a translocation associated with schizophrenia.[62] Post-mortem studies revealed an increase in mGluR1 in the pre-frontal cortex of schizophrenia patients,[63] whereas mGluR1 and mGluR5 knockout mice showed disrupted PPI.[59,64]

Regulator of G protein signaling 4 (RGS4) inhibits signaling by the group I mGluRs;[65] it also has been proposed to be a candidate gene for the development of schizophrenia.[33,66] This provides further support for group I playing a role in

Table 16-3. Pharmacological Evidence Implicating Group I mGluRs in Schizophrenia

Receptor	Modulation	Selectivity	Route	Name	Effect	Test	Reference
mGluR5	Antagonist	mGluR5	Intraperitoneal	MPEP	Increased the PPI disrupting effects of PCP	Pre-pulse inhibition	[58]
mGluR5	Antagonist	mGluR5	Intracerebroventricu larly	MPEP	Increased the PPI disrupting effects of PCP	Pre-pulse inhibition	[59]
mGluR5	Agonist	mGluR5	Intracerebroventricularly	CHPG	Increased the PPI disrupting effects of amphetamine	Pre-pulse inhibition	[59]
mGluR1	Antagonist	mGluR1	Intravenous	BAY 36–7620	Attenuated some of the stereotypy behavior induced by NMDA antagonists	Stereotypy behavior	[60]
mGluR5	Agonist	mGluR5	i.v. Intravenous	CDPPB	Decreased the PPI disrupting effects of amphetamine	Pre-pulse inhibition	[61]

schizophrenia. It has also been suggested that a deficit in PPI may be linked to a decrease in expression of mGluR5 and its signaling protein PLC.[67]

GROUP I MGLURS IN OTHER PSYCHIATRIC DISORDERS

Depression

Glutamate was originally implicated in having a possible role in depression in the 1960s when it was noticed that depressed patients treated with the NMDA receptor antagonist D-cycloserine showed improvement in their depressive symptoms.[68] Since then, other NMDA modulators have been shown to possess antidepressant properties.[69] As previously mentioned, the group I mGluRs can modulate the activity of the iGluRs, providing the scope for mGluR1 and mGluR5 to play a role in depression. Studies have shown that chronic treatment with antidepressants has led to change in both signaling and expression of the group I mGluRs.[70–73] As shown in Table 16.4, selective mGluR5 antagonists produced antidepressant-like effects in mice.[24,74] More recently, specific mGluR1 and mGluR5 antagonists have produced antidepressant-like activity in rats.[75]

Fragile X syndrome

Fragile X is the most common inherited form of mental retardation. It is caused by a tri-nucleotide repeat expansion in the X-linked gene Fmr1. This expansion inhibits the expression of the encoded protein fragile X mental retardation protein (FMRP). The loss of FMRP results in improper brain development leading to the mental retardation seen in fragile X syndrome.[76] The mGluRs are involved in the plasticity-altering process, long-term depression (LTD),[77] and, as mentioned previously, LTP. This alteration of plasticity by the group I mGluRs involves the synthesis of new proteins at the synapse, including the translation of existing Fmr1 mRNA into FMRP protein.[78] Therefore, signaling by the group I mGluRs influences the production of FMRP, the key protein absent in fragile X syndrome. An FMRP knockout animal was anticipated to be defective in hippocampal LTD; however, it turned out to have increased LTP.[79] The authors proposed an explanation for the result by suggesting that FMRP is not directly involved in LTP per se but rather functions to repress the translation of other proteins responsible for the induction of LTP. A nonfunctioning FMRP could result in exaggerated effects of mGluR1 and 5 signaling that could partly contribute to the mental retardation seen in fragile X syndrome. Supporting this hypothesis, mGluR5 antagonists have been shown to reverse some of the behavioral symptoms in the FMRP mouse.[80] Current treatment for fragile X syndrome focuses on treating the symptoms using a variety of psychoactive compounds; it has been proposed that pharmacological manipulation of the group I mGluRs may be a more suitable treatment strategy because it will potentially treat the underlying pathophysiology.[81]

Table 16-4. Pharmacological Evidence Implicating Group I mGluRs in Depression

Receptor	Modulation	Selectivity	Route	Name	Effect	Test	Reference
mGluR5	Antagonist	mGluR5	Intraperitoneal	MPEP	Antidepressant like effect	Tail suspension test	[24]
mGluR5	Antagonist	mGluR5	Intraperitoneal	MTEP	Antidepressant like effect	DRL-72 s paradigm	[74]
mGluR5	Antagonist	mGluR5	Intraperitoneal	MTEP	Antidepressant like effect	Tail suspension test	[75]
mGluR1	Antagonist	mGLuR1	Intraperitoneal	EMQMCM	Antidepressant like effect	Tail suspension test	[75]

GROUP I MGLURS IN PARKINSON'S DISEASE

The commonly accepted hypothesis behind the pathophysiology of Parkinson's disease is progressive destruction of nigrostriatal dopamine-producing neurons leading to alteration in neural circuits that regulate movement.[82] The testing of the potential role of mGluRs in Parkinson's disease requires the development of appropriate animal models. Animal models are most commonly achieved through the administration of agents that lead to a depletion of dopamine levels or the disruption of systems that produce dopamine, such as lesion of dopaminergic striatal neurons, and administration of compounds such as reserpine, methamphetamine, 6-hydroxydopamine, and 1-methyl-4-phenyl-1,2,3,6-tetrahydropyridine.[83] In pharmacological analysis, the potential effect of a given antagonist or agonist is based on its effects on experimentally induced motor alterations such as catalepsy, hypokinesia, or muscle rigidity, as well as expression of proenkephalin (PENK) mRNA, which has been shown to increase due to a lesion of dopaminergic striatal neurons.[82] The first evidence suggesting that the glutamatergic system may play a role in the pathophysiology of Parkinson's disease was when the administration of NMDA receptor antagonists led to a reduction in Parkinsonian symptoms.[84] The fact that the group I mGluRs are coupled to NMDA receptors and serve to facilitate glutamatergic neurotransmission puts them in a key position to play a role in the aberrant neurotransmission seen in Parkinson's disease.

There has been a large amount research devoted to the role that the group I mGluRs may play in Parkinson's disease (Table 16.5). Pharmacological group I antagonists have been shown to have the potential to reduce many of the induced Parkinsonian symptoms used in animal models of Parkinson's disease, including haloperidol-induced catalepsy,[82,85,86] haloperidol-induced proenkephalin mRNA expression,[82] and reserpine-induced akinesia.[87] In line with this, the blockade of the group I mGluRs using DCG-IV led to an increase in reserpine-induced akinesia.[87] Subtype-specific pharmacological agents have shown that both mGluR1 and mGluR5 play a role in Parkinson's disease, with mGluR5 probably having a greater role. mGluR5 antagonists have been shown to reduce Parkinsonian symptoms using a variety of models, including haloperidol-induced muscle rigidity,[88–90] 6-OHDA lesioned animals,[91–95] and increased proenkaphelin mRNA expression.[90,96] In addition to motor symptoms, Parkinson's disease also produces cognitive impairments. Research has shown that an mGluR5 antagonist reversed the visuo-spatial deficits induced by 6-hydroxydopamine (6-OHDA) induced lesions of the dorsal striatum in mice.[97] These studies show the role that mGluR5 plays in a wide spectrum of Parkinsonian symptoms ranging from motor impairments to cognitive deficits to levodopa-induced dyskinisias. In line with antagonists reducing Parkinsonian symptoms, an mGluR5 agonist was found to increase proenkaphelin mRNA expression.[96] mGluR 1 antagonists have also been shown to have anti-Parkinsonian effects.[90,98]

mGluR5 modulation works synergistically with NMDA receptors in animal models of Parkinson's disease. Treatment with the mGluR5 antagonist MPEP

Table 16-5. Pharmacological Evidence Implicating Group I mGluRs in Parkinson's Disease

Receptor	Modulation	Selectivity	Route	Name	Effect	Test	Reference
mGluR1 & 5	Agonist	Group I	Intraventricular and intranigral	DCG-IV	Increase in Parkinsonian symptoms	Resepine induced ankinesia	[87]
mGluR1 & 5	Antagonist	Group I (Dose Dependent)*	Bilateral intrastriatal	AIDA	Decrease in Parkinsonian symptoms	Haloperidol-induced proenkephalin mRNA expression	[82]
mGluR1 & 5	Antagonist	Group I	Intraperitoneal	LY 354740	Decrease in Parkinsonian symptoms	Haloperidol induced muscle rigidity	[86]
mGluR1 & 5	Antagonist	Group I	Intraperitoneal	LY 354740	Decrease in Parkinsonian symptoms	Haloperidol induced muscle rigidity	[85]
mGluR1 & 5	Antagonist	Group I	Intraventricular and intranigral	EGLU	Decrease in Parkinsonian symptoms	Resepine induced ankinesia	[87]
mGluRI	Antagonist	mGluR1 (Dose Dependent)*	Bilateral intrastriatal	AIDA	Decrease in Parkinsonian symptoms	Haloperidol induced muscle rigidity	[98]
mGluRI	Antagonist	mGluRI	Bilateral intrastriatal	LY 367385	Decrease in Parkinsonian symptoms	Haloperidol-induced proenkephalin mRNA expression	[90]
mGluR5	Agonist	mGluR5	Intrastriatal	CHPG	Increase in Parkinson induced associated gene expression	Preproenkaphelin mRNA expression	[96]
mGluR5	Antagonist	mGluR5	Per Os (Oral)	MPEP	Decrease in Parkinsonian symptoms	6-OHDA lesion model	[234]
mGluR5	Antagonist	mGluR5	Intraperitoneal	MPEP	Decrease in Parkinson induced associated gene expression	Preproenkaphelin mRNA expression	[96]
mGluR5	Antagonist	mGluR5	Intraperitoneal	MPEP	Decrease in Parkinsonian symptoms	6-OHDA lesion model	[91]

mGluR5	Antagonist	mGluR5	Intraperitoneal	MPEP	Decrease in Parkinsonian symptoms	Haloperidol-induced proenkephalin mRNA expression	[90]
mGluR5	Antagonist	mGluR5	Intraperitoneal	MPEP	Decrease in Parkinsonian symptoms	Haloperidol induced muscle rigidity	[98]
mGluR5	Antagonist	mGluR5	Intraperitoneal	MTEP	Decrease in Parkinsonian symptoms	Haloperidol induced muscle rigidity	[88]
mGluR5	Antagonist	mGluR5	Intraperitoneal.	MPEP	Decrease in Parkinsonian symptoms	6-OHDA induced defecits in visuo-spatial discrimination	[97]
mGluR5	Antagonist	mGluR5	Intraperitoneal	MPEP	Decrease in Parkinsonian symptoms	6-OHDA induced motor defecits	[95]
mGluR5	Antagonist	mGluR5	Intraperitoneal	MPEP	Decrease in Parkinsonian symptoms	6-OHDA lesion model	[93]
mGluR5	Antagonist	mGluR5	Intrasub thalamic nuceus	MPEP	Decrease in Parkinsonian symptoms	6-OHDA lesion model	[94]
mGluR5	Antagonist	mGluR5	Intraperitoneal	MPEP	Decrease in Parkinsonian symptoms	6-OHDA lesion model	[92]
mGluR5	Antagonist	mGluR5	Intraperitoneal	MPEP	Decrease in Parkinsonian symptoms	6-OHDA lesion model	[91]
mGluR5	Antagonist	mGluR5	Intraperitoneal	MTEP	Decrease in Parkinsonian symptoms	L-DOPA induced akinesia	Mela, Marti et al. 2007

led to a reduction in 6-OHDA lesion-induced Parkinson symptoms in rats at a 0.75 mg/kg concentration, whereas 0.375 mg/kg had no effect.[95] Treatment with the NMDA receptor antagonist MK-801 alone also had no effect on induced Parkinson symptoms; however, when MPEP was administered at the noneffective dose of 0.375 mg/kg along with MK-801, a decrease in the induced Parkinson symptoms was observed.[95] A similar result was found for an mGluR5 antagonist (MPEP) and an adenosine A2A receptor antagonist (8-(3-chlorostyryl)caffeine (CSC)). Co-administration of 0.375 mg/kg MPEP and 0.625 mg/kg CSC resulted in a decrease in the 6-OHDA lesion-induced Parkinsonian symptoms.[92]

Radioligand binding studies in primates have shown an increased binding of mGluR5 in models of Parkinson's disease.[99,100] This demonstrates that the group I mGluRs play a role in Parkinson's disease contributing to motor and cognitive symptoms, with mGluR5 probably having a greater role than mGluR1. In November 2008, Novartis announced that the mGluR5 antagonist, AFQ056, has the potential to become the first approved treatment for levodopa-induced dyskinesias in Parkinson's disease (http://www.novartis.com/newsroom/media-releases/en/2008/1271207.shtml). Although further clinical trials are needed to confirm such findings, it highlights the potential for therapeutic intervention in Parkinson's disease by modulating the function of group 1 mGluRs.

GROUP I MGLURS IN EPILEPSY

Epilepsy is a neurological condition affecting the nervous system, which can result in the onset of seizures (www.epilepsy.com). Glutamate has been implicated in epilepsy, because modulation of the glutamatergic system by a wide variety of molecular mechanisms can produce seizures.[101] The anti- and proconvulsant properties of pharmacological iGluR modulators has been established.[101] The mGluRs may affect the onset of epileptic seizure through a variety of excitatory modulating activities (see earlier discussion). The group I mGluRs affect ion Ca^{2+} and Na^+ channels directly and indirectly through the modulation of AMPA and NMDA iGLuRs.

Pharmacological data shows that the group I mGluRs play a role in epilepsy (Table 16.6). Administration of the mGluR group I agonist DHPG induced limbic seizures; these seizures can be attenuated by L-AP3 and dantrolene; L-AP3 and dantrolene have the ability to mobilize Ca^{2+} subsequent to mGluR activation; the authors suggest that the mGluR activation may be due to mGluR receptor activation effect on Ca^{2+} subsequent to PLC activation.[102] The mGluR1 selective agonist DHPG and nonselective agonist ACPD induced seizures in rats and also produced neuronal damage by excitotoxicity.[103] The release of Ca^{2+} due to activation of mGluRs may be partially responsible for the inducement of the seizure and neuronal damage. The mGluR group I agonist DHPG reduced the latency of pentylenetetrazole (PTZ)-induced seizures whereas conversely, the group I antagonist AIDA was found to have anticonvulsant properties and attenuated PTZ-induced seizures.[104] More selective pharmcalogical agents have demonstrated the

Table 16-6. Pharmacological Evidence Implicating Group I mGluRs in Epilepsy

Receptor	Modulation	Selectivity	Route	Name	Effect	Test	Reference
mGluR1	Antagonist	mGluR1 (antagonist) mGluR2 (agonist)	Intracerebroventricularly	(S)-4C3HPG	Protected against seizures	Audio induced seizures	[105]
mGluR1	Antagonist	mGluR1 (antagonist) mGluR2 (agonist)	Bilateral microinjection into the inferior colliculus	(S)-4C3HPG	Protected against seizures	Epilepsy prone rat strain	[106]
mGluR1 & 5	Agonist	Group I	Intracerebroventricularly	DHPG	Induced epileptic seizures		[102]
mGluR1 & 5	Agonist	Group I & Group 2	Intracerebroventricularly	1S,3R-ACPD	Induced epileptic seizures		[102]
mGluR1 & 5	Agonist	Group I	Intracerebroventricularly	DHPG	Induced epileptic seizures		[103]
mGluR1 & 5	Agonist	Group I	Intracerebroventricularly	DHPG	Induced epileptic seizures	PTZ induced seizures	[104]
mGluR1 & 5	Antagonist	Group I & Group 2	Intracerebroventricularly	AIDA	Protected against seizures	PTZ induced seizures	[104]
mGluR1	Antagonist	mGluR1	Intraperitoneal	LY456236	Protected against seizures	Audio induced seizures	[108]
mGluR5	Antagonist	mGluR5	Intraperitoneal	SIB 1893	Protected against seizures	Audio induced seizures	[235]
mGluR5	Antagonist	mGluR5	Intraperitoneal	MPEP	Protected against seizures	Audio induced seizures	[235]

role that the individual group I mGluR subtypes have in epilepsy. The selective mGluR1 antagonist (S)-4C3HPG inhibited the spontaneously evoked epileptic spikes in a cingulate cortex-corpus callosum slice preparation.[105] (S)-4C3HPG also showed anticonvulsant properties when administered to an epilepsy-prone rat strain.[106] However, data produced using (S)-4C3HPG need to be interpreted with the caveat that (S)-4C3HPG is also an mGluR2 agonist at low doses,[107] and mGluR2 activation may contribute to the results seen due to its administration. The mGluR1 selective antagonist LY456236 attenuated audio-induced seizures but had no effect on PTZ-induced seizures.[108] This highlights the different mechanisms utilized by different seizure models and shows why conflicting results may be produced depending on the pharmacological agent and model used. The mGluR5 selective antagonists MPEP and SIB 1893 attenuated seizures equally in a variety of epileptic models, with differences arising depending on the route of administration.[101]

Pharmacological evidence has suggested that mGluR1 and mGluR5 have different roles in seizure induction and maintenance. The mGluR1 antagonist LY367385 was more effective in reducing seizure prolongation than the mGluR5 antagonist MPEP; however, MPEP was shown to have a greater inhibitory effect on the induction of the seizure.[109] Despite the wealth of evidence implicating both mGluR1 and mGluR5 in epilepsy, it is worth noting there is evidence pointing to the contrary. mGluR5 knockout mice displayed no elevated protection to seizures, which would have been expected due to the proconvulsant properties of so many mGluR5 agonists,[110] whereas in an animal model of difficult-to-treat partial epilepsy, mGluR5 antagonists did not display anticonvulsant properties.[111] The discrepancies between the studies may be due to effects such as the route of administration, unforeseen nonspecific effects, or differences in the animal model used. Despite these contrasting results, a wealth of evidence has been collected implicating the group I mGluRs in epilepsy; moreover, data collected also shows that the subtypes in the group I mGluRs may have different roles in the induction and maintenance of epileptic seizures.

GROUP I MGLURS IN PAIN DISORDERS

Pain refers to the unpleasant sensation that occurs due to perceived tissue damage.[112] Pain consists of both a physical element and an unpleasant emotional response. The most common type of pain is that which is induced by a noxious stimulus. This pain serves two functions: It alerts the organism to the tissue damage taking place and also serves to teach the organism to avoid this damaging stimulus in the future.[113] Pathological pain occurs when there is no appropriate noxious stimulus or when there is an exaggerated or sustained response to a stimulus.[112] This type of pain is nonadaptive and can be damaging to the organism due to the severe and exaggerated emotional response experienced, as well as the potential masking of actual tissue damage. There are different classes of persistent pain including neuropathic pain resulting from disease or injury to the spinal cord or peripheral nerves, and joint related pain such as inflammation.[114] Modulation of neuronal activity in nociceptive pathways can result in the up- or down-regulation of the signal created by a noxious stimulus; this modulation is normally adaptive, but oversensitization can lead to a pain disorder.[113] Hypersensitization is often believed to be caused by excess glutamate transmission resulting in overexcitability.[113] This overexcitability is at least partially mediated by the mGluRs, highlighting their potential to play a key role in pain disorders.

Group 1 mGluRs are located pre- and postsynpatically in many nociceptive pathways, including those in the periphery, spinal cord, and brain.[114] Pharmacological modulation of the group I mGluRs has also demonstrated their potential role in pain disorders (Table 16.7). Group I mGluR antagonists (not subtype-specific) have been shown to reduce secondary-phase nociception in a formalin-induced animal model, whereas group I mGluR agonists increased nociception.[115] Knocking down of the mGluR1 receptor using antisense oligonucleotides resulted in a decrease in nociception in the formalin-induced pain model[116] and decreased the nociceptive response to thermal hyperalgesia and mechanical alodynia in a CFA-induced model.[117]

The two subtypes of the group I mGluRs may not be involved in all stages of pain processing, may possess differing roles in different pain stages, and may also be differentially involved in pain produced by different stimuli. It has been reported that an mGluR1 antagonist CPCCPEt had no effect on pain before the inducing of arthritis in an animal pain model, but did have an analgesic effect during the development and arthritis state, whereas an mGluR5 antagonist MPEP had that effect even before the inducing of the pain state.[118] This points to mGluR1 having a role in prolonged pain states, whereas mGluR5 may have a role in the induction as well as the maintenance of pain states.

GROUP II MGLURS

Group II mGluRs consist of the subtypes mGluR2 and mGluR3. These mGluRs are negatively coupled to adenylyl cyclase in artificial recombinant reporter systems.[3] mGluR3 is extensively distributed throughout the CNS and present on both pre- and postsynaptic elements, whereas mGluR2 has a more limited distribution to that of mGluR3 and the group I mGluRs, and is mainly found on presynaptic elements (see Figure 16.3). mGluR2 is largely restricted to expression on neuronal cells, whereas mGluR3 can be found on many non-neuronal cells including glia and astrocytes[4,119] and other cells not part of the CNS.[3] mGluR2 and mGluR3 are present both in the somatodendritic domains of neurons and in axonal domains that are further from the protein synthesizing regions,[3] meaning the neurons need to employ some sort of transport mechanism to move the receptors to their axonal destinations. The group II mGluRs are mainly located in extrasynaptic sites with very little expression detected in the active pre- or postsynaptic sites, this puts them in a position where they may be inactive under

Table 16-7. Pharmacological Evidence Implicating Group I mGluRs in Pain Disorders

Receptor	Modulation	Selectivity	Route	Name	Effect	Test	Reference
mGluR1 & 5	Antagonist	Group I	Intrathecal	LY393053	Increased pain threshold	Kaolin and carrageenan induced knee inflammation	[236]
mGluR1 & 5	Antagonist	Group I	Intrathecal	(S)-4C3HPG	Decreased nociceptive response	Formalin induced nociception	[115]
mGluR1 & 5	Antagonist	Group I	Intrathecal	(S)-4CPG	Decreased nociceptive response	Formalin induced nociception	[115]
mGluRI & 5	Agonist	Group I	Injection into hind paw	DHPG	Decrease in pain threshold	Sensitivity to noxious heat	[237]
mGluRI & 5	Agonist	Group I	Intrathecal	DHPG	Increase in nociceptive response	Formalin induced nociception	[115]
mGluRI & 5	Agonist	Group I	Intracerebroventricularly	DHPG	Increased nociceptive response	Nerve damage induced neuropathy	[238]
mGluR1	Antagonist	mGluR1	Subcutaneous	LY367385	Reduced DHPG induced decreased pain threshold	Sensitivity to noxious heat	[237]
mGluRI	Antagonist	mGluRI	Intrathecal	LY367385	Increased pain threshold	Kaolin and carrageenan induced knee inflammation	[236]
mGluRI	Antagonist	mGluRI	Intrathecal	AIDA	Increased pain threshold	Kaolin and carrageenan induced knee inflammation	[236]
mGluR1	Antagonist	mGluR1 (and mGluR5 at higher doses)	Intraamygdlar	CPCCOEt	Decreased nociceptive response	Kaolin and carrageenan induced knee inflammation	[239]
mGluR1	Antagonist	mGluR1 (and mGluR5 at higher doses)	Intraamygdlar	CPCCOEt	Decreased nociceptive response	Kaolin and carrageenan induced knee inflammation	[118]
mGluRI	Antagonist	mGluRI	Intrathecal	AIDA	Decreased nociceptive response	Capsaicin induced hypersensitivity	[218]

mGluR1	Antagonist	mGluR1 (and mGluR5 at higher doses)	Intrathecal	CPCCOEt	Decrease in nociceptive response	Nerve damage induced neuropathy	[238]
mGluR1	Antagonist	mGluR1 (and mGluR5 at higher doses)	Subcutaneous	CPCCOEt	Reduced DHPG induced decreased pain threshold	Sensitivity to noxious heat	[237]
mGluR1	Antagonist	mGluR1 (and mGluR5 at higher doses)	Subcutaneous	CPCCOEt	Reduced nocieptive response	Formalin induced nociception	[237]
mGluR1	Antagonist	mGluR3	Intraperitoneal	LY456236	Decreased nocieptive response	Formalin induced nociception and spinal nerve ligation	[25]
mGluR3	Antagonist	mGluR5	Subcutaneous	MPEP	Abolished DHPG induced decreased pain threshold	Sensitivity to noxious heat	[237]
mGluR5	Antagonist	mGluR5	Subcutaneous	MPEP	Reduced nocieptive response	Formalin induced nociception	[237]
mGluR5	Antagonist	mGluR5	Intraamygdalar	MPEP	Decreased nociceptive response	Kaolin and carrageenan induced knee inflammation	[118]
mGluR5	Antagonist	mGluR5	Intrathecal	MPEP	Decreased nociceptive response	Capsaicin induced hypersensitivity	[218]
mGluR5	Antagonist	mGluR5	Intraperitoneal	MPEP	Decreased nociceptive response	Formalin induced nociception and spinal nerve ligation	[25]
mGluR5	Antagonist	mGluR5	Intraperitoneal	MTEP	Decreased nociceptive response	Formalin induced nociception and spinal nerve ligation	[25]
mGluR5	Antagonist	mGluR5	Intracerebroventricularly	MPEP	Decreased nociceptive response	Nerve damage induced neuropathy	[238]

normal physiological conditions, but becoming activated during excessive, possibly pathological glutamate release resulting in a spillover to nonsynaptic sites; thus, Group 2 mGluRs may play a role in reducing glutamate transmission during pathological conditions.[3,119] Many studies support this hypothesis of Group II mediated regulation of glutamate release, with selective pharmacological agonists having been shown to reduce the level of glutamate neurotransmission.[119] Activation of group II mGluRs have been shown to potentiate cAMP increases due to activation of other receptors.[2] Like all mGluRs, group II mGluRs regulate ion channels such as K^+, nonselective cation, and Ca^{2+} channels, resulting in changes to neuronal excitability due to receptor activation, the specific subtypes mediating this regulation depends on the brain region.[2] Group II mGluRs also function as inhibitory heteroreceptors on GABAergic synapses reducing inhibitory neurotransmission.[2] As has been described previously for group I mGluRs, mGluR2 and mGluR3 activation may alter gene expression by modulating the activity of transcription factors; for example, in cultured rat cortical neurons, group II agonist produced increased DNA binding levels of the transcription factor AP-1.[120] Group II mGluRs also modulate the functioning of many different ion channels, including Ca^{2+} and K^+ channels,[15] allowing the mGluRs to regulate neuronal excitability. In the following sections we will summarize some of the main role of group II mGluRs in psychiatric and neurological disorders.

GROUP II MGLURS IN ANXIETY DISORDERS

mGluR2 and mGluR3 function as autoreceptors reducing excitatory glutamate neurotransmission; in principle, this would have the same effect as increasing inhibitory neurotransmission (such as is clinically exploited with benzodiazepine anxiolytics) and may lead to reduced anxiety. Using *in situ* hybridization, it has been shown that mGluR2 and mGluR3 expression levels are high in regions of the brain associated with fear and anxiety processing, including frontal cortex, basolateral amygdala complex, and the hypothalamic paraventricular nucleus.[121,122] As mentioned previously, the group II mGluRs are located perisynaptically on presynaptic elements,[119] and mGluR2/3 agonists have been shown to decrease glutamate release in brain regions associated with fear and anxiety.[119,123–125] Group II mGluRs may be responsible for reducing the amount of glutamate released during a highly excitable event such as a fear-producing or anxiogenic stimulus.

As with the group I mGluRs, the development of selective, orally active pharmacological agents have allowed the identification of the roles mGluR2 and 3 play in anxiety (Table 16.8). Interestingly, there appears to be a role for the GABAergic system in mediating some of these effects as co-treatment with benzodiazepine inverse agonist flumazenil attenuated the anxiolytic effects of the mGluR2/3 agonist LY354740 (eglumegad).[126] LY354740 was also found to reduce stress-induced c-fos expression in the hippocampus, and LY354740 treatment

Table 16-8. Pharmacological Evidence Implicating Group II mGluRs in Anxiety

Receptor	Modulation	Selectivity	Route	Name	Effect	Test	Reference
mGluR 2 & 3	Agonist	Group II	Intraperitoneal	LY354740	Anxiolytic	Defensive withdrawl test	[240]
mGluR 2 & 3	Agonist	Group II	Per os (Oral)	LY354740	Anxiolytic	Elevated plus maze and fear potentiated startle	[159]
mGluR2	Agonist	mGluR2	Per os (Oral)	LY314582	Anxiolytic	Stress induced hyperthermia	[233]
mGluR 2 & 3	Agonist	Group II	Intaperitoneal	LY354740	Anxiolytic	Elevated Plus Maze	[126]
mGluR 2 & 3	Agonist	Group II	Intraperitoneal	LY354740	Anxiolytic	Conflict drinking test	[163]
mGluR 2 & 3	Agonist	Group II	Intrahippocampal	LY354740	Anxiolytic	Conflict drinking test	[163]
mGluR 2 & 3	Agonist	Group II	Intrahippocampal	L-CCG-I	Anxiolytic	Conflict drinking test	[163]
mGluR 2 & 3	Agonist	Group II	Subcutaneous	LY354740	Anxiolytic	Elevated plus maze	[135]
mGluR 2 & 3	Agonist	Group II	Intraperitoneal	LY354740	Anxiolytic	Fear potentiated startle	[102]
mGluR 2 & 3	Agonist	Group II	Intraamygdalar	LY354740	Anxiolytic	Fear potentiated startle	[241]
mGluR 2 & 3	Agonist	Group II	Per Os	LY354740	Anxiolytic	Elevated plus maze	[173]
mGluR 2 & 3	Agonist	Group II	Intrahippocampal	LY354740	Anxiolytic	Conflict drinking test	[23]
mGluR 2 & 3	Agonist	Group II	Intrahippocampal	L-CCG-I	Anxiolytic	Conflict drinking test	[23]
mGluR2	Potentiator	mGluR2	Per os (oral)	4-MPPTS	Anxiolytic	Fear potentiated startle	[130]
mGluR2	Potentiator	mGluR2	Per os (oral)	4-APPES	Anxiolytic	Fear potentiated startle	[130]
mGluR2	Potentiator	mGluR2	Subcutaneous	CBiPES	Anxiolytic	Stress induced hyperthermia	[130]
mGluR 2 & 3	Agonist	Group II	Subcutaneous	LY354740	Anxiolytic	Elevated Plus Maze	[127]
mGluR 2 & 3	Agonist	Group II	Intraperitoneal	LY354740	Anxiolytic	Lactate induced panic attack in an anxious prone rat stra	[127]
mGluR 2 & 3	Antagonist	Group II	Intraperitoneal	MGS0039	Anxiolytic	Marble burying	[131]
mGluR 2 & 3	Antagonist	Group II	Intraperitoneal	MGS0039	Anxiolytic	Conditioned fear stress	[132]
mGluR 2 & 3	Antagonist	Group II	Intraperitoneal	MGS0039	Anxiolytic	Stress induced hyperthermia	[133]
mGluR 2 & 3	Agonist	Group II	Intrahippocampal	L-CCG-I	Anxiolytic	Elevated plus maze	[242]
mGluR 2 & 3	Agonist	Group II	Intraamygdalar	L-CCG-I	Anxiolytic	Elevated plus maze	[206]

alone increased c-fos expression in brain regions associated with stress, such as the amygdala and locus coeruleus.[127]

LY354740 has shown promising results for the treatment of anxiety in human clinical trials; one group showed reduced anxiety using a fear-potentiated startle test,[128] whereas in a separate study, a subset of LY354740-treated patients showed reduced anxiety in a CCK-4-induced anxiety model.[129] However, future clinical development is currently halted due to formulation-specific side effects.

Allosteric potentiators such as APPES and 4-MPPTS, binding to distinct sites on the receptor to agonists, antagonists, or the orthosteric ligand, are being developed as anxiolytics.[130] It should be noted that the mGluR2/3 antagonist MGS0039 has also been reported to have anxiolytic properties in the defensive marble burying,[131] conditioned fear stress,[132] and stress-induced hyperthermia[133] tests. This paradoxically opens up the premise that both group II agonists *and* antagonists may have anxiolytic potential. Reasons for this are currently unclear and may be due to complex downstream effects on other neurotransmitter systems, including dopamine[134], and serotonin.[133]

Knockout mice have shown that the anxiolytic effects of the group II agonist, LY354740, are due to its effects on mGluR2/3 and not nonspecific effects; administration of LY354740 to mGluR2 and mGluR3 knockout mice did not produce any anxiolytic effects in the EPM, whereas a significant anxiolytic effect was seen in their wild-type counterparts.[135] However, due to the lack of pharmacological tools the relative contribution of mGluR2 versus mGluR3 to the mediating anxiety-like effects is unclear; brain activation pattern studies indicate that both subtypes participate in the anxiolytic effects.[136]

All these studies demonstrate that mGluR2 and 3 play a role in anxiety-like behavior. Paradoxical effects have been found with agonists and antagonists; these may be due to the group II mGluR effects on inhibiting both glutamate and GABA neurotransmission, or may also be due to nonspecific effects of theseligands on other receptors such as NMDA receptors. mGluR pharmacological modifiers are thus promising therapeutics with some encouraging early clinical results. Further research needs to be done in this area to fully elucidate the roles these receptors play in anxiety disorders and so that their full potential as therapeutic targets can be realized.

GROUP II MGLURS IN COGNITIVE DISORDERS

Given that the group II mGluRs are largely located perisynaptically and function to reduce glutamate release, they are in a prime position to modify neuronal excitability, which can lead to alterations in LTD and LTP and thus may affect cognitive functions such as memory retention and spatial awareness.

As discussed previously, the mixed group I and II antagonist MCPG has been shown to inhibit information processing, block LTP, produce a deficit in spatial learning, and disrupt memory retention,[38,40–42,137] providing potential scope for mGluR2/3 to play a role in cognition. More specific modulators have helped unravel the role of these receptors in cognition (Table 16.9). The group II mGluR

Table 16-9. Pharmacological Evidence Implicating Group II mGluRs in Cognitive Disorders

Receptor	Modulation	Selectivity	Route	Name	Effect	Test	Reference
mGluR2 & 3	Agonist	Group II	Intraperitoneal	LY354740	Impaired working memory	T-maze	[141]
mGluR2 & 3	Agonist	Group II	Intraperitoneal	LY354740	Improved working memory in a PCP induced deficit model	T-maze	[141]
mGluR2 & 3	Agonist	Group II	Intraperitoneal	LY354740	Impaired working memory	Morris water maze	[143]
mGluR2 & 3	Antagonist	Group II	Intraperitoneal	LY354740	Improved working memory	Morris water maze	[143]
mGluR2 & 3	Agonist	Non-Selective	Intraperitoneal	LY341495	Impaired working memory	Morris water maze	[143]
mGluR2 & 3	Agonist	Group II	Intraperitoneal	APDC	Impaired working memory	Fear conditioning	[145]
mGluR2 & 3	Antagonist	Group II	Intraperitoneal	LY354740	Abolished MCPG induced deficit in memory formation	Lever press task	[243]

specific agonist ACPD has been shown to block the induction of LTP in the CA1 region of the hippocampus in urethane-anesthetized rats,[138] and DCG-IV, another group II agonist, induced LDP in the dentate gyrus.[139] MCPG did not impair the animals during the training portion of the memory test but did severely affect them in the memory recollection portion of the test.[140] This suggests the group II mGluRs are not responsible for the encoding of information but are vital for the consolidation of memory. As with research into mGluR2/3 roles in anxiety, the agonist LY354740 has been at the forefront as a research tool for cognitive function. LY354740 has been shown to impair working memory in a t-maze test.[141] Conversely, in a PCP-induced schizophrenia model, LY354740 was found to improve the memory deficits produced by PCP administration.[142] This suggests that its mechanism of action in PCP-treated subjects is different from that seen in normal subjects. LY354740 also produced a delay-dependent impairment in the Morris water maze; this impairment was absent in mGluR2 knockout mice, suggesting a role for mGluR2 activation as the mechanism behind this impairment.[143] In agreement with this, the nonselective agonist LY341495 improved spatial working memory.[143] The same authors found that LY354740 reduced excitatory postsynaptic potentials in hippocampal slices in wild-type mice but not in mGluR2 knockout mice. This strongly suggests that the effects of LY354740 on working memory are due to its effect on inhibiting glutamate release.

As mentioned earlier in the chapter, LTP and LDP are two forms of synaptic plasticity believed to be heavily involved in learning and memory; mGluR2 knockout mice have been shown to have severely reduced LTD in the CA3 region of the hippocampus, although this did not affect spatial learning in the Morris water maze.[144] The group II specific mGluR agonist ACPD disrupts fear conditioning;[145] however, when the salience of the context was increased, this disruption was abolished. The authors of this study suggest that mGluR2/3 activation produces nonspecific neural activity that prevents the context from being detected and influencing neurotransmission. This data suggest that decreasing glutamate neurotransmission by mGluR2/3 activation may impair learning by nonspecific mechanisms that act to block the registering of external stimuli.

Taken together, data accumulated to date suggest that pharmacological modulation of mGluR2/3 produces conflicting results depending on test and pharmacological agent used. It may be that the group II mGluRs have different roles in different forms of memory depending on factors such as the context, stimulus, and type of learning.

GROUP II MGLURS IN SCHIZOPHRENIA

As mentioned earlier in the chapter, schizophrenia is a psychiatric disorder the symptoms of which include altered perception of the world and cognitive deficits (www.nimh.nih.gov). The modern hypothesis used to explain the pathophysiology of schizophrenia is based on alterations in a large collection of neurotransmitter systems, including the glutamatergic system, leading to impaired functioning of neuronal circuits[49,50] (see the section on schizophrenia earlier in the chapter for more detail). mGluR2 and mGluR3 are presynaptic autoreceptors that negatively regulate the release of glutamate.[119] Since glutamate can influence the release of other neurotransmitters, such as dopamine,[51] mGluR2/3 have the potential to play a role in the altered neurotransmission that may be part of the cause of schizophrenia. Furthermore, mGluR2 can form active complexes with the serotonin 5-HT$_{2A}$ receptor; many hallucinogenic drugs target this complex, producing unique cellular responses.[146] mGluR2 was found to be down-regulated whereas 5-HT$_{2A}$ expression was increased in post-mortem studies of schizophrenia patients.[146] The change in expression of both receptors would lead to altered signaling via the mGluR2–5-HT2A complex, which may contribute to the manifestation of the psychotic symptoms of schizophrenia.

The majority of pharmacological agents used to investigate the role of the group II mGluRs are not subtype specific (Table 16.10), making it difficult to identify the role each individual subtype plays. In general, agonists of the group II mGluRs have been shown to reduce schizophrenia symptoms, likely due to their ability to inhibit glutamate release. This has been advanced recently into clinical trials with the successful utility of the mGluR2 agonist LY404039 in patients with schizophrenia.[147]

GROUP II MGLURS IN ADDICTION

Repeated exposure of neurons to drugs of abuse alters the way those neurons function, and this altered neuronal function leads to the characteristics associated with addiction, including dependence, craving, and tolerance.[148] It was originally discovered by chance that rats would act to electrically self-stimulate regions of their own brain.[149] This intracranial self-stimulation (ICSS) activates brain reward circuitries, and the lowest current that would induce self-stimulation was termed the threshold ICSS. Drugs of abuse were shown to lower ICSS.[150,151] This not only gave an insight into how drugs of abuse mediate their addictive effects, but also provided a mechanism that could be used to study the effects of compounds on addiction. Pharmacological mGluR modulators have been shown to effect the ICSS threshold; LY314582, a group II agonist, elevated the ICSS threshold.[152,153] mGluR2/3 agonists have been shown to reduce drug-seeking behavior; in rats trained to self-administer cocaine, the group II agonist LY379268 was shown to dose-dependently inhibit cocaine[154] and nicotine[155] seeking behavior after withdrawal. Part of the craving properties of drugs of abuse may be mediated by reduced functioning of mGluR2 and 3. Other data supports this hypothesis, with cocaine exposure being shown to decrease mGluR2/3 functioning in the NAc[156] and amygdala.[157] mGluR2/3 agonist induced LTD in the NAc was also found to be attenuated by exposure to morphine.[158] Sensorimotor gating is another technique that has been used to study drug addiction; withdrawal from drugs of abuse have been shown to increase the response to a startling auditory response.[150]LY354740, an mGluR2/3 agonist, attenuated the increase

Table 16-10. Pharmacological Evidence Implicating Group II mGluRs in Schizophrenia

Receptor	Modulation	Selectivity	Route	Name	Effect	Test	Reference
mGluR2 & 3	Agonist	Group II	Intraperitoneal	LY354740	Reduced PCP induced schizophrenia like symptoms	Working memory, stereotypy and locomotion	[142]
mGluR2 & 3	Agonist	Group II	Subcutaneous	LY354740	Reduced PCP induced schizophrenia like symptoms	PCP induced locomotion	[244]
mGluR2 & 3	Agonist	Group II	Subcutaneous	LY379268	Reduced PCP induced schizophrenia like symptoms	PCP induced locomotion	[244]
mGluR2 & 3	Agonist	Group I and II	Intraaccumbens	ACPD	Reduced schizophrenia like symptoms	Pre-pulse inhibition	[245]
mGluR2 & 3	Agonist	Group I and II	Intraaccumbens	MCPG	Increased ACPD induced reduction in schizophrenia like symptoms	Pre-pulse inhibition	[245]
mGluR2 & 3	Agonist	Group II	Subcutaneous	LY379268	Reduced PCP induced schizophrenia like symptoms	PCP induced locomotion	[244]
mGluR2 & 3	Agonist	Group II	Subcutaneous	LY379268	Reduced PCP induced schizophrenia like symptoms	PCP induced locomotion	[4]
mGluR2 & 3	Agonist	Group II	Intraperitoneal	LY314582	Reduced PCP induced schizophrenia like symptoms	PCP induced locomotion	[234]
mGluR2 & 3	Agonist	Group II	Intraperitoneal	LY354740	No effect	PCP induced locomotion	[246]
mGluR2 & 3	Agonist	Group II	Intraperitoneal	LY354740	No effect	Pre-pulse inhibition	[247]
mGluR2 & 3	Agonist	Group II	Intraperitoneal	LY354740	Reduced PCP induced schizophrenia like symptoms	PCP induced locomotion	[248]
mGluR2 & 3	Agonist	Group II	Intraperitoneal	LY354740	No effect	Pre-pulse inhibition	[248]
mGluR2 & 3	Agonist	Group II	Intraperitoneal	LY404039	Reduced PCP induced schizophrenia like symptoms	PCP induced locomotion	[188]

in the startle response after nicotine withdrawal in rats;[159] decreasing glutmatergic neurotransmission using mGluR2/3 agonists may be a method to reduce the irritability experienced during nicotine withdrawal. A similar result was found for cocaine withdrawal where cocaine-dependent rats displayed heightened sensitivity to the anxiolytic effects of LY354740.[160] Withdrawal from certain drugs of abuse produces somatic signs that can easily be observed and used as a method for studying processes underlying drug dependence. The group II agonist DCG-IV attenuated the somatic effects of opiates in rats.[161] Similarly, activation of mGluR2/3 by LY354740 reduced somatic symptoms of morphine withdrawal in rats.[162,163]

Much research has shown that the group II mGluRs play a role in drug addiction. Their ability to negatively regulate glutamate release allows them to modify the excitability of neuronal circuits that may be altered through substance abuse. Their functioning is altered in addictive states, which may contribute to the negative effects of addiction, such as dependence. What is most exciting is that mGluR2/3 agonists have shown great promise in treating addiction to substances such as cocaine and nicotine. More research needs to be done into these receptors and their role in addiction so that their true therapeutic potential can be realized.

GROUP II MGLURS IN OTHER PSYCHIATRIC DISORDERS

Depression

As mentioned earlier in the chapter, glutamate has been shown to play a role in depression, with NMDA modulators having antidepressant properties. mGluR2 and 3 can inhibit the release of glutamate, which may affect NMDA signaling, placing the group II mGluRs in a position where they may influence depression-related physiology and behavior. The 5-hydroxytryptamine(2A) receptor (5-HT(2A)) has been linked to the effects of antidepressants, and mGluR2/3 agonism has been shown to antagonize the glutamate release mediated by 5-HT(2A) receptor agonists. Treatment with the antidepressants imipramine and phenelzine reduced synaptic glutamate levels in the prefrontal cortex; mGluR2/3 agonism could mediate a similar effect.[164] Furthermore, treatment with imipramine produced an up-regulation of mGluR2/3 expression in the rat hippocampus.[164] This up-regulation may be partly how the antidepressant mediates its effects. Preclinical data has shown that mGluR2/3 modulation affects antidepressant-like behavior; LY341495 and MGS0039, mGluR2/3 antagonists, reduced immobility in the forced swim test.[132,165,166] The fact that antagonists produce an antidepressant effect is somewhat unexpected, as antidepressants have been shown to reduce glutamate release, and antagonism of mGluR2/3 would lead to an increase in glutamate neurotransmission; also, as mentioned earlier, mGluR2/3 agonists produce an anxiolytic phenotype. This may be because increasing glutamate release may produce an antidepressant effect, or due to nonspecific effects of the drug, or mGluR-mediated inhibition of GABAergic neurons.

Table 16-11. Pharmacological Evidence Implicating Group II mGluRs in Parkinson's Disease

Receptor	Modulation	Selectivity	Route	Name	Effect	Test	Reference
mGluR2/3	Agonist	mGluR2/3	Intraperitoneal	LY354740	Reduction in Parkinsonian like symptoms	Haloperidol induced catalepsy	[85]
mGluR2/3	Agonist	mGluR2/3	Intraperitoneal	LY354740	Reduction in Parkinsonian like symptoms	Haloperidol induced catalepsy	[86]
mGluR2/3	Agonist	mGluR2/3	Injection into substantia nigra pars reticula	DCG-IV	Reduction in Parkinsonian like symptoms	Reserpine induced movements	[87]

The data presented earlier suggest group II mGluRs play a role in depression; their expression is changed due to antidepressant treatment, and mGluR2/3 modulators have been shown to affect antidepressant-like behavior. However, much more research needs to be done to unravel the true role of mGluR2 and 3 in depression.

Group II mGluRs in Parkinson's disease

The ability of group II mGluRs to inhibit glutamate release means they may be able to influence dopamine release and alter excitability, giving them the potential to play a role in Parkinson's disease. Pharmacological modulators and animal models are helping in the effort to elucidate role of group II mGluRs in Parkinson's disease (Table 16.11). LY354740, a group II agonist, was shown to have the ability to reduce Parkinsonian motor symptoms in both rats and mice.[85,86] Direct injection of DCG-IV, an mGluR2/3 agonist, to the substantia nigra pars reticulate reduced Parkinsonian-like symptoms in reserpine-treated rats.[87] Systematically active agonists and DCG-IV injected directly into the substantia nigra pars reticulate reduced Parkinsonian symptoms; however, 2R,4R-APDC, a selective group II agonist, injected directly into the striatum had no effect on haloperidol-treated animals,[98] suggesting the mGluR group II effects on Parkinsonian symptoms is not mediated by decreasing glutamate transmission in the striatum, but rather in other areas of the brain believed to be involved in the pathophysiology of Parkinson's disease.

Post-mortem studies of Parkinson's disease patients have shown a decrease in mGluR2/3 levels in the internal globus pallidus, an element of the basal ganglia which sends projections into the substantia nigra.[167]

GROUP II MGLURS IN EPILEPSY

As mentioned earlier in the chapter, epilepsy is a neurological condition that can result in the onset of seizures; glutamate has been implicated in epilepsy, with a large number of modulations of the glutamatergic system producing seizures.

Group II mGluRs function to reduce glutamate neurotransmission during time of high excitability; activation of mGluR2/3 may reduce excitability and therefore produce an anticonvulsant effect. For the most part, it is found that mGluR2/3 agonists possess anticonvulsant propertiers (Table 16.12). ((1S,3S)-ACPD), an agonist of mGluR2 and 3 receptors, produced an immediate and prolonged anti-convulsant effect.[102,106,168] Authors suggest the proconvulsant effects may be due to activity at other receptors or by the presynpatic inhibition of GABA neurotran-simission. DCG-IV, an agonist of mGluR2/3, displayed anticonvulsant effects at lower doses while having proconvulsant effects at higher doses;[169] MTPG, an antagonist at group II metabotropic glutamate receptors, blocked the effects of the lower dose of DCG-IV, indicating the anticonvulsant activity is mediated through mGluR2/3. DCG-IV has been shown to be an agonist at NMDA recep-tors at high doses,[170] and MTPG was unable to reverse the effects of high doses of DCG-IV. This suggest that the proconvulsant activity of DCG-IV was mediated through its activity on NMDA receptors. Further supporting the anticonvulsant effects of group II mGluR activation is the fact that LY379268, LY389795 and, LY354740 produced anticonvulsant effects without proconvulsant effects[171–173].

In addition to pharmacological studies, expression studies have also suggested mGluR2/3 may play a role in epilepsy. Immunohistochemical staining found a marked reduction in mGluR2/3 expression in the hippocampus in epileptic patients[174] and preclinically in chronically epileptic rats.[175] This reduction in mGluR2/3 may contribute to the overexcitability that may result in epilepsy.[175]

As can be seen from the data presented earlier in the chapter, mGluR2/3 acti-vation largely produces anticonvulsant effects, with proconvulsant effects pos-sible due to nonspecific effects: if more specific pharmacological modifiers can be developed, they may be valuable therapeutics in epilepsy.

Group II mGluRs in pain disorders

As mentioned in the earlier section on pain, hypersensitization of neurons to noxious stimuli is believed to be partly caused by overexcitability due to excess glutamate transmission.[113] Given that the group II mGluRs are perisynaptic autoreceptors inhibiting the release of glutamate, they are active in times of high glutamate release into the synaptic cleft and act to inhibit further gluta-mate release during periods of high excitability. Expression studies have shown mGluR2/3 to be expressed on synaptic elements in many of the nociceptive ascending and descending pathways in the CNS.[114]

Group II pharmacological modulators have been invaluable in research into group II mGluR function in pain (Table 16.13). Pretreatment with the group II receptor agonists LY379268 and LCCG1 had no effect on noxious stimuli in spi-nothalamic tract cells (STT) under control conditions but did reduce the sen-sitivity of STT neurons to capsaicin treatment.[176] This suggests the mGluR2/3 receptors do not play a role in pain under normal conditions, but may serve an important role during more severe pain states and act to reduce glutamate release. Since the group II mGluRs are also present on GABAergic neurons and

Table 16-12. Pharmacological Evidence Implicating Group II mGluRs in Epilepsy

Receptor	Modulation	Selectivity	Route	Name	Effect	Test	Reference
mGluR2/3	Agonist	Group II	Intracereberal	((1S,3S)-ACPD)	Immediate proconvulsant effect followed by prolonged anticonvulsant effect	mGluR group I agonist induced seizures	[102]
mGluR2/3	Agonist	Group II	Intracereberal	((1S,3S)-ACPD)	Immediate proconvulsant effect followed by prolonged anticonvulsant effect	Audio induced seizures	[106]
mGluR2/3	Agonist	Group II	Intracereberal	DCG IV	Anticonvulsant effect at low doses, proconvulsant at high doses	Homocysteic acid induced seizures	[169]
mGluR2/3	Agonist	Group II	Intraperitoneal	LY389795	Anticonvulsant	Audio and DHPG induced seizures	[76]
mGluR2/3	Agonist	Group II	Intraperitoneal	LY379268	Anticonvulsant	Audio and DHPG induced seizures	[171]
mGluR2/3	Agonist	Group II	Intracereberal	((1S,3S)-ACPD)	Immediate proconvulsant effect followed by prolonged anticonvulsant effect	Audio induced seizures	[168]
mGluR2/3	Agonist	Group II	Intracereberal	DCG IV	Anticonvulsant effect at low doses, proconvulsant at high doses	Electrically induced seizures	[249]
mGluR2/3	Agonist	Group II	Intraperitoneal	LY354740	Anticonvulsant	PTZ and picrotoxin induces seizures	[172]
mGluR2/3	Agonist	Group II	Per os (oral)	LY354740	Anticonvulsant	ACPD induced seizures	[173]
mGluR2/3	Agonist	Group II	Intracereberal	R,4R-APDC	Anticonvulsant	Homocysteic acid induced seizures	[169]

Table 16-13. Pharmacological Evidence Implicating Group II mGluRs in Pain Disorders

Receptor	Modulation	Selectivity	Route	Name	Effect	Test	Reference
mGluR2/3	Agonist	Group II	Microdialysis	LY379268	Reduced caspaicin induced hyperalgesia	Extracellular recordings to mechanical stimuli	[176]
mGluR2/3	Agonist	Group II	Microdialysis	LCCG1	Reduced caspaicin induced hyperalgesia	Extracellular recordings to mechanical stimuli	[176]
mGluR2/3	Agonist	Group II	Intraperitoneal	LY389795	Reduced nociception	Formalin induced persistant pain state	[177]
mGluR2/3	Agonist	Group II	Intraperitoneal	LY379268	Reduced nociception	Formalin induced persistant pain state	[177]
mGluR2/3	Agonist	Group II	Intraperitoneal	LY354740	Reduced nociception	Formalin induced persistant pain state	[177]
mGluR2/3	Agonist	Group II	Intracisternal	LY389795	Reduced nociception	Formalin induced persistant pain state	[177]
mGluR2/3	Agonist	Group II	Intracisternal	LY379268	Reduced nociception	Formalin induced persistant pain state	[131].
mGluR2/3	Agonist	Group II	Intraperitoneal	APDC	Reduced nociception	Peripheral nerve injury induced pain state	[181]
mGluR2/3	Agonist	Group II	Intrathecal	DCG-IV	Increased nociception	Formalin induced persistant pain state	[115]
mGluR2/3	Agonist	Group II	Intraplantar	APDC	Reduced nociception	Formalin induced persistant pain state	[179]
mGluR2/3	Agonist	Group II	Microdialysis	LY354740	Reduced nociception	Kaolin/carrageenan-induced knee-joint arthritis	[180]

355

inhibit the release of this inhibitory neurotransmitter, there may be no net effect on excitation under resting conditions; however, under exaggerated painful conditions, the agonists may produce an effects where excitability is lowered due to decreased glutamate release. In support of this hypothesis, group II agonists LY354740, LY379268, and LY389795 all reduced nociception in a formalin-induced model of persistent pain.[177] This analgesic effect was also reversed by the group II antagonist LY341495,[177] suggesting the nociceptive effects are mediated through the agonists effects on mGluR2/3 and not nonspecific effects. Similar results were found for LY389795 and LY379268[178] and for APDC.[179] In another persistent pain model, arthritis, the mGluR2/3 agonist LY354740 was found to have a significant analgesic effect.[180] In a neuropathic model of pain in the rat, the group II agonist APDC was found to be involved in reducing nociception in the induction phase of pain,[181] as opposed to the maintenance phase as was found with inflammatory pain. The analgesic effects of different mGluR2/3 may depend on the site of action. An i.t. injection of the mGluR2/3 agonist DCG-IV resulted in increased nociception.[115] This pro-nociceptive effect may be due to DCG-IVs effects on NMDA receptors, or it may be DCG-IV is activating mGluR2/3 receptors on GABAergic neurons and reducing the amount inhibitive neurotransmission, resulting in an overall excitatory effect. Supporting this is the finding of increased levels of mGluR3 in the spinal cord of animals used as a model of persistent inflammation and hyperalgesia.[182]

Together, a growing corpus of evidence collected to date demonstrates that group II mGluRs play a key role in nociception; in general, agonists possess analgesic properties probably as a result of their negative effects on glutamate release. Group II mGluR modulators may prove to be very valuable therapeutics for the treatment of pain disorders; clinical validation is now warranted.

GROUP III MGLURS

The third group of mGluRs is the largest group, consisting of mGluR4, 6, 7, and 8. However, they have been the least investigated due to a lack of selective pharmacological tools. More recently, selective ligands have emerged and are very important in the continuing elucidating of the effects of these receptors in CNS disorders. Typically these receptors are located presynaptically and function as autoreceptors to inhibit glutamate and GABA release; however, they are also located postsynaptically and function to modulate neuronal excitability[4] (see Figure 16.3). The ability of these receptors to modulate the release of both the main excitatory and inhibitory neurotransmitters in the CNS provides a large scope for the functioning of these receptors. The group III receptors are extensively distributed throughout the CNS (with the exception of mGluR6, which is restricted to the retina) and have been shown to be located on limbic system nuclei. (For a more extensive review on group III mGluR expression patterns, see Ferraguti and Shigemoto.[3]) These receptors also are negatively coupled to adenylyl cyclase in artificial reporter systems; however, their actual functioning in vivo may be far more complex.[2] mGluR7 is the most extensively distributed and shows the highest degree of evolutionary conservation of all the mGluRs,[183] suggesting an important role for this receptor in proper CNS functioning. mGluR7 is presynaptically located and has a relatively low affinity for glutamate;[184] these receptors function as an inhibitory autoreceptor and therefore may be highly relevant in a pathological condition where there is excessive glutamate neurotransmission.

GROUP III MGLURS IN ANXIETY DISORDERS

As mentioned earlier in the chapter, group III mGluRs are mainly located pre-synaptically, modulate the release of glutamate, and function to modulate neuronal excitability, giving these receptors the large potential to regulate excitation levels, which places them in a position where they may contribute in regulating anxiety levels. Less is known about the roles of the group III mGluRs roles in anxiety, largely due to a lag in the development of selective pharmacological modulators for this group. However, some data have been collected, which helps to shed some light on the role these receptors play.

Work done using non-subtype-specific group III pharmacological modulators show the group III mGluRs do play some role in anxiety (Table 16.14). A group III agonist L-SOP produced anxiolytic effects in the conflict drinking test after intra-hippocampal administration.[23] Similarly, the group III agonists, HomoAMPA and ACPT-I, produced anxiolytic effects in the conflict drinking test.[185] These effects could be reversed by the group III antagonist CPPG,[185] showing the anxiolytic effects are due to activation of the group III mGluRs. In apparent conflict with these results, the same research group demonstrated that the group III antagonists MSOP and CPPG also produced anti-anxiety-like effects.[22,186] These apparently conflicting results may have arisen due to different modulators having varying affinities for different group III receptors, which may have varying contribution, both positive and negative, to the anxiolytic effects. In the case of CPPG, it produced an anxiolytic effect when directly injected into the amygdala.[186] In the amygdala, CPPG may antagonize mGluRs on GABAergic neurons, leading to an increase in the amount of GABA release, and this would lead to an overall lowering of excitation, which may be responsible for the anxiolytic effects. Subtype-specific pharmacological modulators have helped unravel the roles of the group III mGluR subtypes; an mGluR4 postive allosteric modulator produced anxiolytic effects after intra-amygdala administration.[187] The mGluR8 agonist DCPGG attenuated the effects of stress-induced hyperthermia.[188] Early results show that mGluR7 activation may help reduce anxiety levels: the mGluR7 selective agonist, AMN082, facilitated the extinction of acquired fearful memories, whereas mGluR7 knockdown using siRNA attenuated the extinction of fearful memories.[189]

Knockout studies have shown the role group III mGluRs play in anxiety, mGluR8 deficient mice were shown to have increased anxiety like behaviour in the elevated plus maze and the open-field test,[190,191] yet also show reduced

Table 16-14. Pharmacological Evidence Implicating Group III mGluRs in Anxiety

Receptor	Modulation	Selectivity	Route	Name	Effect	Test	Reference
mGluR group III	Agonist	group III	Intrahippocampal	HomoAMPA	Anxiolytic	Conflict drinking test	[185]
mGluR group III	Agonist	group III	Intrahippocampal	ACPT-I	Anxiolytic	Conflict drinking test	[185]
mGluR group III	Antagonist	group III	Intrahippocampal	CPPG	AnxiogenicAttenuated anxiolytic effect of ACPT-I andHomoAMPA	Conflict drinking test	[185]
mGluR group III	Agonist	group III	Intrahippocampal	L-SOP	Anxiolytic	Conflict drinking test	[23]
mGluR group III	Antagonist	group III	Intrahippocampal	MSOP	Anxiolytic	Conflict drinking test	[22]
mGluR group III	Antagonist	group III	Intraamygdalar	CPPG	Anxiolytic	Conflict drinking test	[186]
mGluR4	Postive alloster	mGluR4	Intraamygdalar	PHCCC	Anxiolytic	Conflict drinking test	[187]
mGluR8	Agonist	mGluR8	Intraperitoneal	DCPG	Anxiolytic	Stress induced hyperthermia	[188]
mGluR7	Agonist	mGLuR7	Intraperitoneal	AMN082	Anxiolytic	Fear potentiated startel and conflict drinking test	[189]
mGluR group III	Agonist	group III	Intraperitoneal	ACPT-I	Anxiolytic	Stress induced hyperthermia, elevated plus maze, cnflic	[187]

contextual fear.[192] In contrast to this, mGluR7 knockout mice display anxiolytic-like behavior in the elevated plus maze, light dark box, the staircase test, and stress-induced hyperthermia,[193,194] as well as deficits in fear extinction.[195] These apparently contrasting results from different knockout animals may be due to differing localization of mGluR7 and 8 on GABAergic and glutamatergic neurons, with mGluR7 reducing GABA release and mGluR8 reducing glutamate release. This may also contribute to the reason why mGluR7 knockout animals display an anxious phenotype and mGluR8 animals are less anxious. More research in this area, with more selective pharmacological agents, needs to be done to unravel the specific role each receptor subtype plays in anxiety.

GROUP III MGLURS IN COGNITION

Given that group III mGluRs are mainly located presynpatically and thus can function to inhibit glutamate release, they modulate cellular excitability, which can lead to changes in LTP and LDP that are believed to be involved in learning and memory. Similar to studies on anxiety, cognitive studies focusing on the group III mGluRs are not as abundant as other subtypes, possibly due to a lack of selective agents available. There is some data available on mGluR group III pharmacological modulators effects on neuronal plasticity. The group III antagonist CPPG has been shown to impair LTD and produce a deficit in long-term spatial memory.[196,197] mGluR4 knockout mice showed no impairment in spatial acquisition but did have impairment to their long-term memory.[198] As mentioned earlier in the chapter, mGluR7 has been shown to be essential for fear extinction,[189] which is believed to be a form of new learning. mGluR7 knockout animals showed impaired memory performance in the conditioned taste aversion test and delayed freezing in a footshock test,[199] as well as an impairment to performance in the radial-arm maze.[195]

Together, early data show the group III mGluRs do play a role in learning and memory. More research needs to be done to improve our knowledge of the role these receptors play in cognition.

GROUP III MGLURS IN SCHIZOPHRENIA

As presynaptic autoreceptors inhibiting the release of glutamate, group III mGluRs have the potential to modify cellular excitability and dopamine release, producing the possibility they may play a role in the molecular basis of schizophrenia (see discussion earlier in the chapter). Research into the role of group III mGluRs is also lacking; mGluR8 has been shown to be expressed in regions of the brain associated with schizophrenia.[200] Polymorphisms in both the mGluR7 and the mGluR8 gene have been shown to be associated with schizophrenia.[201,202] More research needs to be done into this subgroup to unravel their role in schizophrenia and their potential as therapeutic targets to treat schizophrenia.

GROUP III MGLURS IN OTHER PSYCHIATRIC DISEASES

Depression

As mentioned previously, the glutamatergic system has been implicated in depression. Antidepressants have been shown to inhibit glutamate release – a phenomenon mirrored by activation of group III mGluRs.[203] It is currently unclear if this property of antidepressants could be achieved via mGluR group III modulation. The group III agonists ACPT-I and *RS*-PPG have been shown to reduce depression-like behavior in the forced swim test.[185,204] mGluR7 knockout mice have been shown to display an antidepressant-like phenotype.[193] mGluR7 knockouts also have a disrupted hypothalamus-pituitary-axis axis.[205] mGluR7 expression was shown to be reduced in the hippocampus and prefrontal cortex after chronic administration of the antidepressant citalopram.[206] Activation of AMN082 was found to reduce immobility in the forced swim test (FST) and tail suspension test (TST).[207] Further research needs to be done to help discover the role of these receptors in depression and whether they are potential therapeutic targets.

GROUP III MGLURS IN PARKINSON'S DISEASE

The ability of group III mGluRs to inhibit glutamate release means they may be able to influence dopamine release and alter excitability, giving them the potential to play a role in Parkinson's disease. Immunohistochemical studies have shown that the group III mGluRs are located in regions associated with Parkinson's disease.[208] Like other disorders, research into the group III mGluRs is lacking compared to the other two groups. Administration of the group III agonists L-AP4 and ACPT-I produced differing effects on Parkinson-like symptoms depending on the brain regions they are delivered to; in the global pallidus (GP), the agonists led to a reduction in induced Parkinson symptoms, whereas administration to the substantia nigra pars reticulate (SNr) induced Parkinsonian symptoms.[209] In the GP, the anti-Parkinsonian effects may be due to mGluR4 mediated inhibition of GABA release, whereas in the SNr, it may be due to inhibitory effects on glutamate release. The mGluR4 allosteric modulator PHCCC reversed reserpine-induced akinesia in rats.[210] The same modulator also attenuated nigrostriatal degeneration after administration of MPTP, a compound used to induce Parkinosnian-like symptoms.[211] Other work has also shown the neuroprotective effect of group III modulators; L-AP4, a group III agonist, provided neuroprotection against 6-OHDA, a neurotoxin used to induce Parkinsonian-like symptoms.[212] Another study showed that L-AP4 influenced behavior in 6-OHDA-treated rats but not in controls.[213] This could mean that mGluR group III function may be altered during Parkinson's disease and may contribute to the observed pathology.

As can be seen from the presented data, group III mGluRs do play a role in Parkinson's disease; they likely inhibit glutamate release, which can influence

Parkinsonian-like symptoms; their function may be altered in a Parkinson's disease, contributing to the symptoms. Early work has also shown the group III mGluRs, particularly mGluR4, may be promising therapeutic targets for the treatment of Parkinson's disease. More research needs to be done into the role that these receptors play in Parkinson's disease.

GROUP III MGLURS IN EPILEPSY

As mentioned previously, glutamate has been implicated in epilepsy, because modulation of the glutamatergic system by a wide variety of molecular mechanisms can produce seizures.[101] The group III mGluRs can inhibit glutamate and GABA release, which influences neuronal excitability and consequently may play a role in epileptic seizures. Increased glutamate release has been shown both clinically in patients with epilepsy and in animal models.[101] Early work showed that group III agonists can be proconvulsant.[214] Later work showed that, like the group II agonists, mGluR group III agonists also have a mixed proconvulsant and anticonvulsant properties; L-SOP produced an immediate proconvulsant effect followed by a prolonged anticonvulsant effect in audio-induced seizures.[102,106] Like the group II mGluRs, this seemingly paradoxical effect may be due to action on both glutamatergic and GABAergic neurons. The mGluR4 agonist ACPT-1, and mGluR8 agonist PPG both display anticonvulsant properties in an animal models of epilepsy.[215] In general, group III mGluRs are anticonvulsant, likely due to their ability to negatively regulate the release of glutamate (Table 16.15).

Changes in group III mGluR expression have been shown in epileptic animal models; mGluR4 expression was up-regulated in an animal model of temporal lobe epilepsy.[216] In human patient studies, mGluR4 and mGluR8 expression was shown to be changed in patients with medial temporal lobe epilepsy.[217]

The data presented earlier show that the group III mGluRs do play a role in epilepsy. They can control the release the glutamate and GABA, leading to altered excitability that may influence epileptic seizures. Expression differences in animal and models of epilepsy and patients with epileptic syndromes have also been shown; it is likely this change in expression contributes to the epileptic phenotype. Furthermore, the fact that mGluR group III agonists are usually anticonvulsant presents these receptors as promising therapeutic targets that warrant further research.

GROUP III MGLURS IN PAIN DISORDERS

Given the role of glutamate in gating nociceptive responses to noxious stimuli, group III mGluRs, as negative regulators of glutamate release, have the potential to regulate pain processing. The group III agonist, LAP4, reduced responses to cutaneous mechanical stimuli both before and after sensitization with capsaicin.[176] LAP4 administered i.t. also reduced capsaicin-induced sensitivity to

Table 16.15. Pharmacological Evidence Implicating Group III mGluRs in Epilepsy

Receptor	Modulation	Selectivity	Route	Name	Effect	Test	Reference
mGluR4, 6, 7&8	Agonist	Group III	Intracereberal	L-SOP	Immediate proconvulsant effect followed by prolonged anticonvulsant effect	mGluR group I agonist induced seizures	[102]
mGluR4, 6, 7&8	Agonist	Group III	Intracereberal	L-SOP	Immediate proconvulsant effect followed by prolonged anticonvulsant effect	Audio induced seizures	[106]
mGluR4, 6, 7&8	Agonist	Group III	Intracereberal	LAP4	Proconvulsant	Epilepsy prone mouse strain	[214]
mGluR4, 6, 7&8	Agonist	Group III	Intracereberal	L-SOP	Proconvulsant	Epilepsy prone mouse strain	[214]
mGluR4	Agonist	mGluR4	Intracereberal	ACPT-1	Anticonvulsant	Audio induced seizures and epilepsy prone rats	[215]
mGluR8	Agonist	mGluR8	Intracereberal	PPG	Anticonvulsant	Audio induced seizures and epilepsy prone rats	[215]
mGluR8	Agonist	mGluR8	Intracereberal	PPG	Anticonvulsant	Audio induced seizures	[235]
mGluR8	Agonist	mGluR8	Intracereberal	PPG	Anticonvulsant	Audio induced seizures	Gasparini, Bruno et al. 1999
mGluR8	Agonist	mGluR8	Intracereberal	3,4-DCPG	Anticonvulsant	Audio induced seizures	[168]
mGluR8	Agonist	mGluR8	Intracereberal	3,4-DCPG	Anticonvulsant	DL-homocysteic acid induced seizures	Folbergrova, Druga et al. 2008
mGluR4, 6, 7&8	Agonist	Group III	Intracereberal	((R,S)-PPG)	Anticonvulsant	DL-homocysteic acid induced seizures	[169]

mechanical stimuli.[218] ACPT-I, a group III agonist, reduced nociception in a formalin-induced model of pain but not in control animals,[219] suggesting the mGluRs are mainly involved in nociception in the sensitized state. LAP4 was also found to be effective in reducing nociception after spinal cord injury[220] In contrast to this, the group III agonist L-SOP administered directly to the periaqueductal gray matter resulted in a pro-nociceptive effect in a formalin-induced model of a persistent pain state.[221] The specific injection site of group III modulators appears to be crucial in determining whether its net effect is pro- or antinociceptive. This may be due to differing distribution of glutamtergic and GABAergic neurons. LAP4 reduced synpatic glutamtergic neurotransmission in a neuropathic model of pain; this reduction in glutamatergic neurotransmission is likely to contribute to the reduction in nociception caused by group III agonists.[222] Recent work has shown the different subtypes of the group III mGluRs participate in different pain states; the mGluR7 agonist AMN082 administered into the amygdala reduced nociception in control animals while having no effect in a kaolin/carrageenan induced model of arthritis.[223] However, AMN082 was effective as an analgesic effect in a neuropathic pain model,[224] while in the arthritic model the mGluR 8 agonist, S-3,4-DCPG, reduced nociception but had no effect on control animals.[223] Thus different mGluR subtypes located on different nociceptive pathways appear to be responsible for specific types of pain processing.

Generally mGluR group III agonists have an antinociceptive effect, which is likely due to their effect on the inhibition of glutamate release. This reduction in glutamate leads to reduced excitability in nociceptive pathways containing mGluRs, which may contribute to a lowering of the pain experienced. Different mGluRs seem to be located on different pain processing pathways and therefore contribute to the processing of different types of pain. More research needs to be conducted in this area, because these are promising therapeutic targets for the treatment of different types of pain disorders.

GROUP III MGLURS IN RETINAL DYSFUNCTION

All eight mGluRs are found in the retina, with mGluR6 expression being solely confined to the retina.[225] mGluR6 has been identified as a key receptor in photoreceptor signaling; it is present on the ON bipolar cells of rods and cones and is essential for transmitting the dark-induced glutamate release from photoreceptors into hyperpolarization of bipolar cells.[226] mGluR6 has been shown to be essential for visual processing; mice lacking mGluR6 showed a loss of sensitivity of ON bipolar cells to light.[227]. The other mGluR group III subtypes have also been shown to be involved in visual processing, with mGluR7 being the only glutamate receptor to be present on presynaptic bipolar cell where it likely functions as an autoreceptor [226] The same group went on to show that mGluR6 knockout mice display impaired behavioral suppression to light.[228] This work has shown that mGluR6 is essential for photoprocessing, and impaired mGluR6 function could lead to retinal dysfunction. Studies examining patients with retinal dysfunction have shown mutations in mGluR6 can lead to night blindness.[229,230]

The other members of the group III mGluRs have also been implicated in retinal dysfunction; in an animal model of glaucoma, mGluR 4, 6, and 7 mRNA expression was found to be up-regulated while mGluR8 mRNA was down-regulated,[231] suggesting impaired regulation of the expression of these receptors in glaucoma.

CONCLUSIONS AND PERSPECTIVES

The past twenty years have been very exciting in glutamate research, especially with regard to the discovery, characterization, and functionalization of mGluRs. Advances in the molecular pharmacology of mGluRs have led to a greater understanding of the complex roles these receptors play in maintaining homeostatic functioning of the CNS. There is much data available demonstrating the role of the group I and II mGluRs in the CNS, whereas research on the group III mGluRs has lagged somewhat. The advent of new more specific pharmacological agents and advances in in vivo gene silencing provides the tools by which these receptors can be investigated and should represent a high priority research area in the coming years. The glutamatergic system has long been considered a promising target for therapeutic intervention for many CNS disorders. However, in largely exploiting ionotropic receptors, a number of false dawns have been encountered in areas ranging from cerebral ischaemia to cognitive enhancement. Thus there is a level of cautious optimism that pharmacological modulators of the mGluRs, which show such promising results in much preclinical work, can translate into therapeutic modalities for CNS disorders. Given that many of these have already reached the clinical-trial stage for disorders such as Parkinsons's disease anxiety disorders and schizophrenia it is an exciting time in the mGluR field. There are still many issues to resolve in terms of how both activation and inhibition of the same receptors produce potentially therapeutic effects as in the case of Group II and III receptors in anxiety/depression. Moreover, the complex interactions of mGluRs with other neurotransmitter systems and receptors are being slowly unlocked and may offer further avenues for therapeutic intervention.

Overall, mGluRs offer themselves as a diverse group of therapeutic targets that can be potentially exploited to treat a host of disorders, where glutamatergic transmission is dysregulated in an effective and safe fashion.

SUGGESTED READING

1. Niederberger, E., Schmidtko, A., Coste, O., Marian, C., Ehnert, C. & Geisslinger, G. (2006) The glutamate transporter GLAST is involved in spinal nociceptive processing. *Biochem Biophys Res Commun*, 346, 393–9.
2. Conn, P. J. & Pin, J. P. (1997) Pharmacology and functions of metabotropic glutamate receptors. *Annu Rev Pharmacol Toxicol*, 37, 205–37.

3. Ferraguti, F. & Shigemoto, R. (2006) Metabotropic glutamate receptors. *Cell Tissue Res*, 326, 483–504.

4. Swanson, C. J., Bures, M., Johnson, M. P., Linden, A. M., Monn, J. A. & Schoepp, D. D. (2005) Metabotropic glutamate receptors as novel targets for anxiety and stress disorders. *Nat Rev Drug Discov*, 4, 131–44.

5. Bockaert, J. & Pin, J. P. (1999) Molecular tinkering of G protein-coupled receptors: an evolutionary success. *Embo J*, 18, 1723–9.

6. Meldrum, B. S. (2000) Glutamate as a neurotransmitter in the brain: review of physiology and pathology. *J Nutr*, 130, 1007S-15S.

7. Mckenna, M. C. (2007) The glutamate-glutamine cycle is not stoichiometric: fates of glutamate in brain. *J Neurosci Res*, 85, 3347–58.

8. Azbill, R. D., Mu, X. & Springer, J. E. (2000) Riluzole increases high-affinity glutamate uptake in rat spinal cord synaptosomes. *Brain Res*, 871, 175–80.

9. Schousboe, A., Svenneby, G. & Hertz, L. (1977) Uptake and metabolism of glutamate in astrocytes cultured from dissociated mouse brain hemispheres. *J Neurochem*, 29, 999–1005.

10. Danbolt, N. C. (2001) Glutamate uptake. *Prog Neurobiol*, 65, 1–105.

11. Conn, P. J. (2003) Physiological roles and therapeutic potential of metabotropic glutamate receptors. *Ann N Y Acad Sci*, 1003, 12–21.

12. Parmentier, M. L., Galvez, T., Acher, F., Peyre, B., Pellicciari, R., Grau, Y., Bockaert, J. & Pin, J. P. (2000) Conservation of the ligand recognition site of metabotropic glutamate receptors during evolution. *Neuropharmacology*, 39, 1119–31.

13. Riedel, G., Platt, B. & Micheau, J. (2003) Glutamate receptor function in learning and memory. *Behav Brain Res*, 140, 1–47.

14. Valenti, O., Conn, P. J. & Marino, M. J. (2002) Distinct physiological roles of the Gq-coupled metabotropic glutamate receptors Co-expressed in the same neuronal populations. *J Cell Physiol*, 191, 125–37.

15. Anwyl, R. (1999) Metabotropic glutamate receptors: electrophysiological properties and role in plasticity. *Brain Res Brain Res Rev*, 29, 83–120.

16. O'Riordan K. J., Huang I. C., Pizzi M. Spano P. Boroni F., Egli R., Desai P., Fitch O., Malone L., Ahn H. J., Liou H. C., Sweatt J. D., & Levenson J. M., (2006) Regulation of nuclear factor kappaB in the hippocampus by group I metabotropic glutamate receptors. *J Neurosci*, 26, 4870–9.

17. Yang, L., Mao, L., Chen, H., Catavsan, M., Kozinn, J., Arora, A., Liu, X. & Wang, J. Q. (2006) A signaling mechanism from G alpha q-protein-coupled metabotropic glutamate receptors to gene expression: role of the c-Jun N-terminal kinase pathway. *J Neurosci*, 26, 971–80.

18. Kingston, A. E., Burnett, J. P., Mayne, N. G. & Lodge, D. (1995) Pharmacological analysis of 4-carboxyphenylglycine derivatives: comparison of effects on mGluR1 alpha and mGluR5a subtypes. *Neuropharmacology*, 34, 887–94.

19. Andlin-Sobocki, P., Jonsson, B., Wittchen, H. U. & Olesen, J. (2005) Cost of disorders of the brain in Europe. *Eur J Neurol*, 12 Suppl 1, 1–27.

20. Cryan, J. F. & Holmes, A. (2005) The ascent of mouse: advances in modelling human depression and anxiety. *Nat Rev Drug Discov*, 4, 775–90.

21. Schoepp, D. D., Wright, R. A., Levine, L. R., Gaydos, B. & Potter, W. Z. (2003) LY354740, an mGlu2/3 receptor agonist as a novel approach to treat anxiety/stress. *Stress*, 6, 189–97.

22. Chojnacka-Wojcik, E., Tatarczynska, E. & Pilc, A. (1997) The anxiolytic-like effect of metabotropic glutamate receptor antagonists after intrahippocampal injection in rats. *Eur J Pharmacol*, 319, 153–6.

23. Tatarczynska, E., Klodzinska, A., Kroczka, B., Chojnacka-Wojcik, E. & Pilc, A. (2001b) The antianxiety-like effects of antagonists of group I and agonists of group II and III metabotropic glutamate receptors after intrahippocampal administration. *Psychopharmacology (Berl)*, 158, 94–9.

24. Tatarczynska, E., Klodzinska, A., Chojnacka-Wojcik, E., Palucha, A., Gasparini, F., Kuhn, R. & Pilc, A. (2001a) Potential anxiolytic- and antidepressant-like effects of MPEP, a potent, selective and systemically active mGlu5 receptor antagonist. *Br J Pharmacol*, 132, 1423–30.

25. Varty, G. B., Grilli, M., Forlani, A., Fredduzzi, S., Grzelak, M. E., Guthrie, D. H., Hodgson, R. A., Lu, S. X., Nicolussi, E., Pond, A. J., Parker, E. M., Hunter, J. C., Higgins, G. A., Reggiani, A. & Bertorelli, R. (2005) The antinociceptive and anxiolytic-like effects of the metabotropic glutamate receptor 5 (mGluR5) antagonists, MPEP and MTEP, and the mGluR1 antagonist, LY456236, in rodents: a comparison of efficacy and side-effect profiles. *Psychopharmacology (Berl)*, 179, 207–17.

26. Koch, M. (1993) Microinjections of the metabotropic glutamate receptor agonist, trans-(+/-)-1-amino-cyclopentane-1,3-dicarboxylate (trans-ACPD) into the amygdala increase the acoustic startle response of rats. *Brain Res*, 629, 176–9.

27. Lima, V. C., Molchanov, M. L., Aguiar, D. C., Campos, A. C. & Guimaraes, F. S. (2008) Modulation of defensive responses and anxiety-like behaviors by group I metabotropic glutamate receptors located in the dorsolateral periaqueductal gray. *Prog Neuropsychopharmacol Biol Psychiatry*, 32, 178–85.

28. Rodrigues, S. M., Bauer, E. P., Farb, C. R., Schafe, G. E. & Ledoux, J. E. (2002) The group I metabotropic glutamate receptor mGluR5 is required for fear memory formation and long-term potentiation in the lateral amygdala. *J Neurosci*, 22, 5219–29.

29. Schulz, B., Fendt, M., Gasparini, F., Lingenhohl, K., Kuhn, R. & Koch, M. (2001) The metabotropic glutamate receptor antagonist 2-methyl-6-(phenylethynyl)-pyridine (MPEP) blocks fear conditioning in rats. *Neuropharmacology*, 41, 1–7.

30. Riedel, G., Casabona, G., Platt, B., Macphail, E. M. & Nicoletti, F. (2000) Fear conditioning-induced time- and subregion-specific increase in expression of mGlu5 receptor protein in rat hippocampus. *Neuropharmacology*, 39, 1943–51.

31. Muller, N. & Schwarz, M. J. (2007) The immune-mediated alteration of serotonin and glutamate: towards an integrated view of depression. *Mol Psychiatry*, 12, 988–1000.

32. Page, M. E., Szeliga, P., Gasparini, F. & Cryan, J. F. (2005) Blockade of the mGlu5 receptor decreases basal and stress-induced cortical norepinephrine in rodents. *Psychopharmacology (Berl)*, 179, 240–6.

33. Chowdari, K. V., Mirnics, K., Semwal, P., Wood, J., Lawrence, E., Bhatia, T., Deshpande, S. N., B, K. T., Ferrell, R. E., Middleton, F. A., Devlin, B., Levitt, P., Lewis, D. A. & Nimgaonkar, V. L. (2002) Association and linkage analyses of RGS4 polymorphisms in schizophrenia. *Hum Mol Genet*, 11, 1373–80.

34. Brann, D. W. (1995) Glutamate: a major excitatory transmitter in neuroendocrine regulation. *Neuroendocrinology*, 61, 213–25.

35. Johnson, M. P., Kelly, G. & Chamberlain, M. (2001) Changes in rat serum corticosterone after treatment with metabotropic glutamate receptor agonists or antagonists. *J Neuroendocrinol*, 13, 670–7.

36. Pecknold, J. C., Mcclure, D. J., Appeltauer, L., Wrzesinski, L. & Allan, T. (1982) Treatment of anxiety using fenobam (a nonbenzodiazepine) in a double-blind standard (diazepam) placebo-controlled study. *J Clin Psychopharmacol*, 2, 129–33.

37. Porter, R. H., Jaeschke, G., Spooren, W., Ballard, T. M., Buttelmann, B., Kolczewski, S., Peters, J. U., Prinssen, E., Wichmann, J., Vieira, E., Muhlemann, A., Gatti, S., Mutel, V. & Malherbe, P. (2005) Fenobam: a clinically validated nonbenzodiazepine anxiolytic is a potent, selective, and noncompetitive mGlu5 receptor antagonist with inverse agonist activity. *J Pharmacol Exp Ther*, 315, 711–21.

38. Riedel, G., Casabona, G. & Reymann, K. G. (1995a) Inhibition of long-term potentiation in the dentate gyrus of freely moving rats by the metabotropic glutamate receptor antagonist MCPG. *J Neurosci*, 15, 87–98.

39. Riedel, G., Wetzel, W. & Reymann, K. G. (1994b) (R,S)-alpha-methyl-4-carboxyphenylglycine (MCPG) blocks spatial learning in rats and long-term potentiation in the dentate gyrus in vivo. *Neurosci Lett*, 167, 141–4.

40. Riedel, G. & Reymann, K. (1993) An antagonist of the metabotropic glutamate receptor prevents LTP in the dentate gyrus of freely moving rats. *Neuropharmacology*, 32, 929–31.

41. Richter-Levin, G., Errington, M. L., Maegawa, H. & Bliss, T. V. (1994) Activation of metabotropic glutamate receptors is necessary for long-term potentiation in the dentate gyrus and for spatial learning. *Neuropharmacology*, 33, 853–7.

42. Balschun, D. & Wetzel, W. (1998) Inhibition of group I metabotropic glutamate receptors blocks spatial learning in rats. *Neurosci Lett*, 249, 41–4.

43. Balschun, D., Manahan-Vaughan, D., Wagner, T., Behnisch, T., Reymann, K. G. & Wetzel, W. (1999) A specific role for group I mGluRs in hippocampal LTP and hippocampus-dependent spatial learning. *Learn Mem*, 6, 138–52.

44. Balschun, D. & Wetzel, W. (2002) Inhibition of mGluR5 blocks hippocampal LTP in vivo and spatial learning in rats. *Pharmacol Biochem Behav*, 73, 375–80.

45. Riedel, G., Manahan-Vaughan, D., Kozikowski, A. P. & Reymann, K. G. (1995b) Metabotropic glutamate receptor agonist trans-azetidine-2,4-dicarboxylic acid facilitates maintenance of LTP in the dentate gyrus in vivo. *Neuropharmacology*, 34, 1107–9.

46. Riedel, G., Wetzel, W., Kozikowski, A. P. & Reymann, K. G. (1995c) Block of spatial learning by mGluR agonist tADA in rats. *Neuropharmacology*, 34, 559–61.

47. Conquet, F., Bashir, Z. I., Davies, C. H., Daniel, H., Ferraguti, F., Bordi, F., Franz-Bacon, K., Reggiani, A., Matarese, V., Conde, F. & ET AL. (1994) Motor deficit and impairment of synaptic plasticity in mice lacking mGluR1. *Nature*, 372, 237–43.

48. Lu, Y. M., Jia, Z., Janus, C., Henderson, J. T., Gerlai, R., Wojtowicz, J. M. & Roder, J. C. (1997) Mice lacking metabotropic glutamate receptor 5 show impaired learning and reduced CA1 long-term potentiation (LTP) but normal CA3 LTP. *J Neurosci*, 17, 5196–205.

49. Carlsson, A., Hansson, L. O., Waters, N. & Carlsson, M. L. (1997) Neurotransmitter aberrations in schizophrenia: new perspectives and therapeutic implications. *Life Sci*, 61, 75–94.

50. Kim, J. S., Kornhuber, H. H., Schmid-Burgk, W. & Holzmuller, B. (1980) Low cerebrospinal fluid glutamate in schizophrenic patients and a new hypothesis on schizophrenia. *Neurosci Lett*, 20, 379–82.

51. Carlsson, A., Waters, N., Holm-Waters, S., Tedroff, J., Nilsson, M. & Carlsson, M. L. (2001) Interactions between monoamines, glutamate, and GABA in schizophrenia: new evidence. *Annu Rev Pharmacol Toxicol*, 41, 237–60.

52. Allen, R. M. & Young, S. J. (1978) Phencyclidine-induced psychosis. *Am J Psychiatry*, 135, 1081–4.

53. Dulawa, S. C. & Geyer, M. A. (1996) Psychopharmacology of prepulse inhibition in mice. *Chin J Physiol*, 39, 139–46.

54. Collier, D. A. & Li, T. (2003) The genetics of schizophrenia: glutamate not dopamine? *Eur J Pharmacol*, 480, 177–84.

55. Harrison, P. J., Lyon, L., Sartorius, L. J., Burnet, P. W. & Lane, T. A. (2008) The group II metabotropic glutamate receptor 3 (mGluR3, mGlu3, GRM3): expression, function and involvement in schizophrenia. *J Psychopharmacol*, 22, 308–22.

56. Krivoy, A., Fischel, T. & Weizman, A. (2008) The possible involvement of metabotropic glutamate receptors in schizophrenia. *Eur Neuropsychopharmacol*, 18, 395–405.

57. Geyer, M. A. & Braff, D. L. (1987) Startle habituation and sensorimotor gating in schizophrenia and related animal models. *Schizophr Bull*, 13, 643–68.

58. Henry, S. A., Lehmann-Masten, V., Gasparini, F., Geyer, M. A. & Markou, A. (2002) The mGluR5 antagonist MPEP, but not the mGluR2/3 agonist LY314582, augments

PCP effects on prepulse inhibition and locomotor activity. *Neuropharmacology*, 43, 1199–209.

59. Kinney, G. G., Burno, M., Campbell, U. C., Hernandez, L. M., Rodriguez, D., Bristow, L. J. & Conn, P. J. (2003) Metabotropic glutamate subtype 5 receptors modulate locomotor activity and sensorimotor gating in rodents. *J Pharmacol Exp Ther*, 306, 116–23.

60. De Vry, J., Horvath, E. & Schreiber, R. (2001) Neuroprotective and behavioral effects of the selective metabotropic glutamate mGlu(1) receptor antagonist BAY 36–7620. *Eur J Pharmacol*, 428, 203–14.

61. Kinney, G. G., O'Brien J. A., Lemaire W.,Burno M., Bickel D. J., Clements M. K., Chen T. B., Wisnoski D. D., Lindsley C. W., Tiller P. R., Smith S. Jacobson M. A., Sur C., Duggan M. E., Pettibone D. J., Conn P. J. & Williams D. L., JR. (2005) A novel selective positive allosteric modulator of metabotropic glutamate receptor subtype 5 has in vivo activity and antipsychotic-like effects in rat behavioral models. *J Pharmacol Exp Ther*, 313, 199–206.

62. Devon, R. S., Anderson, S., Teague, P. W., Muir, W. J., Murray, V., Pelosi, A. J., Blackwood, D. H. & Porteous, D. J. (2001) The genomic organisation of the metabotropic glutamate receptor subtype 5 gene, and its association with schizophrenia. *Mol Psychiatry*, 6, 311–4.

63. Gupta, D. S., Mccullumsmith, R. E., Beneyto, M., Haroutunian, V., Davis, K. L. & Meador-Woodruff, J. H. (2005) Metabotropic glutamate receptor protein expression in the prefrontal cortex and striatum in schizophrenia. *Synapse*, 57, 123–31.

64. Brody, S. A., Conquet, F. & Geyer, M. A. (2003) Disruption of prepulse inhibition in mice lacking mGluR1. *Eur J Neurosci*, 18, 3361–6.

65. Saugstad, J. A., Marino, M. J., Folk, J. A., Hepler, J. R. & Conn, P. J. (1998) RGS4 inhibits signaling by group I metabotropic glutamate receptors. *J Neurosci*, 18, 905–13.

66. Williams, N. M., Preece, A., Spurlock, G., Norton, N., Williams, H. J., Zammit, S., O'Donovan M. C. & Owen M. J. (2003) Support for genetic variation in neuregulin 1 and susceptibility to schizophrenia. *Mol Psychiatry*, 8, 485–7.

67. Grottick, A. J., Bagnol, D., Phillips, S., Mcdonald, J., Behan, D. P., Chalmers, D. T. & Hakak, Y. (2005) Neurotransmission- and cellular stress-related gene expression associated with prepulse inhibition in mice. *Brain Res Mol Brain Res*, 139, 153–62.

68. Javitt, D. C. (2004) Glutamate as a therapeutic target in psychiatric disorders. *Mol Psychiatry*, 9, 984–97, 979.

69. Paul, I. A. & Skolnick, P. (2003) Glutamate and depression: clinical and preclinical studies. *Ann N Y Acad Sci*, 1003, 250–72.

70. Pilc, A., Branski, P., Palucha, A., Tokarski, K. & Bijak, M. (1998) Antidepressant treatment influences group I of glutamate metabotropic receptors in slices from hippocampal CA1 region. *Eur J Pharmacol*, 349, 83–7.

71. Pilc, A. & Legutko, B. (1995b) The influence of prolonged antidepressant treatment on the changes in cyclic AMP accumulation induced by excitatory amino acids in rat cerebral cortical slices. *Neuroreport*, 7, 85–8.

72. Pilc, A. & Legutko, B. (1995a) Antidepressant treatment influences cyclic AMP accumulation induced by excitatory amino acids in rat brain. *Pol J Pharmacol*, 47, 359–61.

73. Zahorodna, A. & Bijak, M. (1999) An antidepressant-induced decrease in the responsiveness of hippocampal neurons to group I metabotropic glutamate receptor activation. *Eur J Pharmacol*, 386, 173–9.

74. Molina-Hernandez, M., Tellez-Alcantara, N. P., Perez-Garcia, J., Olivera-Lopez, J. I. & Jaramillo, M. T. (2006) Antidepressant-like and anxiolytic-like actions of the mGlu5 receptor antagonist MTEP, microinjected into lateral septal nuclei of male Wistar rats. *Prog Neuropsychopharmacol Biol Psychiatry*, 30, 1129–35.

75. Belozertseva, I. V., Kos, T., Popik, P., Danysz, W. & Bespalov, A. Y. (2007) Antidepressant-like effects of mGluR1 and mGluR5 antagonists in the rat forced swim and the mouse tail suspension tests. *Eur Neuropsychopharmacol*, 17, 172–9.

76. Bear, M. F., Huber, K. M. & Warren, S. T. (2004) The mGluR theory of fragile X mental retardation. *Trends Neurosci*, 27, 370–7.

77. Oliet, S. H., Malenka, R. C. & Nicoll, R. A. (1997) Two distinct forms of long-term depression coexist in CA1 hippocampal pyramidal cells. *Neuron*, 18, 969–82.

78. Weiler, I. J., Irwin, S. A., Klintsova, A. Y., Spencer, C. M., Brazelton, A. D., Miyashiro, K., Comery, T. A., Patel, B., Eberwine, J. & Greenough, W. T. (1997) Fragile X mental retardation protein is translated near synapses in response to neurotransmitter activation. *Proc Natl Acad Sci U S A*, 94, 5395–400.

79. Huber, K. M., Gallagher, S. M., Warren, S. T. & Bear, M. F. (2002) Altered synaptic plasticity in a mouse model of fragile X mental retardation. *Proc Natl Acad Sci U S A*, 99, 7746–50.

80. Yan, Q. J., Rammal, M., Tranfaglia, M. & Bauchwitz, R. P. (2005) Suppression of two major Fragile X Syndrome mouse model phenotypes by the mGluR5 antagonist MPEP. *Neuropharmacology*, 49, 1053–66.

81. Dolen, G. & Bear, M. F. (2008) Role for metabotropic glutamate receptor 5 (mGluR5) in the pathogenesis of fragile X syndrome. *J Physiol*, 586, 1503–8.

82. Ossowska, K., Wardas, J., Pietraszek, M., Konieczny, J. & Wolfarth, S. (2003) The strio-pallidal pathway is involved in antiparkinsonian-like effects of the blockade of group I metabotropic glutamate receptors in rats. *Neurosci Lett*, 342, 21–4.

83. Betarbet, R., Sherer, T. B. & Greenamyre, J. T. (2002) Animal models of Parkinson's disease. *Bioessays*, 24, 308–18.

84. Ossowska, K. (1994) The role of excitatory amino acids in experimental models of Parkinson's disease. *J Neural Transm Park Dis Dement Sect*, 8, 39–71.

85. Bradley, S. R., Marino, M. J., Wittmann, M., Rouse, S. T., Awad, H., Levey, A. I. & Conn, P. J. (2000) Activation of group II metabotropic glutamate receptors inhibits synaptic excitation of the substantia Nigra pars reticulata. *J Neurosci*, 20, 3085–94.

86. Konieczny, J., Ossowska, K., Wolfarth, S. & Pilc, A. (1998) LY354740, a group II metabotropic glutamate receptor agonist with potential antiparkinsonian properties in rats. *Naunyn Schmiedebergs Arch Pharmacol*, 358, 500–2.

87. Dawson, L., Chadha, A., Megalou, M. & Duty, S. (2000) The group II metabotropic glutamate receptor agonist, DCG-IV, alleviates akinesia following intranigral or intraventricular administration in the reserpine-treated rat. *Br J Pharmacol*, 129, 541–6.

88. Ossowska, K., Konieczny, J., Wolfarth, S. & Pilc, A. (2005) MTEP, a new selective antagonist of the metabotropic glutamate receptor subtype 5 (mGluR5), produces antiparkinsonian-like effects in rats. *Neuropharmacology*, 49, 447–55.

89. Ossowska, K., Konieczny, J., Wolfarth, S., Wieronska, J. & Pilc, A. (2001) Blockade of the metabotropic glutamate receptor subtype 5 (mGluR5) produces antiparkinsonian-like effects in rats. *Neuropharmacology*, 41, 413–20.

90. Wardas, J., Pietraszek, M., Wolfarth, S. & Ossowska, K. (2003) The role of metabotropic glutamate receptors in regulation of striatal proenkephalin expression: implications for the therapy of Parkinson's disease. *Neuroscience*, 122, 747–56.

91. Breysse, N., Amalric, M. & Salin, P. (2003) Metabotropic glutamate 5 receptor blockade alleviates akinesia by normalizing activity of selective basal-ganglia structures in parkinsonian rats. *J Neurosci*, 23, 8302–9.

92. Coccurello, R., Breysse, N. & Amalric, M. (2004) Simultaneous blockade of adenosine A2A and metabotropic glutamate mGlu5 receptors increase their efficacy in reversing Parkinsonian deficits in rats. *Neuropsychopharmacology*, 29, 1451–61.

93. Oueslati, A., Breysse, N., Amalric, M., Kerkerian-Le Goff, L. & Salin, P. (2005) Dysfunction of the cortico-basal ganglia-cortical loop in a rat model of early parkinsonism is reversed by metabotropic glutamate receptor 5 antagonism. *Eur J Neurosci*, 22, 2765–74.

94. Phillips, J. M., Lam, H. A., Ackerson, L. C. & Maidment, N. T. (2006) Blockade of mGluR glutamate receptors in the subthalamic nucleus ameliorates motor asymmetry in an animal model of Parkinson's disease. *Eur J Neurosci*, 23, 151–60.

95. Turle-Lorenzo, N., Breysse, N., Baunez, C. & Amalric, M. (2005) Functional interaction between mGlu 5 and NMDA receptors in a rat model of Parkinson's disease. *Psychopharmacology (Berl)*, 179, 117–27.

96. Parelkar, N. K. & Wang, J. Q. (2003) Preproenkephalin mRNA expression in rat dorsal striatum induced by selective activation of metabotropic glutamate receptor subtype-5. *Synapse*, 47, 255–61.

97. De Leonibus, E., Manago, F., Giordani, F., Petrosino, F., Lopez, S., Oliverio, A., Amalric, M. & Mele, A. (2009) Metabotropic glutamate receptors 5 blockade reverses spatial memory deficits in a mouse model of Parkinson's disease. *Neuropsychopharmacology*, 34, 729–38.

98. Ossowska, K., Konieczny, J., Pilc, A. & Wolfarth, S. (2002) The striatum as a target for anti-rigor effects of an antagonist of mGluR1, but not an agonist of group II metabotropic glutamate receptors. *Brain Res*, 950, 88–94.

99. Samadi, P., Gregoire, L., Morissette, M., Calon, F., Hadj Tahar, A., Dridi, M., Belanger, N., Meltzer, L. T., Bedard, P. J. & DI PAOLO, T. (2008) mGluR5 metabotropic glutamate receptors and dyskinesias in MPTP monkeys. *Neurobiol Aging*, 29, 1040–51.

100. Sanchez-Pernaute, R., Wang, J. Q., Kuruppu, D., Cao, L., Tueckmantel, W., Kozikowski, A., Isacson, O. & Brownell, A. L. (2008) Enhanced binding of metabotropic glutamate receptor type 5 (mGluR5) PET tracers in the brain of parkinsonian primates. *Neuroimage*, 42, 248–51.

101. Chapman, A. G. (2000) Glutamate and epilepsy. *J Nutr*, 130, 1043S-5S.

102. Tizzano, J. P., Griffey, K. I. & Schoepp, D. D. (1995) Induction or protection of limbic seizures in mice by mGluR subtype selective agonists. *Neuropharmacology*, 34, 1063–7.

103. Camon, L., Vives, P., De Vera, N. & Martinez, E. (1998) Seizures and neuronal damage induced in the rat by activation of group I metabotropic glutamate receptors with their selective agonist 3,5-dihydroxyphenylglycine. *J Neurosci Res*, 51, 339–48.

104. Thomsen, C. & Dalby, N. O. (1998) Roles of metabotropic glutamate receptor subtypes in modulation of pentylenetetrazole-induced seizure activity in mice. *Neuropharmacology*, 37, 1465–73.

105. Thomsen, C., Klitgaard, H., Sheardown, M., Jackson, H. C., Eskesen, K., Jacobsen, P., Treppendahl, S. & Suzdak, P. D. (1994) (S)-4-carboxy-3-hydroxyphenylglycine, an antagonist of metabotropic glutamate receptor (mGluR) 1a and an agonist of mGluR2, protects against audiogenic seizures in DBA/2 mice. *J Neurochem*, 62, 2492–5.

106. Tang, E., Yip, P. K., Chapman, A. G., Jane, D. E. & Meldrum, B. S. (1997) Prolonged anticonvulsant action of glutamate metabotropic receptor agonists in inferior colliculus of genetically epilepsy-prone rats. *Eur J Pharmacol*, 327, 109–15.

107. Hayashi, Y., Sekiyama, N., Nakanishi, S., Jane, D. E., Sunter, D. C., Birse, E. F., Udvarhelyi, P. M. & Watkins, J. C. (1994) Analysis of agonist and antagonist activities of phenylglycine derivatives for different cloned metabotropic glutamate receptor subtypes. *J Neurosci*, 14, 3370–7.

108. Shannon, H. E., Peters, S. C. & Kingston, A. E. (2005) Anticonvulsant effects of LY456236, a selective mGlu1 receptor antagonist. *Neuropharmacology*, 49 Suppl 1, 188–95.

109. Merlin, L. R. (2002) Differential roles for mGluR1 and mGluR5 in the persistent prolongation of epileptiform bursts. *J Neurophysiol*, 87, 621–5.

110. Witkin, J. M., Baez, M., Yu, J. & Eiler, W. J., 2ND (2008) mGlu5 receptor deletion does not confer seizure protection to mice. *Life Sci*, 83, 377–80.

111. Loscher, W., Dekundy, A., Nagel, J., Danysz, W., Parsons, C. G. & Potschka, H. (2006) mGlu1 and mGlu5 receptor antagonists lack anticonvulsant efficacy in rodent models of difficult-to-treat partial epilepsy. *Neuropharmacology*, 50, 1006–15.

112. Wieseler-Frank, J., Maier, S. F. & Watkins, L. R. (2004) Glial activation and pathological pain. *Neurochem Int*, 45, 389–95.

113. Verne, G. N., Himes, N. C., Robinson, M. E., Gopinath, K. S., Briggs, R. W., Crosson, B. & Price, D. D. (2003) Central representation of visceral and cutaneous hypersensitivity in the irritable bowel syndrome. *Pain*, 103, 99–110.

114. Bleakman, D., Alt, A. & Nisenbaum, E. S. (2006) Glutamate receptors and pain. *Semin Cell Dev Biol*, 17, 592–604.

115. Fisher, K. & Coderre, T. J. (1996) The contribution of metabotropic glutamate receptors (mGluRs) to formalin-induced nociception. *Pain*, 68, 255–63.

116. Noda, K., Anzai, T., Ogata, M., Akita, H., Ogura, T. & Saji, M. (2003) Antisense knockdown of spinal-mGluR1 reduces the sustained phase of formalin-induced nociceptive responses. *Brain Res*, 987, 194–200.

117. Fundytus, M. E., Osborne, M. G., Henry, J. L., Coderre, T. J. & Dray, A. (2002) Antisense oligonucleotide knockdown of mGluR1 alleviates hyperalgesia and allodynia associated with chronic inflammation. *Pharmacol Biochem Behav*, 73, 401–10.

118. Li, W. & Neugebauer, V. (2004) Differential roles of mGluR1 and mGluR5 in brief and prolonged nociceptive processing in central amygdala neurons. *J Neurophysiol*, 91, 13–24.

119. Cartmell, J. & Schoepp, D. D. (2000) Regulation of neurotransmitter release by metabotropic glutamate receptors. *J Neurochem*, 75, 889–907.

120. Sugiyama, C., Nakamichi, N., Ogura, M., Honda, E., Maeda, S., Taniura, H. & Yoneda, Y. (2007) Activator protein-1 responsive to the group II metabotropic glutamate receptor subtype in association with intracellular calcium in cultured rat cortical neurons. *Neurochem Int*, 51, 467–75.

121. Ohishi, H., Shigemoto, R., Nakanishi, S. & Mizuno, N. (1993a) Distribution of the messenger RNA for a metabotropic glutamate receptor, mGluR2, in the central nervous system of the rat. *Neuroscience*, 53, 1009–18.

122. Ohishi, H., Shigemoto, R., Nakanishi, S. & Mizuno, N. (1993b) Distribution of the mRNA for a metabotropic glutamate receptor (mGluR3) in the rat brain: an in situ hybridization study. *J Comp Neurol*, 335, 252–66.

123. Dube, G. R. & Marshall, K. C. (1997) Modulation of excitatory synaptic transmission in locus coeruleus by multiple presynaptic metabotropic glutamate receptors. *Neuroscience*, 80, 511–21.

124. Marek, G. J., Wright, R. A., Schoepp, D. D., Monn, J. A. & Aghajanian, G. K. (2000) Physiological antagonism between 5-hydroxytryptamine(2A) and group II metabotropic glutamate receptors in prefrontal cortex. *J Pharmacol Exp Ther*, 292, 76–87.

125. Neugebauer, V., Keele, N. B. & Shinnick-Gallagher, P. (1997) Epileptogenesis in vivo enhances the sensitivity of inhibitory presynaptic metabotropic glutamate receptors in basolateral amygdala neurons in vitro. *J Neurosci*, 17, 983–95.

126. Ferris, P., Seward, E. & Dawson, G. R. (2001) Interactions between LY354740, a group II metabotropic agonist and the GABA(A)-benzodiazepine receptor complex in the rat elevated plus-maze. *J Psychopharmacol*, 15, 76–82.

127. Linden, A. M., Greene, S. J., Bergeron, M. & Schoepp, D. D. (2004) Anxiolytic activity of the MGLU2/3 receptor agonist LY354740 on the elevated plus maze is associated with the suppression of stress-induced c-Fos in the hippocampus and increases in c-Fos induction in several other stress-sensitive brain regions. *Neuropsychopharmacology*, 29, 502–13.

128. Grillon, C., Cordova, J., Levine, L. R. & Morgan, C. A., 3RD (2003) Anxiolytic effects of a novel group II metabotropic glutamate receptor agonist (LY354740) in the fear-potentiated startle paradigm in humans. *Psychopharmacology (Berl)*, 168, 446–54.

129. Kellner, M., Muhtz, C., Stark, K., Yassouridis, A., Arlt, J. & Wiedemann, K. (2005) Effects of a metabotropic glutamate(2/3) receptor agonist (LY544344/LY354740) on panic anxiety induced by cholecystokinin tetrapeptide in healthy humans: preliminary results. *Psychopharmacology (Berl)*, 179, 310–5.

130. Johnson, M. P., Barda, D., Britton, T. C., Emkey, R., Hornback, W. J., Jagdmann, G. E., Mckinzie, D. L., Nisenbaum, E. S., Tizzano, J. P. & Schoepp, D. D. (2005) Metabotropic glutamate 2 receptor potentiators: receptor modulation, frequency-dependent synaptic activity, and efficacy in preclinical anxiety and psychosis model(s). *Psychopharmacology (Berl)*, 179, 271–83.

131. Shimazaki, T., Iijima, M. & Chaki, S. (2004) Anxiolytic-like activity of MGS0039, a potent group II metabotropic glutamate receptor antagonist, in a marble-burying behavior test. *Eur J Pharmacol*, 501, 121–5.

132. Yoshimizu, T., Shimazaki, T., Ito, A. & Chaki, S. (2006) An mGluR2/3 antagonist, MGS0039, exerts antidepressant and anxiolytic effects in behavioral models in rats. *Psychopharmacology (Berl)*, 186, 587–93.

133. Iijima, M., Shimazaki, T., Ito, A. & Chaki, S. (2007) Effects of metabotropic glutamate 2/3 receptor antagonists in the stress-induced hyperthermia test in singly housed mice. *Psychopharmacology (Berl)*, 190, 233–9.

134. Greenslade, R. G. & Mitchell, S. N. (2004) Selective action of (-)-2-oxa-4-aminobicyclo[3.1.0]hexane-4,6-dicarboxylate (LY379268), a group II metabotropic glutamate receptor agonist, on basal and phencyclidine-induced dopamine release in the nucleus accumbens shell. *Neuropharmacology*, 47, 1–8.

135. Linden, A. M., Shannon, H., Baez, M., Yu, J. L., Koester, A. & Schoepp, D. D. (2005) Anxiolytic-like activity of the mGLU2/3 receptor agonist LY354740 in the elevated plus maze test is disrupted in metabotropic glutamate receptor 2 and 3 knock-out mice. *Psychopharmacology (Berl)*, 179, 284–91.

136. Hetzenauer, A., Corti, C., Herdy, S., Corsi, M., Ferraguti, F. & Singewald, N. (2008) Individual contribution of metabotropic glutamate receptor (mGlu) 2 and 3 to c-Fos expression pattern evoked by mGlu2/3 antagonism. *Psychopharmacology (Berl)*, 201, 1–13.

137. Riedel, G., Wetzel, W. & Reymann, K. G. (1994a) Computer-assisted shock-reinforced Y-maze training: a method for studying spatial alternation behaviour. *Neuroreport*, 5, 2061–4.

138. Holscher, C., Anwyl, R. & Rowan, M. J. (1997) Activation of group-II metabotropic glutamate receptors blocks induction of long-term potentiation and depotentiation in area CA1 of the rat in vivo. *Eur J Pharmacol*, 322, 155–63.

139. Huang, L., Killbride, J., Rowan, M. J. & Anwyl, R. (1999) Activation of mGluRII induces LTD via activation of protein kinase A and protein kinase C in the dentate gyrus of the hippocampus in vitro. *Neuropharmacology*, 38, 73–83.

140. Bianchin, M., Da Silva, R. C., Schmitz, P. K., Medina, J. H. & Izquierdo, I. (1994) Memory of inhibitory avoidance in the rat is regulated by glutamate metabotropic receptors in the hippocampus. *Behav Pharmacol*, 5, 356–359.

141. Aultman, J. M. & Moghaddam, B. (2001) Distinct contributions of glutamate and dopamine receptors to temporal aspects of rodent working memory using a clinically relevant task. *Psychopharmacology (Berl)*, 153, 353–64.

142. Moghaddam, B. & Adams, B. W. (1998) Reversal of phencyclidine effects by a group II metabotropic glutamate receptor agonist in rats. *Science*, 281, 1349–52.

143. Higgins, G. A., Ballard, T. M., Kew, J. N., Richards, J. G., Kemp, J. A., Adam, G., Woltering, T., Nakanishi, S. & Mutel, V. (2004) Pharmacological manipulation of mGlu2 receptors influences cognitive performance in the rodent. *Neuropharmacology*, 46, 907–17.

144. Yokoi, M., Kobayashi, K., Manabe, T., Takahashi, T., Sakaguchi, I., Katsuura, G., Shigemoto, R., Ohishi, H., Nomura, S., Nakamura, K., Nakao, K., Katsuki, M. & Nakanishi, S. (1996) Impairment of hippocampal mossy fiber LTD in mice lacking mGluR2. *Science*, 273, 645–7.

145. Riedel, G., Harrington, N. R., Kozikowski, A. P., Sandager-Nielsen, K. & Macphail, E. M. (2002) Variation of CS salience reveals group II mGluR-dependent and -independent forms of conditioning in the rat. *Neuropharmacology*, 43, 205–14.

146. Gonzalez-Maeso, J., Ang, R. L., Yuen, T., Chan, P., Weisstaub, N. V., Lopez-Gimenez, J. F., Zhou, M., Okawa, Y., Callado, L. F., Milligan, G., Gingrich, J. A., Filizola, M.,

Meana, J. J. & Sealfon, S. C. (2008) Identification of a serotonin/glutamate receptor complex implicated in psychosis. *Nature*, 452, 93–7.

147. Patil, S. T., Zhang, L., Martenyi, F., Lowe, S. L., Jackson, K. A., Andreev, B. V., Avedisova, A. S., Bardenstein, L. M., Gurovich, I. Y., Morozova, M. A., Mosolov, S. N., Neznanov, N. G., Reznik, A. M., Smulevich, A. B., Tochilov, V. A., Johnson, B. G., Monn, J. A. & Schoepp, D. D. (2007) Activation of mGlu2/3 receptors as a new approach to treat schizophrenia: a randomized Phase 2 clinical trial. *Nat Med*, 13, 1102–7.

148. Nestler, E. J. & Aghajanian, G. K. (1997) Molecular and cellular basis of addiction. *Science*, 278, 58–63.

149. Olds, J. & Milner, P. (1954) Positive reinforcement produced by electrical stimulation of septal area and other regions of rat brain. *J Comp Physiol Psychol*, 47, 419–27.

150. Kenny, P. J. & Markou, A. (2004) The ups and downs of addiction: role of metabotropic glutamate receptors. *Trends Pharmacol Sci*, 25, 265–72.

151. Cryan, J. F., Hoyer, D. & Markou, A. (2003) Withdrawal from chronic amphetamine induces depressive-like behavioral effects in rodents. *Biol Psychiatry*, 54, 49–58.

152. Harrison, A. A., Gasparini, F. & Markou, A. (2002) Nicotine potentiation of brain stimulation reward reversed by DH beta E and SCH 23390, but not by eticlopride, LY 314582 or MPEP in rats. *Psychopharmacology (Berl)*, 160, 56–66.

153. Kenny, P. J., Gasparini, F. & Markou, A. (2003) Group II metabotropic and alpha-amino-3-hydroxy-5-methyl-4-isoxazole propionate (AMPA)/kainate glutamate receptors regulate the deficit in brain reward function associated with nicotine withdrawal in rats. *J Pharmacol Exp Ther*, 306, 1068–76.

154. Baptista, M. A., Martin-Fardon, R. & Weiss, F. (2004) Preferential effects of the metabotropic glutamate 2/3 receptor agonist LY379268 on conditioned reinstatement versus primary reinforcement: comparison between cocaine and a potent conventional reinforcer. *J Neurosci*, 24, 4723–7.

155. Liechti, M. E., Lhuillier, L., Kaupmann, K. & Markou, A. (2007) Metabotropic glutamate 2/3 receptors in the ventral tegmental area and the nucleus accumbens shell are involved in behaviors relating to nicotine dependence. *J Neurosci*, 27, 9077–85.

156. Xi, Z. X., Ramamoorthy, S., Baker, D. A., Shen, H., Samuvel, D. J. & Kalivas, P. W. (2002) Modulation of group II metabotropic glutamate receptor signaling by chronic cocaine. *J Pharmacol Exp Ther*, 303, 608–15.

157. Neugebauer, V., Zinebi, F., Russell, R., Gallagher, J. P. & Shinnick-Gallagher, P. (2000b) Cocaine and kindling alter the sensitivity of group II and III metabotropic glutamate receptors in the central amygdala. *J Neurophysiol*, 84, 759–70.

158. Robbe, D., Bockaert, J. & Manzoni, O. J. (2002) Metabotropic glutamate receptor 2/3-dependent long-term depression in the nucleus accumbens is blocked in morphine withdrawn mice. *Eur J Neurosci*, 16, 2231–5.

159. Helton, D. R., Tizzano, J. P., Monn, J. A., Schoepp, D. D. & Kallman, M. J. (1997) LY354740: a metabotropic glutamate receptor agonist which ameliorates symptoms of nicotine withdrawal in rats. *Neuropharmacology*, 36, 1511–6.

160. Aujla, H., Martin-Fardon, R. & Weiss, F. (2008) Rats with extended access to cocaine exhibit increased stress reactivity and sensitivity to the anxiolytic-like effects of the mGluR 2/3 agonist LY379268 during abstinence. *Neuropsychopharmacology*, 33, 1818–26.

161. Fundytus, M. E. & Coderre, T. J. (1997) Attenuation of precipitated morphine withdrawal symptoms by acute i.c.v. administration of a group II mGluR agonist. *Br J Pharmacol*, 121, 511–4.

162. Vandergriff, J. & Rasmussen, K. (1999) The selective mGlu2/3 receptor agonist LY354740 attenuates morphine-withdrawal-induced activation of locus coeruleus neurons and behavioral signs of morphine withdrawal. *Neuropharmacology*, 38, 217–22.

163. Klodzinska, A., Chojnacka-Wojcik, E., Palucha, A., Branski, P., Popik, P. & Pilc, A. (1999) Potential anti-anxiety, anti-addictive effects of LY 354740, a selective group

II glutamate metabotropic receptors agonist in animal models. *Neuropharmacology*, 38, 1831–9.

164. Matrisciano, F., Storto, M., Ngomba, R. T., Cappuccio, I., Caricasole, A., Scaccianoce, S., Riozzi, B., Melchiorri, D. & Nicoletti, F. (2002) Imipramine treatment up-regulates the expression and function of mGlu2/3 metabotropic glutamate receptors in the rat hippocampus. *Neuropharmacology*, 42, 1008–15.

165. Bespalov, A. Y., Van Gaalen, M. M., Sukhotina, I. A., Wicke, K., Mezler, M., Schoemaker, H. & Gross, G. (2008) Behavioral characterization of the mGlu group II/III receptor antagonist, LY-341495, in animal models of anxiety and depression. *Eur J Pharmacol*, 592, 96–102.

166. Chaki, S., Yoshikawa, R., Hirota, S., Shimazaki, T., Maeda, M., Kawashima, N., Yoshimizu, T., Yasuhara, A., Sakagami, K., Okuyama, S., Nakanishi, S. & Nakazato, A. (2004) MGS0039: a potent and selective group II metabotropic glutamate receptor antagonist with antidepressant-like activity. *Neuropharmacology*, 46, 457–67.

167. Samadi, P., Rajput, A., Calon, F., Gregoire, L., Hornykiewicz, O., Rajput, A. H. & Di Paolo, T. (2009) Metabotropic glutamate receptor II in the brains of Parkinsonian patients. *J Neuropathol Exp Neurol*, 68, 374–82.

168. Moldrich, R. X., Talebi, A., Beart, P. M., Chapman, A. G. & Meldrum, B. S. (2001b) The mGlu(2/3) agonist 2R,4R-4-aminopyrrolidine-2,4-dicarboxylate, is anti- and proconvulsant in DBA/2 mice. *Neurosci Lett*, 299, 125–9.

169. Folbergrova, J., Haugvicova, R. & Mares, P. (2001) Attenuation of seizures induced by homocysteic acid in immature rats by metabotropic glutamate group II and group III receptor agonists. *Brain Res*, 908, 120–9.

170. Wilsch, V. W., Pidoplichko, V. I., Opitz, T., Shinozaki, H. & Reymann, K. G. (1994) Metabotropic glutamate receptor agonist DCG-IV as NMDA receptor agonist in immature rat hippocampal neurons. *Eur J Pharmacol*, 262, 287–91.

171. Moldrich, R. X., Jeffrey, M., Talebi, A., Beart, P. M., Chapman, A. G. & Meldrum, B. S. (2001a) Anti-epileptic activity of group II metabotropic glutamate receptor agonists (–)-2-oxa-4-aminobicyclo[3.1.0]hexane-4,6-dicarboxylate (LY379268) and (–)-2-thia-4-aminobicyclo[3.1.0]hexane-4,6-dicarboxylate (LY389795). *Neuropharmacology*, 41, 8–18.

172. Klodzinska, A., Bijak, M., Chojnacka-Wojcik, E., Kroczka, B., Swiader, M., Czuczwar, S. J. & Pilc, A. (2000) Roles of group II metabotropic glutamate receptors in modulation of seizure activity. *Naunyn Schmiedebergs Arch Pharmacol*, 361, 283–8.

173. Monn, J. A., Valli, M. J., Massey, S. M., Wright, R. A., Salhoff, C. R., Johnson, B. G., Howe, T., Alt, C. A., Rhodes, G. A., Robey, R. L., Griffey, K. R., Tizzano, J. P., Kallman, M. J., Helton, D. R. & Schoepp, D. D. (1997) Design, synthesis, and pharmacological characterization of (+)-2-aminobicyclo[3.1.0]hexane-2,6-dicarboxylic acid (LY354740): a potent, selective, and orally active group 2 metabotropic glutamate receptor agonist possessing anticonvulsant and anxiolytic properties. *J Med Chem*, 40, 528–37.

174. Tang, F. R., Chia, S. C., Chen, P. M., Gao, H., Lee, W. L., Yeo, T. S., Burgunder, J. M., Probst, A., Sim, M. K. & Ling, E. A. (2004) Metabotropic glutamate receptor 2/3 in the hippocampus of patients with mesial temporal lobe epilepsy, and of rats and mice after pilocarpine-induced status epilepticus. *Epilepsy Res*, 59, 167–80.

175. Pacheco Otalora, L. F., Couoh, J., Shigamoto, R., Zarei, M. M. & Garrido Sanabria, E. R. (2006) Abnormal mGluR2/3 expression in the perforant path termination zones and mossy fibers of chronically epileptic rats. *Brain Res*, 1098, 170–85.

176. Neugebauer, V., Chen, P. S. & Willis, W. D. (2000a) Groups II and III metabotropic glutamate receptors differentially modulate brief and prolonged nociception in primate STT cells. *J Neurophysiol*, 84, 2998–3009.

177. Simmons, R. M., Webster, A. A., Kalra, A. B. & Iyengar, S. (2002) Group II mGluR receptor agonists are effective in persistent and neuropathic pain models in rats. *Pharmacol Biochem Behav*, 73, 419–27.

178. Jones, C. K., Eberle, E. L., Peters, S. C., Monn, J. A. & Shannon, H. E. (2005) Analgesic effects of the selective group II (mGlu2/3) metabotropic glutamate receptor agonists

LY379268 and LY389795 in persistent and inflammatory pain models after acute and repeated dosing. *Neuropharmacology*, 49 Suppl 1, 206–18.

179. Du, J., Zhou, S. & Carlton, S. M. (2008) Group II metabotropic glutamate receptor activation attenuates peripheral sensitization in inflammatory states. *Neuroscience*, 154, 754–66.

180. Li, W. & Neugebauer, V. (2006) Differential changes of group II and group III mGluR function in central amygdala neurons in a model of arthritic pain. *J Neurophysiol*, 96, 1803–15.

181. Jang, J. H., Kim, D. W., Sang Nam, T., Se Paik, K. & Leem, J. W. (2004) Peripheral glutamate receptors contribute to mechanical hyperalgesia in a neuropathic pain model of the rat. *Neuroscience*, 128, 169–76.

182. Dolan, S., Kelly, J. G., Monteiro, A. M. & Nolan, A. M. (2003) Up-regulation of metabotropic glutamate receptor subtypes 3 and 5 in spinal cord in a clinical model of persistent inflammation and hyperalgesia. *Pain*, 106, 501–12.

183. Flor, P. J., Van Der Putten, H., Ruegg, D., Lukic, S., Leonhardt, T., Bence, M., Sansig, G., Knopfel, T. & Kuhn, R. (1997) A novel splice variant of a metabotropic glutamate receptor, human mGluR7b. *Neuropharmacology*, 36, 153–9.

184. Okamoto, N., Hori, S., Akazawa, C., Hayashi, Y., Shigemoto, R., Mizuno, N. & Nakanishi, S. (1994) Molecular characterization of a new metabotropic glutamate receptor mGluR7 coupled to inhibitory cyclic AMP signal transduction. *J Biol Chem*, 269, 1231–6.

185. Palucha, A., Tatarczynska, E., Branski, P., Szewczyk, B., Wieronska, J. M., Klak, K., Chojnacka-Wojcik, E., Nowak, G. & Pilc, A. (2004) Group III mGlu receptor agonists produce anxiolytic- and antidepressant-like effects after central administration in rats. *Neuropharmacology*, 46, 151–9.

186. Stachowicz, K., Chojnacka-Wojcik, E., Klak, K. & Pilc, A. (2007) Anxiolytic-like effect of group III mGlu receptor antagonist is serotonin-dependent. *Neuropharmacology*, 52, 306–12.

187. Stachowicz, K., Klak, K., Klodzinska, A., Chojnacka-Wojcik, E. & Pilc, A. (2004) Anxiolytic-like effects of PHCCC, an allosteric modulator of mGlu4 receptors, in rats. *Eur J Pharmacol*, 498, 153–6.

188. Rorick-Kehn, L. M., Hart, J. C. & Mckinzie, D. L. (2005) Pharmacological characterization of stress-induced hyperthermia in DBA/2 mice using metabotropic and ionotropic glutamate receptor ligands. *Psychopharmacology (Berl)*, 183, 226–40.

189. Fendt, M., Schmid, S., Thakker, D. R., Jacobson, L. H., Yamamoto, R., Mitsukawa, K., Maier, R., Natt, F., Husken, D., Kelly, P. H., Mcallister, K. H., Hoyer, D., Van Der Putten, H., Cryan, J. F. & Flor, P. J. (2008) mGluR7 facilitates extinction of aversive memories and controls amygdala plasticity. *Mol Psychiatry*, 13, 970–9.

190. Duvoisin, R. M., Zhang, C., Pfankuch, T. F., O'Connor, H., Gayet-Primo, J., Quraishi, S. & Raber, J. (2005) Increased measures of anxiety and weight gain in mice lacking the group III metabotropic glutamate receptor mGluR8. *Eur J Neurosci*, 22, 425–36.

191. Linden, A. M., Johnson, B. G., Peters, S. C., Shannon, H. E., Tian, M., Wang, Y., Yu, J. L., Koster, A., Baez, M. & Schoepp, D. D. (2002) Increased anxiety-related behavior in mice deficient for metabotropic glutamate 8 (mGlu8) receptor. *Neuropharmacology*, 43, 251–9.

192. Fendt, M., Burki, H., Imobersteg, S., Van Der Putten, H., Mcallister, K., Leslie, J. C., Shaw, D. & Holscher, C. (2009) The effect of mGlu(8) deficiency in animal models of psychiatric diseases. *Genes Brain Behav*.

193. Cryan, J. F., Kelly, P. H., Neijt, H. C., Sansig, G., Flor, P. J. & Van Der Putten, H. (2003) Antidepressant and anxiolytic-like effects in mice lacking the group III metabotropic glutamate receptor mGluR7. *Eur J Neurosci*, 17, 2409–17.

194. Stachowicz, K., Branski, P., Klak, K., Van Der Putten, H., Cryan, J. F., Flor, P. J. & Andrzej, P. (2008) Selective activation of metabotropic G-protein-coupled glutamate 7 receptor elicits anxiolytic-like effects in mice by modulating GABAergic neurotransmission. *Behav Pharmacol*, 19, 597–603.

195. Callaerts-Vegh, Z., Beckers, T., Ball, S. M., Baeyens, F., Callaerts, P. F., Cryan, J. F., Molnar, E. & D'Hooge R. (2006) Concomitant deficits in working memory and fear extinction are functionally dissociated from reduced anxiety in metabotropic glutamate receptor 7-deficient mice. *J Neurosci*, 26, 6573–82.

196. Altinbilek, B. & Manahan-Vaughan, D. (2007) Antagonism of group III metabotropic glutamate receptors results in impairment of LTD but not LTP in the hippocampal CA1 region, and prevents long-term spatial memory. *Eur J Neurosci*, 26, 1166–72.

197. Klausnitzer, J., Kulla, A. & Manahan-Vaughan, D. (2004) Role of the group III metabotropic glutamate receptor in LTP, depotentiation and LTD in dentate gyrus of freely moving rats. *Neuropharmacology*, 46, 160–70.

198. Gerlai, R., Roder, J. C. & Hampson, D. R. (1998) Altered spatial learning and memory in mice lacking the mGluR4 subtype of metabotropic glutamate receptor. *Behav Neurosci*, 112, 525–32.

199. Masugi, M., Yokoi, M., Shigemoto, R., Muguruma, K., Watanabe, Y., Sansig, G., Van Der Putten, H. & Nakanishi, S. (1999) Metabotropic glutamate receptor subtype 7 ablation causes deficit in fear response and conditioned taste aversion. *J Neurosci*, 19, 955–63.

200. Robbins, M. J., Starr, K. R., Honey, A., Soffin, E. M., Rourke, C., Jones, G. A., Kelly, F. M., Strum, J., Melarange, R. A., Harris, A. J., Rocheville, M., Rupniak, T., Murdock, P. R., Jones, D. N., Kew, J. N. & Maycox, P. R. (2007) Evaluation of the mGlu8 receptor as a putative therapeutic target in schizophrenia. *Brain Res*, 1152, 215–27.

201. Takaki, H., Kikuta, R., Shibata, H., Ninomiya, H., Tashiro, N. & Fukumaki, Y. (2004) Positive associations of polymorphisms in the metabotropic glutamate receptor type 8 gene (GRM8) with schizophrenia. *Am J Med Genet B Neuropsychiatr Genet*, 128B, 6–14.

202. Ohtsuki, T., Koga, M., Ishiguro, H., Horiuchi, Y., Arai, M., Niizato, K., Itokawa, M., Inada, T., Iwata, N., Iritani, S., Ozaki, N., Kunugi, H., Ujike, H., Watanabe, Y., Someya, T. & Arinami, T. (2008) A polymorphism of the metabotropic glutamate receptor mGluR7 (GRM7) gene is associated with schizophrenia. *Schizophr Res*, 101, 9–16.

203. Klak, K., Palucha, A., Branski, P., Sowa, M. & Pilc, A. (2007) Combined administration of PHCCC, a positive allosteric modulator of mGlu4 receptors and ACPT-I, mGlu III receptor agonist evokes antidepressant-like effects in rats. *Amino Acids*, 32, 169–72.

204. Tatarczynska, E., Palucha, A., Szewczyk, B., Chojnacka-Wojcik, E., Wieronska, J. & Pilc, A. (2002) Anxiolytic- and antidepressant-like effects of group III metabotropic glutamate agonist (1S,3R,4S)-1-aminocyclopentane-1,3,4-tricarboxylic acid (ACPT-I) in rats. *Pol J Pharmacol*, 54, 707–10.

205. Mitsukawa, K., Mombereau, C., Lotscher, E., Uzunov, D. P., Van Der Putten, H., Flor, P. J. & Cryan, J. F. (2006) Metabotropic glutamate receptor subtype 7 ablation causes dysregulation of the HPA axis and increases hippocampal BDNF protein levels: implications for stress-related psychiatric disorders. *Neuropsychopharmacology*, 31, 1112–22.

206. Wieronska, J. M., Klak, K., Palucha, A., Branski, P. & Pilc, A. (2007) Citalopram influences mGlu7, but not mGlu4 receptors' expression in the rat brain hippocampus and cortex. *Brain Res*, 1184, 88–95.

207. Palucha, A., Klak, K., Branski, P., Van Der Putten, H., Flor, P. J. & Pilc, A. (2007) Activation of the mGlu7 receptor elicits antidepressant-like effects in mice. *Psychopharmacology (Berl)*, 194, 555–62.

208. Bradley, S. R., Standaert, D. G., Levey, A. I. & Conn, P. J. (1999) Distribution of group III mGluRs in rat basal ganglia with subtype-specific antibodies. *Ann N Y Acad Sci*, 868, 531–4.

209. Lopez, S., Turle-Lorenzo, N., Acher, F., De Leonibus, E., Mele, A. & Amalric, M. (2007) Targeting group III metabotropic glutamate receptors produces complex behavioral effects in rodent models of Parkinson's disease. *J Neurosci*, 27, 6701–11.

210. Marino, M. J., Williams, D. L., JR., O'Brien, J. A., Valenti, O., Mcdonald, T. P., Clements, M. K., Wang, R., Dilella, A. G., Hess, J. F., Kinney, G. G. & Conn, P. J. (2003) Allosteric modulation of group III metabotropic glutamate receptor 4: a potential approach to Parkinson's disease treatment. *Proc Natl Acad Sci U S A*, 100, 13668–73.

211. Battaglia, G., Busceti, C. L., Molinaro, G., Biagioni, F., Traficante, A., Nicoletti, F. & Bruno, V. (2006) Pharmacological activation of mGlu4 metabotropic glutamate receptors reduces nigrostriatal degeneration in mice treated with 1-methyl-4-phenyl-1,2,3,6-tetrahydropyridine. *J Neurosci*, 26, 7222–9.

212. Vernon, A. C., Palmer, S., Datla, K. P., Zbarsky, V., Croucher, M. J. & Dexter, D. T. (2005) Neuroprotective effects of metabotropic glutamate receptor ligands in a 6-hydroxydopamine rodent model of Parkinson's disease. *Eur J Neurosci*, 22, 1799–806.

213. Feeley Kearney, J. A. & Albin, R. L. (2003) mGluRs: a target for pharmacotherapy in Parkinson disease. *Exp Neurol*, 184 Suppl 1, S30–6.

214. Ghauri, M., Chapman, A. G. & Meldrum, B. S. (1996) Convulsant and anticonvulsant actions of agonists and antagonists of group III mGluRs. *Neuroreport*, 7, 1469–74.

215. Chapman, A. G., Talebi, A., Yip, P. K. & Meldrum, B. S. (2001) Anticonvulsant activity of a mGlu(4alpha) receptor selective agonist, (1S,3R,4S)-1-aminocyclopentane-1-,2,4-tricarboxylic acid. *Eur J Pharmacol*, 424, 107–13.

216. Chen, J., Larionov, S., Pitsch, J., Hoerold, N., Ullmann, C., Elger, C. E., Schramm, J. & Becker, A. J. (2005) Expression analysis of metabotropic glutamate receptors I and III in mouse strains with different susceptibility to experimental temporal lobe epilepsy. *Neurosci Lett*, 375, 192–7.

217. Tang, F. R. & Lee, W. L. (2001) Expression of the group II and III metabotropic glutamate receptors in the hippocampus of patients with mesial temporal lobe epilepsy. *J Neurocytol*, 30, 137–43.

218. Soliman, A. C., Yu, J. S. & Coderre, T. J. (2005) mGlu and NMDA receptor contributions to capsaicin-induced thermal and mechanical hypersensitivity. *Neuropharmacology*, 48, 325–32.

219. Goudet, C., Chapuy, E., Alloui, A., Acher, F., Pin, J. P. & Eschalier, A. (2008) Group III metabotropic glutamate receptors inhibit hyperalgesia in animal models of inflammation and neuropathic pain. *Pain*, 137, 112–24.

220. Mills, C. D., Johnson, K. M. & Hulsebosch, C. E. (2002) Role of group II and group III metabotropic glutamate receptors in spinal cord injury. *Exp Neurol*, 173, 153–67.

221. Maione, S., Oliva, P., Marabese, I., Palazzo, E., Rossi, F., Berrino, L. & Filippelli, A. (2000) Periaqueductal gray matter metabotropic glutamate receptors modulate formalin-induced nociception. *Pain*, 85, 183–9.

222. Zhang, H. M., Chen, S. R. & Pan, H. L. (2009) Effects of activation of group III metabotropic glutamate receptors on spinal synaptic transmission in a rat model of neuropathic pain. *Neuroscience*, 158, 875–84.

223. Palazzo, E., Fu, Y., Ji, G., Maione, S. & Neugebauer, V. (2008) Group III mGluR7 and mGluR8 in the amygdala differentially modulate nocifensive and affective pain behaviors. *Neuropharmacology*, 55, 537–45.

224. Osikowicz, M., Mika, J., Makuch, W. & Przewlocka, B. (2008) Glutamate receptor ligands attenuate allodynia and hyperalgesia and potentiate morphine effects in a mouse model of neuropathic pain. *Pain*, 139, 117–26.

225. Gerber, U. (2003) Metabotropic glutamate receptors in vertebrate retina. *Doc Ophthalmol*, 106, 83–7.

226. Brandstatter, J. H. (2002) Glutamate receptors in the retina: the molecular substrate for visual signal processing. *Curr Eye Res*, 25, 327–31.

227. Masu, M., Iwakabe, H., Tagawa, Y., Miyoshi, T., Yamashita, M., Fukuda, Y., Sasaki, H., Hiroi, K., Nakamura, Y., Shigemoto, R. & ET AL. (1995) Specific deficit of the ON response in visual transmission by targeted disruption of the mGluR6 gene. *Cell*, 80, 757–65.

228. Takao, M., Morigiwa, K., Sasaki, H., Miyoshi, T., Shima, T., Nakanishi, S., Nagai, K. & Fukuda, Y. (2000) Impaired behavioral suppression by light in metabotropic glutamate receptor subtype 6-deficient mice. *Neuroscience*, 97, 779–87.

229. Zeitz, C., Van Genderen, M., Neidhardt, J., Luhmann, U. F., Hoeben, F., Forster, U., Wycisk, K., Matyas, G., Hoyng, C. B., Riemslag, F., Meire, F., Cremers, F. P. & Berger, W. (2005) Mutations in GRM6 cause autosomal recessive congenital stationary night blindness with a distinctive scotopic 15-Hz flicker electroretinogram. *Invest Ophthalmol Vis Sci*, 46, 4328–35.

230. Dryja, T. P., Mcgee, T. L., Berson, E. L., Fishman, G. A., Sandberg, M. A., Alexander, K. R., Derlacki, D. J. & Rajagopalan, A. S. (2005) Night blindness and abnormal cone electroretinogram ON responses in patients with mutations in the GRM6 gene encoding mGluR6. *Proc Natl Acad Sci U S A*, 102, 4884–9.

231. Dyka, F. M., May, C. A. & Enz, R. (2004) Metabotropic glutamate receptors are differentially regulated under elevated intraocular pressure. *J Neurochem*, 90, 190–202.

232. Spooren, W. P., Vassout, A., Neijt, H. C., Kuhn, R., Gasparini, F., Roux, S., Porsolt, R. D. & Gentsch, C. 2000b. Anxiolytic-like effects of the prototypical metabotropic glutamate receptor 5 antagonist 2-methyl-6-(phenylethynyl)pyridine in rodents. *J Pharmacol Exp Ther*, 295, 1267–75.

233. Spooren, W. P., Schoeffter, P., Gasparini, F., Kuhn, R. & Gentsch, C. 2002. Pharmacological and endocrinological characterisation of stress-induced hyperthermia in singly housed mice using classical and candidate anxiolytics (LY314582, MPEP and NKP608). *Eur J Pharmacol*, 435, 161–70.

234. Spooren, W. P., Gasparini, F., Bergmann, R. & Kuhn, R. 2000a. Effects of the prototypical mGlu(5) receptor antagonist 2-methyl-6-(phenylethynyl)-pyridine on rotarod, locomotor activity and rotational responses in unilateral 6-OHDA-lesioned rats. *Eur J Pharmacol*, 406, 403–10.

235. Chapman, A. G., Nanan, K., Williams, M. & Meldrum, B. S. (2000) Anticonvulsant activity of two metabotropic glutamate group I antagonists selective for the mGlu5 receptor: 2-methyl-6-(phenylethynyl)-pyridine (MPEP), and (E)-6-methyl-2-styryl-pyridine (SIB 1893). *Neuropharmacology*, 39, 1567–74.

236. Zhang, L., Lu, Y., Chen, Y. & Westlund, K. N. 2002. Group I metabotropic glutamate receptor antagonists block secondary thermal hyperalgesia in rats with knee joint inflammation. *J Pharmacol Exp Ther*, 300, 149–56.

237. Bhave, G., Karim, F., Carlton, S. M. & Gereau, R. W. T. 2001. Peripheral group I metabotropic glutamate receptors modulate nociception in mice. *Nat Neurosci*, 4, 417–23.

238. Ansah, O. B., Goncalves, L., Almeida, A. & Pertovaara, A. 2009. Enhanced pronociception by amygdaloid group I metabotropic glutamate receptors in nerve-injured animals. *Exp Neurol*, 216, 66–74.

239. Han, J. S. & Neugebauer, V. 2005. mGluR1 and mGluR5 antagonists in the amygdala inhibit different components of audible and ultrasonic vocalizations in a model of arthritic pain. *Pain*, 113, 211–22.

240. Bruijnzeel, A. W., Stam, R. & Wiegant, V. M. 2001. LY354740 attenuates the expression of long-term behavioral sensitization induced by a single session of foot shocks. *Eur J Pharmacol*, 426, 77–80.

241. Walker, D. L. & Davis, M. 2002. The role of amygdala glutamate receptors in fear learning, fear-potentiated startle, and extinction. *Pharmacol Biochem Behav*, 71, 379–92.

242. Smialowska, M., Wieronska, J. M., Domin, H. & Zieba, B. 2007. The effect of intra-hippocampal injection of group II and III metobotropic glutamate receptor agonists on anxiety; the role of neuropeptide Y. *Neuropsychopharmacology*, 32, 1242–50.

243. Mathis, C. & Ungerer, A. 1999. The retention deficit induced by (RS)-alpha-methyl-4-carboxyphenylglycine in a lever-press learning task is blocked by selective agonists of either group I or group II metabotropic glutamate receptors. *Exp Brain Res*, 129, 147–55.

244. Cartmell, J., Monn, J. A. & Schoepp, D. D. 2000. Attenuation of specific PCP-evoked behaviors by the potent mGlu2/3 receptor agonist, LY379268 and comparison with the atypical antipsychotic, clozapine. *Psychopharmacology (Berl)*, 148, 423–9.

245. Grauer, S. M. & Marquis, K. L. 1999. Intracerebral administration of metabotropic glutamate receptor agonists disrupts prepulse inhibition of acoustic startle in Sprague-Dawley rats. *Psychopharmacology (Berl)*, 141, 405–12.

246. Bradford, H. F. 1998. Specific group II metabotropic glutamate receptor activation inhibits the development of kindled epilepsy in rats. *Brain Res*, 787, 286 Ossowska, K., Pietraszek, M., Wardas, J., Nowak, G., Zajaczkowski, W., Wolfarth, S. & Pilc, A. 2000. The role of glutamate receptors in antipsychotic drug action. *Amino Acids*, 19, 87–94.

247. Schreiber, R., Lowe, D., Voerste, A. & De Vry, J. 2000. LY354740 affects startle responding but not sensorimotor gating or discriminative effects of phencyclidine. *Eur J Pharmacol*, 388, R3–4.

248. Schlumberger, C., Schafer, D., Barberi, C., More, L., Nagel, J., Pietraszek, M., Schmidt, W. J. & Danysz, W. 2009. Effects of a metabotropic glutamate receptor group II agonist LY354740 in animal models of positive schizophrenia symptoms and cognition. *Behav Pharmacol*, 20, 56–66.

249. Attwell, P. J., Koumentaki, A., Abdul-Ghani, A. S., Croucher, M. J. & –91.

250. Senkowska, A. & Ossowska, K. (2003) Role of metabotropic glutamate receptors in animal models of Parkinson's disease. *Pol J Pharmacol*, 55, 935–50.

251. Mela, F., Marti, M., Dekundy, A., Danysz, W., Morari, M. & Cenci, M. A. 2007. Antagonism of metabotropic glutamate receptor type 5 attenuates l-DOPA-induced dyskinesia and its molecular and neurochemical correlates in a rat model of Parkinson's disease. *J Neurochem*, 101, 483–97.

252. Simmons, R. M., Webster, A. A., Kalra, A. B. & Iyengar, S. 2002. Group II mGluR receptor agonists are effective in persistent and neuropathic pain models in rats. *Pharmacol Biochem Behav*, 73, 419–27.

17 S1P receptor agonists, a novel generation of immunosuppressants

Rosa López Almagro, Gema Tarrasón, and Nuria Godessart

SPHINGOSINE 1-PHOSPHATE

Sphingosine 1-phosphate (S1P) is a bioactive sphingolipid that regulates different cellular processes including proliferation, survival, migration, cytoskeletal organization, and differentiation. S1P is present in body fluids and tissues at different concentrations and its levels are regulated by the balance between its synthesis and degradation. Some intermediates of this metabolic pathway, such as ceramide, also play a role in diverse physiological and pathological conditions.

S1P can function intracellularly as a second messenger or extracellularly by binding to S1P receptors in an autocrine or paracrine manner. To date, five S1P receptors have been identified. The wide expression and coupling promiscuity of these receptors explains the multiple biological actions of S1P, most of them elucidated from gene deletion and chemical biology studies.

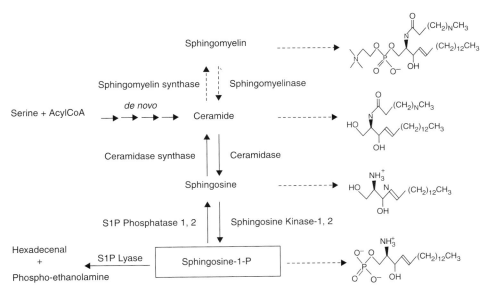

Figure 17-1: Metabolic pathways of S1P.

Biosynthesis and metabolism of S1P

Sphingolipids are structural components of all membranes in eukaryotic cells. This class of lipids presents a sphingoid longchain base (sphingosine) backbone that is linked to a fatty acid molecule through an amide bond. The metabolism of sphingolipids (Figure 17.1) produces bioactive molecules such as ceramide, ceramide-1-phosphate, sphingosine, and S1P that play an important role in both homeostasis and disease.[1] These molecules are produced in a multi-enzymatic pathway in which activation of sphingomyelinase by stress factors and proinflammatory cytokines produces ceramide from sphingomyelin. Ceramide can also be produced by a *de novo* synthesis pathway initiated by a condensation of serine and palmitoyl-CoA catalyzed by serine palmitoyltransferase.

Deacylation of ceramide by ceramidases yields sphingosine, which can be phosphorylated by two ubiquitously expressed protein kinases, sphingosine kinase-1 and 2 (SphK1, Sphk2), to form S1P.[2] As many other signaling molecules, there is a balance between S1P synthesis and degradation. Following S1P synthesis, this sphingolipid can be accumulated in the intracellular space or secreted to the extracellular media, but it can also be reversibly dephosphorylated to sphingosine by nonspecific as well as specific phosphatases, namely sphingosine phosphate phosphatase-1 and 2 (SPP1 and SPP2). S1P is irreversibly degraded to hexadecanal and phosphoethanolamine by S1P lyase.[3]

Sphingolipid rheostat and sphingosine kinases

Ceramide, sphingosine, and S1P are readily interconvertible, and each molecule regulates cell growth and survival in a different manner . This has led to the concept of sphingolipid rheostat. Ceramide and sphingosine are important regulators of stress responses, typically inducing apoptosis and cell-cycle arrest. S1P has opposing effects, promoting cell growth and survival. Suppression of apoptosis seems to be associated with an increase in S1P levels and a decrease in ceramide. Therefore, the balance between S1P and its precursors, each of them controlling opposed signaling pathways, would determine cell fate.[4]

Sphingosine kinases converting sphingosine to S1P play a crucial role in regulating this balance. They are activated by cytokines, immunoglobulin receptors, G protein-coupled receptors (GPCR), and receptor tyrosine kinases, among other stimuli.[5] Although both SphK1 and 2 produce S1P, they show significant differences in expression during embryonic development, subcellular localization, and catalytic properties. Additionally, they appear to have opposing roles regarding cell fate, promoting cell growth in the case of SphK1 and apoptosis by SphK2.[6] They also regulate ceramide biosynthesis in different ways.[7] However, despite these opposing roles, both isoforms can have compensatory and overlapping functions. The single SphK1 or SphK2 knock-out mice have normal S1P levels and develop and reproduce normally, whereas the double knockout animals are totally devoid of S1P and is embryonically lethal due to severe neurogenic and angiogenic defects.[8] SphK1 also plays an important role in tumorigenesis. It is overexpressed in many types of cancers, deregulating the sphingolipid rheostat, and it seems to be involved in tumor angiogenesis and resistance to radio- and chemotherapy.[9]

S1P gradient and cellular sources of S1P

In mammals, a large concentration gradient of S1P exists between vascular and extravascular compartments. This gradient is extremely important for trafficking and maturation of hematopoietic cells, as well as for regulating homing and egress of immune cells in lymphoid organs.[10]

S1P levels in plasma are in the range of 100 nM to 1 µM. Levels in lymph oscillate between 30 and 300 nM, whereas those found in tissues are about one-thousand times lower than in plasma. Although the S1P levels in plasma can reach the low micromolar range, the bioactive concentration of S1P is lower, being close to the K_d for its receptors. S1P in plasma is known to be associated with albumin and lipoproteins. It seems that the fraction bound to albumin and low-density lipoproteins (LDL) may be inactive, but the S1P bound to high-density lipoproteins (HDL) may be biologically active.[11]

There are multiple sources of S1P that contribute to maintaining the high levels in blood. For many years, it was assumed that platelets were the principal source of S1P in plasma due to the fact that upon activation they release S1P in large amounts, and because they show a high SphK1 activity and no S1P lyase activity.[12] Recently, several studies have pointed at erythrocytes as the main source of S1P in blood.[10] These cells have less SphK1 activity than platelets, but they lack S1P lyase and S1P phosphatases.[10,13] Erythrocytes can store S1P produced from other tissues and release it when necessary without previous stimulation. It has been proposed that activated platelets would have a role in pathologies like atherosclerosis, producing a local increase in S1P levels and regulating processes like wound healing and inflammation.[14] Other sources of S1P have been described, like the vascular endothelium, particularly under shear stress,[15] and mast cells, under IgE dependent activation.[16] S1P produced by vascular endothelium could be the main source of lymph S1P.[10]

All these cells that produce S1P need a mechanism to secrete it through the plasma membrane, because S1P has a polar head group. Several reports suggest that the ATP-binding cassette (ABC) family of transporters are involved in secreting these lipids across the plasma membrane. Different ABC members seem to mediate this active transport system in the different cell types that produce S1P; ABCC1 in mast cells,[17] ABCA1 in platelets,[18] and ABCA1 and ABCC1 in endothelial cells.[19]

Several studies have suggested that S1P found in extracellular space is not only derived from intracellular production and subsequent secretion. Biosynthetic enzymes of the S1P metabolism pathway like shpingomyelinase, ceramidase, and SphKs have been reported to exist in the extracellular space.[20–22]

Intracellular and extracellular actions of S1P

S1P plays different roles in many biological processes, acting intracellularly as a second messenger or in an extracellular mode through activation of S1P receptors at the cellular surface in an autocrine or paracrine manner. S1P synthesized inside the cell is transported to the extracellular milieu where it interacts with its receptors at the cellular surface, a mode of action that has been described as "inside-out" signaling.[5]

Several studies show that S1P acting in an intracellular mode can participate in processes like mitogenesis, apoptosis suppression, and calcium mobilization independently of S1P receptors.[23] However, the intracellular targets mediating these effects have not been identified. Given that intracellular S1P is rapidly converted to glycerophospholipid precursors by the action of S1P lyase, it has been proposed that the intracellular actions of S1P are elicited by S1P metabolites instead of S1P itself.[24] Despite the debate about possible intracellular actions of S1P, there is no doubt that the most important roles of S1P are mediated by its binding to specific cell surface receptors.

SPHINGOSINE 1-PHOSPHATE RECEPTORS

Since the discovery of the S1P receptors in the late 1990s, the S1P receptor signaling system has been subjected to intense investigation. The biological effects of S1P are mediated by five high-affinity GPCR, namely S1P1 to S1P5. The differential temporal and spatial pattern of receptor expression, as well as the coupling to different G proteins, explains the variety of responses of S1P (summarized in Figure 17.2 and Table 17.1).

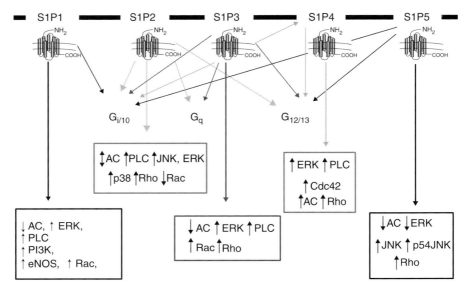

Figure 17-2: G proteins coupled to the S1P receptors and downstream signaling pathways. *See color plate 15.*

S1P1 receptor

S1P1 was the first member of the S1P receptor family to be cloned. It was identified as a gene, edg-1, up-regulated during endothelial cell differentiation in vitro.[25] S1P1 receptor shows a widespread expression, with the highest levels detected in brain, lung, spleen, heart, and liver. It is also expressed in a variety of immune system cell lineages like T cells, B cells, dendritic cells, macrophages, and natural killer (NK) cells, as well as in endothelial cells of most tissues.

S1P1 couples exclusively to the pertussis toxin (PTX)-sensitive G_i pathway.[26] In hematopietic cells (such as lymphocytes), however, $S1P_1$ couples also to G_{16}, a promiscuous member of the G_q protein family. The downstream signaling pathways activated include adenylyl cyclase (AC) inhibition, activation of the small guanosine triphosphatase (GTPase) Ras, and the extracellular signal-regulated kinase (ERK) to promote proliferation. Moreover, an activation of phospholipase C (PLC) by $\beta\gamma$ subunits has been observed that leads to an increase in intracellular calcium also has been observed. This calcium release has a role in some biological responses, such as cytokine secretion. S1P1 also leads to a PI3 kinase-dependent activation of Akt, promoting survival, and a PI3K-dependent activation of Rac, promoting cell migration. Activated S1P1 transactivates growth factor receptor tyrosine kinases (RTKs), such as PDGF receptor.[27] Conversely, after growth factor binding to PDGF receptors SphK1 is activated leading to S1P production, which in turn activates S1P1-Rac signaling to promote cell migration.

The phenotype of S1P1 knockout indicates the important role of this receptor in vascular maturation and neurogenesis.[28] S1P1 knockout embryos die between E12.5 and E14.5 due to massive hemorrhage. Although vasculogenesis and angiogenesis appeared normal in the embryos, vascular maturation was incomplete due to a deficiency of vascular smooth muscle cells (VSMC). Specific deletion of

Table 17-1. S1P Receptors: Knockout Phenotypes and Biological Functions

Receptor	Knock-out phenotype	Main biological functions reported from in vitro and in vivo studies
S1P1	Embryonic lethality: defects in vascular maturation and neurogenesis Failure of null thymocyte egress in radiation chimeras	Angiogenesis Vascular maturation and tone Increase in endothelial barrier integrity Lymphocyte migration Neurogenesis Astrocyte migration
S1P2	Slightly reduced viability Reduced litter size Seizures, deafness Neuronal hyperexcitability Vascular dysfunction in adult mice	Adherens junction disruption Hearing Neuronal excitation Inhibition of cell migration Mast cell degranulation Angiogenesis Vasoconstriction Wound healing in liver injury
S1P3	Slightly reduced viability Reduced litter size Disruption of alveolar epithelial junctions Worsens sepsis outcome	Bradycardia Increased vascular permeability Cardioprotection Vascular smooth muscle contraction
S1P4	Embryonic lethality Behavioral changes in heterozygotes (decreased prepulse inhibition in heterozygotes during startle testing)	T cell proliferation Cytokine secretion
S1P5	Normal phenotype Defect in trafficking of NK cells	Inhibition of oligodendrocyte progenitor migration Oligodendrocyte survival NK trafficking

S1P1 in endothelial cells showed that S1P1 expression is required for proper vessel maturation and migration of VSMC.[29] The phenotype of the S1P1-deficient mice is similar to that of double SphK1 and SphK2 knockout mice.[8]

The important role of S1P1 in immune cell trafficking has been demonstrated in studies in mice with conditional deletion of S1P1 in T cells and by injecting fetal liver cells derived from S1P1deficient mice.[30,31] These mice showed a block in the egress of mature T lymphocytes from the thymus and S1P1-deficient T cells were sequestered into secondary lymphoid organs, disappearing from blood and lymph. On the other hand, the overexpression of S1P1 receptor in T cells translates into a potentiated egress from secondary lymphoid organs and major distribution into blood.[32] Loss of S1P1 in B cells results in a less pronounced homing to lymphoid organs than in T cells but induces a reduced recirculation to blood and a decreased homing to the bone marrow.[33]

A crucial role of the S1P/S1P1 axis in homeostasis is the tightening of the endothelial barrier through the maintenance of adherens junctions. S1P induced recruitment of S1P1 to caveolin-enriched domains promotes PI3 kinase mediated

Tiam/Rac 1 activation required for cortical actin rearrangement and endothelium barrier enhancement.[34]

Recently, using S1P1 silencing by lentiviral infection, an interesting role in neural stem cell migration toward an injured brain area has been described. This biological action suggests a possible therapeutic potential for this receptor in spinal cord injury.[35]

S1P2 and S1P3

Both S1P2 and S1P3 receptors are widely expressed, with S1P2 being more abundant in lung and heart and S1P3 in lung, heart, kidney, spleen, and pancreas.[36] They couple to G_i, G_q, and $G_{12/13}$ but with different preferences, leading to different biological responses.[37] Both activate ERK mitogen-activated protein (MAP) kinase, although S1P3 more strongly, and they induce calcium mobilization in both PTX-sensitive and PTX-insensitive manners, although this is a preferred pathway for S1P3.[36] They also activate Rho by a $G_{12/13}$-dependent mechanism, resulting in stress fiber formation, cell rounding, and activation of a serum response element promoter.[38,39] Analysis of S1P signaling in S1P2-null mouse embryonic fibroblast (MEFs) revealed a defect in Rho activation, which suggests that S1P2 couples preferentially to $G_{12/13}$.[36] This would explain why, whereas S1P1 and S1P3 activate Rac and promote migration, S1P2 inhibits Rac in a Rho-dependent manner and inhibits cell migration. Activation of adenylyl cyclase independent of G_s, and of the MAP kinases p38 and JNK by S1P2 have also been reported.

Similarly to S1P1, S1P2 and S1P3 participate in the cross-talk with PDGF receptors, but with opposite effects. While S1P3 is involved in PDGF receptor activation, S1P2 seems to negatively regulate this receptor, inhibiting its proliferative effects, migration, and SphK1 induction.[40,41]

S1P2 and S1P3 null mice are viable and have no obvious phenotypic defects, except for slightly smaller litter size.[36,42-44] This suggests some kind of compensation between receptors expression and signaling. However, in some strains of mice lacking S1P2, occasional fatal seizure episodes have been reported.[44] Loss of S1P2 leads to an increase in the excitability of neocortical pyramidal neurons demonstrating the important role of S1P2 in development of neuronal excitability. S1P2 also contributes to the development of auditory and vestibular systems because its deficiency results in deafness. In addition, S1P2 seems to have an important and unique role in wound healing during liver injury, because S1P2-null mice presented a reduced accumulation of hepatic myofibroblasts.[45] Other studies with RNA_i in VSMC showed that S1P2 is also involved in VSMC contraction.[46]

S1P1/S1P2-double deletion in mice leads to an exacerbation of the S1P1-null phenotype, with a less mature vasculature and an earlier embryonic lethality. S1P1/S1P2/S1P3 triple knock-out mice show a slightly more severe phenotype than the double S1P1/S1P2 knock-out animals, indicating a cooperative action of these receptors in vascular development.[43]

Deletion of S1P3 has been useful to elucidate its role in the bradycardia observed after in vivo administration of the nonselective S1P receptor agonist FTY720 and to indicate an important role of S1P3 in vascular tone regulation and cardioprotection in rodents.[47,48] This receptor also has pulmonary functions in adult mice because its deletion leads to lung leakage as a result of increased permeability derived from tight junction opening.[49] Therefore, S1P1 and S1P3 seem to have opposing effects on pulmonary epithelial and endothelial barriers. Recently, studies with S1P3-null mice and PAR1-deficient dendritic cells (DCs) have revealed that S1P3 acts as a downstream component in PAR1-mediated septic lethality.[50]

S1P4 and S1P5

The S1P4 and S1P5 receptors show a more restricted expression pattern. S1P4 expression is limited to lymphoid cells and tissues and to airway smooth muscle cells. S1P5 is expressed in the white matter of the brain, mainly in oligodendrocytes, and also in skin, spleen, and peripheral blood leukocytes. S1P signaling through S1P4 and S1P5 is mediated by coupling to G_i and $G_{12/13}$.[51] S1P4 activates ERK, PLC, and the GTPase Cdc42 via G_i coupling and activates Rho via the $G_{12/13}$ pathway, inducing cytoskeletal changes. S1P4 has also been described to activate adenylyl cyclase, but the G protein involved has not been determined. S1P5 inhibits adenylyl cyclase and activates Akt in a PTX-sensitive manner. A PTX-insensitive ERK inhibition and JNK activation has also been described.[52] Via $G_{12/13}$ S1P5 can activate Rho kinase.

Homozygous S1P4 null mice die between E10.5 and E12.5, with embryos appearing partially resorbed by E10.5.[53] Heterozygous mice showed a nervous system phenotype, with a decrease in prepulse inhibition (a weaker prestimulus, prepulse inhibits the reaction of an organism to a subsequent strong startling stimulus, named pulse). This effect could be a consequence of receptor compensation, because S1P4 does not show a pattern of expression in the nervous system.

S1P5-deficient mice do not present an obvious phenotype.[54] Although the expression of S1P5 in oligodendrocytes suggests a role of this receptor in myelination, S1P5-null mice revealed immature oligodendrocytes, but no deficits in myelination were observed. Oligodendrocytes showed a defect in migration when S1P5 was knocked down by siRNA.[55] S1P5 is also highly expressed in NK cells, and its deletion alters the trafficking of these cells in normal and pathological conditions.[56]

S1P RECEPTORS IN IMMUNE FUNCTION

In the last decade, the role of S1P and their receptors in different aspects of the immune response have been elucidated. Cells from both the innate and the adaptive immune systems express S1P receptors, with S1P1 and S1P4 being predominantly expressed in lymphocytes, and S1P1 and S1P2 in cells of the innate immune system.[57] Despite the fact that S1P4 is also expressed in all lymphocyte subsets, all attention has been focused on S1P1. This receptor has been reported to be involved in lymphocyte trafficking during immunosurveillance, control of the immune response, and T cell lineage differentiation.[57]

Lymphocyte trafficking and imunosurveillance

The circulation of lymphocytes between blood and secondary lymphoid tissues plays a role in the establishment of the immune response to foreign antigens and is crucial for generating immunity. Mature naïve B and T lymphocytes travel continuously between different lymph nodes (LN) and other secondary lymphoid organs, surveying for their specific antigen.

Naïve lymphocytes migrate from the blood stream to the LN by extravasation through high endothelial venules (HEV), enter lymph nodes and migrate to the specific compartments to allow the appropriate cell interactions to occur. The role that chemokines play in these two processes is well established.[58] Cells that do not encounter antigen exit LNs via efferent lymphatics and return to circulation through the thoracic duct. When naïve T cells find the appropriate antigen, they become activated, proliferate, and migrate to peripheral tissues.[59]

The mechanisms regulating lymphocytes egress from LN have not been well understood. In the 1990s, Chaffin and colleagues[60] pointed at the requirement of a Gα$_i$-coupled receptor for lymphocyte egress. The involvement of a specific GPCR came from the discovery that a small molecule with immunosuppressant properties, FTY720, caused lymphopenia.[61] The lymphopenia was due to a block in lymphocyte egress from LNs, as demonstrated by histology of lymph node medullary sinuses of FTY-treated mice that were emptied of lymphocytes.[62] Key observation to the understanding of the mode of action of FTY720 was its structural similarity to the lipid sphingosine. Following in vivo administration, FTY720 was phosphorylated by SphK2 and converted into FTY720-P, a potent agonist of S1P1, S1P3, S1P4, and S1P5 receptors.[62]

The specific S1P receptor involved in lymphocyte trafficking was clarified by Matloubian and colleagues,[31] showing that mice whose hematopoietic cells lacked S1P1 receptor had no T cells in the periphery because mature T cells were unable to exit the thymus. Adaptive transfer of S1P1 null lymphocytes to wild-type mice revealed a role for S1P1 in lymphoid organ egress, similar to the phenotype observed upon FTY720 treatment of mice.[31] S1P1 would thus be involved in the egress of mature T lymphocytes from thymus and of naïve T and B cells from LN following the S1P gradient (see the subchapter on S1P gradient and cellular sources) that exist from lymphoid organs to blood.[63]

Mechanism of lymphopenia evoked by S1P1 agonists

Whereas S1P in blood facilitates the egress of lymphocytes during immunosurveillance, synthetic agonists of S1P1 have shown to induce peripheral blood lymphopenia following in vivo administration.[47,62,64] Nevertheless, the exact mechanism leading to lymphopenia is still unknown. There are currently two hypotheses to explain lymphopenia, which differ in the target cell and in the nature of the effect evoked on the receptor.[65,66] The first hypothesis states that S1P1 agonists bind S1P1 expressed on lymphocytes and cause receptor internalization, preventing the cells from following the S1P gradient and sequestering them inside the LN. According to this hypothesis, synthetic S1P1 agonists would act as "functional antagonists" of S1P1 in lymphocytes. The second hypothesis states that S1P1 agonists bind the S1P1 receptor on lymphatic endothelial cells and act as pure agonists, inducing downstream signaling that results in changes in cytoskeleton organization, enhancement of the barrier function, and blockade of egress of lymphocytes from the medullary parenchyma to the draining sinuses.[67]

Although no evidence excludes the fact that agonists may act directly on lymphoid endothelium to reduce egress, many experimental data support the hypothesis of the functional antagonism on lymphocyte S1P1.[65] However, if this hypothesis would be correct, an S1P1 antagonist should also induce lymphopenia in vivo, and it has been reported not to be the case.[68] However, available antagonists show weak potency at the receptor and poor pharmacokinetic profiles in vivo. Therefore, the lack of lymphopenia could be explained by the fact that the levels of the antagonists in vivo are insufficient to counteract the endogenous levels of S1P.[65]

Internalization of S1P1 receptor by S1P and synthetic agonists has been observed using different experimental approaches.[69] S1P and agonists like SEW2871 induce receptor internalization and rapid recycling back to membrane, whereas FTY720 induces a sustained receptor internalization and subsequent degradation.[62] Whether agonists showing different effects on the receptor will behave differentially in vivo remains to be elucidated.

Effect of S1P on Th17 cell generation

Apart from the well-known Th1 and Th2 cell subsets, CD4+ T cells can differentiate into effector Th17 cells. These cells are characterized by the production of the cytokine IL-17 and are involved in the physiopathology of some autoimmune diseases like multiple sclerosis, psoriasis, or intestinal bowel disease, as well as in defense against certain pathogens.[70]

Recent discoveries have pointed at the S1P/S1P1 axis as one of the main stimuli driving terminal differentiation of Th17 cells,[71] together with IL-23. In vitro differentiation of mouse splenic T cells into Th17 cells is driven by TCR stimulation (anti-CD3 plus anti-CD28) and treatment with TGFβ, IL-6 and IL-1. When S1P is added to the medium, a significant increase in IL-17 is observed some days later. It has been hypothesized that S1P and IL-23 represent alternative means of recruiting and activating Th17 cells in different compartments. S1P is thought to be more relevant in acute systemic and vascular inflammation, whereas IL-23 is thought to be involved in amplification of Th17 cells in infections and intestinal inflammatory states. In vivo work using antigen-challenged transgenic S1P1 mice overexpressing CD4 T cells showed amplification of Th17 cells, indicating that the same phenomenon may occur in vivo.[72]

Effect of S1P on regulatory T cells

Regulatory T cells (T_{reg} cells) play a fundamental role in the maintenance of immune tolerance and prevention of autoimmunity.[73] T_{reg} cells develop and mature in the thymus and are then released to the periphery to exert their immunosuppressive function on effector cells. Excessive suppression by T_{reg} cells may render the host susceptible to infection and cancer.

Although T_{reg} cells express the five S1P receptors, a series of elegant studies from Liu and colleagues[74] demonstrated that S1P1 is the receptor subtype that plays an essential role in their development and function. Genetic approaches to either eliminate or enhance S1P1 function selectively in these cells were used to show that the loss of S1P1 translates into enhanced thymic differentiation and suppressive activity of T_{reg} cells. By contrast, higher S1P1 signaling led to diminished development and function of these cells in vitro and in vivo, and to the development of spontaneous autoimmunity. S1P1 appears to be an intrinsic negative regulator of the thymic differentiation, peripheral maintenance, and suppressive activity of T_{reg} cells. The downstream signal evoked by S1P on T_{reg} cells via S1P1 is the activation of the Akt-mTOR pathway.[74]

The authors propose that a dynamic regulation of S1P1 expression contributes to lymphocyte priming and the maintenance of immune homeostasis. At the early stage of immune activation, S1P1 expression is maintained mainly in T_{reg} cells, which results in low suppressive activity and high mobility of these cells. In contrast, S1P1 is down-regulated in conventional T cells to mediate their sequestration into the draining LNs to allow for an efficient interaction with antigen-presenting cells. At a later stage of the immune response, S1P1 is down-regulated in T_{reg} cells to facilitate suppressive activity and preventing an overt immune response due to T cells. Then, S1P1 slowly recovers in conventional T cells to permit them to migrate to peripheral tissues.

THERAPEUTIC EXPLOITATION OF THE S1P SYSTEM

The multiple roles of S1P in biological processes offer the possibility to exploit the system therapeutically for the treatment of inflammation, autoimmunity, and cancer.[75] Experimental work in animal models suggesting that S1P receptor agonists might be effective in human diseases like multiple sclerosis have been confirmed in the clinic with FTY720, also known as fingolimod.

Fingolimod: Clinical validation of S1P receptor agonism in multiple sclerosis

Multiple sclerosis (MS) is an autoimmune disease that affects young adults in which the immune system destroys the myelin sheath of the central nervous system (CNS), leading to progressive neurological disabilities. The relapsing-remitting form of MS is characterized by unpredictable relapses of the disease followed by periods of remission. The social and economic impact of the disease is enormous.

Fingolimod (Novartis) is currently in phase 3 trials for the treatment of relapsing-remitting MS.[76,77] Results from a 24-month extension study of the phase 2 trial showed that 77% of patients treated with fingolimod were free of relapses during the treatment, and 79% to 91% of patients had an improvement in the inflammatory lesions in the CNS.[78] Current treatments of patients with MS are all injectable therapies and offer about 30% of improvement in disease relapses.[77] In addition to having better clinical efficacy than current therapies, fingolimod is an oral drug, which represent a significant improvement in MS treatment.

Fingolimod causes lymphopenia in humans. At therapeutic doses, the mean total lymphocyte count in peripheral blood is reduced by 61%. Both CD4+ and CD8+ naïve and central memory T cells are depleted, whereas effector memory T cells are not affected.[39] The inhibition of the migration of autoreactive T cells in the CNS by fingolimod may prevent destruction of the myelin sheath. Lymphopenia is likely the main mechanism responsible for the efficacy observed with this drug. Additionally, fingolimod has been reported to preserve the integrity of the brain blood barrier,[76] which might also contribute to the efficacy. Other effects of the drug on the immune system could also play a role, since fingolimod has been shown to inhibit the development of Th17 cells in vitro[71] and, in mouse models, to enhance functional activity of T_{reg} cells.[79]

Novel therapeutic approaches targeting S1P system

Fingolimod has paved the way for the search of novel strategies devoted to interfere with the S1P system. Fingolimod is an agonist of all the S1P receptors except S1P2.[62] Pharmaceutical companies like Novartis, Actelion/Roche, Sankyo, or Kyorin Pharmaceuticals are developing a second generation of S1P1 agonists, with a more selective profile versus other S1P receptors. Fingolimod causes bradycardia, an effect that in rodents is associated to S1P3 agonism (47). Thus, developing S1P1 agonist with selectivity against S1P3 was an obvious strategy to avoid this side effect. However, compound ACT-128800, a selective S1P1 agonist from Actelion/Roche (http://www1.actelion.com/en/investors/events/actelion-day-2009.page) induces bradycardia in humans to a degree comparable to fingolimod, suggesting this effect in humans is mediated by S1P1. Similar results were reported for BAF-312, a Novartis compound.[80] These results clearly indicate that species differences in S1P receptor expression pattern exist, and that caution should be taken in extrapolating data from receptor knock-out mice to humans. On the other hand, whether a selective S1P1 agonist will be equally effective to fingolimod remains to be seen, given that receptors other than S1P1 might play a role in the disease, like S1P3 or S1P5, expressed abundantly in the CNS.[54,55]

There are strategies other than S1P receptor agonism with potential therapeutic utility, such as the inhibition of sphingosine kinases,[81] S1P lyase,[82] or monoclonal

Table 17-2. Diseases Described in Laboratory Animals in Which Different Strategies Targeting S1P Have Reported Efficacy

Preclinical disease models	Mechanistic approach and/or Compound	References
Atherosclerosis in the apolipoprotein E-deficient mice	S1P receptor agonism FTY720	[84]
Experimental asthma in rats (local application to the lung)		[85]
Acute pulmonary injury in rats with acute necrotizing pancreatitis		[86]
Rat renal transplantation	S1P receptor agonism KRP-203	[87]
Mouse lupus eryhtematodes		[88]
Rat autoimmune myocarditis		[89]
Wound healing in diabetic mice	S1P	[90]
Colitis in mice	SphK inhibition ABC747080, ABC294640	[91]
Airway inflammation in a mouse model of asthma	SphK inhibition N,N-dimethylsphingosine	[92]
Murine collagen-induced arthritis	SphK inhibition N,N-dimethylsphingosine and SphK1 knock down with small interfering RNA	[93]
Dextran sulfate sodium-induced colitis	SphK1 knock-out	[94]
Laser-induced choroidal neovascularization in mouse	Anti-S1P monoclonal antibody	[95]
Retinal and choroidal neovascularization	Sonepcizumab (humanized anti-S1P monoclonal antibody)	[96]

antibodies against S1P.[83] Table 17.2 depicts one example of the different animal models of disease in which different strategies have shown to be effective. Cancer has been excluded due to space constraints.

CONCLUDING REMARKS

The knowledge of the biological effects of S1P has been extensively studied in the last decade. The discovery of FTY720 has made an extraordinary contribution to the field and has allowed to elucidate the significant roles of the S1P pathway in immunological processes. FTY720 has validated clinically the S1P receptors as therapeutic targets, but has also paved the way to explore the therapeutic potential of other approaches based on this fascinating lipid mediator. To our knowledge, the S1P receptors provide the first example of a GPCR as a target for immunosuppression, and fingolimod the first drug to exploit agonist-induced internalization therapeutically.

REFERENCES

1. Pyne S *et al*. (2000) Sphingosine 1-phosphate signalling in mammalian cells. *Biochem J*, **349** (2), 385–402.
2. Pitson SM *et al*. (2000) Human sphingosine kinase: Purification, molecular cloning and characterization of the native and recombinant enzymes. *Biochemistry Journal*, **350** (2), 429–441.
3. Zhou J *et al*. (1998) Identification of the first mammalian sphingosine phosphate lyase gene and its functional expression in yeast. *Biophys. Res. Commun.* **242**, 502–507.
4. Cuvillier O *et al*. (1996) Supression of ceramide-mediated programmed cell death by sphingosine-1-phosphate. *Nature*, **381**, 800–803.
5. Alvarez SE *et al*. (2007) Autocrine and paracrine roles of sphingosine-1-phosphate. *Trends Endocrinol. Metab.*, **18**, 300–307.
6. Maceyka M *et al*. (2005) Sphk1 and Sphk2: Sphingosine kinase isoenzymes with opposing functions in sphingolipid metabolism. *J. Biol. Chem.*, **280**, 37118–37129.
7. Le Stunff H *et al*. (2007) Recycling of sphingosine is regulated by the concerted actions of sphingosine-1-phosphate phosphohydrolase 1 and sphingosine kinase 2. *J. Biol. Chem.* **282**, 3472–3480.
8. Mizugishi K *et al*. (2005) Essential role for sphingosine kinases in neural and vascular development. *Mol. Cell Biol.*, **25**, 11113–11121.
9. Xia P *et al*. (2000) An oncogenic role of sphingosine kinase. *Curr. Biol.* **10**, 1527–1530.
10. Pappu R *et al* (2007) Promotion of lymphocyte egress into blood and lymph by distinct sources of sphingosine-1-phosphate. *Science*, **316**, 295–298.
11. Okajima F (2002) Plasma lipoproteins behave as carriers of extracellular sphingosine 1-phosphate: is this an atherogenic mediator or an anti-atherogenic mediator? *Biochim. Biophys. Acta*, **1582**, 132–137.
12. Yatomi Y *et al*. (1997) Sphingosine 1-phosphate, a bioactive sphingolipid abundantly stored in platelets, is a normal constituent of human plasma and serum. *J. Biochem.* **121**, 969–973.
13. Hanel P *et al*. (2007) Erythrocytes store and release sphingosine 1-phosphate in blood. *FASEB J.*, **21**, 1202–1209.
14. Kihara A *et al*. (2008) Production and release of sphingosine-1-phosphate and the phosphorylated form of the immunomodulator FTY720. *Biochim. Biophys. Acta*, **1781**, 496–502.
15. Ventakaraman K *et al*. (2008) Vascular endothelium as a contributor of plasma sphingosine 1-phosphate. *Circ. Res.*, **102**, 669–676.
16. Olivera A *et al*. (2006) IgE-dependent activation of sphingosine kinases 1 and 2 and secretion of sphingosine 1-phosphate requires Fyn kinase and contributes to mast cell responses. *J. Biol. Chem.*, **281**, 2515–2525.
17. Mitra P *et al*. (2006) Role of ABCC1 in export of sphingosine-1-phosphate from mast cells. *Proc. Natl. Acad. Sci. U.S.A.*, **103**, 16394–16399.
18. Kobayashi N, *et al*. (2006) Sphingosine 1-phosphate is released from the cytosol of rat platelets in a carrier-mediated manner. *J. Lipid Res.*, **47**, 614–621.
19. Lee YM *et al*. (2007) A novel method to quantify sphingosine 1-phosphate by immobilized metal affinity chromatography (IMAC). *Prostaglandins Other Lipid Mediat.*, **84**, 154–162.
20. Romiti E *et al* (2000) Neutral/alkaline and acid ceramidase activities are activily released by murine endothelial cells. *Biochem. Biophys. Res. Commun.*, **275**, 476–751.
21. Tabas I (1999) Secretory sphingomyelinase. *Chem. Phys. Lipids*, **102**, 123–130.
22. Ancellin N, Colmont C, Su J, Li Q, Mittereder N, Chae SS, Stefansson S, Liau G, Hla T (2002) Extracellular export of sphingosine kinase-1 enzyme. Sphingosine 1-phosphate generation and the induction of angiogenic vascular maturation. *J. Biol. Chem.*, **277**, 6667–6675.
23. Van Brocklyn JR *et al*. (1998) Dual actions of sphingosine-1-phosphate: extracellular through the G_i- coupled receptor Edg-1 and intracellular to regulate proliferation and survival. *J. Cell Biol.*, **142**(1), 229–240.

24. Kariya Y *et al.* (2005) Products by the sphingosine kinase/sphingosine-1-phosphate (S1P) lyase pathway but not S1P stimulate mitogenesis. *Genes Cells*, **10** (6), 605–615.

25. Hla T *et al.* (1990) An abundant transcript induced in differentiating human endothelial cells encodes a polypeptide with structural similarities to G-protein-coupled receptors. *J. Biol. Chem.* **265**, 9308–9313.

26. Okamoto H *et al.* (1998) EDG1 is a functional sphingosine-1-phosphate receptor that is linged via a Gi/o to multiple signalling pathways, including phospholipase C activation, Ca2+ mobilization, Ras-mitogen-activated protein kinase activation, and adenylate cyclase inhibition. *J. Biol. Chem.* **273**, 27104–27110.

27. Alderton F *et al.* (2001) Tethering of the platelet-derived growth factor beta receptor to G-protein coupled receptors: a novel platform for integrative signalling by these receptor classes in mammalian cells. *J. Biol. Chem.* **276**, 28578–28585.

28. Liu Y *et al.*(2000) Edg-1, the G protein-coupled receptor for sphingosine-1-phosphate, is essential for vascular maturation. *J. Clin. Invest.* **106**, 951–961.

29. Allende ML *et al.* (2003) G-protein-coupled receptor S1P1 acts within endothelial cells to regulate vascular maturation. *Blood* **102**, 3665–3667.

30. Allende ML *et al.* (2004) Expression of the sphingosine-1-phosphate receptor, S1P1, on T cells controls thymic emigration. *J. Biol. Chem.* **279** (15), 15396–15401.

31. Matloubian M *et al.* (2004) Lymphocyte egress from thymus and peripheral lymphoid organs is dependent on S1P receptor 1. *Nature* **427** (6972), 355–360.

32. Chi H *et al.* (2005) Cutting edge: Regulation of T cell trafficking and primary immune responses by sphingosine 1-phosphate receptor 1. *J. Immunol.* **174** (5), 2485–2488.

33. Kabashima K *et al.* (2006) Plasma cell S1P1 expression determines secondary lymphoid organ retention versus bone marrow tropism. *J. Exp. Med.* **203**(12), 2683–2690.

34. Singleton PA *et al.* (2005), Regulation of sphingosine 1-phosphate-induced endothelial cytoskeletal rearrangement and barrier enhancement by S1P1 receptor, PI3 kinase, Tiam1/Rac1, and alpha-actinin.*FASEB J.* Oct; **19** (12):1646–56.

35. Kimura A *et al* (2007) Essential roles of sphingosine 1-phosphate/S1P1 receptor axis in the migration of neural stem cells toward a site of spinal cord injury. *Stem Cells* **25**, 115–124.

36. Ishii I *et al.* (2001) Selective loss of sphingosine 1-phosphate signalling with no obvious phenotypic abnormality in mice lacking its G protein-coupled receptor, LP (B3)/EDG-3. *J. Biol. Chem.* **276**, 33697–33704.

37. Ancellin N *et al.*(1999) Differential pharmacological properties and signal transduction of the sphingosine 1-phosphate receptors EDG-1, EDG-3, and EDG-5. *J. Biol. Chem.* **274**, 18997–19002.

38. Okamoto H *et al.* (2000) Inhibitory regulation of Rac activation, membrane ruffling, and cell migration by the G protein-coupled sphingosine-1-phosphate receptor EDG5 but not EDG1 or EDG3. *Mol. Cell. Biol.* **20**, 9247–9261.

39. Paik JH *et al.* (2001) Sphingosine 1-phosphate-induced endothelial cell migration requires the expression of EDG-1 and EDG-3 receptors and Rho-dependent activation of alpha vbeta3- and beta1-containing integrins. *J. Biol. Chem.* **276**, 11830–11837.

40. Goparaju SK *et al.* (2005) The S1P2 receptor negatively regulates platelet-derived growth factor-induced motility and proliferation. *Mol. Cell. Biol.* **25**(10), 4237–4249.

41. Baudhuin LM *et al.* (2004) S1P3-mediated Akt-activation and cross-talk with platelet-derived growth factor receptor (PDGFR). *FASEB J.* **18**(2), 341–343.

42. Ishii I *et al.* (2002) Marked perinatal lethality and cellular signalling deficits in mice null for the two sphingosine 1-phosphate (S1P) receptors, S1P(2)/LP(B2)/EDG-5 and S1P(3)/LP(B3)/EDG-3. *J. Biol. Chem.* **277**, 25152–25159.

43. Kono M *et al.*(2004) The sphingosine-1-phosphate receptors S1P1, S1P2, and S1P3 function co-ordinately during embryonic angiogenesis. *J. Biol. Chem.* **279**, 29367–29373.

44. MacLennan AJ *et al.* (2001) An essential role for the H218/AGR16/Edg-5/LP(B2) sphingosine 1-phosphate receptor in neuronal excitability. *Eur. J. Neurosci.* **14**, 203–209.

45. Serriere-Lanneau V *et al.* (2007) The sphingosine 1-phosphate receptor S1P2 triggers hepatic wound healing. *FASEB J.* **21**, 2005–2013.

46. Hu W *et al.* (2006) Lentiviral siRNA silencing of sphingosine-1-phosphate receptors S1P1 and S1P2 in smooth muscle. *Biochem. Biophys. Res. Commun.* **343**, 1038–1044.

47. Sanna MG *et al.* (2004) Sphingosine 1-phosphate (S1P) receptor subtypes S1P1 and S1P3, respectively, regulate lymphocyte recirculation and heart rate. *J. Biol. Chem.* **279**, 13839–13848.

48. Theilmeier G *et al.* (2006) High-density lipoproteins and their constituent, sphingosine-1-phosphate, directly protect the heart against ischemia/reperfusion injury in vivo via the S1P3 lysophospholipid receptor. *Circulation* **114**, 1403–1409.

49. Gon Y *et al.* (2005) S1P3 receptor-induced reorganization of epithelial tight junctions compromises lung barrier integrity and is potentiated by TNF. *Proc. Natl. Acad. Sci. U.S.A.* **102**, 9270–9275.

50. Niessen F *et al.* (2008) Dendritic cell PAR1-S1P3 signalling couples coagulation and inflammation. *Nature* **452**, 654–658.

51. Graler MH *et al.* (2003) The sphingosine 1-phosphate receptor S1P4 regulates cell shape and motility via coupling to Gi and G12/13. *J. Cell. Biochem.* **89**, 507–519.

52. Malek RL *et al.* (2001) Nrg-1 belongs to the endothelial differentiation gene family of G protein-coupled sphingosine-1-phosphate receptors. *J. Biol. Chem.* **276**: 5692–5699.

53. Deltagen Inc. 2005: http://www.informatics.jax.org/external/ko/deltagen/302

54. Jaillard C *et al.* (2005) Edg8/S1P5: an oligodendroglial receptor with dual function on process retraction and cell survival. *J. Neurosci.* **25**, 1459–1469.

55. Novgorodov AS *et al.* (2007) Activation of sphingosine-1-phosphate receptor S1P5 inhibits oligodendrocyte progenitor migration. *FASEB J.* **21**, 1503–1514.

56. Walzer T *et al.* (2007) Natural killer cell trafficking in vivo requires a dedicated sphingosine 1-phosphate receptor. *Nat. Immunol.* **8**, 1337–1344.

57. Rivera J *et al.* (2008) The alliance of sphingosine 1-phospahte and its receptors in immunity. *Nat. Rev. Immunol.* **8**, 753–763.

58. Cyster JG. (2007) Specifying the patterns of immune cell migration *Novartis Found. Symp.* **281**; 54–61

59. Grigorova IL *et al.* (2009) Cortical sinus probing, S1P1-dependent entry and flow-based capture of egressing T cells. *Nat Immunol.*; **10** (1):58–65.

60. Chaffin KE, and Perlmutter RM (1991) A pertussis toxin-sensitive process controls thymocyte emigration. *Eur J Immunol.*; **21**(10):2565–73.

61. Chiba K, *et al.* (2006) Role of sphingosine 1-phosphate receptor type 1 in lymphocyte egress from secondary lymphoid tissues and thymus. *Cell Mol. Immunol.* **3** (1):11–9

62. Mandala S. *et al* (2002) Alteration of lymphocyte trafficking by sphingosine-1-phosphate receptor agonists. *Science;* **296** (5566):346–9.

63. Hla T *et al.* (2008) The vascular S1P gradient: cellular sources and biological significance. *Biochim Biophys Acta;* **1781**(9):477–82.

64. Vachal P *et al.* (2006) Highly selective and potent agonists of sphingosine-1-phosphate 1 (S1P1) receptor. Bioorg Med Chem Lett. ;**16**(14):3684–7.

65. Schwab SR and Cyster JG.(2007) Finding a way out: lymphocyte egress from lymphoid organs. *Nat Immunol.* **8** (12):1295–301

66. Rosen H, *et al.* (2009) Sphingosine 1-phosphate receptor signaling. *Annu Rev Biochem;* **78**:743–68.

67. Wei SH, *et al.* (2005) Sphingosine 1-phosphate type 1 receptor agonism inhibits transendothelial migration of medullary T cells to lymphatic sinuses. *Nat Immunol.* **6** (12):1228–35.

68. Sanna G *et al.* (2006) Enhancement of capillary leakage and restoration of lymphocyte egress by a chiral S1P1 antagonist in vivo. *Nat Chem Biol.* **2** (8):434–41.

69. Oo ML *et al.* (2007) Immunosuppressive and anti-angiogenic sphingosine 1-phosphate receptor-1 agonists induce ubiquitinylation and proteasomal degradation of the receptor. *J Biol Chem.*; **282** (12):9082–9.

70. Harrington, LE. *et al.* (2005) Interleukin 17-producing CD4+ effector T cells develop via a lineage distinct from the T helper type 1 and 2 lineages. *Nat. Immunol.* **6**:1123–1132.

71. Liao JJ *et al.* (2007) Cutting edge: alternative signalling of Th17 cell development by sphingosine 21-phosphate. *J.Immunol.* **178**, 5425–5428.

72. Huang MC *et al.* (2007) Th17 augmentation in OTII TCR plus T cell-selective type 1 sphingosine 1-phosphate receptor double transgenic mice. *J. Immunol.* **178**, 6806–6813.

73. Piccirillo CA *et al.* (2008) CD4+ Foxp3+ regulatory T cells in the control of autoimmunity: in vivo veritas. *Curr. Opin. Immunol.* **20**, 655–662.

74. Liu G *et al.* (2009) The receptor SP1 overrides regulatory T cell mediated immune suppression through Akt-mTOR. *Nat. Immunol.* Jul;**10** (7):769–77.

75. Huwiler A *et al.* (2008) New players on the center stage: Sphingosine 1-phosphate and its receptors as drug targets. *Biochem.Pharmacol.* **75**: 1893–1900.

76. Baumruker T. *et al.* (2007) FTY720, an immunomodulatory sphingolipd mimetic: translation of a novel mechanism into clinical benefit in multiple sclerosis. *Expert Opin. Investig.Drugs* **16**; 283–289

77. Martini S *et al.* (2007) Current perspectives on FTY720. *Expert Opin. Invest. Drugs* **16**, 505–518.

78. O' Connor P *et al.* (2009) Oral fingolimod (FTY720) in multiple sclerosis, Two year results of a phase II extension study. *Neurology* **72**, 73–79.

79. Sawicka E *et al.* (2005) The sphingosine 1-phosphate receptor agonist FTY720 differentially affects the sequestration of CD4+/CD25+ T-regulatory cells and enhances their functional activity. *J Immunol*; **175** (12):7973–80.

80. Gergely P. *et al.* (2009) Phase I study with the selective S1P1/S1P5 receptor modulator BAF312 indicates that S1P1 rather than S1P3 mediates transient heart rate reduction in humans. Abstract P437. 25th Congress of the European Committee for Treatment and Research in Multiple Sclerosis. Sep 9–12, Düsseldorf, Germany.

81. Shida D *et al.* (2008) Targeting SphK1 as a new strategy against cancer. *Curr Drug Targets.* ;**9**(8):662–73.

82. Pappas, C. *et al.* (2008) LX2931: A Potential Small Molecule Treatment for Autoimmune Disorders. *Presentation # 351*, American College of Rheumatology meeting.

83. O' Brien N *et al.* (2009) Production and characterization of monoclonal anti-sphingosine-1-phosphate antibodies. *J. Lipid Res.* **50**, 2245–2257.

84. Keul P *et al.* (2007) The sphingosine-1-phosphate analogue FTY720 reduces atherosclerosis in apolipoprotein E-deficient mice. *Arterioscler. Thromb. Vasc. Biol.* **27**, 607–613.

85. Idzko M *et al.* (2006) Local application of FTY720 to the lung abrogates experimental asthma by altering dendritic cell function. *J. Clin. Invest.* **116**, 2935–2944.

86. Liu HB *et al.* (2008) Sphingosine-1-phosphate and its analogue FTY720 diminish acute pulmonary injury in rats with acute necrotizing pancreatitis. *Pancreas* **36**, e10–5.

87. Fujishiro J, *et al.* (2006). Use of sphingosine-1-phosphate 1 receptor agonist, KRP-203, in combination with a subtherapeutic dose of cyclosporine A for rat renal transplantation. *Transplantation* **82**, 804–812.

88. Wenderfer SE *et al.* (2008) Increased survival and reduced renal injury in MRL/lpr mice treated with a novel sphingosine-1-phosphate receptor agonist. *Kidney Int.* **74**, 1319–1326.

89. Ogawa R *et al.* (2007) A novel sphingosine-1-phosphate receptor agonist KRP-203 attenuates rat autoimmune myocarditis. *Biochem. Biophys. Res. Commun.* **361**, 621–628.

90. Kawanabe T, *et al.* (2007) Sphingosine 1-phosphate accelerates wound healing in diabetic mice. *J. Dermatol. Sci.* **48**,53–60.

91. Maines LW *et al.* (2008) Suppression of ulcerative colitis in mice by orally available inhibitors of sphingosine kinase. *Dig. Dis. Sci.* **53**, 997–1012.

92. Nishiuma T *et al.* (2008) Inhalation of sphingosine kinase inhibitor attenuates airway inflammation in asthmatic mouse model. *Am. J. Physiol. Lung Cell. Mol. Physiol.* **294**, L1085–93.

93. Lai WQ *et al.* (2008) Anti-inflammatory effects of spohingosine kinase moudulation in inflammatory arthritis. *J. Immunol.* **1**, 8010–8017.

94. Snider AJ *et al.* (2009) A role for sphingosine kinase 1 in dextran sulfate sodium-induced colitis. *FASEB J.* **23**, 143–152.
95. Caballero S *et al.* (2009) Anti-sphingosine-1-phosphate monoclonal antibodies inhibit angiogenesis and sub-retinal fibrosis in a murine model of laser-induced choroidal neovascularization. *Exp. Eye Res.* **88**, 367–377.
96. Xie B *et al.* (2009) Blockade of sphingosine-1-phosphate reduces macrophage influx and retinal and choroidal neovascularization. *J. Cell. Physiol.* **218**, 192–198.

18 Wnt/Frizzled receptor signaling in osteoporosis

Georges Rawadi

INTRODUCTION TO WNT SIGNALING

Wnt proteins are secreted signaling molecules coordinating several aspects of early development and are well studied in the fly, worm, fish, frog, mouse, and human.[1] The intracellular signaling pathway of Wnt is also conserved evolutionally and regulates cellular proliferation, morphology, motility, fate, axis formation, and organ development.[1,2] The overall downstream Wnt signaling can be segregated into three pathways (Figure 18.1): the canonical Wnt pathway, which acts through the protein β-catenin to regulate transcription; the planar cell polarity (PCP) pathway, which regulates *drosophila* development independently of β-catenin; and the calcium (Ca^{2+}) pathway. Among these intracellular cascades, the canonical β–catenin pathway has been most extensively studied. It has been shown that abnormalities of this pathway lead to several human diseases, including tumor formation and bone abnormalities.[3]

Wnt signaling has been linked to several diseases such as cancer, obesity, Type II diabetes, Familial Exudative Vitreoretinopathy, and others. In the last few years, a role for the Wnt/β-catenin signaling pathway has been demonstrated

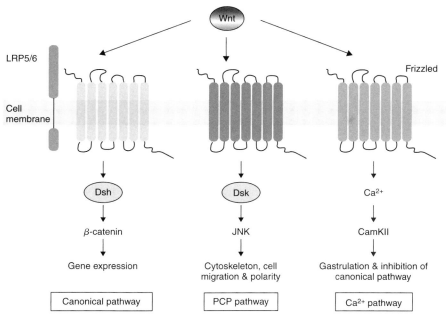

Figure 18-1: Current general view of Frizzled receptor-dependent Wnt signaling. A large body of evidence supports a model in which the Frizzleds are involved at the cell surface to determine the outcome of Wnt activity. In some cases the downstream signaling events following pathway activation such as β-catenin pathways are well understood, and others such as Ca²⁺ pathway are still unclear.

in bone metabolism and shown to be a key regulator of bone accrual during growth.[4,5]. A strong body of human and mouse genetic evidence clearly indicate that pharmacological manipulation of elements in this biological cascade results in stimulation of osteoblast (the bone forming cells) function and subsequent increase in bone mass. This offers new and exciting opportunities for the development of novel bone anabolic drugs for the treatment of osteoporosis and other bone diseases.

This chapter summarizes the current evidence of the role of Wnt/β-catenin cascade in skeletal biology and discuss potential approaches for the identification of Wnt/β-catenin modulators with bone anabolic activity.

FRIZZLED RECEPTORS: CENTRAL ROLE IN WNT SIGNALING

The mammalian genome encodes 10 Frizzled receptors and 19 Wnt protein ligands, which suggests that about 190 potential Wnt/Frizzled distinct combinations are plausible. All members of the Frizzled protein family are characterized by a putative signal peptide, followed by a sequence of 120 amino acids (aa) containing 10 highly conserved cysteine residues (CRD) [6, 7]. Then a highly divergent region of 40–100 aa predicted to form a flexible linker, seven transmembrane segments separated by short extracellular and cytoplasmic loops, and finally a cytoplasmic tail (Figure 18.2). The CRD is the Wnt ligand-binding site of Frizzleds.

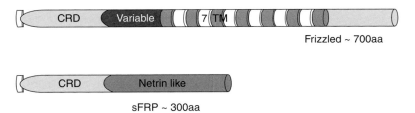

Figure 18-2: Schematic representation of frizzled and secreted frizzled (sFRP) proteins. CRD: cystein rich domain; 7TM: seven transmembranes.

The C-terminus of many Frizzleds contains a S/TXV motif, which confers the capacity to bind PDZ domains. Disheveled protein (Dsh) that displays a PDZ domain was shown to bind to the C-terminus and enable downstream signaling. An additional motif, KTXXXW, in the C-terminus of Frizzled is also required for Wnt signaling.[8]

Although the overall structure of Frizzleds resembles that of G protein-coupled receptors (GPCRs), whether Frizzleds are coupled to small G protein remains controversial. There is some evidence that Frizzled receptors, mainly Frizzled2, involved in the Wnt/Ca^{2+} pathway are coupled to heterotrimeric G protein subunits. Whether this is also the case for Frizzleds acting through the Wnt/β-catenin remains controversial.[9,10]

Remarkably, the Frizzled receptors for Wnts,[6] are related to the Smoothened (Smo) protein that is necessary for Hedgehog signaling.[11,12] Both receptors have seven transmembrane domains and a long N-terminal CRD extension (cysteine-rich domain). As a group, these molecules are more closely related to each other than they are to the other families of serpentine receptors, of which there are many. Although the actual mechanisms of activation of Frizzled and Smo are fundamentally different, both share the same controversy about their capacity to signal through binding to small G proteins.

Cascade downstream Frizzled

Because Wnt proteins have been notoriously difficult to manipulate, the biochemical behavior of these proteins and their Frizzled receptors has been difficult to investigate in details. A large majority of studies designed to address Wnt/Frizzled signaling involved gene transfer experiments with in vivo models (e.g., mouse, *Drosophila*, or *Xenopus*) or in vitro cell culture systems. As indicated above, it has been well established that Wnts activate a number of different signaling pathways,[13,14] each of which has been shown to intersect with numerous other intracellular signal transduction pathways,[15] and all involve Frizzled receptors.

Canonical Wnt signaling

The activation of β-catenin downstream of Frizzled receptor is the signaling pathway that has been characterized as and historically named "canonical Wnt signaling". Wnt/β-catenin signaling involves the binding of Wnts to two

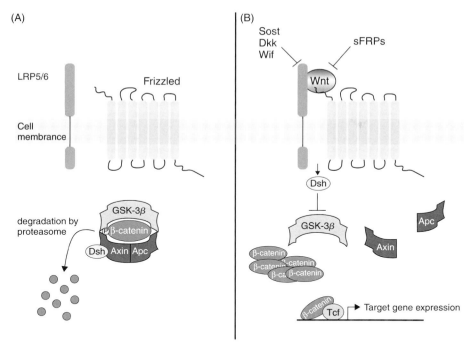

Figure 18-3: Canonical Wnt signaling. (A) In the absence of Wnt proteins, β-catenin is phosphorylated by GSK-3β leading to its degradation and pathway inactivation. The β-catenin phosphorylation/degradation mechanism involves other cytoplasmic proteins mainly Apc, Axin and Dsh. (B) Following Wnt binding to its LRP5/6 and frizzled co-receptors GSK-3β is inactivated. As a consequence, β-catenin is stabilized, accumulates in the cytoplasm and translocates into the nucleus where it triggers target gene expression in concert with TCF/lef transcription factor family members. Secreted proteins such as Dkk, Sost or Wifs inhibit Wnt signaling through binding to LRP5/6, whereas sFRPs inhibit the pathway through interaction with Wnt proteins.

receptors: Frizzled, through CRD, and low density lipoprotein (LDL) receptor-related protein 5 (LRP5) or LRP6[6,7,16,17].

In the absence of Wnt, ligand glycogen synthase kinase-3β (GSK-3β), a serine/threonine kinase, phosphorylates and targets cytoplasmic β-catenin for degradation (Figure 18.3a). As a result, the cytoplasmic β-catenin level is kept at a very low concentration at which it is unable to activate the underlying transcriptional machinery. When an appropriate Wnt, including Wnt1, Wnt2, Wnt3a, or Wnt10a, acts on the cell surface receptors, Frizzled and LRP5 or 6, it triggers an intracellular signaling cascade that results in the inhibition of GSK-3β action (Figure 18.3b). This signaling cascade involves more than ten different proteins, mainly Dsh, Axin, adenomatous polyposis coli (Apc), and GSK binding protein/frequently rearranged in advanced T-cell lymphoma 1 (GBP/Frat1). As a consequence, the phosphorylation and degradation of β-catenin is reduced, resulting in its accumulation in the cytoplasm. When it reaches a certain threshold, accumulated β-catenin translocates into the nucleus, binds to a member of the Tcf/Lef family of transcription factors, and stimulates gene expression.

β-catenin activity and stability is not only regulated by phosphorylation but also by acetylation.[18,19] Moreover, in the nucleus, several proteins that bind to Tcf/Lef (such as Smad proteins or Groucho) also regulate the formation of the β-catenin-Tcf-DNA complex. Thus, in the canonical β-catenin pathway, Wnt increases the stability of β–catenin, thereby stimulating Tcf/Lef mediated gene expression.

Nature has designed a number of Wnt antagonists in order to control and fine-tune the canonical signaling. Frzb is the first reported secreted Wnt antagonist[20-24] that belongs to a family of secreted Frizzled-related proteins (sFRPs) containing CRD motifs similar to those of Frizzleds (Figure 18.2). In addition to CRDs, sFRPs contain a C-terminal region with similarity to netrins: secreted proteins involved in axon guidance. sFRPs can bind Wnt ligands and act as decoy receptors antagonizing Wnt signaling. Today numerous sFRPs have been described in vertebrates.[25] Wnt-inhibitory factor 1 (Wif1) is another secreted Wnt antagonist structurally diverse from sFRPs but also able to antagonize both the canonical and noncanonical Wnt pathways.[26] Although Wnt signaling is widely conserved across species, no orthologes of either sFRPs or Wif1 have been identified in *Drosophila* or *C. elegans*.

Other secreted Wnt signaling antagonists highly specific to the canonical pathway have been described and include Dickkopf (Dkk) connective tissue growth factor (Ctgf), Wise and sclerostin/Sost proteins. Dkk1, Dkk2, and Dkk4, but not Dkk3, are able to bind LRP5 and 6.[27-29] Kremen1, a transmembrane protein, binds Dkks and forms a tertiary complex with LRP5/6 that is then internalized, targeting LRP5/6 for degradation.[30,31]

Sost, a potent Wnt inhibitor, is a secreted protein characterized by a cysteine-knot motif and a significant similarity with the dan family of secreted glycoproteins.[32,33] Sost displays a good homology with the Wnt antagonist Wise, and both were recently shown to inhibit Wnt activity in vitro.[34,35] Sost inhibitory activity is mediated through its binding to LRP5/6 receptor.[36]

Planar cell polarity signaling

The other best-studied variant Wnt pathway and first described in *Drosophila* is the planar cell polarity (PCP), which controls the orientation of hairs, bristles, and ommatidia. The PCP pathway has been the focus of much attention mainly because it has been duly recognized as a critical regulator of many major morphogenetic processes. The current findings point out to the involvement in the PCP pathway of the small GTPases, Rho, Rac, and Cdc42, that lead to regulation of the Jun kinase (JNK) cascade [40,41,38]. Interest in these small GTPases and the JNK cascade has generally been in relation to their roles in organizing the cytoskeleton and regulating changes in cell shape and behavior.[37] Whereas GTPases and JNK can be activated by many different stimuli, there is now a general consensus that they are downstream of Frizzled activation of Dsh and through this pathway they have been shown to be critical for regulating cell polarity and morphogenetic movements in vertebrates.[38] The PCP pathway overlaps to some extent with the canonical Wnt signaling pathway in that it requires Frizzled receptors and the cytoplasmic molecule Dsh. However, neither LRP5/6, β-catenin, nor TCF is required. Most importantly, findings in *Drosophila* strongly suggest that Wnt proteins are not dispensable for the PCP pathway. As a consequence, unlike the Wnt canonical signaling pathway, which can exploit β-catenin reporter models such as TOPgal mice,[39] there are no transcriptional reporters for the PCP pathway and characterization of signaling through this pathway generally relies on localization of, and assessment of activity of its signaling components in various experimental models.

Ca²⁺ signaling

β-catenin and PCP signaling are not the only elucidated signaling pathways that involve Frizzled receptors. A separate body of evidence suggests that some noncanonical Wnt signaling involves intracellular calcium release. In zebrafish blastulae, overexpression of Wnt5a or rat Frizzled2 stimulates the frequency of Ca^{2+} fluxes in the enveloping layer (EVL) cells assessed using a Ca^{2+}-sensitive dye. In *Xenopus* embryos, overexpression of Wnt5a or Wnt11 activate the Ca^{2+}-sensitive kinase protein kinase C (PKC)[42] and Ca^{2+}/calmodulin dependent kinase II (CamKII)[43]. PKC is also upregulated in response to Wnt5a overexpression in melanoma cells UACC 1273.[44] NF-AT, the Ca^{2+} downstream transcription factor, has been shown as a potential target under specific conditions.[45]

Whether Frizzled receptors rely on Wnt binding to trigger Ca^{2+} signaling remains is debatable. Evidence that Ca^{2+} flux and activation of Ca^{2+}-dependent target proteins have been only provided in response to added Wnts to cells. Independently of Wnts, some Frizzleds trigger PKC and CamkII activities without exogenous Wnt.[42,43,46] These data suggest that like Frizzled/PCP signaling, the Frizzled/Ca^{2+} signaling may not require Wnts or at least not in certain cell environment. In addition, because other ligands, such as Norrin, a protein wholly unrelated to Wnts, that were also described to bind to Frizzled4 suggest that Frizzled can potentially activate downstream signaling as mentioned in a Wnt-independent manner.[47,48]

As to the evidence whether Frizzled behave as other GPCR members in activating the Ca^{2+} pathway, there is scattered evidence out in the literature to support this hypothesis. Slusarski and colleagues showed that pertussis toxin inhibits Frizzled signaling, suggesting binding to heterodimeric guanine nucleotide proteins (G proteins).[40] Some other studies suggest that $G_{\beta\gamma}$ subunits may be the relevant signaling G protein subunits for Frizzled signaling.[49] In vitro experiments have incriminated $G_{i/o}$ and family transducin G proteins.[50,51] Chimeric forms of Frizzleds, making use of the cytoplasmic domains of Frizzleds, substituted into the exofacial and transmembrane segments of the prototypic GPCR β_2-adrenoceptor, are functional and display the well-known GTP-shift in receptor affinity. Suppression of specific G protein subunits suppresses the ability of chimeric as well as authentic Frizzled-1 and Frizzled-2 to signal to their canonical pathways upon activation. Today more refined studies are applied using loss-of-function and genetic approaches to systematically and rigorously identify specific heterotrimeric G protein subunits involved and to increase our understanding of these receptors and of their mechanisms of downstream signaling.

Involvement in development and cellular processes

Loss of function mutations in Wnt genes have been extensively characterized in *Drosophila*, *Caenorhabditis elegans*, and mice, and these mutations demonstrate that a single Wnt can have multiple and diverse functions in different developmental contexts. For example, *Drosophila* Wingless (Wg), the most extensively studied Wnt protein, plays an essential role in embryogenesis to maintain segment polarity in the epidermis, define the identities of a subset of developing neuroblasts, and pattern the midgut and heart; in imaginal disc development, Wg acts to pattern the wings, legs, and eyes.[52] Gain of function mutations that lead to Wnt overexpression have also been characterized in *Drosophila* and mice. For example, overexpression of Wnt-1 or Wnt-3a in mice, first identified as a consequence of retroviral insertion, causes hyperproliferation and oncogenic transformation.[53,54] Consistent with this observation, recent work has implicated mutational activation of downstream effectors of Wnt signaling in the pathogenesis of various human cancers.

Knockout mice have been created for Frizzled3, 4, 5, 6, and 9. Loss of Frizzled3 resulted in mice dying within 30 minutes of birth and produced major defects in axonal development and path finding in the central nervous system (CNS).[55, 56] Data had suggested that Frizzled3 expression was largely limited to the CNS. Loss of Frizzled4, another Frizzled member largely limited to CNS expression, produced progressive cerebellar degeneration, esophageal enlargement, atrophy of the stria vascularis in the inner ear, and defective development of the retinal vasculature.[47,57] Mice lacking Frizzled5 die at day 10.75 in utero because of defects in yolk sac and placental angiogenesis.[58] Frizzled6$^{-/-}$ mice have abnormal hair orientation, but detailed analyses of other tissues are not described.[59] Recently Frizzled6 was also suggested as a critical receptor in B cell development.[60] Frizzled9 expression was observed in heart, brain, skeletal muscle, kidney, and testis, and knock-out mice revealed a role for Frizzled9 in lymphoid development and maturation.[61,62]

WNT/FRIZZLED PATHWAY IMPLICATION IN BONE METABOLISM

The search for additional factors that could contribute to our understanding of bone formation biology and provide grounds for the design of new and efficient anabolic drugs for treatment of bone and joint diseases has led investigators to study genetic modification associated with skeletal development in animals or with rare skeletal diseases in human.

A pivotal role of Wnt proteins in the process of skeletal development was established in different knock-out animals and in vitro studies (Figure 18.4). In fact, misexpression of Wnt proteins or players affects chondrocyte differentiation in early mesenchymal condensations, interfering with skeletal maturation and later

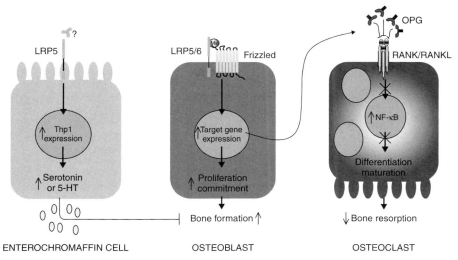

Figure 18-4: Canonical Wnt signaling effect on bone metabolism. Canonical Wnt signaling affects osteoblast proliferation, commitment, differentiation/maturation, function and lifespan. Wnt directly controls the expression of a series of genes playing crucial roles in osteoblast biology such as ALP, Atf4, and others. Importantly Wnt increases OPG and reduces RANKL expression, which translates into a negative impact on osteoclast differentiation and function. Very recently a role of LRP5 in controlling serotonin (5-HT) expression by enterochromaffin cells has been uncovered. High serotonin levels lead to decreased bone formation. Whether this new link between LRP5, serotonin and bone formation is dependent on Wnt binding or on a distinct ligand remains to be understood.

bone growth. These studies have shown that during embryonic development, abnormal expression of Wnt ligands and/or their regulatory proteins can disrupt the Wnt controlled-balance and result in skeleton malformation.[63] Additional investigations have further demonstrated the role of Wnt signaling in postnatal regulation of bone mass.

Studying human genes associated with osteoporosis-pseudoglioma syndrome (OPPG), a rare autosomal recessive disorder affecting bone and vision, and the hereditary high bone mass trait (HBM) led to the identification of LRP5, a Wnt receptor, as the responsible gene in both cases. LRP5 loss of function mutations were linked to OPPG,[64] whereas a gain of function mutation was associated with HBM.[65,66] Additional genetic evidence of the link between Wnt signaling plays and peak bone mass and its maintenance arose from the discovery of the Sost gene as the causing agent in sclerosteosis and Van Buchem disease.[32,67] These two human diseases are characterized by general overgrowth of bone tissue mostly visible in cranial bones and in the diaphysis of the tubular bone.[68] Both, sclerosteosis and Van Buchem disease were associated with loss of the Sost gene product and Wnt inhibitor, sclerostin, or SOST. All these findings clearly demonstrate that activating the Wnt pathway increases bone mass in adult.

Multiple myeloma (MM) is a B cell malignancy that causes a progressive and destructive osteolytic bone lesions, associated with osteoporosis, spinal cord compression, hypercalcemia, and severe bone pain.[69] Under normal conditions, bone resorption is typically reversed when osteoblasts fill lytics lesions with new bone matrix. This process in MM, however, is inhibited even when treated with bisphosphonates. Recent evidence showed that Dkk1 is secreted by MM cells and is responsible for hindering osteoblast differentiation and activity.[70–74] sFRP2, another soluble Wnt inhibitor, is also expressed by MM cells. Both in vitro and in vivo data have demonstrated that Dkk1 and sFRPs are potent inhibitors of bone formation. These together with elevations of RANKL, Dkk1 and sFRP-2 in patients with MM selectively block the normal remodeling process by diminishing bone formation.

In addition to MM, Wnt signaling antagonists are overexpressed in other tumor types such as prostate or breast cancer that metastasizes to bone and causes osteolysis. These results suggest that cancer-produced Dkk1 may be an important mechanistic link between primary tumors and secondary osteolytic bone metastases.[70]

Genetic models

To further investigate the role of the Wnt pathway in skeletal maintenance and bone homeopstatsis in adults, a variety of mouse genetic models have been generated and studied. First of all, knocking out LRP5 gene or overexpressing LRP5 HBM mutant completely reproduce OPPG and high bone mass human phenotypes, respectively. This further confirms that the mouse genetic model is an appropriate tool to investigate bone biology. In these animals, it is mainly osteoblast proliferation, function, and survival that were affected. These findings suggest that the Wnt pathway may influence osteoblast biology at distinct levels. LRP6, the closest homologue to LRP5, is also involved in bone formation and shows a similar function in bone formation. Because complete LRP6 knockout is embryonically lethal, the bone phenotype was only investigated in LRP6 heterozygous mice that display reduced bone mass.[75]

Because Wnt ligands are required during embryogenesis, exploring bone phenotypes in Wnt knock-out mice is quite difficult. Interestingly, Wnt10b does not follow the same rules and its inactivation in mice results in low bone mass in adults. On the opposite, overexpression of Wnt10b results in increased bone mass and strength and induces resistance to aging and hormone-related bone los [76,77]. Although mice harboring deletion for either Frizzled3, Frizzled4, Frizzled5, Frizzled6 or Frizzled9 (see discussion earlier in the chapter) have been described, their skeletal phenotype has not been examined yet and it would be interesting to investigate how the bone phenotypes in these transgenic animals overlap with other transgenics of the Wnt pathway.

Knock-out and transgenic mice of distinct Wnt antagonists were generated. Data show that loss of function of Wnt signaling inhibitors, such as Sfrp1, Dkk1, Kremen, or Sost, results in an increased bone mass.[78–81] Conversely, overexpression of Wnt antagonists such as Wif1, Dkk1, CTGF decreases bone density[82–84] Dkk2 deficiency stands up as a specific case because it resulted in rather a low bone mass phenotype.[85]

As discussed earlier in the chapter, mouse genetic studies clearly show that Wnt pathway directly influences bone formation. These results suggest that osteo-blasts are the main target of Wnt pathway. Dynamic histomorphometry analysis of bone samples from Dkk1, Kremen, and SOST knock-out mice show a strong increase in osteoblast number and function. This effect is confirmed with in vitro and ex vivo studies. In vitro overexpression of several Wnt proteins, including Wnt1, Wnt2, Wnt3a, or Wnt10, increases commitment of mesenchymal stem cells towards osteoblast lineages, thereby supporting the role of Wnt pathway in affecting osteoblast differentiation and thus bone formation.[86] The central role of Wnt signaling was further supported by ex vivo deletion of the β–catenin gene in murine calvaria cells. Unlike wild-type calvaria cells that differentiate into mature osteoblasts, calvaria cells lacking β-catenin expression failed to differenti-ate into osteoblast but rather differentiated into chondrocytes.

Very recently, a study by Yadav and colleagues has uncovered a novel molecu-lar mechanism accounting for the LRP5 regulation of bone formation.[87] This novel mechanism relies on the regulation of serotonin secretion in duodenum by the LRP5 receptor in a β-catenin-independent manner. On the one hand, these findings extend our understanding of bone remodeling and the LRP5 role; on the other hand, however, they do not fit with all the other genetic models reported so far (see earlier discussion). One cannot exclude that LRP5 plays dis-tinct roles, one in the Wnt pathway and others through possibly distinct ligands (Figure 18.4). Future additional studies will be expected to either confirm and build on this exciting discovery or to challenge it and further position LRP5 as a Wnt co-receptor.

TRANSLATING WNT INTO THERAPEUTICS

Pharmacological manipulation of the Wnt cascade affects bone formation rates in adults, and positive modulators of this pathway in vivo are expected to act as bone anabolic drugs. A number of drug design opportunities are offered by some components of the Wnt signaling. In fact, several elements of the Wnt cascade constitute potential targets for pharmacological intervention, either for small chemical compounds or for therapeutic antibodies.

Kinases and enzymes are common targets for drug discovery and design. Specific chemical inhibitors have been documented for targets such as GSK-3β, histone deacetylases (HDAC), and proteasomes.[88] As described earlier, these pro-teins negatively regulate cytosolic and nuclear levels of β-catenin, thus mak-ing them valuable targets to inhibit. In vitro data clearly show that GSK-3β, HDAC or proteasomes inhibitors increase Wnt signaling and enhance osteoblast differentiation.[89–91] More importantly, GSK-3β or proteasomes inhibitors display bone anabolic activity when administered in both control and ovariectomized mice. Although these findings strongly suggest that inhibiting GSK-3β or pro-teasomes may constitute an efficient therapy for osteopenic disorders includ-ing OPPG and senile osteoporosis, these enzymes are not exclusive to the Wnt pathway, but rather involved in other signaling mechanisms, hence raising concerns about possible side effects using such inhibitors for long-term treatment.

The design of inhibitors for natural Wnt antagonists such as Dkk1, Sost, or sFRPs is expected to achieve a more specific effect on the canonical Wnt pathway. The fact that Dkk1 and Sost are specifically expressed in adult bone tissue has strongly stimulated the interest in these two targets for bone anabolic development. Both Dkk1 and Sost bind to LRP5/6 and because protein-protein interaction is addressed, the most appropriate way to identify inhibitors is the neutralizing antibody approach. Furthermore, as our understanding of the involvement of such inhibitors in the pathophysiology of bone disease increases, the concept that anti-Dkk1 or anti-SOST therapies may be an effective treatment in the management of osteoporosis and/or in cancer-associated bone disease is now widely accepted.[92,93] Indeed, two fully human anti-Dkk1 and anti-SOST antibodies are currently under Phase I/II clinical investigation for the treatment of osteoporosis and osteolytic lesions induced by multiple myeloma. Data reported recently clearly show that such antibodies are safe and display strong anabolic bone activities.

FRIZZLED RECEPTORS AS POTENTIAL DRUG TARGET FOR TREATMENT OF BONE DISEASES

An alternative way to stimulate the Wnt pathway is by modulating Frizzled activity since these receptors are classified as GPCRs. Drugs targeting members of this protein superfamily, which transmit extracellular signals into a wide variety of cell types, represent the core of modern medicine. GPCRs are clearly today among the most heavily investigated drug targets by the pharma industry. However, the path to novel Frizzled-targeted medicines is not to be considered as a routine activity. This is mainly because of lack of reliable means to identify Frizzled modulators. Indeed, specific Frizzled signaling is well understood in PCP and Ca^{2+} signaling compared to the canonical pathway, the major target in bone cells. Drug discovery for GPCRs has been proven to be successful in the past years; however, it is not well established whether Frizzled receptors involved in the Wnt canonical pathway signal indeed as true GPCRs. Although some reports have shown that potentially Frizzled1, involved in canonical signaling, could be coupled to small G proteins under specific conditions, we and others have failed to reproduce these data at least in formats appropriate for small chemical compound high throughput screening (unpublished data). Furthermore, in vitro data have suggested that Frizzled1 may antagonize Wnt activity on bone-forming cells, indicating that not all Frizzled, but rather specific ones, can be used as a target to identify bone anabolic compounds.

Alternatively to screen for Frizzled modulators by using GPCR readouts, one can use some other general approaches, such as the "T cell factor" (TCF) reporter system or alkaline phosphatase (ALP) in mensemchymal cell lineages. Stimulation with Wnt3a recombinant protein increases commitment of mesenchymal stem cells towards osteoblast lineages. This activity could be demonstrated with either a Wnt-responsive luciferase reporter or by measuring ALP, a marker of osteoblast activity.[86] Both of these readouts have been used successfully to screen for pathway modulators. For instance Frank-Kamenetsky and colleagues have used the ALP readout in C3H10T1/2 cells to screen for smoothened (Smo) agonists and antagonists.[94] Smo is a seven transmembrane receptor involved in the Hedgehog pathway and has higher similarity to Frizzled. The difficulty in using such readouts that are so much downstream in the signaling pathway is that it is then necessary to filter for every potential target within the pathway, which enourmously increases the number of false-positive hit compounds. In addition, these types of readouts are usually not well accepted by structure activity relationship (SAR) establishment for chemical hits and afterwards for compound's lead optimization. Finally, identifying specific compound hits is highly dependent on the compound diversity of used chemical libraries and one should expect to screen a significant high number of compounds in order to identify specific hits.

A more specific way to modulate Frizzled receptors are antibodies. CRD is the effective domain of Frizzled and can raise specific antibodies that interact with CRDs. These antibodies can be thereafter tested for their agonist or antagonist activities on the Wnt canonical pathway. Such an approach has been previously used to identify GPCR modulators. More specifically, a Frizzled neutralizing antibody has been already reported with proven activity in vivo.[95]

Altogether, Wnt pathway offers an invaluable opportunity for new drug discovery in the field of bone anabolism. At the current time, secreted extracellular inhibitors were successfully used by the pharmaceutical industry in order to identify bone anabolic drugs. Although Frizzled receptors are considered as potential targets, the lack of specific screening assays has either discouraged using these receptors for drug discovery activities or failed to deliver specific hits. Future development in screening technologies applicable to Frizzled receptors for further understanding of Frizzled specific molecular signaling will stimulate the interest in this unique type of receptors and certainly place them high in the list for drug hunting research.

REFERENCES

1. A. Wodarz, R. Nusse, Mechanisms of Wnt signaling in development, Annu Rev Cell Dev Biol (1998) 59–88.
2. P. Polakis, Wnt signaling and cancer, Genes Dev 15 (2000) 1837–1851.
3. R. Nusse, The Wnt gene family in tumorigenesis and in normal development, J Steroid Biochem Mol Biol 1–3 (1992) 9–12.
4. R. Baron, G. Rawadi, Targeting the Wnt/beta-catenin pathway to regulate bone formation in the adult skeleton, Endocrinology. 6 (2007) 2635–2643.
5. G. Rawadi, S. Roman-Roman, Wnt signalling pathway: a new target for the treatment of osteoporosis, Expert Opin Ther Targets. 5 (2005) 1063–1077.
6. P. Bhanot, et al., A new member of the frizzled family from Drosophila functions as a Wingless receptor, Nature 6588 (1996) 225–230.
7. J. Yang-Snyder, J.R. Miller, J.D. Brown, C.J. Lai, R.T. Moon, A frizzled homolog functions in a vertebrate Wnt signaling pathway, Curr Biol 10 (1996) 1302–1306.

8. M. Umbhauer, A. Djiane, C. Goisset, A. Penzo-Mendez, J.F. Riou, J.C. Boucaut, D.L. Shi, The C-terminal cytoplasmic Lys-thr-X-X-X-Trp motif in frizzled receptors mediates Wnt/beta-catenin signalling, Embo J 18 (2000) 4944–4954.

9. C.C. Malbon, H. Wang, R.T. Moon, Wnt signaling and heterotrimeric G-proteins: strange bedfellows or a classic romance?, Biochem Biophys Res Commun 3 (2001) 589–593.

10. C.C. Malbon, Frizzleds: new members of the superfamily of G-protein-coupled receptors, Front Biosci (2004) 1048–1058.

11. J. Alcedo, M. Noll, Hedgehog and its patched-smoothened receptor complex: a novel signalling mechanism at the cell surface, Biol Chem 7 (1997) 583–590.

12. M. van den Heuvel, Hedgehog signalling: off the shelf modulation, Curr Biol 17 (2003) R686–688.

13. M.V. Semenov, R. Habas, B.T. Macdonald, X. He, SnapShot: Noncanonical Wnt Signaling Pathways, Cell. 7 (2007) 1378.

14. M. Semenov, K. Tamai, X. He, SOST is a ligand for LRP5/LRP6 and a WNT signaling inhibitor, J Biol Chem (2005) 20.

15. M.T. Veeman, J.D. Axelrod, R.T. Moon, A second canon. Functions and mechanisms of beta-catenin-independent Wnt signaling, Dev Cell. 3 (2003) 367–377.

16. M. Wehrli, *et al.*, arrow encodes an LDL-receptor-related protein essential for Wingless signalling, Nature 6803 (2000) 527–530.

17. K. Tamai, *et al.*, LDL-receptor-related proteins in Wnt signal transduction, Nature 6803 (2000) 530–535.

18. C. Labalette, C.A. Renard, C. Neuveut, M.A. Buendia, Y. Wei, Interaction and functional cooperation between the LIM protein FHL2, CBP/p300, and beta-catenin, Mol Cell Biol 24 (2004) 10689–10702.

19. L. Levy, Y. Wei, C. Labalette, Y. Wu, C.A. Renard, M.A. Buendia, C. Neuveut, Acetylation of beta-catenin by p300 regulates beta-catenin-Tcf4 interaction, Mol Cell Biol 8 (2004) 3404–3414.

20. S. Wang, M. Krinks, M. Moos, Jr., Frzb-1, an antagonist of Wnt-1 and Wnt-8, does not block signaling by Wnts -3A, -5A, or -11, Biochem Biophys Res Commun 2 (1997) 502–504.

21. S. Wang, M. Krinks, K. Lin, F.P. Luyten, M. Moos, Jr., Frzb, a secreted protein expressed in the Spemann organizer, binds and inhibits Wnt-8, Cell 6 (1997) 757–766.

22. K. Lin, S. Wang, M.A. Julius, J. Kitajewski, M. Moos, Jr., F.P. Luyten, The cysteine-rich frizzled domain of Frzb-1 is required and sufficient for modulation of Wnt signaling, Proc Natl Acad Sci U S A 21 (1997) 11196–11200.

23. L. Leyns, T. Bouwmeester, S.H. Kim, S. Piccolo, E.M. De Robertis, Frzb-1 is a secreted antagonist of Wnt signaling expressed in the Spemann organizer, Cell 6 (1997) 747–756.

24. P.W. Finch, *et al.*, Purification and molecular cloning of a secreted, Frizzled-related antagonist of Wnt action, Proc Natl Acad Sci U S A 13 (1997) 6770–6775.

25. Y. Kawano, R. Kypta, Secreted antagonists of the Wnt signalling pathway, J Cell Sci Pt 13 (2003) 2627–2634.

26. J.C. Hsieh, *et al.*, A new secreted protein that binds to Wnt proteins and inhibits their activities, Nature 6726 (1999) 431–436.

27. B. Mao, W. Wu, Y. Li, D. Hoppe, P. Stannek, A. Glinka, C. Niehrs, LDL-receptor-related protein 6 is a receptor for Dickkopf proteins, Nature 6835 (2001) 321–325.

28. A. Bafico, G. Liu, A. Yaniv, A. Gazit, S.A. Aaronson, Novel mechanism of Wnt signalling inhibition mediated by Dickkopf-1 interaction with LRP6/Arrow, Nat Cell Biol 7 (2001) 683–686.

29. V.E. Krupnik, *et al.*, Functional and structural diversity of the human Dickkopf gene family, Gene 2 (1999) 301–313.

30. B. Mao, *et al.*, Kremen proteins are Dickkopf receptors that regulate Wnt/beta-catenin signalling, Nature 6889 (2002) 664–667.

31. B. Mao, C. Niehrs, Kremen2 modulates Dickkopf2 activity during Wnt/LRP6 signaling, Gene 1–2 (2003) 179–183.

32. W. Balemans, *et al.*, Increased bone density in sclerosteosis is due to the deficiency of a novel secreted protein (SOST), Hum Mol Genet 5 (2001) 537–543.

33. M.E. Brunkow, *et al.*, Bone dysplasia sclerosteosis results from loss of the SOST gene product, a novel cystine knot-containing protein, Am J Hum Genet 3 (2001) 577–589.

34. M.C. Steitz, J.K. Wickenheisser, E. Siegfried, Overexpression of zeste white 3 blocks wingless signaling in the Drosophila embryonic midgut, Dev Biol 2 (1998) 218–233.

35. D.G. Winkler, *et al.*, Sclerostin inhibition of Wnt-3a-induced C3H10T1/2 cell differentiation is indirect and mediated by bone morphogenetic proteins, J Biol Chem 4 (2005) 2498–2502.

36. X. Li, *et al.*, Sclerostin Binds to LRP5/6 and Antagonizes Canonical Wnt Signaling, J Biol Chem 20 (2005) 19883–19887.

37. A. Hall, Rho GTPases and the control of cell behaviour, Biochem Soc Trans. Pt 5 (2005) 891–895.

38. R. Habas, X. He, Activation of Rho and Rac by Wnt/frizzled signaling, Methods Enzymol. (2006) 500–511.

39. R. DasGupta, E. Fuchs, Multiple roles for activated LEF/TCF transcription complexes during hair follicle development and differentiation, Development 20 (1999) 4557–4568.

40. D.C. Slusarski, V.G. Corces, R.T. Moon, Interaction of Wnt and a Frizzled homologue triggers G-protein-linked phosphatidylinositol signalling, Nature 6658 (1997) 410–413.

41. D.C. Slusarski, J. Yang-Snyder, W.B. Busa, R.T. Moon, Modulation of embryonic intracellular Ca2+ signaling by Wnt-5A, Dev Biol 1 (1997) 114–120.

42. L.C. Sheldahl, M. Park, C.C. Malbon, R.T. Moon, Protein kinase C is differentially stimulated by Wnt and Frizzled homologs in a G-protein-dependent manner, Curr Biol 13 (1999) 695–698.

43. M. Kuhl, L.C. Sheldahl, C.C. Malbon, R.T. Moon, Ca(2+)/calmodulin-dependent protein kinase II is stimulated by Wnt and Frizzled homologs and promotes ventral cell fates in Xenopus, J Biol Chem 17 (2000) 12701–12711.

44. A.T. Weeraratna, Y. Jiang, G. Hostetter, K. Rosenblatt, P. Duray, M. Bittner, J.M. Trent, Wnt5a signaling directly affects cell motility and invasion of metastatic melanoma, Cancer Cell. 3 (2002) 279–288.

45. T. Saneyoshi, S. Kume, Y. Amasaki, K. Mikoshiba, The Wnt/calcium pathway activates NF-AT and promotes ventral cell fate in Xenopus embryos, Nature 6886 (2002) 295–299.

46. P. Pandur, D. Maurus, M. Kuhl, Increasingly complex: new players enter the Wnt signaling network, Bioessays 10 (2002) 881–884.

47. Q. Xu, *et al.*, Vascular development in the retina and inner ear: control by Norrin and Frizzled-4, a high-affinity ligand-receptor pair, Cell 6 (2004) 883–895.

48. H. Clevers, Wnt signaling: Ig-norrin the dogma, Curr Biol 11 (2004) R436–437.

49. A. Penzo-Mendez, M. Umbhauer, A. Djiane, J.C. Boucaut, J.F. Riou, Activation of Gbetagamma signaling downstream of Wnt-11/Xfz7 regulates Cdc42 activity during Xenopus gastrulation, Dev Biol 2 (2003) 302–314.

50. H.Y. Wang, C.C. Malbon, Wnt signaling, Ca2+, and cyclic GMP: visualizing Frizzled functions, Science 5625 (2003) 1529–1530.

51. J. Dejmek, K. Dib, M. Jonsson, T. Andersson, Wnt-5a and G-protein signaling are required for collagen-induced DDR1 receptor activation and normal mammary cell adhesion, Int J Cancer 3 (2003) 344–351.

52. J. Klingensmith, R. Nusse, Signaling by wingless in Drosophila, Dev Biol 2 (1994) 396–414.

53. R. Nusse, H.E. Varmus, Many tumors induced by the mouse mammary tumor virus contain a provirus integrated in the same region of the host genome, Cell 1 (1982) 99–109.

54. H. Roelink, E. Wagenaar, S. Lopes da Silva, R. Nusse, Wnt-3, a gene activated by proviral insertion in mouse mammary tumors, is homologous to int-1/Wnt-1 and is normally expressed in mouse embryos and adult brain, Proc Natl Acad Sci U S A 12 (1990) 4519–4523.

55. A.I. Lyuksyutova, *et al.*, Anterior-posterior guidance of commissural axons by Wnt-frizzled signaling, Science 5652 (2003) 1984–1988.

56. Y. Wang, N. Thekdi, P.M. Smallwood, J.P. Macke, J. Nathans, Frizzled-3 is required for the development of major fiber tracts in the rostral CNS, J Neurosci 19 (2002) 8563–8573.

57. Y. Wang, D. Huso, H. Cahill, D. Ryugo, J. Nathans, Progressive cerebellar, auditory, and esophageal dysfunction caused by targeted disruption of the frizzled-4 gene, J Neurosci 13 (2001) 4761–4771.

58. T. Ishikawa, Y. Tamai, A.M. Zorn, H. Yoshida, M.F. Seldin, S. Nishikawa, M.M. Taketo, Mouse Wnt receptor gene Fzd5 is essential for yolk sac and placental angiogenesis, Development 1 (2001) 25–33.

59. N. Guo, C. Hawkins, J. Nathans, Frizzled6 controls hair patterning in mice, Proc Natl Acad Sci U S A 25 (2004) 9277–9281.

60. Q.L. Wu, C. Zierold, E.A. Ranheim, Dysregulation of Frizzled 6 is a critical component of B-cell leukemogenesis in a mouse model of chronic lymphocytic leukemia, Blood. 13 (2009) 3031–3039.

61. E.A. Ranheim, H.C. Kwan, T. Reya, Y.K. Wang, I.L. Weissman, U. Francke, Frizzled 9 knock-out mice have abnormal B-cell development, Blood. 6 (2005) 2487–2494.

62. C. Zhao, C. Aviles, R.A. Abel, C.R. Almli, P. McQuillen, S.J. Pleasure, Hippocampal and visuospatial learning defects in mice with a deletion of frizzled 9, a gene in the Williams syndrome deletion interval, Development 12 (2005) 2917–2927.

63. V.L. Church, P. Francis-West, Wnt signalling during limb development, Int J Dev Biol 7 (2002) 927–936.

64. Y. Gong, *et al.*, LDL receptor-related protein 5 (LRP5) affects bone accrual and eye development, Cell 4 (2001) 513–523.

65. R.D. Little, *et al.*, A Mutation in the LDL Receptor–Related Protein 5 Gene Results in the Autosomal Dominant High–Bone-Mass Trait, American Journal of Human Genetic (2002) 11–19.

66. L.M. Boyden, *et al.*, High Bone Density Due to a Mutation in LDL-Receptor-Related Protein 5, N Engl J Med 20 (2002) 1513–1521.

67. W. Balemans, *et al.*, Identification of a 52 kb deletion downstream of the SOST gene in patients with van Buchem disease, J Med Genet 2 (2002) 91–97.

68. H. Hamersma, J. Gardner, P. Beighton, The natural history of sclerosteosis, Clin Genet 3 (2003) 192–197.

69. G.D. Roodman, Pathogenesis of myeloma bone disease, Leukemia. 3 (2009) 435–441.

70. J.J. Pinzone, B.M. Hall, N.K. Thudi, M. Vonau, Y.W. Qiang, T.J. Rosol, J.D. Shaughnessy, Jr., The role of Dickkopf-1 in bone development, homeostasis, and disease, Blood. 3 (2009) 517–525.

71. Y.W. Qiang, *et al.*, Myeloma-derived Dickkopf-1 disrupts Wnt-regulated osteoprotegerin and RANKL production by osteoblasts: a potential mechanism underlying osteolytic bone lesions in multiple myeloma, Blood. 1 (2008) 196–207.

72. J. Qian, *et al.*, Dickkopf-1 (DKK1) is a widely expressed and potent tumor-associated antigen in multiple myeloma, Blood. 5 (2007) 1587–1594.

73. S. Yaccoby, W. Ling, F. Zhan, R. Walker, B. Barlogie, J.D. Shaughnessy, Jr., Antibody-based inhibition of DKK1 suppresses tumor-induced bone resorption and multiple myeloma growth in vivo, Blood. 5 (2007) 2106–2111.

74. E. Tian, F. Zhan, R. Walker, E. Rasmussen, Y. Ma, B. Barlogie, J.D. Shaughnessy, Jr., The role of the Wnt-signaling antagonist DKK1 in the development of osteolytic lesions in multiple myeloma, N Engl J Med. 26 (2003) 2483–2494.

75. S.L. Holmen, *et al.*, Decreased BMD and limb deformities in mice carrying mutations in both Lrp5 and Lrp6, J Bone Miner Res 12 (2004) 2033–2040.

76. C.N. Bennett, *et al.*, Wnt10b increases postnatal bone formation by enhancing osteoblast differentiation, J Bone Miner Res. 12 (2007) 1924–1932.

77. C.N. Bennett, K.A. Longo, W.S. Wright, L.J. Suva, T.F. Lane, K.D. Hankenson, O.A. MacDougald, Regulation of osteoblastogenesis and bone mass by Wnt10b, Proc Natl Acad Sci U S A 9 (2005) 3324–3329.

78. X. Li, *et al.*, Targeted deletion of the sclerostin gene in mice results in increased bone formation and bone strength, J Bone Miner Res. 6 (2008) 860–869.

79. F. Morvan, *et al.*, Deletion of a single allele of the Dkk1 gene leads to an increase in bone formation and bone mass, J Bone Miner Res. 6 (2006) 934–945.

80. K. Ellwanger, *et al.*, Targeted disruption of the Wnt regulator Kremen induces limb defects and high bone density, Mol Cell Biol. 15 (2008) 4875–4882.

81. P.V. Bodine, *et al.*, The Wnt antagonist secreted frizzled-related protein-1 is a negative regulator of trabecular bone formation in adult mice, Mol Endocrinol 5 (2004) 1222–1237.

82. B.T. MacDonald, D.M. Joiner, S.M. Oyserman, P. Sharma, S.A. Goldstein, X. He, P.V. Hauschka, Bone mass is inversely proportional to Dkk1 levels in mice, Bone. 3 (2007) 331–339.

83. A. Smerdel-Ramoya, S. Zanotti, L. Stadmeyer, D. Durant, E. Canalis, Skeletal overexpression of connective tissue growth factor impairs bone formation and causes osteopenia, Endocrinology. 9 (2008) 4374–4381.

84. J. Li, *et al.*, Transgenic Mice Over-Expressing Dkk-1 in Osteoblasts Develop Osteoporosis, J Bone Miner Res (2004) S6.

85. X. Li, *et al.*, Dkk2 has a role in terminal osteoblast differentiation and mineralized matrix formation, Nat Genet. 9 (2005) 945–952.

86. G. Rawadi, B. Vayssiere, F. Dunn, R. Baron, S. Roman-Roman, BMP-2 controls alkaline phosphatase expression and osteoblast mineralization by a Wnt autocrine loop, J Bone Miner Res 10 (2003) 1842–1853.

87. V.K. Yadav, *et al.*, Lrp5 controls bone formation by inhibiting serotonin synthesis in the duodenum, Cell. 5 (2008) 825–837.

88. M.D. Gordon, R. Nusse, Wnt signaling: multiple pathways, multiple receptors, and multiple transcription factors, J Biol Chem. 32 (2006) 22429–22433.

89. I.R. Garrett, *et al.*, Selective inhibitors of the osteoblast proteasome stimulate bone formation in vivo and in vitro, J Clin Invest. 11 (2003) 1771–1782.

90. H.W. Lee, J.H. Suh, A.Y. Kim, Y.S. Lee, S.Y. Park, J.B. Kim, Histone deacetylase 1-mediated histone modification regulates osteoblast differentiation, Mol Endocrinol. 10 (2006) 2432–2443.

91. P. Clement-Lacroix, *et al.*, Lrp5-independent activation of Wnt signaling by lithium chloride increases bone formation and bone mass in mice, Proc Natl Acad Sci U S A. 48 (2005) 17406–17411.

92. D.J. Heath, *et al.*, Inhibiting Dickkopf-1 (Dkk1) removes suppression of bone formation and prevents the development of osteolytic bone disease in multiple myeloma, J Bone Miner Res. 3 (2009) 425–436.

93. M. Fulciniti, *et al.*, Anti-DKK1 mAb (BHQ880) as a potential therapeutic agent for multiple myeloma, Blood (2009) 5.

94. M. Frank-Kamenetsky, *et al.*, Small-molecule modulators of Hedgehog signaling: identification and characterization of Smoothened agonists and antagonists, J Biol 2 (2002) 10.

95. M. Sen, M. Chamorro, J. Reifert, M. Corr, D.A. Carson, Blockade of Wnt-5A/frizzled 5 signaling inhibits rheumatoid synoviocyte activation, Arthritis Rheum 4 (2001) 772–781.

Index